THE ART OF MODELLING STARS IN THE 21ST CENTURY

IAU SYMPOSIUM No. 252

COVER ILLUSTRATION: Turbulent Shell Burning Convection

As the collection of papers in this proceedings reveals, the art of modeling stars in the 21st century touches on a diverse range of topics, from population synthesis and galactic chemical evolution modeling to the mechanisms underlying supernova and gamma ray burst explosions. The common thread is that almost all of this work relies on our ability to accurately model an individual star. And understanding an individual star is deeply connected to how well we understand the turbulent motions which take place deep in the stellar interior. Today, the exponential growth in computing technology, which is expressed so concisely by Moore's law, is heralding in a new era of sophistication in our ability to model the plasma dynamics within stars. These *numerical experiments* are providing strong constraints on our theories, and inspiring new ways to understand and describe the interiors of the stars.

In this cover figure, an example of one such numerical experiment is presented. The turbulent flow field in an oxygen burning shell is shown for a large angular domain ($120° \times 120°$), 3D simulation. The domain, which is described by spherical polar coordinates, is oriented so that the polar direction is aligned roughly in the up-down direction. The mass fraction of ^{32}S is visualized in order to give a sense of the topology and the complex, multi-scale, turbulent nature of the flow field. Material with a high mass fraction of ^{32}S is being entrained into the turbulent oxygen burning convective shell from the underlying silicon and sulfur rich core of a 23 $\mathcal{M}_\odot$ star, and illustrates a mixing process which is not included in the standard treatment of stars used today. The computational domain for this calculation contains 17 million cells. Evolving the flow for 5 convective turnover times requires $\sim$1 million cpu-hours on a cluster equipped with state of the art processors.
Casey Meakin

IAU SYMPOSIUM PROCEEDINGS SERIES
2008 EDITORIAL BOARD

Chairman
I.F. CORBETT, IAU Assistant General Secretary
European Southern Observatory
Karel-Schwarzschild-Strasse 2
D-85748 Garching-bei-München
Germany
icorbett@eso.org

Advisers
K.A. VAN DER HUCHT, IAU General Secretary,
SRON Netherlands Institute for Space Research, Utrecht, the Netherlands
E.J. DE GEUS, *Dynamic Systems Intelligence B.V., Assen, the Netherlands*
U. GROTHKOPF, *European Southern Observatory, Germany*
M.C. STOREY, *Australia Telescope National Facility, Australia*

Members
IAUS251
SUN KWOK, *Faculty of Science, University of Hong Kong, Hong Kong, China*
IAUS252
LICAI DENG, *National Astronomical Observatories, Chinese Academy of Sciences, Beijing, China*
IAUS253
FREDERIC PONT, *Geneva Observatory, Sauverny, Switzerland*
IAUS254
JOHANNES ANDERSEN, *Astronomical Observatory, Niels Bohr Institute, Copenhagen University, Denmark*
IAUS255
LESLIE HUNT, *INAF - Istituto di Radioastronomia, Firenze, Italy*
IAUS256
JACOBUS Th. van LOON, *Astrophysics Group, Lennard-Jones Laboratories, Keele University, Staffordshire, UK*
IAUS257
NATCHIMUTHUK GOPALSWAMY, *Solar Systems Exploration Div., NASA Goddard Space Flight Center, MD, USA*
IAUS258
ERIC E. MAMAJEK, *Radio and Geoastronomy Division, Harvard Smithsonian CfA, Cambridge, MA, USA*
IAUS259
KLAUS G. STRASSMEIER, *Astrophysics Institute Potsdam, Potsdam, Germany*

INTERNATIONAL ASTRONOMICAL UNION
UNION ASTRONOMIQUE INTERNATIONALE

THE ART OF MODELLING STARS IN THE 21ST CENTURY

PROCEEDINGS OF THE 252th SYMPOSIUM OF THE
INTERNATIONAL ASTRONOMICAL UNION
HELD IN SANYA, HAINAN PROVINCE, CHINA
APRIL 6–11, 2008

Edited by

LICAI DENG
National Astronomical Observatories, Beijing, China

and

KWING-LAM CHAN
The University of Science and Technology, Hongkong, China

CAMBRIDGE UNIVERSITY PRESS
The Edinburgh Building, Cambridge CB2 8RU, United Kingdom
32 Avenue of the Americas, New York, NY 10013-2473, USA
477 Williamstown Road, Port Melbourne, VIC 3207, Australia
Ruiz de Alarcón 13, 28014 Madrid, Spain
Dock House, The Waterfront, Cape Town 8001, South Africa

© International Astronomical Union 2008

This book is in copyright. Subject to statutory exception
and to the provisions of relevant collective licensing agreements,
no reproduction of any part may take place without
the written permission of the International Astronomical Union.

First published 2008

Printed in the United Kingdom at the University Press, Cambridge

Typeset in System LaTeX 2_ε

A catalogue record for this book is available from the British Library

Library of Congress Cataloguing in Publication data

ISBN 9780521889834 hardback
ISSN 1743–9213

Table of Contents

Preface xiii

Organizing committee xv

Conference photograph xvi

Conference participants xix

Address by the SOC & LOC xxi

Session I: Updates of physical ingredients of stellar models

Chair: Norbert Langer

Changing Abundances, Changing Opacities................................. 1
    J. Ferguson & A. Dotter

Is the Suns chemical composition unusual? 13
    M. Asplund

Rigorous and Phenomenological Equations of State 27
    W. Däppen & D. Mao

3D model atmospheres and the solar photosphere oxygen abundance........ 35
    E. Caffau & H.-G. Ludwig

Poster

SMART – a computer program for modelling stellar atmospheres 41
    A. Aret, A. Sapar, R. Poolamäe & L. Sapar

Proxies for overshooting above a convection zone........................ 43
    Kwing L. Chan & Harinder P. Singh

Some discussion of the nonlocal treatment of the dissipation in the Reynolds stress models.. 45
    T. Cai

Session II: Physical processes in stars

Chair: Kwing-Lam Chan

Instabilities and mixing in stellar radiation zones 47
    J.-P. Zahn

Lithium depletion in late-type dwarfs as probe of stellar convection........ 61
    D. R. Xiong & L. Deng

Angular momentum and overshooting: two as yet unsolved problems in stellar mixing.. 67
    V. M. Canuto & Y. Cheng

Radiation-hydrodynamics simulations of surface convection in low-mass stars: connections to stellar structure and asteroseismology 75
    H.-G. Ludwig, E. Cauffau & A. Kučinskas

How extended is convective overshooting? 83
    L. Deng & D.R. Xiong

A new two-columns description for convective transport in stars 89
    A. Stökl

Thermohaline Convection in Main Sequence Stars 97
    S. Vauclair

Thermohaline mixing in low-mass giants 103
    M. Cantiello & N. Langer

Poster

Large-scale Numerical Simulations:Convection in an annular channel rotating about a vertical axis with side-walls 111
    Y. Chang & Y. Liu

Comparing the Nucleosynthesis Parameters of s+r Stars and Ba Stars 113
    W. Cui, D. Cui & B. ZHang

Asteroseismic study of solar-like stars: A method of estimating stellar age 115
    Y. K. Tang, S. L. Bi & N. Gai

The evolution of 'the moment of inertia' of stars 117
    Y.-C. Kim & S. Barnes

The spectral properties of a large sample of low surface brightness disk galaxies 119
    Y. C. Liang, G. H. Zhong, L. Deng & B. Zhang

Observations of clusters with ages from 0.01 to 1.0 Gyr in the Large Magellanic Cloud 121
    Q. Liu, R. de Grijs, L. C. Deng, Y. Hu & I. Baraffe

Asteroseismic constraints on the OPAL opacity interpolation 123
    W. M. Yang & M. Li

The Electric Currents from Viscosity in Differentially Moved Plasma 125
    Z. L. Yang & H. B. Wang

Solar scandium abundance 127
    H. W. Zhang, T. Gehren & G. Zhao

Radio Star Candidates from FIRST and 2MASS Databases 129
    Y. X. Zhang & Y. H. Zhao

A large sample of low surface brightness disk galaxies from SDSS 131
    G. H. Zhong, Y. C. Liang, L. Deng & B. Zhang

Researching turbulent convection models and the density gradient reversing.... 133
    Q. S. Zhang & Y. Li

Session III: From physics to stars, the progenitors of white dwarfs

Chair: Vittorio Canuto

Helio- and asteroseismology. 135
    J. Christensen-Dalsgaard

Progress report on solar age calibration. 149
    G. Houdek & D. O. Gough

Rate of change of the pulsation periods in the PG 1159 star PG 0122+200. 157
    G. Vauclair, J.-N. Fu, J.-E. Solheim, S.-L. Kim, M. Chevreton, N. Dolez, L. Chen, M. A. Wood & I. M. Silver

Deep inside low-mass stars 163
    Corinne Charbonnel & Suzanne Talon

Hydrodynamical Simulations of Turbulent Convection in a Rotating Red Giant Star 175
    A. Palacios & A. S. Brun

Mixing-length parameter from binaries and clusters 183
    M. Yıldız

Chair: Jørgen Christensen-Dalsgaard

Deathzones and exponents: A different approach to incorporating mass loss in stellar evolution calculations. 189
    L. A. Willson

Stellar Evolution from AGB to Planetary Nebulae. 197
    S. Kwok

AGB Star Models. 205
    F. Herwig

Hydrodynamic simulations of the core helium flash 215
    M. Mocák & E. Müller

Non-LTE Spectral Analysis of Extremely Hot Post-AGB Stars: Constraints for Evolutionary Theory 223
    T. Rauch, K. Werner, M. Ziegler, L. Koesterke & J. W. Kruk

Dust-driven Winds Beyond Spherical Symmetry. 229
    P. Woitke

Low-Mass Extremely Metal-Poor Stellar Models: Yields, Uncertainties and the Galactic Halo Stars 235
    S. W. Campbell & J. C. Lattanzio

Poster

Solar-like oscillations in red giant ε Ophiuchi 243
    S. L. Bi & N. Gai

Thermohaline mixing and fossil magnetic fields in red giant stars.............. 245
 C. Charbonnel & J.-P. Zahn

The Structure and Kinematics of Envelope around Red Supergiant AH Sorpii Traced by SiO Masers 247
 X. Chen & Z.Q. Shen

A New, Efficient Stellar Evolution Code for Calculating Complete Evolutionary Tracks 249
 A. Kovetz, O. Yaron & D. Prialnik

Lithium isotopes in halo dwarfs 251
 L. Piau

Surface convection in Population II stars............................. 253
 L. Piau & R. Stein

On MHD rotational transport, instabilities and dynamo action in stellar radiation zones 255
 S. Mathis, J.-P. Zahn & A.-S. Brun

The Rotation of the Solar Radiative zone 257
 S. Turck-Chieze

Mass Loss from Pulsating Cool Stars............................. 259
 Q. Wang, L. A. Willson & S. Kawaler

"Hot Helium Flashers" – The Road to Extreme Horizontal Branch Stars 261
 O. Yaron, A. Kovetz & D. Prialnik

Dust Size Effect On IR Colors Of AGB Stars 263
 H. Wang, B. W. Jiang & R. Szczerba

The age-metallicity relation in the thin disk 265
 J. Li, B. Liang & W. S. Fan

The Variability Of RSG : HV2576 267
 M. Yang & B. W. jiang

The Dynamics Of Galactic Globular Cluster............................. 269
 D. Chen

Session IV: From physics to stars, the progenitors of neutron stars and black holes

Chair: Lee Anne Willson

Mass loss and evolution of hot massive stars............................. 271
 J. S. Vink

The influence of inhomogeneities on hot star wind model predictions.......... 283
 J. Krtička, L. Muijres, J. Puls, J. Kubát & Alex de Koter

Stellar evolution models with mass loss and turbulence 289
 M. Vick, G. Michaud & O. Richard

The most massive AGB stars	297
L. Siess	
Interation of massive stars with their surroundings	309
G. Hensler	

Chair: Martin Asplund

Massive star evolution: from the early to the present day Universe	317
G. Meynet, S. Ekström, C. Georgy, A. Maeder & R. Hirschi	
Around the Pair Instability Valley - Massive SN Progenitors	329
R. Waldman	
^{60}Fe and Massive Stars	333
W. Wang	

Poster

Exploring for the Nucleosynthesis Region of Metal-Poor Stars	339
Y.-Y. Geng, D.-N. Cui, J.Zhang & B. Zhang	
Two-fluid models for the winds of OB stars	341
J. H. Guo	
Tidal disruption of stripped red giants by massive black holes	343
Y. Lu, Y. F. huang, S. N. Zhang & P. Lu	
Study on the spectrum of the injected relativistic protons	345
Y. P. Wang, Y. Lu & L. Chen	
Advanced test of the model stellar atmospheres: the nature of the light variability of magnetic chemically peculiar stars	347
J. Krtička, Z. Mikulášek, J. Zverko, J. Žižňovský & P. Zvěřina	

Session V: Physics of stars in close binaries

Chair: Ph. Podsiadlowski

Binary Evolutionary Models	349
Z. Han & Ph. Podsiadlowski	
The role of binary stars in stellar population synthesis	359
Z. Li & Z. Han	
Rotational mixing in close binaries	365
S. E. de Mink, M. Cantiello, N. Langer, S.-Ch. Yoon, I. Brott, E. Glebbeek, M. Verkoulen & O. R. Pols	
Close binary evolution and blue straggler formation	371
P. Lu & L. Deng	

Chair: Licai Deng

The single degenerate channel for the progenitor of type Ia supernovae	379
X.C. Meng, X. F. Chen & Z. Han	

Modelling the evolution and nucleosynthesis of C-enhanced metal poor stars ... 383
   O. Pols, R. G. Izzard, M. Lugaro & S.E. de Mink

A Spectroscopic Study of Blue Stragglers in M67 391
   G. Q. Liu, L. Deng, M. Chávez & E. Bertone

The Missing Population of Be+Black Hole X-Ray Binaries 399
   A. Sądowski, J. Ziółkowski, K. Belczyński & T. Bulik

Stellar radii from long-baseline interferometry 405
   P. Kervella

The Y^2 Isochrones getting a new dimension 413
   S.K. Yi, Y.-C. Kim, P. Demarque & Y.-W. Lee

Poster

Blue Stragglers from Primordial Binary Evolution 417
   X.F. Chen & Z. han

A Model for Adiabatic Mass-loss 419
   H. Ge, R. Webbink & Z. Han

The history of KZ Hya and its unseen companions 421
   S. Y. Jiang

Structure and evolution of W UMa-type systems 423
   L. F. Li, F. H. Zhang, Z. Han, D. K. Jiang & T. Y. Jiang

A Spectroscopic Study of Barium Stars 425
   G.Q. Liu, Y. C. Liang & L. Deng

Evolutionary scenario for W UMa-type stars 427
   K. Stępień & K. Gazeas

The eclipsing binary IU Per and its intrinsic oscillations 429
   X. B. Zhang

The duration properties of Swift Gamma-Ray Bursts 431
   Z. B. Zhang & C.-S. Choi

The energy transfer in W UMa binary stars 433
   D. K. Jiang, J. C. Wang, Z. han & T. Y. Jiang

The Effect of Binary Interactions on Infrared Passbands 435
   F. H. Zhang, L. F. Li & Z. Han

Secular period decreasing of 17 detached chromospherically active binaries 437
   C. Q. Luo, Y. P. Luo, X. B. Zhang, L. Deng, Z. Q. Luo & S. Z. Yang

Session VI: New tools and future perspective

Chair: Werner Däppen

Hydrodynamic Processes in Massive Stars 439
   C. Meakin

New numerical simulations and the role of coherent structures 451
    F. Kupka

Poster

Analysing the Contributions in Moment Equations of Reynolds Stress Models of
    Convection with Numerical Simulations 463
    F. Kupka & H. J. Muthsam

SONG – Stellar Observations Network Group 465
    F. Grundahl, J. Christensen-Dalsgaard, H. Kjeldsen, S. Frandsen, T. Arentoft, P. Kjaergaard & U. G. Jørgensen

Summary

Conference Summary 467
    N. Langer

Author index 475

Subject index 477

Preface

Stars are at the heart of astrophysics and constitute the link between the microphysical world and large scale structures of the universe. A proper understanding of stellar evolution is required to address questions such as: the origin of re-ionization in the early universe, the chemical and photometric evolution of galaxies, the dynamical evolution of the gas in galaxies, the age of remote starbursts, the origin of dust, the nature of the progenitors of gamma ray bursts, supernovae of different types, neutron stars, black holes and white dwarfs, to cite only a few topical questions. With the advent of new very large telescopes like ALMA, ELT, JWST, still farther remote regions of the Universe will be explored, allowing us to observe stars in conditions which are different from those encountered in the present Universe. Stars might have been much more massive, had much lower metallicities or were even metal free, had different rotational velocities, magnetic fields; the binary fraction might also be different. The results of the theory of stellar structure and evolution are among the most important ingredients embedded in almost all research fields of modern astrophysics. The current overall picture of the Universe and the formation and evolution of structures in different scales depend very much on the outputs of stellar theory. The currently available one-dimensional (1D) stellar models are able to follow most of the hydrostatic phases of stellar evolution, from the pre-main-sequence to the Debye phase of the cooling of WDs. These models also prepare structures as initial inputs to hydrodynamics studies of dynamical phases such as supernova events. This does not mean that the theory of stellar evolution is fully matured. In many applications of current astrophysical interests, the subject models are based only on parameterized, educated guesses of the outcome of the most difficult phases of stellar evolution. With advances in computational technology, the construction of stellar models beyond 1D and the inclusion of more elaborated physical processes (rotation, realistic atmospheres, magnetic fields, mass loss and binary interactions) that have been previously neglected are now becoming possible. At te same tiem, helio-seismology and more recently astero-seismology are offering a new, independent tool for probing stellar interiors, and imposing more stringent constraints on the theoretical models.

To face the new challenges, stellar research needs to advance in two directions: Firstly, the physics of the stellar models should be improved. This is the requirement for extending and consolidating our understanding of stars in the wider HR diagram. Secondly, predictions of stellar models should be further tested with the more accurate and complete data from the new observations. The present symposium aims at discussing the progress made so far and the development to be expected along these two directions. The emphasis is on the hydrostatic phases of evolution for which the observational statistics are more abundant. The following topics were covered during the 5 day meeting:

- Improvements of the physical ingredients of stellar models (opacities, nuclear reaction rates, neutrinos, equation of state, initial compositions;
- Evolution of low and intermediate mass stars;
- Evolution of massive stars;
- Close binary evolution;
- Stellar physics in the era of very large telescopes.

Correspondingly, the symposium was organized into 5 sessions covering the above topics. Starting with reviews by the invited speakers, each session includes oral and poster presentations in the respective area. The specific questions addressed include the following:

- What are the current status of stellar opacity and equation of state? Are the accuracies adequate?
- How can we handle convection? One-dimensional moment approach or 3D simulation?
- How does rotation affect the evolution of stars?
- What do hydrodynamics models tell us about the effects of mass loss?
- What do observations of the Sun tell us about the evolution of low mass stars?
- How much do we know about the physics of AGB stars?
- What do we know about the extremely blue horizontal branch stars?
- What do we know about low- and intermediate- mass stars beyond AGB?
- Do models of massive stars fit both photometric and spectroscopic observations of supergiants?
- What have we learn from Helio- and Astero-seismology about the structures of stars at different stages?
- What can we say about the properties of first stars?
- What do we know about the evolution of close binaries?
- What are the consequences of the interacting binary populations?
- Three-dimensional models are now available, how can we make good use of them?

In recent years, major progresses have been made in answering these questions, but most of them were noted to remain as major challenges for the future.

This symposium is dedicated to Professor XIONG Darun for the celebration of his 70th birthday. Professor Xiong is a pioneer of stellar astrophysics in China. He has been working on stellar convection theory, stellar evolution, stellar oscillation and helio-seismology for over 40 years (including the hard times during the Culture Revolution period between the 1960s and 70s). His theory of stellar convection has made an important impact on the modern way to treat this complicated stellar process.

Financial supports from the International Astronomical Union, the Chinese Academy of Sciences, the National Natural Science Foundation of China, and the National Astronomical Observatories, which made it possible for the organizing committees to bring a lot of young people from all over the world to this symposium, are gratefully acknowledge. We would also like to take this opportunity to thank the young students (some of them are in LOC, especially Mr. Yi Hu) who have been very friendly and helpful to all the participants during the symposium. Without their help, the symposium cannot be so pleasant and fruitful.

Licai Deng and Cesare Chiosi, co-chairs SOC,
Yanchun Liang, chair LOC
Beijing, Padova, July 11, 2008

THE ORGANIZING COMMITTEE

Scientific

D. Arnett (US)
F. Allard (Fr)
I. Baraffe (fr)
K.L. Chan (HK, Cn)
C. Chiosi (co-chair, Italy)
W. Dappen (US)
L. Deng (co-chair, Cn)
F. Kupka (De)

H.J.G.L. Lammers (Nl)
N. Langer (Nl)
J. Lattanzio (AU)
G. Meynet (Ch)
P. Ventura (It)
A. Weiss (De)
L.A. Willson (US)
D.R. Xiong (Cn)

Local

Y.C. Liang (chair)
Y. Lu
Y. Hu
K.L. Chan
J. Yan
X.S. Yun

Y. Xin
S.L. Bi
F.H. Zhang
C.L. Tian
S.J. Xie

Acknowledgements

The symposium is sponsored and supported by the IAU Division IV (Stars), the primary sponsoring commission is No. 35 (Stellar Constitution); and by the IAU Commissions No. 25 (Stellar Photometry & Polarimetry), No. 26 (Binary and Multiple Stars), No. 27 (Variable stars), No. 36 (Theory of Stellar Atmospheres), No. 37 (Star Clusters and Associations).

The Local Organizing Committee operated under the auspices of the
National Astronomical Observatories

Funding by the
International Astronomical Union,
The National Natural Science Foundation of China
The Chinese Academy of Sciences
and
The National Astronomical Observatories

Conference photograph

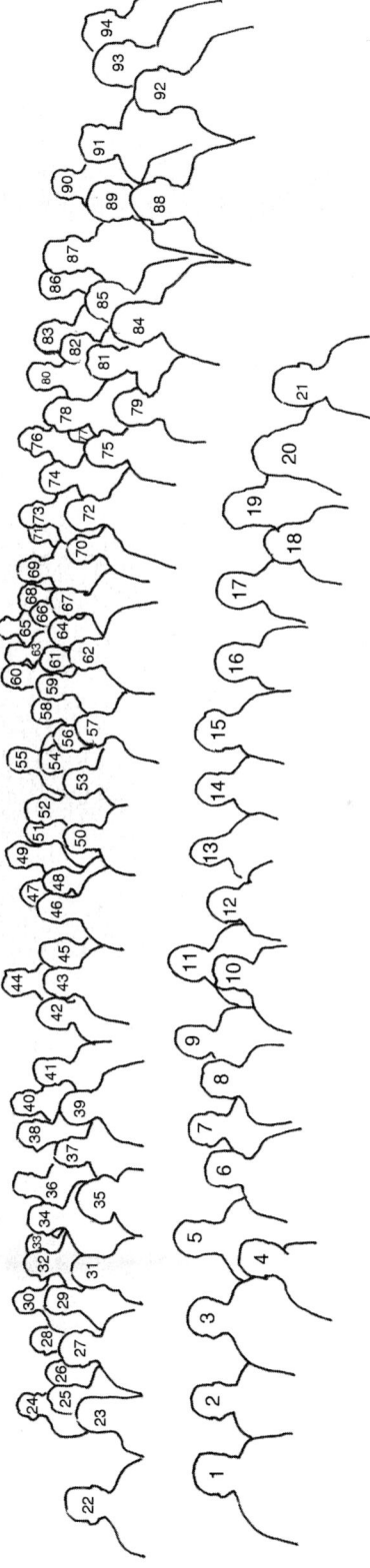

1. Guohu Zhong
2. Wuming Yang
3. Hongwei Ge
4. Xiaoshan Yun
5. Yanke Tang
6. Licai Deng
7. Yangping Luo
8. Ping Lu
9. Xuefei Chen
10. Guoqing Liu
11. Yanping Wang
12. Shunfang Liu
13. Huali Li
14. Huijuan Wang
15. Tao Cai
16. Cuihua Du
17. Yayuan Wen
18. Jundan Nie
19. Corinne Charbonnel
20. Ye Lu
21. Yuanyuan Geng
22. Zhongmu Li
23. Changqing Luo
24. Huawei Zhang
25. Jianheng Guo
26. Fenghui Zhang
27. Xiangcun Meng
28. Lifang Li
29. Stephone Mathis
30. Dengkai Jiang
31. Shaolan Bi
32. Kazimierz Stepien
33. Qiansheng Zhang
34. Jean-Paul Zahm
35. Dina Prialnik
36. Jorick Vink
37. Attany Kovetz
38. Jiri Krticka
39. Xi Chen
40. Philipp Podsiadlowski
41. Mathieu Vick
42. Anna Aret
43. Janusz Ziolkowski
44. Thomas Raoch
45. Ana Palacios
46. Qian Wang
47. Miroslav Mocak
48. Lee Anne Willson
49. Pierre Kervella
50. Sylvie Vauclair
51. Sukyoung Yi
52. Martin Asplund
53. Gunter Houdek
54. Gerard Vauclair
55. Jason Ferguson
56. Y.-C. Kim
57. Sylvaine Turck-Chieze
58. Friedrich Kupka
59. Kwing L Chan
60. Hans-G Ludwig
61. Wenyuan Cui
62. Werner Dappen
63. Huan Wang
64. Sun Kowk
65. Ming Yang
66. Ji Li
67. Ofer Yaron
68. Yitzchak Tuchman
69. Peter Woitke
70. Laurent Piau
71. Xiaobin Zhang
72. Lionel Siess
73. Falk Herwig
74. Casey Meakin
75. Roni Waldman
76. Jiang Zhang
77. Elisabetta Caffau
78. Joergen Christensen-Dalsgaard
79. Selma de Mink
80. Zhiliang Yang
81. Gerhard Hensler
82. Alexander Stoekl
83. Mutlu Yildiz
84. Matteo Cantiello
85. Evert Glebbeek
86. Joao Fernandes
87. Onno Pols
88. Shiyang Jiang
89. Simon Campbell
90. Jiulin Du
91. Georges Meynet
92. Fuyuan Zhao
93. Hongsheng Zhang
94. Norbert Langer

Participants absent from the conference photo:
Chris Belczynski Yingli Chang Zhanwen Han Ligang Li Chunlin Tian Weihua Wang Da Run Xiong Shuzheng Yang
Vittorio Canuto Ding Chen Yi Hu Qiang Liu Wei Wang Yu Xin Zhiquan Luo

Participants

Anna **Aret**, Tartu Observatory, Estonia		aret@aai.ee
Martin **Asplund**, Max Planck Institute for Astrophysics, Garching, Germany		asplund@mpa-garching.mpg.de
Chris **Belczynski**, Los Alamos National Lab, USA		kbelczynski@nmsu.edu
Shaolan **Bi**, Department of Astronomy Beijing Normal University, Beijing, China		bisl@bnu.edu.cn
Elisabetta **Caffau**, GEPI, Observatoire de Paris-Meudon, Meudon, France		Elisabetta.Caffau@obspm.fr
Tao **Cai**, Department of Mathematics, HKUST, Hong Kong, China		ctust@ust.hk
Simon W. **Campbell**, Academia Sinica Institute of Astronomy and Astrophysics, Taipei, China		simcam@asiaa.sinica.edu.tw
Matteo **Cantiello**, Astronomical Institute Utrecht, Utrecht, The Netherlands		M.Cantiello@uu.nl
Vittrio M. **Canuto**, NASA, Goddard Institute for Space Studies, New York, USA		vcanuto@giss.nasa.gov
Yingli **Chang**, Shanghai Fisheries University, Shanghai, China		ylchang@shfu.edu.cn
Corinne **Charbonnel**, Geneva Observatory, University of Geneva, Switzerland		Corinne.Charbonnel@obs.unige.ch
Kwing L. **Chan**, Department of Mathematics and Center for Space Science Research, HKUST, Hong Kong, China		maklchan@ust.hk
Ding **Chen**, Shanghai Astronomical Observatory, SHanghai, China		cding@shao.ac.cn
Xi **Chen**, Shanghai Astronomical Observatory, SHanghai, China		chenxi@shao.ac.cn
Xuefei **Chen**, Yunnan observatory, CAS, Kunming, CHina		xuefeichen717@hotmail.com
Jørgen **Christensen-Dalsgaard**, Danish AsteroSeismology Centre, University of Aarhus, Aarhus, Denmark		jcd@phys.au.dk
Wenyuan **Cui**, Department of Physics, Hebei Normal University, Shijiazhuang, China		wenyuancui@126.com
Licai **Deng**, National Astronomical Observatories, CAS, Beijing, China		licai@bao.ac.cn
Werner **Däppen**, Department of Physics and Astronomy, University of Southern California, Los Angeles, USA		dappen@usc.edu
Selma E. **de Mink**, Astronomical Institute Utrecht, Utrecht, The Netherlands		S.E.deMink@uu.nl
Cuihua **Du**, Graduate University of Chinese Academy of Sciences, Beijing, China		ducuihua@gucas.ac.cn
Jiulin **Du**, Department of Physics, School of Science, Tianjin University, Tianjin, China		jiulindu@yahoo.com.cn
Jason W. **Ferguson**, Department of Physics, Wichita State University, Wichita, USA		jason.ferguson@wichita.edu
João **Fernandes**, Astronomical Observatory, University of Coimbra, Portugal		jmfernan@mat.uc.pt
Ning **Gai**, Yunnan Observatory, CAS, Kunming, China		gaining@ynao.ac.cn
Hongwei **Ge**, Yunnan Observatory, CAS, Kunming, China		hongwei.ge@gmail.com
Yuanyuan **Geng**, Department of Physics, Hebei Normal University, Shijiazhuang, China		gengyuanyuan1982@126.com
Evert **Glebbeek**, Sterrekundig Instituut Utrecht, Utrecht, The Netherlands		e.glebbeek@phys.uu.nl
Jianheng **Guo**, Yunnan Observatory, CAS, Kunming, China		guojh@ynao.ac.cn
Zhanwen, **Han**, Yunnan Observatory, CAS, Kunming, China		zhanwenhan@hotmail.com
Gerhard **Hensler**, Institute of Astronomy, University of Vienna, Vienna, Austria		hensler@astro.univie.ac.at
Falk **Herwig**, Keele University, Staffordshire, UK		fherwig@btinternet.com
Günter **Houdek**, Institute of Astronomy, Cambridge, UK		hg@ast.cam.ac.uk
Yi **Hu**, National Astronomical Observatories, CAS, Beijing, China		huyi@bao.ac.cn
Dengkai **Jiang**, Yunnan Observatory, CAS, Kunming, China		jiangdengkai@hotmail.com
Shiyang **Jiang**, National Astronomical Observatories, CAS, Beijing, China		jiangsy@bao.ac.cn
Pierre **Kervella**, LESIA, Observatoire de Paris, Université Paris Diderot, Meudon, France		Pierre.Kervella@obspm.fr
Yong-Cheol **Kim**, Astronomy Department, Yonsei University, Seoul, Korea		yckim@yonsei.ac.kr
Attay **Kovetz**, Department of Geophysics and Planetary Sciences, Sackler Faculty of Exact Sciences		attay@etoile.tau.ac.il
Jiří **Krtička**, Institute of Theoretical Physics and Astrophysics, Masaryk University, Brno, Czech		krticka@monoceros.physics.muni.cz
Sun **Kwok**, Department of Physics, University of Hong Kong, Hong Kong, China		sunkwok@hku.hk
Friedrich **Kupka**, Max-Planck-Institute for Astrophysics, Garching, Germany		fk@mpa-garching.mpg.de
Norbert **Langer**, Astronomical Institute Utrecht, Utrecht, The Netherlands		langer@phy.uu.nl
Huali **Li**, National Astronomical Observatories, CAS, Beijing, China		lhl@bao.ac.cn
Ji **Li**, College of Physics and Information Engineering, Hebei Normal University, Shijiazhaung, China		liji@mail.hebtu.edu.cn
Lifang **Li**, Yunnan Observatory, CAS, Kunming, Yunnan Province, China		llf@ynao.ac.cn
Ligang **Li**, Shanghai Astronomical Observatory, SHanghai, China		llg@shao.ac.cn
Zhongmu **Li**, Yunnan Observatory, CAS, Kunming, China		zhongmu.li@gmail.com
Guoqing **Liu**, National Astronomical Observatories, CAS, Beijing, China		lgq@bao.ac.cn
Qiang **Liu**, National Astronomical Observatories, CAS, China		liuq@bao.ac.cn
Shunfang **Liu**, National Astronomical Observatories, CAS, Beijing, China		liushunfang@126.com
Pin **Lu**, National Astronomical Observatories, CAS, Beijing, China		lupin@bao.ac.cn
Ye **Lu**, National Astronomical Observatories, CAS, Beijing, China		ly@bao.ac.cn
Hans-G. **Ludwig**, Observatoire de Paris-Meudon, Meudon, France		Hans.Ludwig@obspm.fr
Changqing **Luo**, Institute of Theoretical Physics, China West Normal University, Nanchong, China		changqingluo@126.com
Yangping **Luo**, Institute of Theoretical Physics, China West Normal University, Nanchong, China		luoyangping789@163.com
Zhiquan **Luo**, Institute of Theoretical Physics, China West Normal University, Nanchong, China		zqluo@tom.com
Stéphane **Mathis**, Laboratoire AIM, CEA/DSM-CNRS-Université Paris Diderot, Cedex, France		stephane.mathis@cea.fr
Casey A. **Meakin**, Astronomy Department, University of Arizona, Tucson, USA		casey.meakin@gmail.com
Xiangcun **Meng**, Yunnan Observatory, CAS, Kunming, China		conson859@msn.com
Georges **Meynet**, Observatory of Geneva University, Switzerland		georges.meynet@obs.unige.ch
Miroslav **Mocák**, Max-Planck-Institut für Astrophysik, Garching, Germany		mmocak@mpa-garching.mpg.de
Jundan **Nie**, Department of Astronomy, Beijing Normal University, Beijing, China		niejundan@mail.bnu.edu.cn
Ana **Palacios**, Université Montpellier II - GRAAL, Montpellier, France		palacios@graal.univ-montp2.fr
Laurent **Piau**, CEA-Saclay DSM/IRFU/SAp, France		laurent.piau@cea.fr
Philipp **Podsiadlowski**, Oxford University, Oxford, Uk		podsi@astro.ox.ac.uk
Onno **Pols**, Astronomical Institute Utrecht, Utrecht, The Netherlands		o.r.pols@astro.uu.nl
Dina **Prialnik**, Department of Geophysics and Planetary Sciences, Sackler Faculty of Exact Sciences		dina@planet.tau.ac.il
Thomas **Rauch**, Institute for Astronomy, Eberhard Karls University, Tübingen, Germany		rauch@astro.uni-tuebingen.de
Lionel **Siess**, Institut d'Astronomie et d'Astrophysique, Université libre de Bruxelles, Bruxelles, Belgium		siess@astro.ulb.ac.be
Kazimierz **Stępień**, Warsaw University Observatory, Warsaw, Poland		kst@astrouw.edu.pl
Alexander **Stökl**, CRAL, Université de Lyon, CNRS, École Normale Supérieure de Lyon, Lyon, France		alexander.stoekl@ens-lyon.fr
Yanke **Tang**, Yunnan Observatory, CAS, Kunming, China		tangyanke@ynao.ac.cn
Yitzchak **Tuchman**, Racah Inst. of Physics. Hebrew University in Jerusalem, Jerusalem, Israel		tuchma@vms.huji.ac.il

Participants

Sylvaine **Turck-Chieze**, IRFU/CEA CE Saclay, cedex, France — cturck@cea.fr
Gérard **Vauclair**, Laboratoire d'Astrophysique de Toulouse-Tarbes, Toulouse, France — gerardv@ast.obs-mip.fr
Sylvie **Vauclair**, Laboratoire d'Astrophysique de Toulouse-Tarbes, Toulouse, France — sylvie.vauclair@ast.obs-mip.fr
Mathieu **Vick**, Département de Physique, Université de Montréal, Montréal, — mathieu.vick@umontreal.ca
Jorick S. **Vink**, Armagh Observatory, College Hill, Armagh, Northern Ireland, UK — jsv@arm.ac.uk
Roni **Waldman**, Racah Institute of Physics, Hebrew University, Jerusalem, Israel — waldman@cc.huji.ac.il
Huan **Wang**, Department of Astronomy, Beijing Normal University, Beijing, China — whbnu@mail.bnu.edu.cn
Huijuan **Wang**, National Astronomical Observatories, CAS, Beijing, China — Wanghj@bao.ac.cn
Qian **Wang**, Department of Physics and Astronomy, Iowa State University, Ames, USA — wqinisu@iastate.edu
Wei **Wang**, National Astronomical Observatories, CAS, Beijing, China — wangwei@bao.ac.cn
Weihua **Wang**, Shanghai Astronomical Observatory, Shanghai, China — whwang@shao.ac.cn
Yanping **Wang**, National Astronomical Observatories, CAS, Beijing, China — wangyanping@mail.bnu.edu.cn
Yayuan **Wen**, National Astronomical Observatories, CAS, Beijing, China — wenyy@bao.ac.cn
Lee Anne **Willson**, Department of Physics and Astronomy, Iowa State University, Ames, USA — lwillson@iastate.edu
Peter **Woitke**, UK Astronomy Technology Centre, Edinburgh, Scotland, UK — ptw@roe.ac.uk
Yu **Xin**, National Astronomical Observatories, CAS, Beijing, China — xinyu@bao.ac.cn
Darun **Xiong**, Purple mountain Observatory, Nanjing, China — xiongdr@mail.pmo.ac.cn
Ming **Yang**, Department of Astronomy, Beijing Normal University, Beijing, China — myang@mail.bnu.edu.cn
Shuzheng **Yang**, Institute of Theoretical Physics, China West Normal University, Nanchong, China — szyangcwnu@126.com
Wuming **Yang**, Department of Physics and Chemistry, Henan Polytechnic University, Jiaozuo China — yang.wuming@yahoo.com.cn
Zhiliang **Yang**, Astronomy Department, Beijing Normal University, Beijing, China — lyang@bnu.edu.cn
Ofer **Yaron**, Department of Geophysics and Planetary Sciences, Sackler Faculty of Exact Sciences — oyaron@gmail.com
Mutlu **Yıldız**, Ege University, Dept. of Astronomy and Space Sciences, Izmir, Turkey — mutlu.yildiz@ege.edu.tr
Xiaoshan **Yun**, National Astronomical Observatories, CAS, Beijing, China — xsyun@bao.ac.cn
Jean-Paul **Zahn**, LUTH, Observatoire de Paris, Meudon, France — Jean-Paul.Zahn@obspm.fr
Feihui **Zhang**, Yunnan Observatory, CAS, Kunming, China — zhangfh@ynao.ac.cn
Hongsheng **Zhang**, Korea Astronomy and Space Science Institute, Daejeon, Korea — Hongsheng@kasi.re.kr
Huawei **Zhang**, Department of Astronomy, School of Physics, Peking University, Beijing, China — zhw@bac.pku.edu.cn
Jiang **Zhang**, Department of Physics, Hebei Normal University, Shijiazhuang, China — zhangjiang@sjzue.edu.cn
Qiansheng **Zhang**, Yunnan Observatory, Kunming, China — mail_aoe_1@163.com
Xiaobin **Zhang**, National Astronomical Observatories, CAS, Beijing, China — xzhang@bao.ac.cn
Zhibin **Zhang**, Korea Astronomy and Space Science Institute, Daejon, Korea — z-b-zhang@163.com
Fuyuan **Zhao**, National Astronomical Observatories, CAS, Beijing, China — fyzhao@bao.ac.cn
Guohu **Zhong**, National Astronomical Observatories, CAS, Beijing, China — ghzhong@bao.ac.cn
Janusz **Ziółkowski**, Copernicus Astronomical Center, ul. Bartycka 18, 00-716 Warsaw, Poland — jz@camk.edu.pl

Address by the SOC & LOC

Dear colleagues,

We warmly welcome you to come to Sanya, the only tropical city in China! As you are already here, I wish you find the place nice in all aspects. We believe that you will enjoy this setting as well as a fruitful scientific meeting. Sanya was only a small fishing village 20 years ago; it is now a modern city and is turning more and more international. The city of Sanya can actually represent the fast developments in this country in last 20 years.

China is one of the few countries in the world whose history can trace back 5000 years. China also has a very long history of Astronomy which is witnessed by the very famous ancient instrument Armillary Sphere (illustrated below) invented in Han dynasty ($\sim$ 100 BC). This state of art instrument actually made it possible for ancient astronomers to measure positions of stars at arcsecond level. That may be regarded as the starting point of stellar astronomy in human history. This instrument is used as the logo of the Chinese Astronomical Society.

The Armillary Shpere. By moving the complicated circular parts and pointing arms, one can measure stellar coordinates in equatorial, ecliptic systems, and more. This is the original piece made in Ming dynasty around 1300, now collected at the old observing site of Purple Mountain Observatory in Nanjing.

The modern epoch of Chinese astronomy was started at the beginning of the 20st century at nearly the same time by western preachers in Shanghai, and by Chinese astronomers coming back from the west in Nanjing; Chinese astronomy started with stellar physics, and the tradition has been well kept. You can always find your Chinese colleagues actively working in all subjects related with this symposium. As one of the goals of such an IAU event is to stimulate scientific collaborations, we sincerely welcome you to visit the Chinese astronomical institutions after the meeting, and we are ready to lend our helping hands.

To highlight future developments in Astronomy in China, we would like to mention a few projects that may have important impact on stellar physics (subjects in the brackets), therefore we hope to draw your attention:

(a) Astronomy in Dome A: the path finder, CSTAR, is now in operation (*variable objects, stellar oscillations*);

(b) LAMOST: 4000 spectra (stars/galaxies/QSOs) per shot. The first light will be in the end of this year (*stellar spectroscopy, the structure of the Galaxy*);

(c) FAST: The world's largest 500 meter aperture single dish radio telescope, project approved 2008 (*final stages of stellar evolution, compact stellar objects.*);

(d) Astronomy in Tibetian Plateau: site survey for future large telescopes (Optical/IR/sub-mm, *general in stellar physics.*).

We would like to take this opportunity to express our sincere gratitudes to all SOC members who helped a lot during preparation stage of this meeting. We also thank all Chinese students who have been working so hard in the past month, and will still be working around you during this week, the meeting would never be made so pleasant without their help.

Thank you very much, and welcome.

Licai Deng (co-Chair, SOC) & Y.C. Liang (Chair, LOC
Sanya, 6 April 2008

Changing Abundances, Changing Opacities

Jason W. Ferguson[1], and Aaron Dotter[2]

[1] Department of Physics, Wichita State University, Wichita, KS 67260-0032, USA
email: jason.ferguson@wichita.edu

[2] Department of Physics and Astronomy, Dartmouth College, 6127 Wilder Laboratory,
Hanover, NH 03755, USA
email:aaron.l.dotter@dartmouth.edu

Abstract. With ever changing solar abundances being reported the equation of state and opacities needed for stellar evolution models also change. A discussion of those changes in mean molecular opacities will be presented with a discussion on the effect on evolution models. Aside from changing the abundances of the base mixture the enrichment changes too. Traditionally mean opacity tables are produced for oxygen-rich mixtures, however stars will often become carbon-rich. A discussion of carbon-rich opacities tables will also be presented.

Keywords. atomic processes, equation of state, molecular processes, stars: abundances, stars: evolution

1. Background

For the caluculation of the transfer of radiation in a stellar evolution model or any astrophysical environment, it is necessary to understand and include the interaction of radiation with matter. The physics of attenuation in a medium is called the opacity of material and is a function of the properties of the material (temperature, density or pressure, chemical composition, etc.) and a function of the wavelength of light. Computation of opacities require detailed knowledge of the atomic, molecular, and dust physics over a wide range of wavelengths. A technique that is often used to simplify the computation of radiative transfer is to take a wavelength averaged or mean opacity.

Tables of mean opacities as functions of temperature and density have been compiled for decades (see Iglesias & Rogers(1991) and Iglesias & Rogers(1996), OPAL hereafter; Seaton et al.(1994), OP hereafter; Alexander & Ferguson(1994) and Ferguson et al.(2005), F05 hereafter). The advantage of a grey opacity is the savings of hours of computation time. For example, in a stellar evolution model the age of the star must be advanced in small timesteps at certain stages in the star's evolution and the additional time sink of computing the detailed opacity with the evolution is currently computationally impossible in a reasonable time. The disadvantage of using mean opacity tables is the loss of chemical composition changes for which opacity tables are not available and the closest "set" has to be used. As outlined in F05 the typical mean opacity table set includes over 155 X and Z number fractions each with a large range of temperature and density. A full set can take more than four days of computer time on a small cluster.

This review summarizes the recent work in low temperature (less than 10,000 K) opacities at Wichita State University (WSU). A summary of input physics is discussed first, followed by a section on how new solar abundances mixtures affect the opacity and the effect of changing C/O ratios on the mean opacity. The effect of changing elemental abundances on stellar evolution models is discussed and the last section looks at future directions.

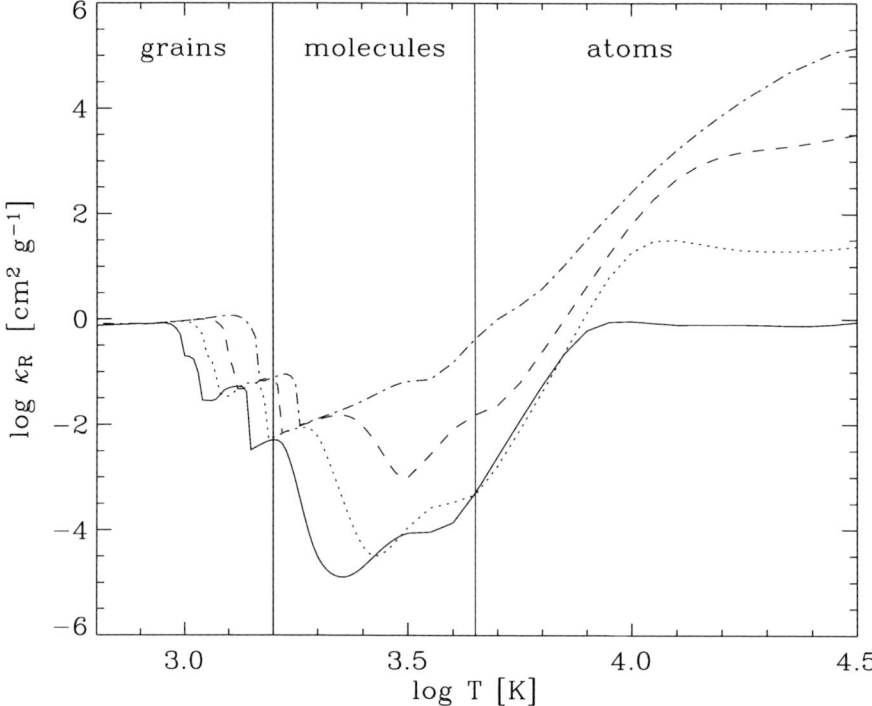

Figure 1. A plot of mean opacity as a function of temperature for four values of *log R*. From bottom to top the solid line is for a *log R* = −5.0, then is *log R* = −3.0 (dotted), *log R* = −1.0 (dashed), and *log R* = +1.0 (dash-dot) with the highest mean opacity.

2. Input Physics

A thourogh discussion of the input physics included in WSU opacity tables is given in F05. Figure 1 shows what the Rosseland mean opacity looks like as a function of temperature for the typical quasi-solar mixture (X = 0.7 and Z = 0.02) with four densities included. Density is often parameterized as R which is defineded as ρ/T_6^3 with the temperature in millions of Kelvin.

The Rosseland mean is defined as

$$\frac{1}{\kappa_R} \equiv \frac{\int_0^\infty \frac{1}{\kappa_\lambda} \frac{\partial B_\lambda}{\partial T} d\lambda}{\int_0^\infty \frac{\partial B_\lambda}{\partial T} d\lambda} \tag{2.1}$$

where κ_λ is the monochromatic opacity, B_λ is the Planck function and $\partial B_\lambda/\partial T$ is therefore the weighting function of the Rosseland mean. For the computations from F05, we integrate Eq. 1 over 24,000 wavelengths. The wavelengths are not evenly spaced and the details are given in F05. What is important to note is that the Rosseland mean is a harmonic mean, not a straight mean. With a straight mean additional sources can effectively be added to the total opacity, but with a harmonic mean it is necessary to compute κ_λ for each different or new composition that might occur.

For illustration purposes the dominate phase of material that controls the strength of the mean opacity is indicated in the figure. Note that the dividing lines are approximate and move in temperature with density, which is espcecially evident that the molecular/grain line at about 1500 K.

2.1. Atoms

Atoms, including free-free and bound-free sources, dominate the mean opacity at temperatures above about 4,000 K as shown in Fig. 1. As the temperature of the gas decreases the opacity also decreases as the atoms become more neutral and the electrons tend to settle in the lowest levels. The most important opacity sources include the most abundant species, hydrogen and helium. Hydrogen lines are very important for temperatures above $\sim$10,000 K. The continuous source H^- is important as the temperature falls below 10,000 K. F05 includes some discussion showing the mean opacity with some sources removed to see the overall importance of them.

For higher values of gas density ($log\ R$) the mean opacity is generally larger. This is due primarily to pressure ionization and collision effects of the atoms and pressure broadening of the lines. However, it is not the rule. For example, between about 4,000 K and 10,000 K the mean opacity of the the $log\ R = -3$ is lower than the mean opacity of $log\ R = -5$. This effect is due to more H^- opacity at the lower density.

2.2. Molecules

As a gas cools molecules form under equilibrium conditions. This is reflected in the mean opacity in two ways. In Fig. 1 at the higher end of the "molecules" region is a bump in the mean opacity that is due to the beginning formation of strongly bonded molecules such as CO. While the abundance is quite small at these temperatures there are enough CO lines to begin to affect the mean opacity in small ways.

At even lower temperatures the mean opacity rises as H_2O and TiO begin to appear. Both of these molecules, especially water, have huge numbers of lines in the region of the weighting function of the Rosseland mean that amplifies the effect they have on the opacity. The rise in the mean opacity can be as much as 2 orders of magnitude depending upon the gas density peaks, and then begins to fall away again as the temperature lowers and the rotational/vibrational modes begin to settle down.

Higher gas densities result in larger values of the mean opacity. In general the water bump appears washed out due to higher gas density. The peak of the water bump moves to higher temperature with increasing density as well.

2.3. Dust

At even lower tempreatures the gas can condense and solid precipitates exist in equilibrium which greatly affect the mean opacity. In general the first condensate to form is either $CaTiO_3$ or Al_2O_3 and the mean opacity jumps by an order of magnitude in a very small temperature range. Other species then form, including $MgSiO_3$, $FeMgSiO_4$ and Fe among others, and begin to generate a mean opacity that is roughly constant with temperature.

A larger value of gas density moves the condensation temperature of the grains. This is evident in Fig. 1 as the line between the grains/molecule regions shifts towards higher temperature. Once the grains exist at lower temperatures the mean opacity is roughly constant.

It should be noted that all of the opacities shown in this paper come from the routines discussed in F05 and include the assumptions of dust physics outlined in that paper. Ferguson *et al.* (2007) discusses the affects of the basic grain physics assumptions (including size distribution, porosity, aggregates, etc.) on the mean opacity. In general these effects are not currently needed for the computation of stellar evolution models as models today do not fall below the grain condensation temperature.

3. Abundance Effects

In light of recent changes in the metallicity mixture of the sun, especially the amount of solar oxygen, an illustration of how these changes affect the mean opacity is given in Figure 2. In the top portion of the figure the mean opacity is given for the abundance mixtures of Grevesse & Noels (1993 GN93 hereafter), Grevesse & Sauval (1998; GS98 hereafter), Seaton (private communication; S92 hereafter), Asplund et al. (2005; AGS05 hereafter) and Lodders (2003; L03 hereafter) for the same scaled abundance set (X=0.7 and Z=0.02).

In the lower portion of the figure the differences in the logarithm of the opacity is shown indicating how much of a true change abundances have on the mean opacity. The GN93 mixture is used as a baseline in the plot for illustration purposes. At higher temperatures where the mean opacity is due mostly to hydrogen the difference between the five abundance sets is neglibible. However, at cooler temperatures when molecules become important the differences become quite large. At temperatures where water is the dominate opacity, differences can be as large for as 25% for the difference between AGS05 and GN93. This is sensible because the AGS05 abundances have an decrease in the amount of oxygen and the amount of water is also less.

At even cooler temperatures the differences in the opacities are generally due to the refractory elements. It is clear from Fig. 1 that the temperature of the condensation of grains move a bit in temperature for the different sets by observing the discontinuities in the lower portion of the figure.

Another feature in opacity calculations that is needed by stellar evolution modellers is the effect of α-element enhancement. Figure 3 shows the effect of a typical alpha enhancement based on the abundances of S92. The α-element abundances are from Achim Weiss (private communication; S92ae hereafter) and clearly show the effect of more metals, particularly oxygen. At the higher temperatures where hydrogen controls the opacity the difference between the GN93 and the α-enhanced mix is nearly neglible as there is a bit more atomic opacity in the S92ae mix compared with the comparison mixture.

There are clear differences in the water bump at about $log\ T \sim 3.25$ as much more oxygen is available and there is more water in the s92ae mix compared with the GN93 mix. At cooler temperatures the dust opacity is actually lower due to the scaling of the total metallicity Z, which is kept constant for both s92ae and GN93.

3.1. Changing the C/O ratio

Of the many consequences of stellar evolution one that is often overlooked in terms of opacity is the changing from an oxygen rich to a carbon rich environment. For high temperature opacities OPAL has produced tables for enhanced carbon and oxygen abundances (see Rogers & Iglesias 1993). Known as OPAL type 2 tables there are hundreds of combinations of X, Z, dX_C and dX_O which have been computed. Essentially the abundance of carbon is enhanced as helium is reduced as the triple-alpha process ensues. Then, material from the star's core is then dredged up causing the outer layers to become carbon rich.

Carbon-rich opacities for low temperatures are not as readily aviaible. AF94 and F05 do not provide carbon-enhanced tables. Extrapolations based on oxygen rich opacities are avaible from Marigo (2002) and more recently tables from Lederer & Aringer (2008) are among those tables that are publically available.

Figure 4 and Figure 5 shows the effect of raising the amount of carbon in the mix. Each of the mean opacities shown in the figures are for the scaled ($X = 0.70, Z = 0.02$) set scaled from GN93, $log\ R = -3.0$ and without dust grains for ease of comparison. In

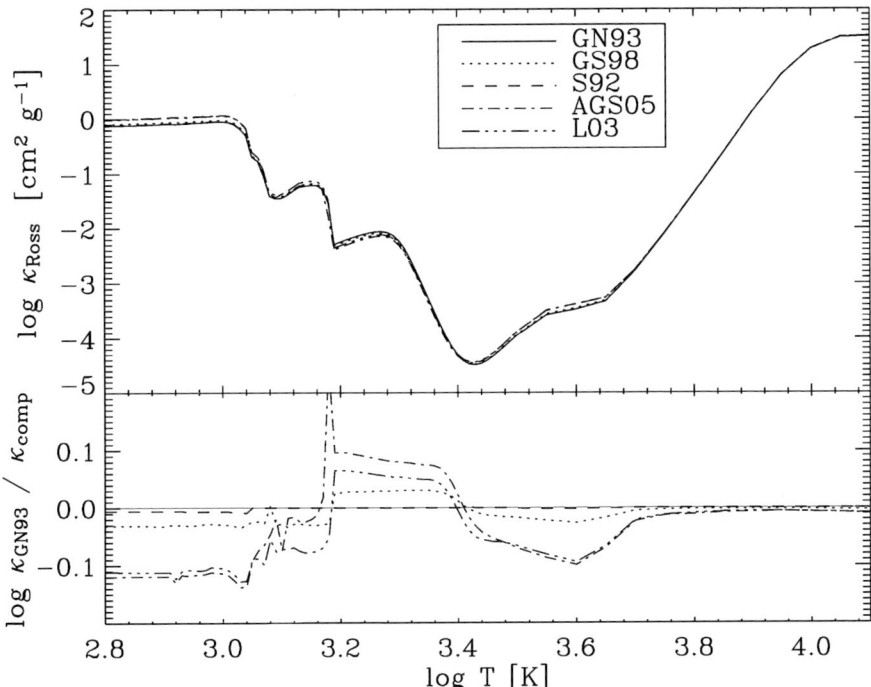

Figure 2. Shown is the mean opacity for five different abundance sets. In the top portion of the figure the solid line represents the mean opacity based on GN93 abundacnes, the dotted line from GS98, the dashed line from S92, the dash-dot line from AGS05, and the dash-dot-dot line is from L03. In the bottom portion the logarithm of the ratio of the opacity from GN93 to the mean opacity being compared to. The thin solid line is a ratio of one, the dotted line compares GN93 with GS98, the dashed line (GN93 to S92), the dash-dot line (GN93 to AGS05), and the dash-dot-dot line represents (GN93 to L03).

Fig. 4 the mean opacity for mixes with C/O ratios less than 1.0 are shown. It is easy to see the water bump at $log\ T \sim 3.25$ becomes smaller as the C/O ratio rises to unity. What is interesting is that the value of $C/O = 1.0$ is not the lowest mean opacity. The lowest value of the mean opacity is (for most temperatures) $C/O = 0.98$. In Fig. 5 the mean opacity is shown for $C/O \geqslant 1.0$. The mean opacity rises at temperatures of $log\ T \sim 3.3$ as molecular CN begins to dominate the opacity. In the last section of this paper, the validity of the carbon opacities used here are discussed.

Overall, the mean opacity from solar values to a C/O ratio of unity to carbon rich lowers as water is removed from the mixture, then rises as cyanide begins to dominate. Similar plots have been made for the OPAL type 2 abundances with F05, however the dX_C and dX_O enhancements have such large stepsizes that most of the physics involved as the opacity changes is lost. Large stepsizes may be reasonable at higher temperatures as hydrogen and other ionized species dominate the opacity. The smaller step sizes as the C/O ratio is slowly increased retains more of the changing opacity due due to the changing abundances.

3.2. Stellar evolution models

With current computational facilities it is not yet possible to compute an entire stellar evolution model for an entire lifetrack of a star and include the detailed opacity computation as well. This reasoning explains why opacity tables are used. Dotter *et al.* (2007) explores the effect of changes in the abundance of individual species on the mean opacity,

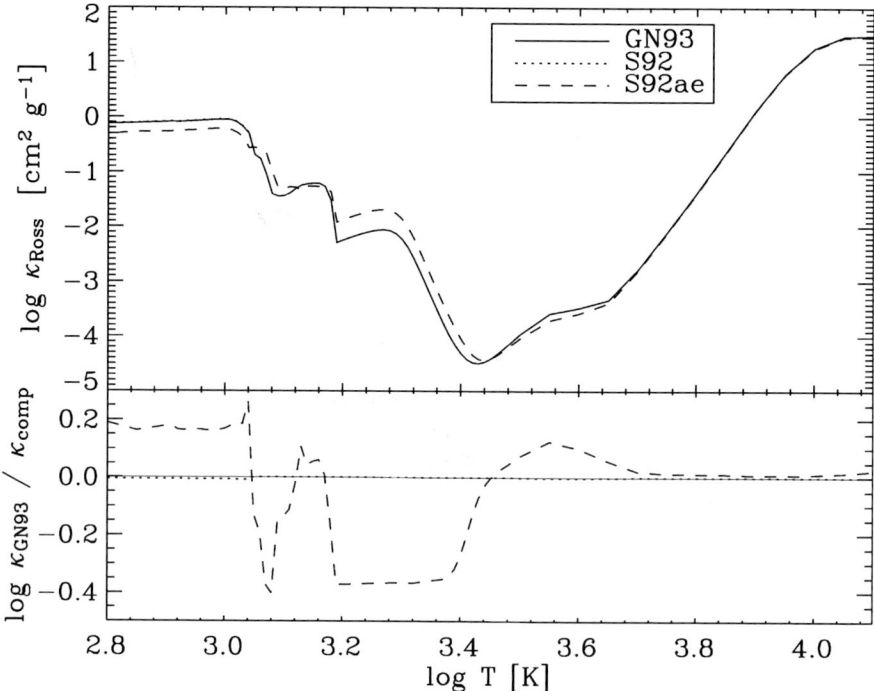

Figure 3. Similar to Fig. 1 the mean opacity is shown for GN93 abundances (solid line), S92 abundances (dotted line) and S92 with the α-elements enhanced (dashed line). The bottom portion represents the logarithm of the ratio of the opacity from the GN93 mix compared to S92 (dotted line). The dashed line represents the ratio of the GN93 mix to S92 with α-enhancement. Note that the scale of the lower portion is different from that shown in Fig. 1.

lifetracks of stars on the H-R diagram and on the mean main sequence lifetimes of stars as well. In that study the H-R diagram life tracks were computed for a series of mixtures each with individually enhanced elements. For example, what happens to a star if the initial fraction of C, N, or O is enhanced by a small fraction?

Figure 6 shows the effect of changing the abundance of individual elements on the mean opacity. Both N- and Ne-enhanced abundances do not have much of an effect on the mean opacity as those elements do not exist in molecules that have large numbers of lines. However, both the O- and C-enhanced changes greatly affect the mean opacity, mostly in the $log\ T \sim 3.3$ regime where water is an important molecular source. One noticeable feature in the figure is the rather significant change to the mean opacity caused by enhancing magnesium in the $log\ T \sim 3.55$ region. This bump in the opacity is due mostly to the very broad Mg II line at λ 2798, the fact that Mg$^+$ is very abundadant in that temperature region, and the weighting function of the Rosseland mean opacity peak is near the Mg-line. All of these features of Mg help in changing the mean opacity significantly.

Stellar evolution tracks have been computed by Dotter *et al.* (2007) and show significant differences for individually increased elements. Figure 7 shows the effect of enhancing individual elemental abundances on H-R digram isochrones for C- and O-enhanced compositions. The isochrones were computed with the Dartmouth Stellar Evolution Program (see Dotter *et al.* (2007) for details).

The conclusions of Dotter *et al.* (2007) basically boil down to the fact that low temperature opacities of various metal elements affect the lifetracks of stars in non-linear ways.

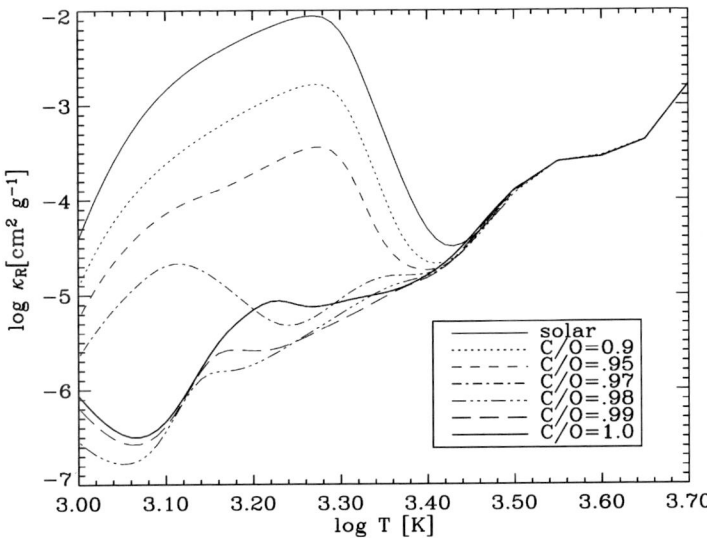

Figure 4. Shown is the mean opacity based upon calculations of varying C/O ratios including a solar mixture and upto a C/O ratio of unity. The legend gives the values of the C/O ratio and the text discusses the details of the models.

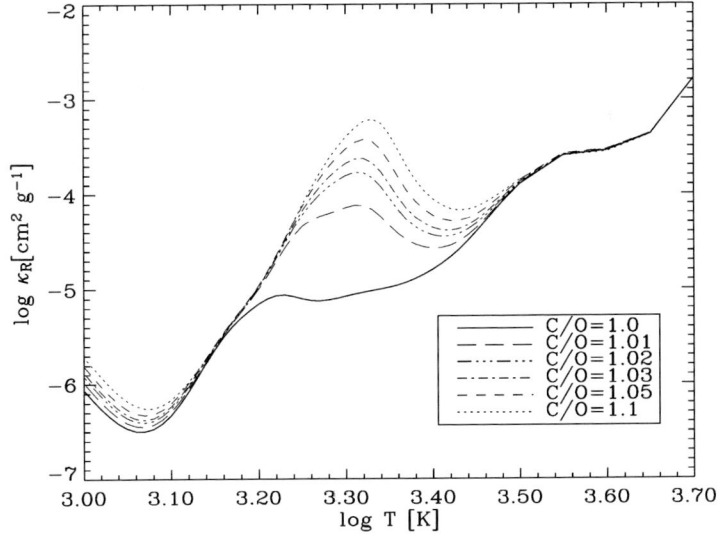

Figure 5. Similar to Fig. 4 with C/O ratios greater than unity.

For example enhancing the lighter elements (C, N, O) tends to lead to hotter shorter lived stars. However, enhancing heavier elements are not as easily constrained. For example, enhancing Fe tends to lead to cooler, longer lived stars, but enhancing other refractory elements lead to cooler stars, but the main sequence lifetimes vary in non-determinate ways.

Figure 8 shows the influence on main sequence lifetimes due to enhancing individual elements for four different stellar masses. Most elements only affect the lifetimes in small ways much less than 5% of the total, however a few elements, O and Fe, make much larger contributions. It should be noted that all of the stellar evolution results shown here are for models with constant Z rather than constant [Fe/H]. Enhancing an element

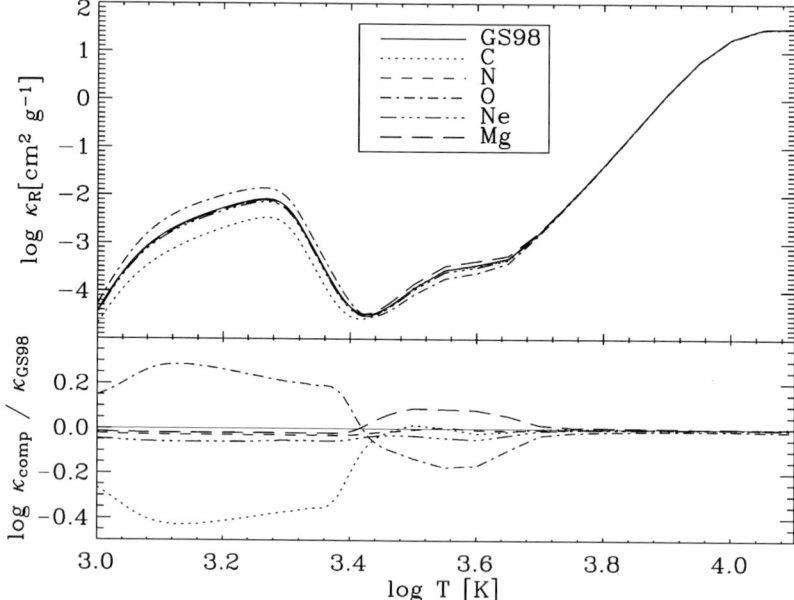

Figure 6. In the top portion of the plot are mean opacities based upon six different mixtures. The baseline consists of mean opacities computed assuming the abundance set of GS98 (solid line). Additionally mean opacities are also shown with individually enhanced abundances carbon (dotted), nitrogen (dashed), oxygen (dash-dot), neon (dash-dot-dot) and magnesium (long dash). In the bottom portion are the difference plots similar to Fig. 1. The dotted line represents the ratio of the baseline to the carbon enhanced opacities. Also shown is the GS98 to N-enhanced (dashed), GS98 to O-enchanced (dash-dot), GS98 to Ne-enhanced (dash-dot,) and GS98 to Mg-enhanced (long dash).

at constant Z requires other metals to be reduced. For example, as Dotter et al. (2007) points out enhancing the light elements mean less heavy elements and the opposite too: more Fe means there is now less C, N and O available.

4. Future Directions

With regard to section 3.1 on the opacity of different values of the C/O ratio there are several items that need updating in F05.

Comparisons of our carbon-rich mean opacities with those of Lederer & Aringer (2008) shows that our opacities may be missing a some important molecular sources. Our molecular data for HCN and for C_2H_2 are out-of-date and do not at this time include a sufficent number of lines for the wavelength coverage necessary for the computation of mean opacities. Comparisions of PHOENIX based carbon star models with observations (Peter Woitke, private communication) show the same lack of appropriate carbon molecule coverage.

A short list of near term future needs include:

• Update carbon molecular line lists to more modern lists using the open electronic databases as outlined in Tennyson et al. (2007). The Tennyson et al. paper discusses only work done on molecules at UCL including HCN/HNC, C_3, H_3^+, and HDO. Other important carbon molcules that need updating include C_2H_2 and CN.

• Explore the effects of carbon rich grains on the mean opacity. As outlined above for the C/O ratio comparisons the effects of grains were turned off in the equation of state and hence the opacity computations. Carbon grains tend to have a much higher

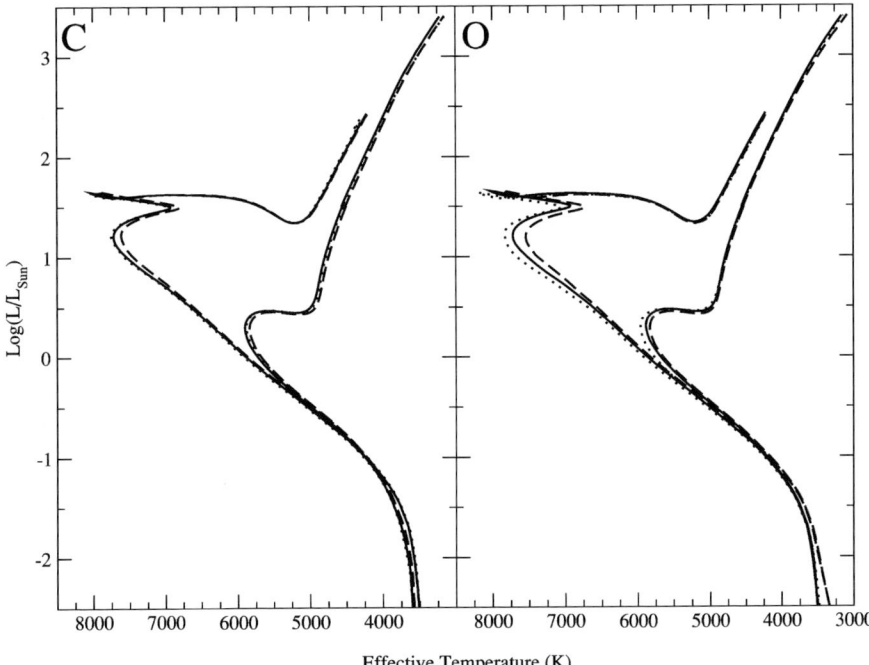

Figure 7. Plots of isochrones for 1 and 8 Gyr are shown for both C-enhanced (left) and O-enhanced (right) computations. The solid line represents the scaled solar abundance (normal computation), dotted lines are enhanced elemental abundances at constant Z and dashed lines are for the enhanced abundances with constant [Fe/H]. See Dotter et al. (2007) for color versions of these plots.

mean opacity than silicate grains. The exploration of the physical properties needed to compute the opacity of carbon grains as they are very different from silicate grains will need to be accomplished as well. For example, the size distribution that is typically used in F05 is not appropriate for carbon grains which tend to be much smaller.

Acknowledgements

Low temperature astrophysics at Wichita State University is supported by NSF grant AST-0239590 with matching support from the State of Kansas. We also acknowledge the support from the National Science Foundation under Grant No. EIA-0216178 and Grant No. EPS-0236913, matching support from the State of Kansas and the Wichita State University High Performance Computing Center. JF thanks D. Alexander for many continuing fruitful discussions.

References

Alexander, D. R. & Ferguson, J. W. 1994, ApJ, 437, 879
Asplund, M., Grevesse, N., Sauval, A. J., Allende Prieto, C., & Blomme, R. 2005, A&A, 431, 693
Dotter, A., Chaboyer, B., Ferguson, J. W., Lee, H.-c., Worthey, G., Jevremović, D., & Baron, E. 2007, ApJ, 666, 403
Ferguson, J. W., Alexander, D. R., Allard, F., Barman, T., Bodnarik, J. G., Hauschildt, P. H., Heffner-Wong, A., & Tamanai, A. 2005, ApJ, 623, 585
Ferguson, J. W., Heffner-Wong, A., Penley, J. J., Barman, T. S., & Alexander, D. R. 2007, ApJ, 666, 261

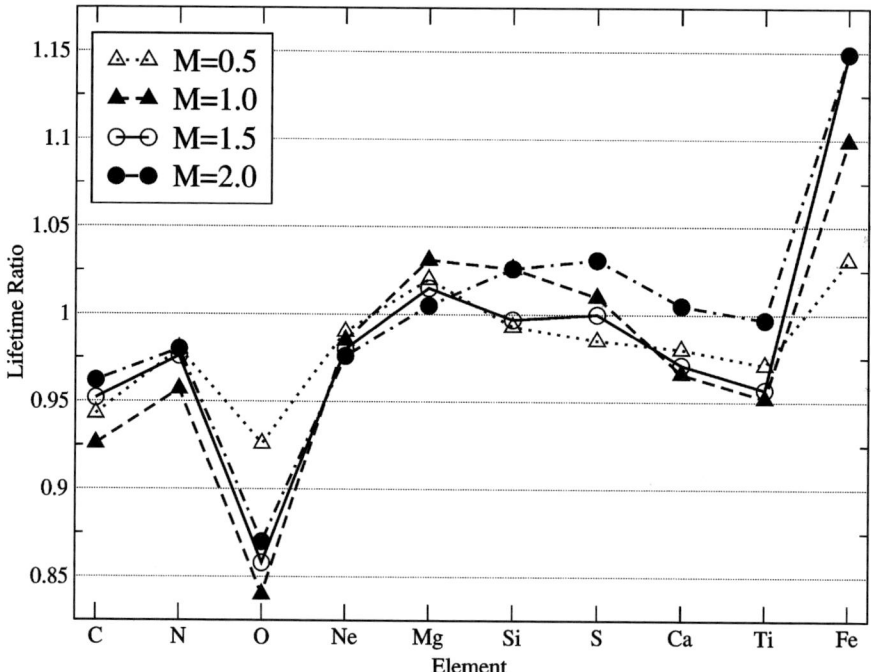

Figure 8. Plots of the ratio of main sequence lifetime due to a model with an enhanced element to a scaled-solar model to probe the effects of each element on the lifetime. Masses in the legend are given in units of solar masses and there does not appear to be any significant trend with mass. See Dotter *et al.* (2007) for a color version of this plot.

Grevesse, N. & Noels, A. 1993, Origin and Evolution of the Elements
Grevesse, N. & Sauval, A. J. 1998, Sp.Sci.Rev, 85, 161
Iglesias, C. A. & Rogers, F. J. 1991, ApJ, 371, 408
—. 1993, ApJ, 412, 752
—. 1996, ApJ, 464, 943
Lederer, M. T. & Aringer, B. 2008, AIPC, 1001, 11
Lodders, K. 2003, ApJ, 591, 1220
Marigo, P. 2002, A&A, 387, 507
Seaton, M. J., Yan, Y., Mihalas, D., & Pradhan, A. K. 1994, MNRAS, 266, 805
Tennyson, J., Harris, G. J., Barber, R. J., La Delfa, S., Voronin, B. A., Kaminsky, B. M., & Pavlenko, Ya. V. 2007, Mol. Phys., 105, 701

Discussion

HERWIG: Can you comment on molecular and dust opacities at very low Z with the primary contribution from AGB stars, like O, Al, Mg etc? What could be implications for mass loss?

FERGUSON: Even at $Z = 0.0001$ the grain opacity can be as much as 2–3 orders of magnitude above the $Z = 0$ case. If mass loss depends on radiation pressure due to grains then there may be enough opacity – but that would be environment dependent.

WILLSON: Your grain opacity is an integral over the grain distribution – what distribution did you use?

FERGUSON: The fiducial size distribution is Mathis, Rumpl + Nordsieck 1977. However, changing the distribution to another does not change the mean opacity much.

WOITKE: Comment: dust size distribution f(a) matters!
— χ_λ depends strongly on $\langle a \rangle$! $\langle a \rangle \to \infty \Rightarrow \chi_\lambda \sim \frac{1}{\langle a \rangle}$ (AGB star winds $\to$ small grains; protoplanetary disks $\to$ large grains).
— in one object, $\langle a \rangle$ may depend strongly on position, e.g., brown dwarfs: $\langle a \rangle$ small in high layers ($0.01\,\mu$m); $\langle a \rangle$ large in deep layers ($100\,\mu$m).

FERGUSON: Absolutely correct! And I want to point out that we can make opacity tables with (almost) any grain distribution you'd like.

KWOK: There are many observational overtone transitions of data in the visible part of the spectrum. The oscillation strengths of these highly excited lines are poorly known as hardly any laboratory data exist. Are these lines included in your opacity calculations?

FERGUSON: I am not sure of this specific case. Our water list is tuned for brown dwarfs. We have 3 H_2O line lists and they do differ slightly in the mean opacity they produce.

The speaker, J. Ferguson (right), is talking to P. Woitke.

Does the Sun have a subsolar metallicity?

Martin Asplund

Max Planck Institute for Astrophysics, Postfach 1317, D-85741 Garching b. München, Germany
email: asplund@mpa-garching.mpg.de

Abstract. The solar chemical composition has recently undergone a drastic revision, in particular in terms of the C, N, O and Ne abundances that have been lowered by almost a factor of two. In this invited review I will describe the different compounding reasons for this change (3D model atmospheres, non-LTE line formation, improved atomic/molecular data) and discuss some astrophysical implications thereof, which fall under both good (solar neighborhood) and bad (helioseismology) news. The most recent literature regarding the solar O abundance is surveyed and a critical evaluation whether or not these support the low solar abundance scale is presented. Finally I venture to make some predictions to what the real solar O abundance may be.

Keywords. Convection, line: formation, radiative transfer, Sun: abundances, Sun: atmosphere, Sun: granulation

1. Introduction

The solar chemical composition is a fundamental yardstick in astronomy, to which the elemental abundances of essentially all cosmic objects, be it planets, stars, nebulae or galaxies, are anchored. The importance of having accurate solar elemental abundances can thus not be overstated. From the pioneering efforts of Russell (1929) and Suess & Urey (1956) to the more recent works of Anders & Grevesse (1989), Grevesse & Sauval (1998), Lodders (2003) and Asplund *et al.* (2005a), compilations of the solar system abundances have had, and will no doubt continue to have, an extremely wide-ranging use in astronomy and cosmology. To illustrate this obvious point, it here suffices to mention that Anders & Grevesse (1989) is currently the fourth most cited astronomy article of all time according to the ADS database.

There are two independent and complementary ways of determining the solar system abundances, each with its pros and cons. Through mass spectroscopy of meteorites in terrestrial laboratories it is possible to directly measure the abundances of essentially all elements, including their various isotopes, with remarkable precision. Even in the most pristine meteorites – the so-called C1 chondrites that have been the least modified by various physical and chemical processes over the past 4.5 Gyr – the volatile elements, including the most abundant elements hydrogen, helium, carbon, nitrogen, oxygen and neon, have been depleted to various degrees. As a consequence, it is not possible to rely on meteorites to determine the primordial solar system abundances for these elements. This also implies that one must measure all meteoritic abundances relative to another element than hydrogen, traditionally chosen to be silicon, and prior knowledge of the solar photospheric Si abundance is therefore required in order to place the meteoritic abundances on the same absolute scale as the Sun (Asplund 2000). With the exception of lithium and possibly beryllium depletion and a general ∼10% modification due to diffusion and gravitational settling, the solar photospheric abundances today reflect those present at the time of the birth of the solar system. On the other hand, the solar abundances can not obviously be determined with the same accuracy as for meteorites, in particular very little isotopic abundance information is available. Furthermore, the abundances are not *observed* but *inferred*: the solar spectrum must be interpreted using realistic models of the solar atmosphere and the spectrum formation process.

Recently the recommended solar atmospheric abundances of in particular C, N, O and Ne – the four most abundant elements next to H and He – have undergone a dramatic downward revision (Asplund et al. 2004, 2005a,b) While very welcome news in many, if not most, areas of astronomy, this lowering of the solar metal content from the canonical 2% to just over 1% (thus provoking the oxymoron in the title of this article) has caused a great deal of consternation for model builders of the solar interior and helioseismologists (e.g. Basu & Antia 2008, and references therein). In this review I will briefly describe the underlying reasons for the revised abundances and present arguments in favour of the new analyses as well as discuss some lingering doubts about their reliability. I will also focus on a number of papers related to the solar oxygen abundance that have appeared very recently before making some cautionary predictions what the real solar abundances may be.

2. Ingredients for solar spectroscopic analyses

The most basic ingredient for any solar abundance analysis is an observed solar spectrum, either intensity at the center of the solar disk (Neckel & Labs 1984) or flux (Kurucz et al. 1984). To minimize the effects of magnetic activity, the disk-center intensity spectrum used is obtained for the quiet Sun, i.e. for the typical solar granulation; for disk-integrated flux spectra the influence of solar activity, sunspots etc is still minimal in terms of derived abundances. Somewhat surprisingly, the available solar atlases sometimes differ and therefore yield slightly different abundances for some lines (e.g. Caffau et al. 2008). Whether the reasons for this can be traced to actual temporal variations in the solar atmosphere over the area averaged for the solar atlas or to subtle differences in for example the terrestrial absorption and data reduction is not yet clear. A renewed effort in acquiring new high-quality solar disk-center and flux atlases addressing these issues should be undertaken (Kurucz 2006).

To model the solar spectrum one then needs a solar model atmosphere. Traditionally various 1D hydrostatic models have been employed for the purpose. These come in two flavours, theoretical models such as the Kurucz (1993), PHOENIX (Hauschildt et al. 1999) and MARCS (Gustafsson et al. 2008) grids, and semi-empirical models where perhaps the best known ones are the Holweger & Mueller (1974) and VAL3C (Vernazza et al. 1976) versions. The Holweger & Müller model has generally been the favoured model atmosphere of solar abundance aficionados, since its inferred temperature stratification from observed Fe I lines and continuum center-to-limb variation has been believed to be an accurate representation of the photospheric structure. All 1D modelling requires the use of the ad-hoc free parameters micro- and macroturbulence to obtain reasonable spectral line profiles; for theoretical models one also need to specify the mixing length parameters.

More recently 3D hydrodynamical models of the solar surface convection and atmosphere have become available (e.g. Stein & Nordlund 1998; Asplund et al. 2000a; Freytag et al. 2002; Nordlund et al. 2008) and applied for solar abundance purposes (e.g. Asplund et al. 2000b, 2004, 2005b; Caffau & Ludwig 2007; Caffau et al. 2008; Ayres 2008). In this modelling the standard hydrodynamical equations of conservation of mass, momentum and energy are solved together with the 3D radiative transfer equation in a small representative volume of the solar atmosphere typically covering ~10 granules at any given time. The 3D spectral line formation can then be computed with these models and the resulting spatially and temporally averaged line profiles compared with observations (see Asplund 2005, and references therein). While allowing for a self-consistent hydrodynamical description of convection makes for more *sophisticated* modelling, it does not necessarily imply more *realistic* outcomes, since approximations have to be made for computational reasons, in particular regarding the radiative transfer in which until now opacity binning has been used instead of opacity sampling. Therefore predictions from the 3D models have been confronted with a large number of observational diagnostics, in particular for the Sun. The 3D model successfully reproduce the solar granulation pattern in terms of length- and time-scales, brightness contrast, velocities etc (Nordlund *et al.* 2008). Perhaps the most impressive agreement comes from a comparison of the detailed line shapes: without invoking any micro- and macroturbulence, the theoretical 3D line profiles are in essentially perfect agreement with the observed lines for a wide range of elements and transitions (Asplund *et al.* 2000a). In fact even the observed line shifts and asymmetries are extremely well reproduced in general. The line profiles are the intricate result of the atmospheric inhomogeneities and (anti-)correlations between temperature, density and velocity and are therefore very sensitive probes of the atmospheric conditions.

A problem with the 3D model of Asplund *et al.* (2000a) it that it appears to predict a slightly too steep temperature gradient as inferred from continuum center-to-limb variation (Ayres *et al.* 2006; Koesterke *et al.* 2008). As seen in Fig. 2 performs better than any available 1D theoretical model but not quite as well as for the Holweger & Mueller (1974) model, which is not surprising since the latter was after all constructed in order to fulfill this observational constraint. The 3D CO5BOLD model employed by Caffau *et al.* (2008) performs as well as the Holweger & Mueller (1974) model in this regard but it has not yet been extensively tested in respect to for example line profiles and asymmetries. The center-to-limb variation is however not the only probe of the atmospheric temperature stratification. The wings of the hydrogen Balmer lines are formed in roughly similar depths but these seemingly paint a conflicting picture. The 3D model of Asplund *et al.* (2000a) yield slightly too strong wings when assuming LTE, which may be rectified when accounting for departures from LTE (Barklem 2007b). The Holweger & Mueller (1974) model, however, yields much too weak wings of Hβ even in LTE, which implies that its temperature gradient is too shallow in these layers (Pereira *et al.*, in preparation). Which model atmosphere has the most realistic temperature structure is thus not yet settled.

In addition to a model of the solar atmosphere, one also needs a model for how the spectrum formation proceeds, in particular the line formation. For late-type stars such as the Sun, this basically boils down whether or not local thermodynamic equilibrium (LTE) is a valid assumption. In general, the collisional rates do not dominate sufficiently over the corresponding radiative rates in the solar atmosphere to ensure that LTE holds. While most elements and transitions do show departures from LTE in the Sun, in terms of abundance corrections the non-LTE effects are typically rather modest ($\leqslant 0.1\,\mathrm{dex}$) (see Asplund 2005, and references therein). Some well-used solar lines, however, are significantly more affected, such as most O I lines.

Finally, all solar analyses rely heavily on having reliable input physics data. For the construction of model atmospheres, essential ingredients are equation-of-state and continuous and line opacities. For the line formation the most obvious data needed are oscillator strengths for the lines in question but a large number of other transition properties are also required which can affect the derived abundances, including the line broadening, partition functions, molecular dissociation energies etc. Moving to non-LTE vastly increases

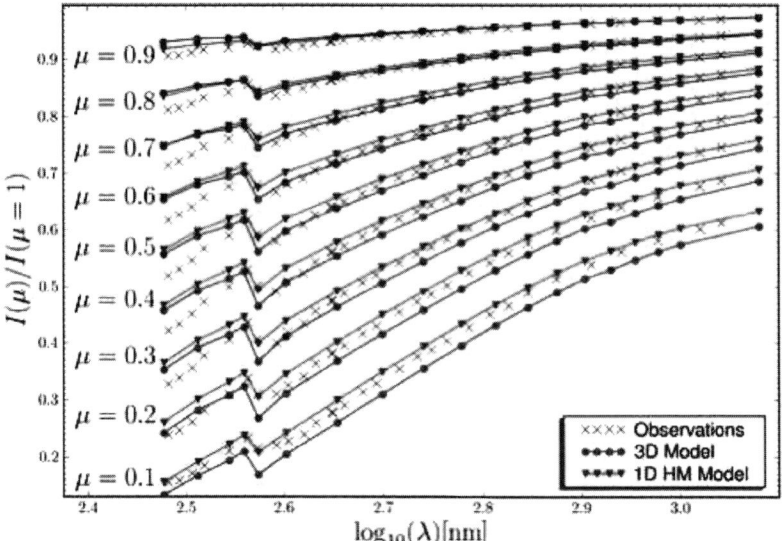

Figure 1. A comparison of the observed solar continuum center-to-limb variation (Neckel & Labs 1984) (crosses) with the predictions for the Holweger & Mueller (1974) 1D semi-empirical model (triangles) and the 3D hydrodynamical model of Asplund et al. (2000a) (circles). It can be seen that the observations fall in between the two theoretical curves, although with a slight preference for the Holweger & Mueller (1974) model for $\mu \leqslant 0.4$. The shortcomings at the shortest wavelengths are probably related to the observations including the effects of lines.

the data requirements, since these need to be specified for all radiative transitions in the element under consideration as well as photoionization and collisional cross-sections. The situation is continuously improving through the heroic efforts of a relatively small number of atomic physicists, especially in regards to transition probabilities (e.g. Kurucz 1993; Plez 1998; Johansson et al. 2003; Lawler et al. 2006) and photoionization rates (e.g. Badnell et al. 2005) The progress for the important collisional data is less satisfactory in spite of some recent advances (Barklem et al. 2003; Barklem 2007a). Non-LTE calculations therefore still normally employ various recipes based on classical physics for electron and hydrogen collisions, often with a scaling factor somehow calibrated. For late-type stars arguably the biggest piece missing is reliable estimates for the inelastic hydrogen collisions (see discussion in Asplund 2005).

3. The new solar chemical composition and implications thereof

The 3D hydrodynamical solar model atmosphere has been the foundation for a reanalysis of the solar chemical composition that has been detailed in a series of papers (e.g. Asplund et al. 2000b, 2004, 2005a,b; Asplund 2000, 2004; Allende Prieto et al. 2001, 2002; Scott et al. 2006). The most noteworthy result of this work is no doubt the significant downward revision of the present-day solar photospheric C, N, O and Ne abundances by almost a factor of two compared with the widely used recommendations by Anders & Grevesse (1989): $\log \epsilon_C = 8.39 \pm 0.05$, $\log \epsilon_N = 7.78 \pm 0.06$, $\log \epsilon_O = 8.66 \pm 0.05$, and $\log \epsilon_{Ne} = 7.84 \pm 0.06$. Besides employing a 3D rather than a 1D model, departures from LTE have been considered for transitions suspected to be affected, including performing full 3D non-LTE calculations for elements like O I. Furthermore, the best possible atomic and molecular data have been utilized and a careful evaluation of the solar spectrum was carried out to only include the most reliable lines the least affected by blends. It is important to remember that the driving force behind the new low solar abundances is not only the application of a 3D solar model but also allowance for non-LTE effects, identification of significant blends and improved line transition data. Indeed for some lines the 3D effects are very small yet a consistent low abundance is obtained from them for other reasons.

Allende Prieto et al. (2001) and Asplund et al. (2004) employed a wide range of abundance indicators to determine the solar O abundance: low excitation forbidden [O I] lines, high-excited permitted O I lines and OH lines in the infrared from both the vibration-rotation and pure rotation bands; Scott et al. (2006) added CO lines to conclude that the ^{16}O/^{18}O isotopic ratio is identical to within the uncertainties to the terrestrial value. Since the various transitions have vastly different sensitivities to the atmospheric conditions and the line formation processes, achieving consistent abundances from them is a very strong test of the accuracy of the analysis. Traditionally 1D-based analyses have implied a significantly higher O abundance from the OH and [O I] lines than for O I when non-LTE effects are considered for the latter (Asplund et al. 2004). The presence of temperature inhomogeneities and a cooler mean temperature stratification in the 3D model compared with for example the Holweger & Mueller (1974) model result in significant reduction in the OH-based results. Furthermore, Allende Prieto et al. (2001) demonstrated that the crucial [O I] 630 nm line is blended by a Ni I line, which was subsequently confirmed experimentally (Johansson et al. 2003). The end result is that finally all O indicators agree, which is a strong argument in favour of the new value. The solar Ne and Ar abundances are similarly affected, since they are based on the coronal O/Ne and O/Ar ratios together with the photospheric O abundance.

In the case of C, even more diagnostics are available: [C I], C I, CH vibration-rotation, CH electronic, C_2 and CO lines. Again, with any 1D model these imply C abundances differing by 0.2−0.3 dex, while the 3D-based result are the same to within 0.1 dex (Allende Prieto et al. 2002; Asplund et al. 2005a,b; Scott et al. 2006). Given the very different formation depths and temperature and pressure sensitivities for the various transitions, this excellent agreement is very gratifying. Compared with the recommendation of Anders & Grevesse (1989) the new value is 0.17 dex lower. The reasons for the downward revision are manifold also for C: 3D effects in particular for the molecules, non-LTE effects for C I, blends and better atomic/molecular data. Using CO lines in the infrared, Scott et al. (2006) found that the solar photospheric ^{12}C/^{13}C ratio is in agreement with the telluric value.

Given that C, N, O and Ne are the most abundant elements in the Universe besides H and He, it is not surprising that the new solar abundances of Asplund et al. (2005a) have had a wide-ranging impact in astronomy. Naturally this affects the abundances of other cosmic objects when using the customary square-bracket notation in astronomy, such as [C/H] and [O/Fe]. For the Sun, the mass fraction of metals Z decreases to from 0.0194 to 0.0122. One of the good news is that the Sun is no longer peculiar in terms of its apparent metal-richness compared to its surroundings: the new solar abundances are in excellent agreement with what is measured for nearby OB stars (e.g. Przybilla et al. 2008) and the local interstellar medium (e.g. Esteban et al. 2005), in particular when factoring in the overall ≈ 0.04 dex reduction in the photospheric metal abundances due to diffusion (Turcotte et al. 1998) and the expected ≈ 0.05 dex chemical enrichment of the interstellar medium over the past 4.5 Gyr (Chiappini et al. 2003).

The bad news is that the revised solar chemical composition messes up the predicted solar interior structure, since C, N, O and Ne are significant contributors to the opacity. The predicted sound speed as a function of depth is therefore altered and no longer agree as nicely with observations as inferred from helioseismology (e.g. Basu & Antia 2008, and references therein). This discrepancy has been named the solar modelling problem (or even the oxygen crisis, which is clearly not an appropriate labelling) and has received a great deal of attention lately. A large number of solutions have been put forward but most have already been ruled out, leaving only very few as viable explanations. Initially it was believed by some that missing opacity would do the trick but the very significant amount required (10–20%) have not been forthcoming in more recent calculations (Badnell et al. 2005). Another possibility is that a ~ 0.5 dex increase in the solar Ne abundance compensates for in particular the lower O abundance, a suggestion that has received some empirical support (Drake & Testa 2005). More recent works, however, suggest that although the solar neighborhood Ne abundance may be higher than the solar value of Asplund et al. (2005a), it is not enough by itself to solve the problem (Cunha et al. 2006; Morel & Butler 2008; Przybilla et al. 2008; Robrade et al. 2008). Perhaps the only proposed explanation still standing is the effects of internal gravity waves, which may act as an effective opacity source and provide additional mixing (Arnett et al. 2005; Charbonnel & Talon 2005). No quantitative estimate of this effect has of yet appeared in the literature.

The stubbornness with which this solar modelling problem has refused to yield a solution has lead some to question the new solar abundances of Asplund et al. (2005a), rather than the physical ingredients of the solar interior models, in spite of the successes of the new 3D-based spectroscopic analyses briefly outlined above. A significantly better agreement with helioseismology is obtained when instead using the (preliminary) abundances recommended by Grevesse & Sauval (1998) obtained from an empirical and somewhat ad-hoc additional adjustment of the temperature structure of the Holweger & Mueller (1974); in this context it is perhaps worth noting that even then the discrepancy is much larger than expected purely from the helioseismic statistical and systematic errors and that the largest problem occur at the same location, namely immediately below the solar convection zone. A further revisit of the solar abundance issue is, however, clearly justified given the current state of affairs.

4. Recent developments

In this section I will review some very recent works on the solar oxygen abundance and discuss how they agree or disagree with the low solar O abundance advocated by Asplund and collaborators; unfortunately the solar C and N abundances have been largely overlooked in this respect.

4.1. Ayres (2008)

Ayres (2008) has carried out an analysis of the [O I] 630 nm line using one snapshot of a 3D CO5BOLD solar atmosphere model. He followed the same procedure as Allende Prieto et al. (2001) and Asplund et al. (2004) for this line, namely to allow the Ni blend to vary freely in obtaining the best overall line profile fit. In many ways this new study agrees well with that of Allende Prieto et al. but differs in one critical respect: the derived O abundance, which here is found to be $\log \epsilon_O = 8.81 \pm 0.04$, or 0.12 dex higher. Part of this difference must stem from the different temperature gradients in the line-forming region of the two 3D models (Caffau et al. 2008) but it is not yet clear if this is the whole explanation. A serious problem is presented by the implied solar Ni abundance, since Ayres find that the best fit is obtained with a Ni line strength of only 70% of the theoretical line strength based on the accurate experimental gf-value of Johansson et al. (2003) and the Ni abundance of Grevesse & Sauval (1998). This difference exceeds the quoted uncertainties in these two values. In terms of abundance, a 0.15 dex lower Ni abundance corresponds to an increase of the derived O abundance by ≈ 0.09 dex for disk-center intensity (Allende Prieto et al. 2001; Caffau et al. 2008), which would reduce the suggested value of Ayres (2008) to ≈ 8.72 but at the expense of a deteriorated profile fit. It remains to be determined if this is an acceptable trade-off.

4.2. Caffau et al. (2008)

Caffau et al. (2008) have studied both [O I] and O I lines using the same 3D CO5BOLD model as employed by Ayres (2008) but covering in total 19 snapshots. Their recommended solar O abundance from the weighted mean of all lines from both flux and disk-center intensity spectra is $\log \epsilon_O = 8.76 \pm 0.07$, i.e. roughly half-way between the values of Grevesse & Sauval (1998) ($\log \epsilon_O = 8.83$) and Asplund et al. (2005a) ($\log \epsilon_O = 8.66$).

For the [O I] 630 nm line, Caffau et al. includes the Ni blend at full strength in contrast to Ayres (2008) and consequently find a much smaller O abundance: $\log \epsilon_O = 8.68$. The value itself is in excellent agreement with Allende Prieto et al. (2001) but is partly due to their larger adopted contribution of the Ni line to the 630 nm feature. They also found that the even weaker [O I] line at 636 nm, which is blended by CN lines and a Ca autoionization line, imply a 0.1 dex higher abundance; Asplund et al. (2004) found the two lines to give identical abundances. The reason for this discrepancy in the Caffau et al. study has not yet been fully identified.

For the O I lines, they carry out 3D LTE calculations to which non-LTE abundance corrections computed with 1D models are applied, in contrast to Asplund et al. (2004) who performed a full 3D non-LTE study; the 1D and 3D non-LTE effects are, however, quite similar so this simplification should not bias the results significantly. The O I-based result of Caffau et al. is 0.12 dex higher than the corresponding value in Asplund et al.. Most of this difference can be traced to different assumptions regarding the poorly known inelastic H collisions for the non-LTE calculations and the adopted equivalent widths with a smaller contribution coming from selection and weighting of lines. In fact when the same input data are used the two studies agree remarkably well for most of the O I lines, in spite of two completely independent 3D models and line formation codes have been used. As a corollary it follows that the outcome for these lines is quite insensitive to the choice of model atmosphere, be it 3D or 1D, but is driven more by other factors, in particular departures from LTE. This is also demonstrated by Socas-Navarro & Norton (2007) who derived a 3D semi-empirical solar model from inversion of spatially resolved observations of two Fe I line profiles akin to the Holweger & Mueller (1974) model constructed from spatially averaged spectra; their O I 777 nm-based abundance is in excellent agreement with the value of Asplund et al. (2004), since neither considered H collisions in their non-LTE studies based on the available laboratory and quantum mechanical evidence for other species (see discussion in Asplund 2005). Caffau et al. on the other hand make use of the classical Drawin (1968) formula with a scaling factor $S_H = 1/3$, apparently largely adopted as a middle-ground between the "extremes" $S_H = 0$ and 1.

Allende Prieto et al. (2004) argued on the basis of the observed center-to-limb variation in the solar O I 777 nm line profiles that $S_H = 1$ is slightly preferable over having no H collisions ($S_H = 0$) while the LTE case could be ruled out at very high significance; the $S_H = 1/3$ case was not considered. It is important to realize that such empirical calibrations of S_H is certainly no substitute to having real quantum mechanical or laboratory estimates, since the necessary thermalization may well come from other atomic processes than H collisions, as in the case of charge transfer reactions in Li (Barklem et al. 2003). Furthermore, the observations of Allende Prieto et al. were not optimal. Pereira et al. (in preparation) have tried to remedy this by observing closer to the limb

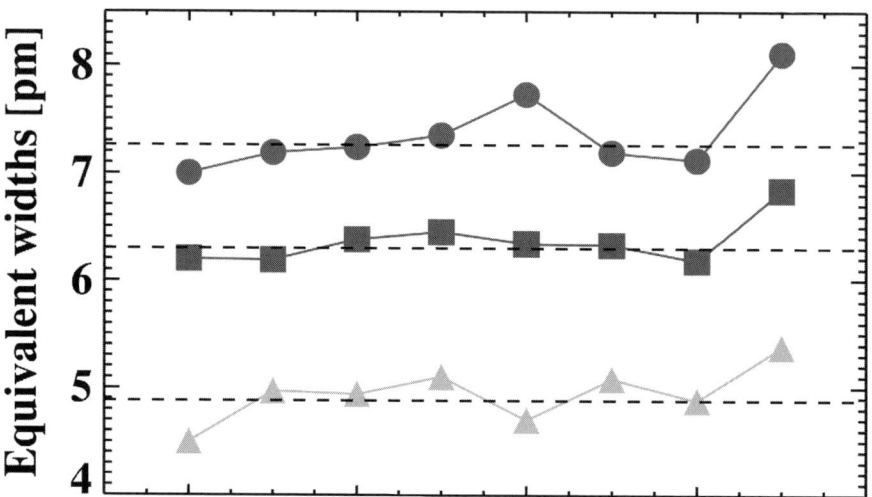

Figure 2. The published flux equivalent widths of the O I 777 nm triplet given in the eight influential O abundance analyses over the past three decades. From left to right the symbols correspond to in chronological order: Lambert (1978), Biemont et al. (1991), Kiselman (1993), Reetz (1999), Holweger (2001), Asplund et al. (2004), Allende Prieto et al. (2004) and Caffau et al. (2008); for those studies only presenting disk-center intensity equivalent widths they have been converted to corresponding flux values using the ratio of intensity to flux line strengths. The Caffau et al. equivalent widths are significantly higher than the other published values for unknown reasons. The dashed lines denote the mean for the three lines when excluding the Caffau et al. values.

and using a more suitable spectrograph. In addition they have obtained spatially resolved spectra of both the 777 nm and other O lines to test the 3D models and line formation; the analysis is currently ongoing. H. Ludwig et al. (private communication) have carried out a similar study using their own observations and find a preference for $S_H = 0$. In terms of abundance, having no H collisions would lower the Caffau et al. O abundance by ≈ 0.03 dex.

Recently, new cross-sections have been computed for electron collisions with O I (Barklem 2007a). Fabbian et al. (2008) find that the non-LTE effects become slightly more severe for the Sun when including the new collisional data, which corresponds to a lowering of the derived abundance by ≈ 0.02 dex.

Surprisingly the largest difference between the Caffau et al. and Asplund et al. studies can be traced to the adopted line strengths. In contrast to Asplund et al., Caffau et al. did not carry out full 3D non-LTE calculations and therefore had to rely on equivalent widths to derive the O abundances (the quoted equivalent widths given in Asplund et al. are the theoretical values from the best fitting 3D non-LTE profiles). Measuring equivalent widths is often tricky and subjective, in particular for the strong, partly saturated 777 nm triplet with continuum placement, exact line shape and which wavelength region to integrate over some of the key factors to worry about. Furthermore, Caffau et al. found a disconcerting difference between the available solar atlases already alluded to above (which of course would also equally affect any attempts to derive abundances from profile fitting). The equivalent widths of the 777 nm lines adopted by Caffau et al. are much larger than any other values in the literature. As demonstrated in Fig. 4.2, the Caffau et al. values differ by $\geqslant 3\sigma$ compared with the seven most recent studies. By itself this does not obviously invalidate their adopted equivalent widths but it is paramount to try to understand why the values are systematically higher than previously found. Independent new remeasurements have yielded equivalent widths in between the Caffau et al. values and the mean shown in Fig. 4.2 (N. Grevesse and J. Meléndez, private communication). More work is needed to resolve this issue. In the meantime, it is clear that profile fitting offers a superior method in determining abundances since the detailed line shape, including the asymmetries, are automatically accounted for. In terms of abundance, adopting instead the Asplund et al. predicted line strengths would lower the Caffau et al. abundances for the 777 nm triplet by ≈ 0.09 dex.

In summary, Caffau et al. (2008) found a slightly higher O abundance than Asplund et al. (2004) but the differences basically vanish when the same input data are adopted. It remains to be convincingly demonstrated that their choices in this respect are preferable with some answers expected in the near future.

4.3. Centeno & Socas-Navarro (2008)

Centeno & Socas-Navarro (2008) have invented a novel method, which to my knowledge has not previously been employed in deriving abundances. Instead of using a normal solar spectrum they analyse spectropolarimetric sunspot observations of the blend of [O I] and Ni I at 630 nm. They estimated an *atomic* ratio of O/Ni = 210 ± 24 from the line asymmetry of the Stokes V profile using their own semi-empirical model atmosphere of the sunspot obtained from an inversion of nearby Fe I lines; adopting the more traditional sunspot model of Maltby et al. (1986) would lower the ratio to ≈ 175. The total O abundance was found to be $\log \epsilon_O = 8.86 \pm 0.07$ after applying the Ni abundance of Grevesse & Sauval (1998) and correcting for the $\sim$50% of O locked up in CO in the cool environments of sunspots, i.e. even slightly higher than the recommended value of Grevesse & Sauval (1998).

A few factors that could influence the final derived abundance should be borne in mind though. Firstly, Centeno & Socas-Navarro (2008) adopted an outdated value for the [O I] gf-value that is 0.06 dex too low. Secondly, converting from atomic O to total O content requires prior knowledge of the C abundance. If it is true as they argue that all other molecules other than CO can be ignored, the maximum correction is obtained when assuming that all C is in the form of CO, which means that adopting instead the C abundance of (Asplund et al. 2005b) would imply a lowering of the derived O abundance. Thirdly, Centeno & Socas-Navarro made use of the solar Ni abundance of Grevesse & Sauval (1998), while a more recent, 3D-based analysis suggest a value ≈ 0.08 dex lower (Scott et al., in preparation). Taken together, the Stokes V profile of the [O I] 630 nm line could thus equally well support an O abundance of $\log \epsilon_O = 8.71$ (or 8.69 had the Maltby et al. sunspot model been used).

While the current observations and analysis are quite inconclusive, this new and interesting alternative approach clearly has a great deal of potential and should be studied more. In this context, it would be very valuable if a similar spectropolarimetric analysis could be performed for features coming from other elements, whose solar photospheric abundances are less in dispute than O in order to confirm that the method yields consistent results with traditional solar spectroscopy.

4.4. Meléndez & Asplund (2008)

Meléndez & Asplund (2008) have employed a forbidden O line previously not used in solar spectroscopy, namely the [O I] 557.7 nm transition. Because the line is heavily blended by two C_2 lines (Fig. 4.4), it has before not been considered a reliable abundance indicator.

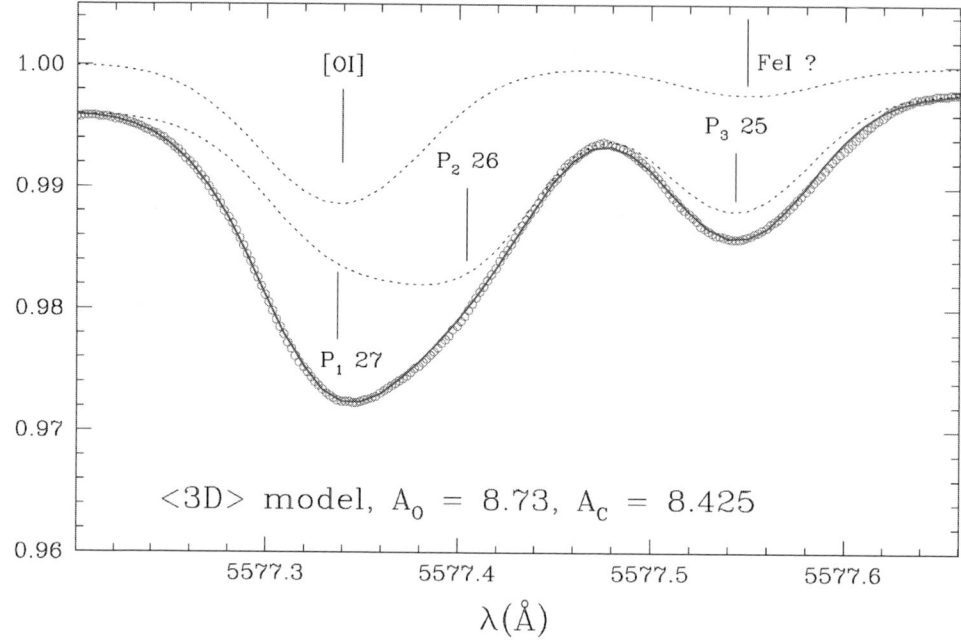

Figure 3. The solar disk-center forbidden [O I] 557.7 nm line (circles) together with the theoretical profile using the horizontally averaged 3D model assuming $\log \epsilon_O = 8.73$ and $\log \epsilon_C = 8.42$. Also shown with dotted lines are the contributions from the [O I] and C_2 lines individually. Figure taken from Meléndez & Asplund (2008).

Meléndez & Asplund have shown, however, that the strength of the C_2 lines can be accurately constrained by numerous other nearby C_2 lines from the same molecular band with essentially identical excitation potential, line strengths and line formation properties; the expected uncertainties in relative line strengths between these C_2 lines are very small indeed. Another advantage with this line is that it originates not from the O I ground state as the [O˙I] lines at 630 and 636 nm but from a level at 2 eV, which makes it less sensitive to the atmospheric structure and thus the details of the model atmosphere Asplund (2005).

Relying on the same 3D hydrodynamical solar model as in Asplund et al. (2004), Meléndez & Asplund derived $\log \epsilon_O = 8.70 \pm 0.08$ with very similar results coming from 1D models (e.g. 8.73 with the Holweger & Mueller (1974) model) The mean abundance of the tested 3D and 1D models was found to be $\log \epsilon_O = 8.71 \pm 0.02$, i.e. in very good agreement with the Asplund et al. (2004) results for the other two [O I] lines.

5. Does the Sun really have a subsolar metallicity?

From the previous section, it is clear that there is currently no clear consensus what the solar O abundance, and by extension what the overall solar metal content, is. It is probably fair to say that nowadays there is no support for the high abundance $\log \epsilon_O = 8.93$ recommended by Anders & Grevesse (1989) but the question is how much lower the real value is: is it as low as $\log \epsilon_O = 8.66$ as advocated by Asplund et al. (2004) or is it closer to $\log \epsilon_O = 8.83$ as the uncompleted study of of Grevesse & Sauval (1998) suggested? It would be premature to here give a definite answer, since a great deal of work still remains to be done before the final tally is in. Nevertheless I will venture to here make some, not entirely well-grounded, speculations and extrapolations based on the available evidence presented above.

The three forbidden [O I] lines seem to suggest $\log \epsilon_O \approx 8.7 - 8.75$, i.e. slightly larger than Asplund *et al.* (2004) due to a 3D model that even better reproduces the continuum center-to-limb variation but partly offset by an increased Ni contribution to the 630 nm line. The O I lines may be ~ 0.05 dex higher than in Asplund *et al.* following adjustments to the non-LTE abundance corrections and/or adopted line strength, which would result in $\log \epsilon_O \approx 8.7$; the exact 3D atmospheric structure is not particularly important in dictating the outcome. Similarly one would expect the molecular-based abundances to increase with a slightly warmer temperature structure, which is consistent with the fact that in Asplund *et al.* the OH lines implied slightly lower O abundances than the atomic lines. An upper limit to the abundance is set by the Holweger & Mueller (1974) model, since it has a slightly too shallow temperature gradient (Fig. 2) and the presence of temperature inhomogeneities inevitably lead to a lowering of the derived O abundance. Given that Asplund *et al.* found that the overall 3D effects for the OH lines are in roughly equal parts a combination of mean stratification and atmospheric inhomogeneities, the OH-based abundance may be something like $\log \epsilon_O \approx 8.7 - 8.75$.

In summary, it seems reasonable to expect the solar O abundance to be $\log \epsilon_O \approx 8.7 - 8.75$, i.e. a slight upward revision of the value presented in Asplund *et al.* (2004, 2005a,b). Such an O abundance would ease but not remove the discrepancy between solar interior models and helioseismology. It should be borne in mind though that this is nothing more than my own personal expectation, which have not yet been confirmed or otherwise by actual calculations. We are currently working on constructing a new 3D solar model based on an improved treatment of radiative transfer, including the first ever models employing opacity sampling rather than opacity binning to treat the line-blanketing. Whether or not this model will fulfill the constraints from center-to-limb variation as well as the other tests such as detailed line profiles remains to be seen. A complete re-analysis of not only the solar C, N and O abundances but of most elements accessible by solar spectroscopy is currently underway, with the results presented elsewhere (Asplund *et al.*, in preparation).

Acknowledgements

It is a pleasure to acknowledge the wonderful work done by my main collaborators on the topic of solar modelling and abundance analysis: Carlos Allende Prieto, Mats Carlsson, Remo Collet, Nicolas Grevesse, Wolfgang Hayek, Dan Kiselman, David Lambert, Åke Nordlund, Jacques Sauval, Patrick Scott, Bob Stein and Regner Trampedach. Finally, the author is grateful to the editors for showing a great deal of patience when waiting for this contribution.

References

Allende Prieto, C., Asplund, M., & Fabiani Bendicho, P. 2004, A&A, 423, 1109
Allende Prieto, C., Lambert, D. L., & Asplund, M. 2001, ApJL, 556, L63
Allende Prieto, C., Lambert, D. L., & Asplund, M. 2002, ApJL, 573, L137
Anders, E. & Grevesse, N. 1989, Geochim. Cosmochim., 53, 197
Arnett, D., Meakin, C., & Young, P. A. 2005, in ASP Conf. Series, Vol. 336, 235
Asplund, M. 2000, A&A, 359, 755
Asplund, M. 2004, A&A, 417, 769
Asplund, M. 2005, ARA&A, 43, 481

Asplund, M., Grevesse, N., & Sauval, A. J. 2005a, in ASP Conf. Series, Vol. 336, 25
Asplund, M., Grevesse, N., Sauval, A. J., Allende Prieto, C., & Blomme, R. 2005b, A&A, 431, 693
Asplund, M., Grevesse, N., Sauval, A. J., Allende Prieto, C., & Kiselman, D. 2004, A&A, 417, 751
Asplund, M., Nordlund, Å., Trampedach, R., Allende Prieto, C., & Stein, R. F. 2000a, A&A, 359, 729
Asplund, M., Nordlund, Å., Trampedach, R., & Stein, R. F. 2000b, A&A, 359, 743
Ayres, T. R. 2008, ApJ, in press
Ayres, T. R., Plymate, C., & Keller, C. U. 2006, ApJS, 165, 618
Badnell, N. R., Bautista, M. A., Butler, K., et al. 2005, MNRAS, 360, 458
Barklem, P. S. 2007a, A&A, 462, 781
Barklem, P. S. 2007b, A&A, 466, 327
Barklem, P. S., Belyaev, A. K., & Asplund, M. 2003, A&A, 409, L1
Basu, S. & Antia, H. M. 2008, Physics Reports, 457, 217
Biemont, E., Hibbert, A., Godefroid, M., Vaeck, N., & Fawcett, B. C. 1991, ApJ, 375, 818
Caffau, E. & Ludwig, H.-G. 2007, A&A, 467, L11
Caffau, E., Ludwig, H.-G., Steffen, M., et al. 2008, ArXiv astro-ph/0805.4398
Centeno, R. & Socas-Navarro, H. 2008, ApJL, 682, L61
Charbonnel, C. & Talon, S. 2005, Science, 309, 2189
Chiappini, C., Romano, D., & Matteucci, F. 2003, MNRAS, 339, 63
Cunha, K., Hubeny, I., & Lanz, T. 2006, ApJL, 647, L143
Drake, J. J. & Testa, P. 2005, Nature, 436, 525
Drawin, H.-W. 1968, Zeitschrift fur Physik, 211, 404
Esteban, C., García-Rojas, J., Peimbert, M., et al. 2005, ApJL, 618, L95
Fabbian, D., Asplund, M., Carlsson, M., & Kiselman, D. 2008, A&A, in press
Freytag, B., Steffen, M., & Dorch, B. 2002, Astronomische Nachrichten, 323, 213
Grevesse, N. & Sauval, A. J. 1998, Space Science Reviews, 85, 161
Gustafsson, B., Edvardsson, B., Eriksson, K., et al. 2008, A&A, 486, 951
Hauschildt, P. H., Allard, F., Ferguson, J., Baron, E., & Alexander, D. R. 1999, ApJ, 525, 871
Holweger, H. 2001, in American Institute of Physics Conference Series, Vol. 598, , 23
Holweger, H. & Mueller, E. A. 1974, Solar Physics, 39, 19
Johansson, S., Litzén, U., Lundberg, H., & Zhang, Z. 2003, ApJL, 584, L107
Kiselman, D. 1993, A&A, 275, 269
Koesterke, L., Allende Prieto, C., & Lambert, D. L. 2008, ApJ, 680, 764
Kurucz, R. L. 1993, in ASP Conf. Series, Vol. 44, IAU Colloq. 138, ed. M. M. Dworetsky, F. Castelli, & R. Faraggiana, 87
Kurucz, R. L. 2006, ArXiv astro-ph/0605029
Kurucz, R. L., Furenlid, I., Brault, J., & Testerman, L. 1984, Solar flux atlas from 296 to 1300 nm (National Solar Observatory)
Lambert, D. L. 1978, MNRAS, 182, 249
Lawler, J. E., Den Hartog, E. A., Sneden, C., & Cowan, J. J. 2006, ApJS, 162, 227
Lodders, K. 2003, ApJ, 591, 1220
Maltby, P., Avrett, E. H., Carlsson, M., et al. 1986, ApJ, 306, 284
Meléndez, J. & Asplund, M. 2008, A&A, submitted
Morel, T. & Butler, K. 2008, ArXiv astro-ph/0806.0491
Neckel, H. & Labs, D. 1984, Solar Physics, 90, 205
Nordlund, A., Stein, R., & Asplund, M. 2008, Living Reviews in Solar Physics, in press
Plez, B. 1998, A&A, 337, 495
Przybilla, N., Nieva, M. F., Heber, U., & Butler, K. 2008, A&A, submitted
Reetz, J. 1999, Astroph. & Space Science, 265, 171
Robrade, J., Schmitt, J. H. M. M., & Favata, F. 2008, A&A, 486, 995
Russell, H. N. 1929, ApJ, 70, 11
Scott, P. C., Asplund, M., Grevesse, N., & Sauval, A. J. 2006, A&A, 456, 675

Socas-Navarro, H. & Norton, A. A. 2007, ApJL, 660, L153
Stein, R. F. & Nordlund, A. 1998, ApJ, 499, 914
Suess, H. E. & Urey, H. C. 1956, Reviews of Modern Physics, 28, 53
Turcotte, S., Richer, J., Michaud, G., Iglesias, C. A., & Rogers, F. J. 1998, ApJ, 504, 539
Vernazza, J. E., Avrett, E. H., & Loeser, R. 1976, ApJS, 30, 1

Discussion

S. TURCK-CHIEZE: Among lot of other questions, I would like to know if the magnetic field and the variation are taken into account in the 3D solar atmosphere?

M. ASPLUND: In the 3D solar atmosphere model I described here that forms the basis of our solar abundance analysis, magnetic fields have not been considered, although the code can handle that. For spectral line formation in the quiet Sun magnetic fields are unimportant and thus this omission does not influence the conclusions regarding the chemical composition of the Sun.

J. CHRISTENSEN-DALSGAARD: Two comments: (1) The slope of $\delta c/c$ is easy to understand: $\delta c/c \simeq 0$ in the convection zone (c^2 is determined by the surface gravity), and in the radiation region is largest at the base of convection zone where the opacity change is largest. (2) The agreement between the new solar abundance and the abundance in the solar neighborhood is impressive, but the solar birth place is far from the present location and the composition could have been different there.

M. ASPLUND: (1) My point was mainly that even with the old, high solar abundances there is a discrepancy, although admittedly not nearly as large, immediately below the convection zone. This may be a coincidence of course, but it may also suggest that there are processes related to convection not yet taken into account in standard models (convection overshoot, internal gravity waves etc.)

(2) It is true of course that the Sun has migrated over the last 4.5 Gyr, but given its essentially perfectly circular orbit today it seems unlikely that it originated at a very different Galactocentric radius. Furthermore, all solar twins with the same stellar parameters as the Sun also have the same O abundance within the uncertainties. The same argument would then require that all of them also have migrated similarly.

HANS-G. LUDWIG: (1) What is the new abundance of nickel which you obtain? (2) What is your "best guess" value of S_H?

M. ASPLUND: (1) Our new 3D-based solar Ni abundance determination (Scott et al., in preparation) gives $\log \epsilon_{Ni} = 6.17 \pm 0.07$ using some ten Ni I lines.

(2) At this stage I don't have a preferred value for the scaling factor S_H to the Drawin (1968) formula for inelastic H collisions. The available evidence from laboratory and quantum mechanical calculations for other elements imply that the Drawin formula overestimates the collisional cross-sections by several order of magnitudes, which suggests that $S_H = 0$ also for O. The available center-to-limb observations for O I 777 nm on the other hand seem to suggest $S_H \simeq 1$. We clearly need quantum mechanical calculations for O I + H I collisions to resolve the issue but in the meantime we have obtained new and better UV observations for several O I and [O I] lines as well as spatially resolved spectra at different viewing angles μ.

S. KWOK: I just want to mention that interstellar medium lines have their own problems. In addition to dynamics and non-LTE effects, there are also uncertainties in excitation and ionization effects. In general, absorption lines are much easier to interpret than emission lines and their corresponding abundances are much more accurate.

M. ASPLUND: I fully agree that ISM abundance analyses have their own problems and systematic errors, which only due to limited time I did not have the opportunity to discuss in any detail.

The Art of Modelling Stars in the 21st Century
Proceedings IAU Symposium No. 252, 2008
L. Deng & K.L. Chan, eds.

Rigorous and Phenomenological Equations of State

Werner Däppen and Dan Mao

Department of Physics and Astronomy, University of Southern California
Los Angeles, CA 90089-1342, USA
email: dappen@usc.edu & dmao@usc.edu

Abstract. For solar and stellar modeling, a high-quality equation of state is crucial. But the inverse is also true: the astrophysical data (helioseismic today, asteroseismic tomorrow) put constraints on the physical formalisms, making the Sun and the stars laboratories for plasma physics. One of the main astrophysical benefits from a good equation of state is an improved abundance determination. Recent theoretical progress in the equation of state has involved both rigorous and phenomenological approaches, giving the user a considerable choice.

Keywords. Equation of state, Plasmas, Atomic processes, Stars: evolution, Sun: helioseismology

1. Introduction

A simple ideal-gas model of the plasma of the solar interior was adequate before helioseismology, but helioseismic and asteroseismic equation-of-state analyses require more sophisticated physical models. In the case of the Sun, the need to go beyond the ideal-gas approximation for helioseismic applications had been recognized in the early 1980s (see Berthomieu *et al.* 1980; Ulrich 1982; Noels *et al.* 1984). With the better data that became available towards the end of the 1980s, a clearer picture began to emerge. Christensen-Dalsgaard *et al.* (1988, 1996) demonstrated that it was essential to include the Coulomb correction. The relative Coulomb pressure correction peaks in the outer part of the convection zone (about –8 per cent) and in the solar core (about –1 per cent). For solar conditions, the Debye-Hückel (DH) theory is a good approximation for the leading term of the Coulomb correction.

Helioseismic equation-of-state studies typically use solar models based on sophisticated new equations of state, in particular, the ones underlying the two ongoing major opacity re-computation efforts. One of these efforts is the international Opacity Project (OP; see the books by Seaton 1995; Berrington 1997); it contains the so-called Mihalas-Hummer-Däppen equation of state (Hummer & Mihalas 1988; Mihalas *et al.* 1988; Däppen *et al.* 1988; Nayfonov *et al.* 1999; Trampedach *et al.* 2006); hereinafter MHD) and it deals with *heuristic* concepts about the modification of atoms and ions in a plasma. The other effort is being pursued at Lawrence Livermore National Laboratory by the OPAL group (Iglesias and Rogers 1996; Rogers *et al.* 1996; Rogers & Nayfonov 2002); its equation of state is based on a detailed *systematic* method to include density effects in a plasma.

Although approximate asymptotic techniques (see Christensen-Dalsgaard *et al.* 1985; Gough 1993) exist to invert solar oscillation frequencies for the internal sound speed, for an accurate analysis of the observations, a fully-fledged, non-asymptotic numerical treatment of the oscillations is mandatory (see Gough *et al.* 1996). Figure 1 is a typical result of such a numerical inversion (Basu & Christensen-Dalsgaard 1997). It shows the relative difference (in the sense sun – model) between the squared sound speed obtained from inversion of oscillation data and that of a two standard solar models. The two solar models

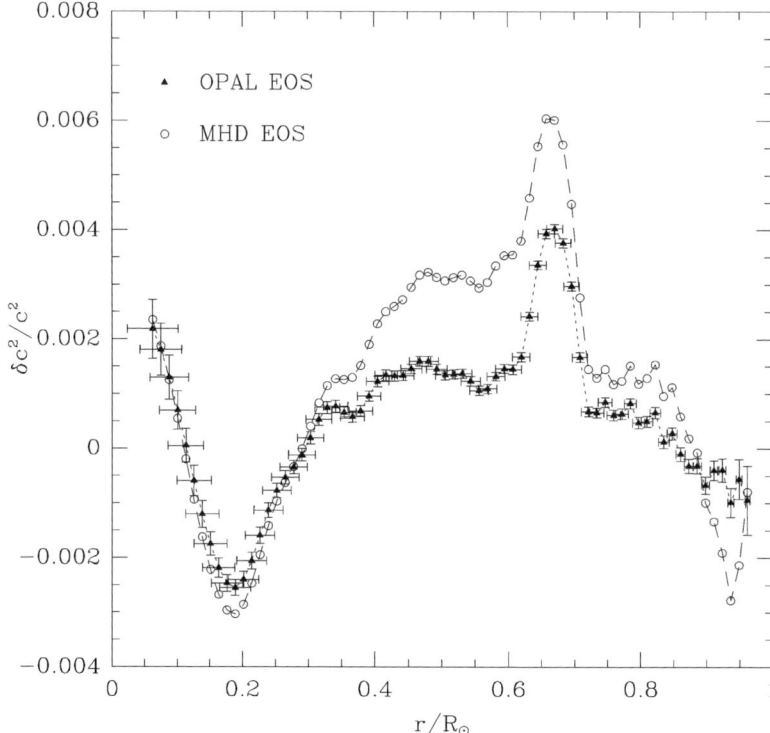

Figure 1. Difference between squared sound speed from inversion and two solar models, which only differ in their equation of state (see text). Figure by S. Basu.

used are identical in all respects except for their equation of state, MHD (circles) and OPAL (triangles), respectively. For the present purpose, we can consider inversion results such as Figure 1 as the *data* of helioseismology, disregarding the procedure through which they were actually obtained from solar oscillation frequencies. It follows from Figure 1 that in most parts of the sun the OPAL equation of state is a better approximation to reality than MHD, but OPAL needs to be improved as well.

2. Various equations of state

With the exception of neutron stars, in stellar interiors the thermal de Broglie wavelength of nuclei, atoms, ions and molecules is always tiny and for them a classical treatment is sufficient. Only electrons have to be treated according to quantum mechanics. At this point there is a bifurcation into two distinct classes of approach, the "chemical picture" and the "physical picture" (see, *e.g.*, the book by Kraeft *et al.* 1986). While in the more conventional chemical picture bound configurations (atoms, ions and molecules) are introduced and treated as new and independent species, only *fundamental* particles (electrons and nuclei) appear in the physical picture. In the chemical picture, reactions between the various species occur, and thus the thermodynamical equilibrium must be sought among the stoichiometrically allowed set of concentration variables by means of a maximum entropy (or minimum free-energy) principle. In contrast, the physical picture has the esthetic advantage that there is no need for a minimax principle; the question of bound states is dealt with implicitly through the Hamiltonian describing the interaction between the fundamental particles.

As mentioned in the Introduction, in most parts of the Sun, OPAL fares better than MHD. However, all OPAL results are so far only available in the form of pre-computed tables, and they have been made for certain chemical compositions of astrophysical relevance (Rogers *et al.* 1996). In 2005, the most recent OPAL equation of state tables were released. These tables allow low stellar mass down to 0.1 $M_\odot$ (Rogers & Nayfonov 2002). However, it is important to note that the OPAL computer code is proprietary and belongs exclusively to the Livermore group. Therefore, the community cannot use OPAL without recourse to tables. If the goal is to have more flexibility than is possible with mere tables, one solution consist in retrofitting chemical-picture formalisms such that they agree with OPAL. Indeed, Liang (2004) and Liang & Däppen (2003, 2004) have successfully emulated the OPAL equation of state. A recent extension of that work (Mao 2008) allowed the OPAL emulation for a mixture of hydrogen and helium (see Section 4).

Alternatively, progress can be made by developing more rigorous equations of state. One example is based on the path-integral formalism in the framework of the so-called Feynman-Kac (FK) representation. The FK formalism leads to a virial expansion of the thermodynamic functions in power of the total density of a Coulomb plasma (Alastuey & Perez 1992; Alastuey *et al.* 1994, 1995). Like OPAL, this approach was developed in the physical picture. Specific to the Feynman-Kac representation is the equivalence between a point-charge quantum system and a classical one made of extended filaments. The explicit calculation of the thermodynamic function has been carried out up to order $\rho^{5/2}$ in total density for general plasmas (Alastuey & Perez 1996). The finite truncation limits its domain of validity to a temperature and density regime close to the conditions of full ionization (see next section). It is planned to achieve the implementation of FK in solar and stellar models (see Section 5).

3. Activity versus virial expansions

In the previous section, we have mentioned that virial expansions limit their domain of validity to a temperature and density regime close to the conditions of full ionization, a restriction to which activity are not subject. We illustrate this different behavior using a simple but typical example. The essential technique can be understood by considering the exact equation of the ideal (non-relativistic) Fermi gas. It is an *implicit* equation of state, obtained from the grand-canonical partition function (see, e.g.)

$$\frac{p}{kT} = \frac{1}{\lambda^3} f_{5/2}(\tilde{z}), \quad \frac{N}{V} = \frac{1}{\lambda^3} f_{3/2}(\tilde{z}). \tag{3.1}$$

Here, p, T, N, V, k are pressure, temperature, particle number, volume, and the Boltzmann constant, respectively. The thermal de Broglie length λ, and the fugacity $\tilde{z}$ are defined based on the Fermi integrals $f_{5/2}$ and $f_{3/2}$

$$\lambda = \frac{h}{\sqrt{2\pi mkT}}; \tilde{z} = e^{\frac{\mu}{kT}}, \quad f_{5/2}(\tilde{z}) = \frac{4}{\sqrt{\pi}} \int_0^\infty dx\, x^2 \ln(1 + \tilde{z} e^{-x^2}), \quad f_{3/2}(\tilde{z}) = \tilde{z} \frac{d}{d\tilde{z}} f_{5/2}(\tilde{z}). \tag{3.2}$$

Expanding the Fermi integrals as a series in the high-temperature limit, and eliminating the fugacity $\tilde{z}$, one obtains the density (virial) expansion

$$\frac{pV}{NkT} = 1 + \frac{1}{2^{5/2}} \frac{N\lambda^3}{V} + \dots, \tag{3.3}$$

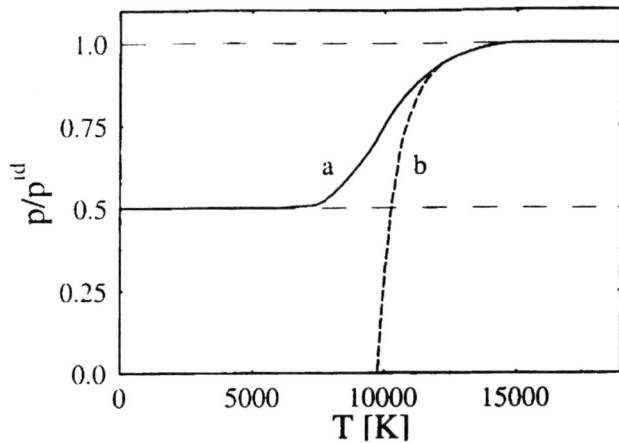

Figure 2. Pressure in a simple reacting system of neutral hydrogen H, recombining to hydrogen molecules H_2. P_{id} refers to fully dissociated hydrogen. The activity expansion (a) correctly describes recombination with just one non-ideal term (equations 3.4,3.5). Without further high-order terms, the corresponding virial expansion (b) (*i.e.* the reacting analog to the single-species equation 3.3) exhibits catastrophic non-physical behavior (negative pressure!) (Ebeling *et al.* 1976).

This is a nice and important result, since it gives not only the familiar classical limit of a rigorous quantum result, but it also states the precise physical conditions when that limit is attained. Note that this equation is a convenient *density* expansion. However, I now show that not eliminating the fugacity also has its rewards. One can write

$$\frac{p}{kT} = \frac{1}{\lambda^3}\left[\tilde{z} - \frac{1}{2^{5/2}}\tilde{z}^2 + ...\right] = z - \frac{\lambda^3}{2^{5/2}}z^2 + ... , \qquad (3.4)$$

$$\frac{N}{V} = \frac{1}{\lambda^3}\left[\tilde{z} - \frac{1}{2^{3/2}}\tilde{z}^2 + ...\right] = z - \frac{\lambda^3}{2^{3/2}}z^2 + ... , \qquad (3.5)$$

with the definition of the *activity* z

$$z = \frac{1}{\lambda^3}\tilde{z} . \qquad (3.6)$$

Note that while these expressions are still for the ideal Fermi gas, from the structure of the formal grand-canonical partition function it is obvious that any real gas has an activity expansion (no need to evaluate the grand-canonical partition function, merely to look at it)

$$\frac{p}{kT} = z - b_2 z^2 + ... , \qquad (3.7)$$

$$\frac{N}{V} = z - 2b_2 z^2 + ... , \qquad (3.8)$$

It is the task of a theory for a specific system to deliver the expansion coefficients. Nevertheless, already before doing that one can convince oneself of an important advantage of the activity formalism when applied to *reacting gases* (the system of neutral hydrogen H recombining to hydrogen molecules H_2 is the simplest case; a system of protons and electrons recombining to atomic hydrogen is another relatively simple case) (Ebeling *et al.* 1976). Figure 2 demonstrates that already the lowest order of an activity expansion describes the low-temperature recombination correctly. Although as *infinite*

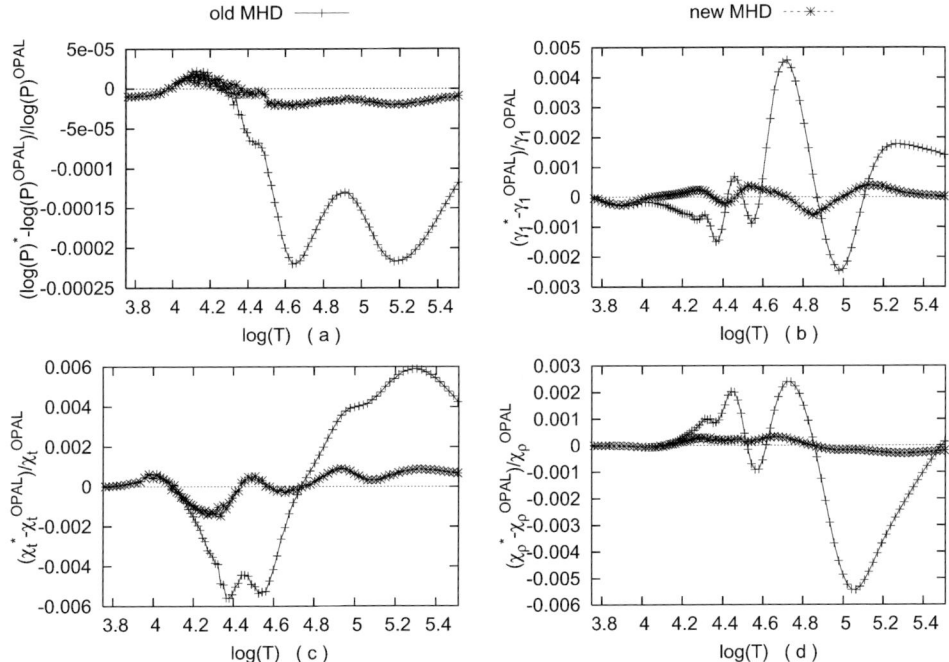

Figure 3. OPAL emulator: relative difference of thermal quantities (a) $\log p$, (b) Γ_1, (c) χ_T, and (d) χ_ρ in a solar model, in the sense old MHD – OPAL and new MHD – OPAL, respectively. OPAL is thus represented by the zero line. The quality of the emulation is reflected by the closeness of the new MHD values (large asterisks) to zero.

series, activity and density expansions are strictly equivalent, a density expansion would have to go to very high order merely to describe recombination (or ionization).

The chemical picture can do without the introduction of the activity variable; it is therefore more intuitive. However, the chemical picture is plagued with fundamental weaknesses, such as the heuristic but never rigorous assumption of the existence of atoms in a plasma (for an introduction to current issues see, *e.g.*, Däppen 2004, 2006). Activity expansions (such as OPAL) are not limited by these weaknesses. The OPAL equation of state has demonstrated that activity expansions combine the conceptual advantage of the physical picture with an accurate practical realization, needing to involve only a relatively small number of expansion terms.

4. Progress with the phenomenological approach: the OPAL emulator

As part of her PhD thesis, Mao (2008) has upgraded the OPAL emulator mentioned in Section 2, by extending it from pure hydrogen to a hydrogen-helium mixture. As before, the MHD equation of state has been appropriately modified, by adding Coulomb-scattering terms and a modified internal partition function (the so-called "Planck-Larkin partition function" PLPF, see e.g. Ebeling *et al.* 1976). The resulting emulator gives excellent agreement with OPAL, as shown in Fig. 3 (here, "old" MHD denotes the usual MHD, "new" MHD the OPAL emulator). It is obvious that the remaining differences between the modified MHD equation of state have become much smaller than those between the original MHD equation of state and OPAL. Work is in progress to release a user-friendly version of the emulator to the community.

5. The rigorous FK equation of state

Although the MHD and OPAL equations of state can be very successful in describing the structure of the Sun and stars, there is still room in improving its precision. Among all the attempts to improve the equation of state, the formalism of path integrals in the framework of the Feynman-Kac (FK) representation promises the most rigorous results, because it contains all non-ideal contributions to the equation of state, systematically, exactly and analytically.

As we have seen in Section 3, the fact that FK is a density (virial) expansion causes a severe limitation of its domain of validity. The plasma has to be nearly fully ionized in order for the method to be valid. Despite this limitation, the FK approach can still be applied to large parts of stellar interiors (see Perez *et al.* 2008). In the locations where FK can be used, it is especially suited to study effects of screening, bound states, the location of the transition to full ionization, and diffraction and exchange effects. Incidentally, follow-up papers of Perez *et al.* (2008) will be dedicated to the extension of the FK calculations to higher orders, including the order ρ^3 in density. This will allow us to take into account exactly the three-body effects occurring in a fairly ionized plasma (for example, in the case of the Sun, the recombination of He^{++}).

6. Conclusions

While currently available equations of state give reasonable accuracy for solar and stellar modelers, the observational data have the potential to aim higher, not only for better models, but especially for a maximum use of the Sun and stars to serve as plasma-physics experiments. These experiments are complementary to on-going efforts in terrestrial laboratories.

In the equation of state work, there has been recent progress both on the phenomenological and rigorous side. On the phenomenological side, the result is a useful OPAL emulator. It is based on new terms put in MHD. These terms were directly borrowed from OPAL and have no natural foundation in the chemical picture; therefore they were missing in the original MHD equation of state. (It is unclear if they could ever be found consistently within the MHD approach.) Physically, the upgrade of MHD reflects the fact that MHD, as any typical chemical-picture formalism, describes ionized electrons in the plane-wave approximation. The main added ingredient is based on Coulomb wave functions for the free (scattering) states and using the Planck-Larkin partition function. We have demonstrated that such an upgrade of MHD does indeed bring it very close to OPAL. The resulting OPAL emulator will be a useful tool for solar and stellar modelers, who will be able to use the OPAL equation of state directly, without recourse to the OPAL tables distributed by the Lawrence Livermore National Laboratory. (Currently there is no alternative to the official OPAL equation of state tables, since their source code is not publicly available.)

On the rigorous side, there has been recent progress with an equation of state based on the Feynman-Kac path-integral formalism. Although it is expressed as a density expansion, which limits its application (Section 3), the FK formalism can still be applied to most regions of the solar interior. The localizing power of helioseismology will allow us to test the FK equation of state in the regions of the solar interior where it works well. And since FK is so far the most exact and systematic equation-of-state approach, it will provide a new way to study the influence of various fine physical effects.

Acknowledgements

We thank Asher Perez, Forrest Rogers and Hugh DeWitt for stimulating discussions. This work was supported by the grant AST-0708568 of the National Science Foundation.

References

Alastuey, A. & Perez, A. 1992, Europhys. Lett. 20, 19
Alastuey, A., Cornu, F., & Perez, A. 1994, Phys. Rev. E, 49, 1077
Alastuey, A., Cornu, F., & Perez, A. 1995, Phys. Rev. E, 51, 1725
Alastuey, A. & Perez, A. 1996, Phys. Rev. E, 53, 5714
Basu, S. & Christensen-Dalsgaard, J. 1997, Astron. Astrophys., 322, L5
Berrington, K. A. 1997, *The Opacity Project*, vol. II, Institute of Physics Publishing. Bristol
Berthomieu, G., Cooper, A. J., Gough, D. O., Osaki, Y., Provost, J., & Rocca, A. 1980, in Hill, H. A., Dziembowski, W., eds., *Lecture Notes in Physics* **125**, Springer, Heidelberg, p. 307
Christensen-Dalsgaard, J., Gough, D. O., & Toomre, J. 1985, *Science*, 229, 923
Christensen-Dalsgaard, J., Däppen, W. & Lebreton, L. 1988, Nature, 336, 634
Christensen-Dalsgaard, J., Däppen, W., and the GONG Team 1996, Science, 272, 1286
Däppen W., 2004, in Danesy D., ed., Proc. SOHO 14/GONG 2004: Helio- and Asteroseismology: Towards a Golden Future. ESA SP-559, Noordwijk, p. 261
Däppen W., 2006, J. Phys. A: Math. Gen., 39, 4441
Däppen, W., Mihalas, D., Hummer, D. G. & Mihalas, B. W. 1988, Astrophys. J., 332, 261
Ebeling W., Kraeft W. D., & Kremp D., 1976, Theory of Bound States and Ionization Equilibrium in Plasmas and Solids. Akademie-Verlag, DDR-Berlin
Gough, D. O., Kosovichev, A. G., Toomre, J., and the GONG Team 1996, Science, 272, 1296
Gough, D. O. 1993, in Zahn, J.-P., Zinn-Justin, J., eds., *Astrophysical Fluid Dynamics*, North-Holland, Amsterdam, p. 399
Huang K., 1963, Statistical Mechanics. John Wiley, New York, Chapt. 14
Hummer, D. G. & Mihalas, D. 1988, Astrophys. J., 331, 794
Iglesias, C. A. and Rogers, F. J. 1996, Astrophys. J., 464, 943
Kraeft W. D., Kremp D., Ebeling W., & Röpke G. 1986 *Quantum Statistics of Charged Particle Systems*, (New York: Plenum)
Liang, A., 2004, "Emulating the OPAL equation of state in the chemical picture formalism", in Equation-of-State and Phase-Transition Issues in Models of Ordinary Astrophysical Matter, edited by V. Celebonovic, W. Däppen, & D. Gough, AIP Conference Proceedings 731, Melville, New York, 2004, p. 106
Liang, A. & Däppen, W. 2003, "Modifications of the Equation of State to Achieve Desired Changes in Thermodynamic Quantities," in Proc. SOHO 12/GONG+ 2002 Workshop (ESA SP-517, Noordwijk, The Netherlands), p. 333–336.
Liang, A. & Däppen, W. 2004, "Emulating the OPAL equation of state", in *"Helio- and Asteroseismology: Towards a Golden Future"*, SOHO14–GONG2004 Meeting held July 12-16 2004 at Yale University, New Haven, CT, USA (ESA Publications Division: Noordwijk, The Netherlands), p. 548.
Mao, D. 2008, *PhD Thesis, USC, Los Angeles*
Mihalas, D., Däppen, W., & Hummer, D. G. 1988, Astrophys. J., 331, 815
Nayfonov, A., Däppen, W., Hummer, D. G., & Mihalas, D. M. 1999, Astrophys. J., 526, 451-464.
Noels, A., Scuflaire, R., & Gabriel, M. 1984, Astron. Astrophys., 130, 389
Perez, A., Mussack, K., Däppen, W., & Mao, D. 2008, Astron. Astrophys., submitted.
Rogers F. J. & Nayfonov A., 2002, Astrophys. J., 576, 1064
Rogers, F. J., Swenson, F. J., & Iglesias, C. A. 1996, Astrophys. J., 456, 902
Seaton, M. J. 1995, *The Opacity Project* Vol. I, Institute of Physics Publishing. Bristol
Trampedach R., Däppen W., & Baturin V. A., 2006, Astrophys. J., 646, 560
Ulrich, R. K. 1982, Astrophys. J., 258, 404

Discussion

CHRISTENSEN-DALSGAARD: What is the potential for determining the abundances of O, C, N etc. from their effect on the equation of state in the convection zone, using helioseismology.

DÄPPEN: The size of the modulation of r_1 by the heavy elements is (marginally) within the power of present-day helioseismology. A preliminary study by C.-H. Lin (2004) has looked at this effect and it was found that Z weight indeed below the Anders-Grevesse value.

3D model atmospheres and the solar photospheric oxygen abundance

E. Caffau[1] and H.-G. Ludwig[1,2]

[1] GEPI, Observatoire de Paris-Meudon; 92195 Meudon, France
email: Elisabetta.Caffau@obspm.fr

[2] CIFIST Marie Curie Excellence Team
Observatoire de Paris-Meudon, 92195 Meudon, France
email: Hans.Ludwig@obspm.fr

Abstract. In recent years the photospheric solar oxygen abundance experienced a significant downward revision. However, a low photospheric abundance is incompatible with the value in the solar interior inferred from helioseismology. For contributing to the dispute whether the solar oxygen abundance is "high" or "low", we re-derived its photospheric abundance independently of previous analyses. We applied 3D (CO5BOLD) as well as 1D model atmospheres. We considered standard disc-centre and disc-integrated spectral atlases, as well as newly acquired solar intensity spectra at different heliocentric angles. We determined the oxygen abundances from equivalent width and/or line profile fitting of a number of atomic lines. As preliminary result, we find an oxygen abundance in the range 8.73–8.79, encompassing the value obtained by Holweger (2001), and somewhat higher than the value obtained by Asplund et al. (2005).

Keywords. Sun: abundances – Sun: photosphere – Line: formation – hydrodynamics

1. Introduction

After hydrogen and helium oxygen is the most abundant element in the Universe, and its abundance has been extensively studied in the Galaxy and beyond. In such studies the solar oxygen abundance serves as natural reference. However, the spectroscopic determination of the solar oxygen abundance is not an easy task, and unfortunately its meteoritic abundance cannot be used for guidance since oxygen – as rather volatile element – was incompletely condensed during the cooling of the proto-solar nebula. Much effort has been devoted to spectroscopic determination of the photospheric oxygen abundance without having led to a convergence on a definite value. This has repercussions on solar and stellar physics. For instance, oxygen is a major contributor to the opacity in the convective envelope of the Sun. Solar oxygen abundance has a direct impact on the internal structure and evolution of the Sun and solar-like stars.

Over the last twenty years the photospheric oxygen abundance has experienced a downward trend, brought only to a halt by a recent analysis of Ayres and collaborators of both CO molecular and [OI] forbidden lines. Some milestone results that we would like to point out are: 8.93 ± 0.035 in Anders & Grevesse (1989), 8.83 ± 0.06 in Grevesse & Sauval (1998), 8.736 ± 0.078 in Holweger (2001), and 8.66 ± 0.06 in Asplund et al. (2004). Despite oxygen is such an abundant element, very few atomic lines are available to determine the photospheric oxygen abundance.

2. Data analysis

We considered various high-resolution, high signal-to-noise (S/N) solar atlases for the oxygen abundance determination. Notably, the difference between them is higher than

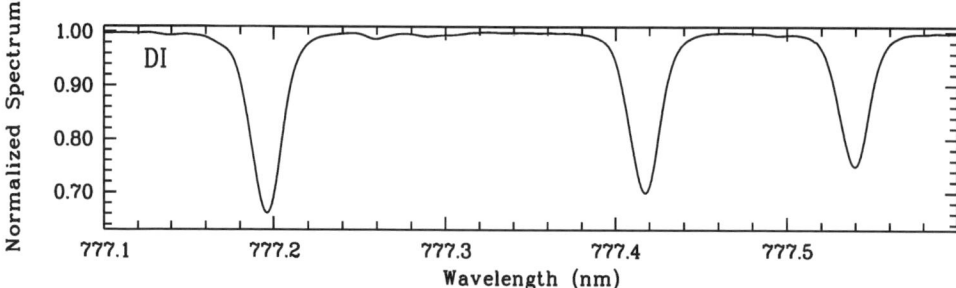

Figure 1. Comparison of two observed disc-centre spectra.

Table 1. Oxygen lines considered in this analysis

Wavelength (nm)	Characteristics	Number of spectra
630.0	blended with NiI line	4
636.3	on CaI autoionisation line, blended with CN	4
777.1	affected by strong NLTE correction	4
777.4	affected by strong NLTE correction	4
777.5	affected by strong NLTE correction	4
615.8	weak, next to a strong line	4
844.6	blended with a stronger FeI line	4
926.6	blended with a telluric absorption	1
1130.2	blended with a telluric absorption	1
1316.5	blended with a telluric absorption	1

expected according to the nominal signal-to-noise level, usually on the order of a few thousand. An example is shown in figure 1 where the grating disc-centre spectrum of Delbouille *et al.* (1973) (solid black line) is compared to the Fourier-transform spectrum of Neckel & Labs(1984) (solid green/grey line) for the triplet lines. The differences in the EW of the three lines between the two spectra are 4, 7, and 3 % in order of increasinf wavelength, which translates in abundance differences of 0.05, 0.07, and 0.03 dex, respectively.

The lines we considered are listed in Table 1.

As model of the solar atmosphere we used a 3D radiation-hydrodynamical model atmosphere (computed with the CO^5BOLD code) which provides a statistical realisation of the evolution of the plasma flow in a volume located at the solar surface comprising optically thick and thin regions. CO^5BOLD solves simultaneously the hydrodynamical equation coupled to radiative transfer in an external gravity field. The particular solar model we used in the present analysis covers a time interval of 4300 s represented by a selection of 19 snapshots.

For spectral-synthesis purposes we used the code Linfor3D. Linfor3D is restricted to local thermodynamic equilibrium (LTE) conditions. Since departures from LTE are important for the determination of the solar oxygen abundance, in addition we performed NLTE calculations with the Kiel code (Steenbock & Holweger (1984)) using the oxygen model atom from Paunzen *et al.* (1999). This Kiel code, however, is restricted to 1D structures. For this reason we used as input model the time and horizontal average of the 3D model over surfaces of equal optical depth. It has been shown (Asplund *et al.* (2004)) that the NLTE problem in the Sun is essentially a 1D problem, i.e., horizontal inhomogeneities play a minor role so that we consider the 1D approximation sufficient.

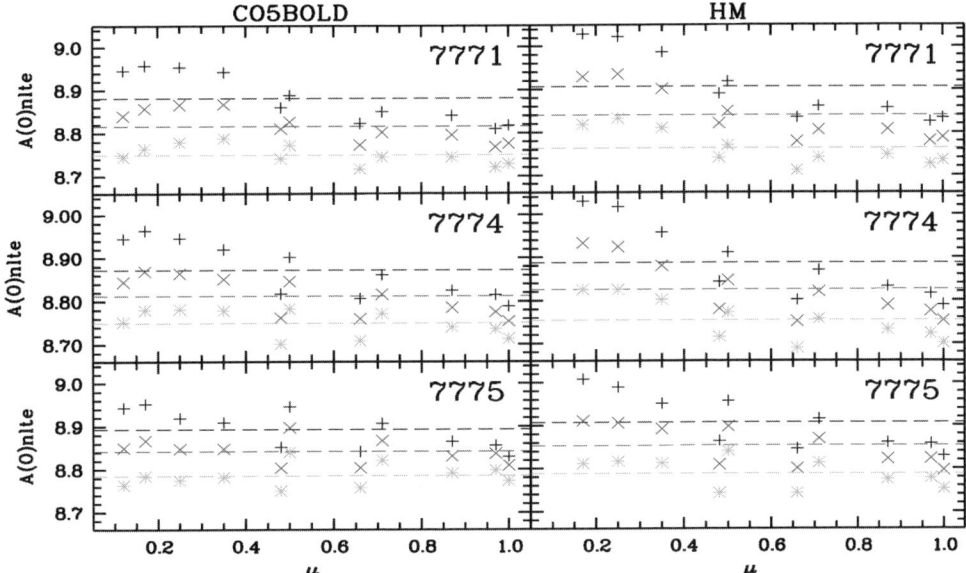

Figure 2. Oxygen abundance from the 777 nm triplet of centre-limb data.

In the Kiel code the excitation and ionisation of oxygen atoms by collisions with neutral hydrogen atoms is taken into account within the approximation provided by the Drawin (Drawin 1969) formula. A scaling factor, S_H, permits to take into account these collision, ($S_H = 1$), to switch them off, ($S_H = 0$), or to consider a fraction of them. No atomic physics calculations or laboratory measurements are available to constrain S_H. We tried to estimate its value by considering the centre-to-limb variation of the infrared triplet lines, similar to the work of Allende Prieto *et al.* (2004). Figure 2 depicts the oxygen abundance as a function of the cosine of the heliocentric angle μ for the three triplet components. We think we can exclude $S_H = 1$ since it is the one that gives the much higher scatter of the oxygen abundance for various μ than the lower test values.

Considering 31 line/observed spectrum combinations, using the CO5BOLD 3D model, with the choice of $S_H = 1/3$ in the NLTE computation we obtain $A(O) = 8.76 \pm 0.07$. The error includes random error related to scatter between the results obtained from different spectra, uncertainties in the continuum placement, the oscillator strength uncertainty, blends, and the choice of S_H in the NLTE computation. This oxygen abundance translates into $Z = 0.014$ if we consider our value for oxygen abundance and the Asplund *et al.* (2005) abundances for the other elements, $Z = 0.016$ when the other abundances are from Grevesse & Sauval (1998). By comparison, the Asplund *et al.* (2005) abundances, including oxygen, imply $Z = 0.012$.

To quantify in how far 3D effects are important we performed a comparison with abundances obtained from a 1D model. We choose a 1D model computed with the code LHD which is a standard 1D model atmosphere code, sharing the same micro-physics with CO^{5}BOLD. The convective energy transport is treated with mixing-length theory. For the mixing-length parameter, α, we used a value of 1.0. In the 1D spectral synthesis we adopted a micro-turbulence of 1 km/s. The choice of the mixing-length parameter is not important except for the 615.8 nm line, since all other oxygen lines are formed in layers whose temperature structure is virtually unaffected by differences in the assumed efficiency of convective energy transport.

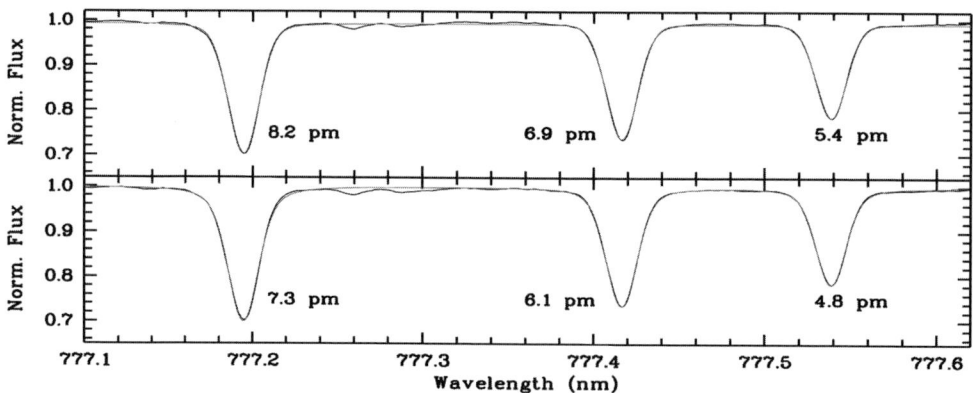

Figure 3. Differences in EW for the 777 nm triplet.

We define 3D correction the difference $A(O)_{3D}-A(O)_{LHD}$, as in Caffau & Ludwig (2007). The average 3D correction we found is $+0.05\,\text{dex}$. For all lines except the 615.8 nm line, the 3D correction is positive. We conclude that 3D effects are not responsible for lowering the oxygen abundance. This conclusion obviously depends on our choice of the 1D model atmosphere we used as reference. We belief, however, that our particular choice of a differentially comparable model in terms of the intrinsic micro-physics provides a well-defined reference. Our higher oxygen abundance in comparison to Asplund et al. (2004) is in part related to the EWs measurements. The triplet lines are important in the A(O) determination, and their EWs are lower in Asplund et al. (2004) with respect to what we measure. In figure 3 we show the subjective component that is present in the placement of the continuum for the 777 nm triplet lines. The upper panel shows the Voigt profile to compute the EW as we did in this work, in the lower panel we tried to reproduce the EWs published in Asplund et al. (2004). We think that in this latter case part of the wings are lost.

We compared the oxygen abundance of the Sun with these of stars with and without detected planets. Ecuvillon et al. (2006) compared the oxygen abundance in a sample of 96 stars with planets to a volume limited sample of 59 stars without planets. The solar oxygen is very close to the mean oxygen abundance of the sample with planets (0.01 dex lower). The Asplund et al. (2004) value is very close to the mean oxygen abundance of the sample without planets. But both determinations of the solar oxygen abundance are compatible with the A(O) for stars hosting planets.

References

Allende Prieto, C., Asplund, M., & Fabinani Bendicho, P. 2004, *A&A* 423, 1109
Anders, E. & Grevesse, N. 1989, *Geochim. Cosmochim. Acta* 53, 197
Asplund, M. et al. 2004, *A&A* 417, 751
Asplund, M., Grevesse, N., & Sauval, A. J. 2005, *ASP Conf. Ser.* 336, 25
Caffau, E. & Ludwig, H.-G. 2007, *A&A* 467, L11
Delbouille, L., Roland, G., & Neven, L. 1973, *Liege: Universite de Liege, IA* 1973
Drawin, H. W., 1969, *Z. Physik* 225, 483
Ecuvillon, A., et al. 2006, *A&A* 445, 633
Grevesse, N. & Sauval, A. J. 1998, *Space Science Reviews* 85, 161
Holweger, H. 2001, AIP Conf. Proc., 598, 23
Neckel, H. & Labs, D. 1984, *Sol. Phys.* 90, 205
Paunzen, E., et al. 1999, *A&A* 345, 597
Steenbock, W. & Holweger, H. 1984, *A&A* 130, 319

Discussion

ASPLUND: Two comments: (1) In Asplund *et al.* (2004) we don't use equivalent widths but line profile fitting in 3D non-LTE. (2) One has to be careful when comparing the Sun with the distribution of stars "with" and "without" planets. We don't know yet whether the "without planets" sample really don't have planets, only that they don't have "hot Jupiters" which the Sun of course does not have.

CAFFAU: (1) Thank you for clarification. (2) Yes. I do not think that the Solar oxygen abundance respect to the histogram of stars with or without "hot Jupiters" can anyway solve the problem.

The Speaker, E. Caffau (right), is talking to A. Aret at the poster area.

SMART – a computer program for modelling stellar atmospheres

A. Aret, A. Sapar, R. Poolamäe and L. Sapar

Tartu Observatory, 61602 Tõravere, Tartumaa, Estonia
email: aret@aai.ee

Abstract. Program SMART (Spectra and Model Atmospheres by Radiative Transfer) has been composed for modelling atmospheres and spectra of hot stars (O, B and A spectral classes) and studying different physical processes in them (Sapar & Poolamäe 2003, Sapar *et al.* 2007). Line-blanketed models are computed assuming plane-parallel, static and horizontally homogeneous atmosphere in radiative, hydrostatic and local thermodynamic equilibrium. Main advantages of SMART are its shortness, simplicity, user friendliness and flexibility for study of different physical processes. SMART successfully runs on PC both under Windows and Linux.

Keywords. Radiative transfer, stars: atmospheres

1. Main features of the program

Model atmospheres are calculated iteratively varying only temperature and pressure dependency on column density. Flux constancy about 0.1–0.5 % is achieved by about 10 iterations, using Kurucz ATLAS9 models as input. Number of atmospheric layers can be multiplied up if necessary. Line absorption has been completely taken into account with spectral resolution 300 000.

Programming language is Fortran 90, the program is compiled using Intel Fortran Compiler and runs both on Windows and Linux computers. Code is extensively commented on right-hand margin. Graphical interface (written in C++) has been composed for visualizing results of calculation.

Radiative transfer has been calculated using integration by parts, yielding series of exponential integrals. The scattering processes are computed by simple Λ-iteration. Radiative transfer calculations give radiative flux $F_\nu(\lambda, \tau)$ in all layers of atmosphere.

Capabilities of program SMART include also computations of evolution of diffusive separation of isotopes in atmospheres of CP stars, relaxational formation of NLTE in line spectra, accelerations of clumps in stellar wind, computation of detailed spectral limb darkening and hence the spectra of rotating stars and non-irradiated eclipsing binaries. Pan-spectral method for determining element abundances from high-quality observed spectra have been developed and implemented as an extension to SMART.

The basic restriction is the assumption of plane-parallel, static and horizontally homogeneous atmosphere with no convection and no molecular absorption ($T_{\text{eff}} > 9\,000$ K). Present version assumes also LTE. NLTE calculations are not yet included into model computations. There are several simplifying assumptions reducing accuracy of modelling. Multiple light scattering is treated using simple Λ-iterations. A simplified treatment of Stark broadening of H and He lines has been used. Problems are also an instability of algorithms near Eddington limit and incompleteness of atomic data.

Typical running times on a PC with CPU 3.2 GHz and 2 GB RAM are several hours for model atmosphere computation with spectral resolution 300 000 and 64 layers of atmosphere. One time step in evolutionary computations of separation of mercury isotopes in atmospheres of CP stars with resolution 5 000 000 takes approximately 15 min.

2. Special tasks

SMART enables to compute evolution of chemical composition in atmospheres of CP stars due to diffusive separation of elements and isotopes driven by radiative acceleration, light-induced drift and gravity (Aret & Sapar 2002, Sapar et al. 2005). Extensive high-precision line list and collision cross-sections are necessary. Currently the calculations have been made for mercury isotopes, similar calculations for calcium are in preparation.

Relaxational formation of NLTE in spectral lines can be calculated by rapidly converging iterations. Equilibrium quantum state populations of ion states are found from the equations of unbalanced statistical equilibrium treated as an relaxational initial value problem from LTE to NLTE populations.

Detailed spectral limb darkening has been computed for some model stellar atmospheres and used thereafter for finding spectra of rotating stars and non-irradiated eclipsing binaries. Codes also accounting for stellar surface distortion and gravitational darkening are currently prepared.

To enlighten the problem of stellar wind triggering in stellar atmospheres, the radiative acceleration of moving clumps with Doppler shifted spectral lines has been studied and found to give hopeful results.

Pan-spectral method for determining element abundances (Sapar et al. 2008) aimed for the automatic processing of high-quality stellar spectra has been elaborated. The method is based on weighted cumulative line-widths Q_λ defined as

$$Q_\lambda = \int_{\lambda_0}^{\lambda} \left| \frac{dR_\lambda}{dZ} \right| (1 - R_\lambda) d\lambda ,$$

where R_λ is the residual flux (intensity) and $Z = \log(N_{elem}/N_{tot})$ is the abundance of studied element or isotope. The derivative of residual flux R_λ with respect to abundance Z automatically excludes spectral regions insensitive to changes of the abundance of studied element and gives a large contribution in the most sensitive regions, i.e. in the centres of non-saturated lines and in the steep wings of strong lines of the element. Best fit of quantities Q_λ found from synthetic and observed spectra gives final abundance taking duly into account all lines of studied element including blended ones. Abundances can be found simultaneously for many elements. The method can also be used to find corrections of effective temperature and gravity.

Acknowledgements

We are grateful to Estonian Science Foundation for financial support by grant ETF 6105.

References

Sapar A. & Poolamäe R. 2003, ASP Conf. Series, 288, 95
Sapar A., Poolamäe R., Sapar L., & Aret, A. 2007, Spectroscopic Methods in Modern Astrophysics, INASAN, Moscow, 236 (in Russian)
Aret A. & Sapar A. 2002, Astr. Nachr., 323, 1, 21
Sapar A., Aret A., & Poolamäe R. 2005, EAS Publ. Series, 17, 341
Sapar A., Aret A., Sapar L., & Poolamäe R. 2008, Precision Spectroscopy in Astrophysics: ESO Astrophysics Symposia, Springer Verlag, 145

Proxies for overshooting above a convection zone

Kwing L. Chan[1] and Harinder P. Singh[2]

[1]Department of Mathematics and Center for Space Science Research,
Hong Kong University of Science and Technology, Hong Kong, China
email: maklchan@ust.hk

[2]Department of Physics & Astrophysics, University of Delhi, Delhi, India
email: singh@iucaa.ernet.in

Abstract. Due to the up/down asymmetry caused by stratification, overshooting above differs from overshooting below a convection zone. The flux of kinetic energy, frequently used as a proxy of overshooting below a convection zone, cannot be used for the upward problem.

Keywords. Convection, turbulence, waves, methods: numerical

In many numerical studies, the flux of kinetic energy, F_k, is used as an indicator of the presence of turbulence. The extent of its penetration into the stable layer is taken as the depth of overshooting. Below a convection zone, this is understandable, as F_k tracks the continuation of the down-flow columns into the region below. Above the convection zone, however, this interpretation does not hold. Upward flows disperse and do not form narrow coherent columns. Thus, other proxies for overshooting need to be found. Based on a set of recently computed numerical models, we discuss the difficulties that arise.

Three cases with different input energy fluxes (F) are discussed here. While all other parameters are the same, $F(\text{Case 3}) = 2F(\text{Case 2}) = 4F(\text{Case 1})$. The domain is a $3 \times 3 \times 1$ rectangular box ($162 \times 162 \times 122$ grids). The depths of the convection zone and the stable zone above are approximately 2.5 and 3.9 pressure scale heights, respectively.

$\underline{F_k \text{ does not turn positive across the unstable/stable boundary}}$

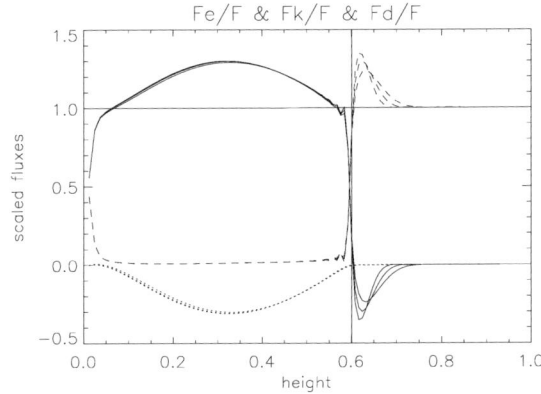

Figure 1. The scaled mean (horizontally and temporally averaged) enthalpy flux (F_e/F), kinetic energy flux (F_k/F), and diffusive flux (F_d/F) are shown by the solid, dotted, and dashed curves, respectively. The value of the total flux F depends on the case. The case with the largest (smallest) F produces maximum (minimum) overshoots in the stable region (height > 0.6).

Different overshooting quantities decay with different spatial scales

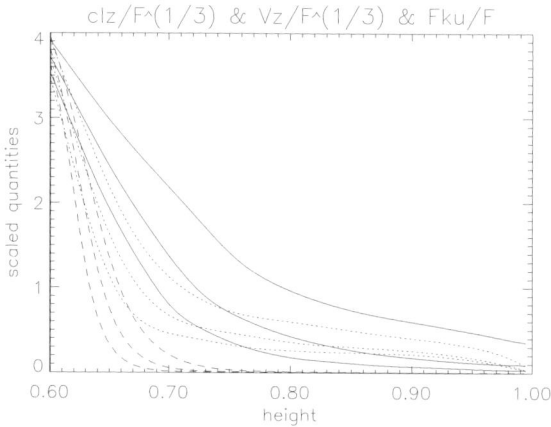

Figure 2. The solid, dotted, and dashed curves show the scaled root-mean-square (rms) vertical vorticity, $(\mathrm{curl}\,V)''_z/F^{1/3}$, the scaled rms vertical velocity, $24.3V''_z/F^{1/3}$, and the scaled upward conditional mean (i.e. average over grid points with positive V_z) kinetic energy flux, $10F_{ku}/F$, respectively. The scalings are to assist visual comparisons only. It is clear that the rates of decay are very different. The case with larger flux generally has slower rates of decay.

Some quantities do decay at similar rates

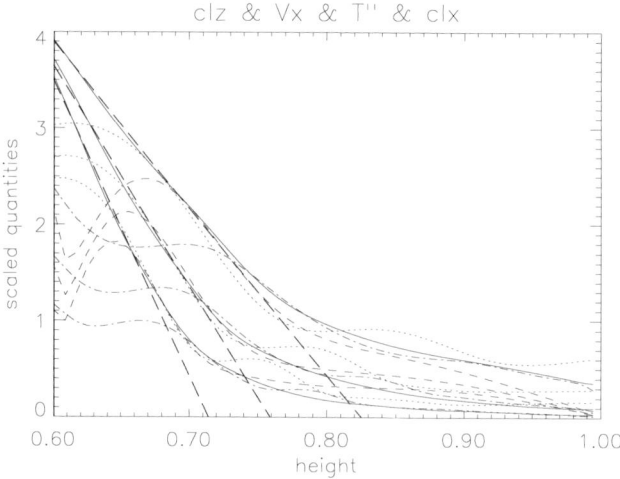

Figure 3. The solid, dotted, dashed, dot-dashed curves show the scaled rms vertical vorticity (same as the previous figure), the scaled rms horizontal velocity, the scaled rms temperature variation, and the scaled rms horizontal vorticity, respectively. The long dashed straight lines show that the initial decay rates of these quantities are similar, over certain portions of the overshoot region. These lines intersect the zero value at distances 0.115, 0.16, and 0.225 from the unstable/stable boundary. The distances scale roughly as $1:2^{1/2}:4^{1/2}$. If they are taken as the 'overshoot distance', the scaling with the flux is $F^{1/2}$.

Acknowledgements

KLC thanks the Hong Kong Research Grants Council for support.

Some discussion of the nonlocal treatment of the dissipation in the Reynolds stress models

Tao Cai

Department of Mathematics, HKUST, Clear Water Bay, Hong Kong
email: ctust@ust.hk

Abstract. We investigate the weakness of the present turbulence model with the nonlocal treatment of dissipation rate. A revised version is well tested for the solar convection. The suggestion of constant mixing length parameter of MLT could not hold any more if we refer to the nonlocal description of the dissipation rate, especially in the region of overshooting zone.

Keywords. Convection, hydrodynamics, sun:interior

1. Introduction

Discarding of the traditional local mixing length theory(MLT) (Böhm-Vitense (1958)), Xiong developed a nonlocal turbulence model by using the Reynolds stress methods for studying the stellar convection (Xiong (1979)). However, Canuto argued that Xiong's model is not a fully nonlocal turbulence model since it still uses a local description of dissipation rate, hence he developed a fully nonlocal turbulence model by describing the dissipation rate in a nonlocal way (Canuto & Dubovikov (1998)). By using this fully nonlocal model, Kupka and his coworkers have calculated the cases of A-star and DA/DB white dwarfs with thin convection zones (Kupka & Montgomery (2002), Montgomery & Kupka (2004)). Meanwhile the more attractive case of the sun with a deep convection zone is still unclear. The purpose of this paper is to try to explain where the difficulty comes from and what's the difference if we introduce the nonlocal description of the dissipation rate.

2. Results and discussion

The fully nonlocal turbulence model with the nonlocal treatment of dissipation rate is given by (Canuto & Dubovikov (1998))

$$\frac{\partial K}{\partial t} + D_f = g\alpha J - \epsilon \tag{2.1}$$

$$\frac{\partial}{\partial t}\frac{1}{2}\overline{\theta^2} + D_f = \beta J - \tau_\theta^{-1}\overline{\theta^2} + \frac{1}{2}\frac{\partial}{\partial r}(\chi\frac{\partial}{\partial r}\overline{\theta^2}) \tag{2.2}$$

$$\frac{\partial}{\partial t}J + D_f = \beta\overline{\omega^2} + (1-\gamma_1)g\alpha\overline{\theta^2} - \tau_{p\theta}^{-1}J + \frac{1}{2}\frac{\partial}{\partial r}(\chi\frac{\partial}{\partial r}J) \tag{2.3}$$

$$\frac{\partial}{\partial t}\frac{1}{2}\overline{\omega^2} + D_f = -\tau_{pv}^{-1}(\overline{\omega^2} - \frac{2}{3}K) + \frac{1}{3}(1+2\beta_5)g\alpha J - \frac{1}{3}\epsilon \tag{2.4}$$

$$\frac{\partial \epsilon}{\partial t} + D_f = 2c_1\tau^{-1}g\alpha J - 2c_2\tau^{-1}\epsilon + c_3\epsilon N \tag{2.5}$$

$$N = (g\alpha|\beta|)^{\frac{1}{2}} \tag{2.6}$$

$$\tau = \frac{2}{\epsilon}K \tag{2.7}$$

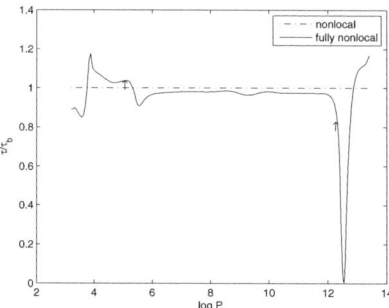

Figure 1. The ratio of the dissipative timescale over the buoyant timescale of the fully nonlocal (solid curve) and nonlocal model (dash-dotted curve) in the solar envelope. The arrows indicate the boundaries of the convective instable zone where $\nabla = \nabla_{ad}$. The ratio in the lower overshooting zone is very small due to the nonlocal treatment of the dissipation rate.

When K^2/ϵ is very small compared with $1/\tau$, the diffusion term in the equations can be neglected. Then suppose that $(K, \overline{\theta^2}, J, \overline{w^2}, \epsilon)$ is the solution of the system. It is not difficult to check that $\lambda(K, \overline{\theta^2}, J, \overline{w^2}, \epsilon)$ is also the solution by substituting it into the equations if we notice that the timescales remain constant at the same level and the system can be viewed as a linear system, here λ is a constant. This tells us that the solution is not unique and that's the reason why steady solution is hard to be obtained by solving these equations. To overcome this weakness, a new formula of Brunt-Väisälä frequency is introduced for replacing the former one. It can be defined by

$$N = \frac{c_\epsilon}{\alpha_m H_p} K^{1/2}. \qquad (2.8)$$

$\tau_b = 2/N$ could be viewed as the buoyant characteristic timescale describing the travelling time of the bulb to go through the mixing length. We know that τ is the dissipative characteristic timescale for the bulb to consume all of its kinetic energy. If $\tau/\tau_b > 1$, the bulb would reserve some energy when finishing travelling the mixing length. If $\tau/\tau_b < 1$, the bulb would use up all of its kinetic energy within the mixing length. If $\tau/\tau_b = 1$, which is the local description of dissipation rate, the kinetic energy of the bulb is exactly used up after the bulb goes through the mixing length. This may provide us information to explain why the mixing length parameter α_m is supposed to have different values at different levels when using the local mixing length model. This disadvantage will disappear when we refer back to the nonlocal description. If $\tau/\tau_b > 1$, an adjustment of increasing the mixing length parameter would balance the timescales, hence it is corresponding to a larger mixing length parameter of MLT. While $\tau/\tau_b < 1$ is corresponding to a smaller mixing length parameter of MLT.

Acknowledgements

The author thanks Prof. Chan,K.L. and Prof. Xiong,D.R. for their valuable suggestions.

References

Böhm-Vitense E. 1958, *ZfA* 46,108
Canuto, V. M. & Dubovikov, M. 1998, *ApJ* 493,834
Kupka, F. & Montgomery, M. H. 2002, *MNRAS*(letters) 330,L6
Montgomery, M. H. & Kupka, F. 2004, *MNRAS* 350,267
Xiong, D. R. 1979, *Acta Astron. Sinica* 20,238

Instabilities and mixing in stellar radiation zones

Jean-Paul Zahn

LUTH, Observatoire de Paris, 92195 Meudon, France
email: Jean-Paul.Zahn@obspm.fr

Abstract. The standard model of stellar structure is unable to account for various observational facts, and there is now a large consensus that some 'extra mixing' must occur in the radiation zones. The possible causes for such mixing are briefly reviewed. The most efficient among them is probably the shear-turbulence generated by the differential rotation, which itself results from the transport of angular momentum that can be mediated through the large-scale circulation induced by structural adjustments or by the applied torques (stellar wind, accretion, tides). In solar-type stars this angular momentum transport is ensured mainly by internal gravity waves that are excited at the boundary with convection zones. Another cause of mixing manifests itself in the red giant phase, namely the thermohaline instability due to an inversion of the molecular weight gradient. The implementation of these processes in stellar evolution codes is giving rise to a new generation of stellar models, which are in much better agreement with the observational constraints.

Keywords. Instabilities, turbulence, stars: evolution, interiors, rotation

1. Mixing in stellar radiation zones: the observational evidence

All along the 20th century, considerable progress has been achieved in modeling stellar interiors, thanks to our great pioneers: Eddington, Chandrasekhar, Bethe, Schwarzschild, Kippenhahn, Iben, and many others. Presently the standard model consists of well-mixed convection zones, whose boundaries are defined by the Schwarzschild criterion, and of stable radiation zones that do not allow for any mixing, except some convective penetration. The most recent improvement was to include the atomic processes that are responsible for the microscopic diffusion of the chemical species, and also for their gravitational settling and radiative acceleration. This standard model proves very successful in predicting the main features of stellar evolution, such as the ascent of the giant branch after the exhaustion of hydrogen in the stellar core. However, some properties of this standard model are in disagreement with a number of observational facts.

Stars of intermediate mass – A-type stars in particular – display anomalies in their surface composition, when they are compared to other, similar stars that are regarded as 'normal'. These anomalies are due to atomic processes, as was shown by Michaud (1970) and his collaborators. Indeed, the radiative acceleration experienced by a given ion depends strongly on its spectral properties, i.e. on its ability of absorbing or diffusing photons, whereas its gravitational settling increases with atomic mass. Depending on which process wins over the other, the considered species appear overabundant or underabundant. But it turns out that these atomic processes are so efficient that they would produce surface anomalies which are much more pronounced than those observed. One consequence, for instance, is that helium would disappear from the surface of A-type stars in about one million years, as was pointed out already by Vauclair *et al.* (1974) and

Michaud et al. (1976). Since this is not observed, Vauclair et al. (1978) suggested that some mild turbulence operates near the surface to smooth the composition gradients.

Another sign that some mixing must occur in stellar atmospheres is the striking flatness of the celebrated Spite plateau, with the ^{7}Li abundance in halo stars depending little on effective temperature (Spite & Spite 1982). If element segregation were alone to operate, the Li depletion would increase with effective temperature, as was recognized by Deliyannis et al. (1990) and Proffitt & Michaud (1991), and was confirmed recently again by Richard et al. (2002). Therefore one must invoke again some mild mixing occurring below the surface that will enforce the flatness of this plateau.

Furthermore the overabundance, at the surface of massive stars, of chemical elements (such as He, N) that are synthesized in the nuclear core, can only be explained if the radiative interior undergoes some mixing (cf. Meynet & Maeder 2000). Interestingly, these overabundances seem to be correlated with the rotation velocity of the stars (Herrero et al. 1992, 2000).

A final example is provided by the surface abundances of Li, C and N of low-mass red giants, after they reach the so-called bump in the luminosity function; these would not appear if there were no extra mixing.

To sum it up, there are many indications that stellar radiation zones do not behave as inert regions, but that they allow for some mild mixing.

2. How shall one treat that mixing?

Why should one bother with that mixing? One reason is that it injects hydrogen-rich material in the nuclear core of stars, thus extending their life-time and changing the age estimates based on cluster diagrams. Furthermore, since it modifies the profile of chemical composition within the star, mixing has a strong impact on the later stages of stellar evolution. Also, the matter that is carried away by winds has a non-standard composition, which will orient differently the chemical evolution of the host galaxy.

One way to treat that mixing is what may be considered as the minimalist approach: it consists in introducing a suitably parametrized turbulent diffusivity, and to adjust the parameter(s) to fit the observations. Recently this was done rather successfully by Korn et al. (2006), who showed that a unique parametric prescription for this turbulent diffusivity could explain both the slow rise of the abundance of Fe with evolutionary age in a metal-poor globular cluster, and that of ^{7}Li (followed by its destruction as the convection zone digs deeper). But there is no guarantee that this prescription would work in other circumstances. That is why we prefer the physical approach, namely to strive to implement the physical processes that are likely to cause the mixing.

3. Mixing processes in the radiation zone of rotating stars

The main mixing processes that operate in the radiation zone of rotating stars have been identified some time ago (cf. Zahn 1992). These are the large scale circulation which is induced by external torques or structural changes, and the turbulence generated by the shear of differential rotation; their combined action is called *rotational mixing*.

3.1. Meridional circulation

Until recently, the large-scale meridional circulation which occurs in a stellar radiation zone was ascribed to the fact that the isothermal surfaces are distorted by the centrifugal force, with the consequence that the radiative flux is no longer divergence-free (Von Zeipel 1924). In the original treatment (Eddington 1925, Vogt 1925, Sweet 1950), the meridional circulation was thus linked to the state of rotation, i.e. the circulation speed and hence the amount mixing scaled proportionally to the square of the angular velocity. The characteristic time was derived by Sweet, and it has since been named the Eddington-Sweet time $t_{\rm ES}$:

$$t_{\rm ES} = t_{\rm KH} \frac{GM}{\Omega^2 R^3}, \qquad (3.1)$$

with $t_{\rm KH} = GM^2/RL$ being the thermal adjustment time (or Kelvin-Helmholtz time). R, M, L designate respectively the radius, mass and luminosity, Ω the angular velocity and G the gravitational constant. Sweet's result implied that rapidly rotating stars should be well mixed by this circulation, and therefore that they would be prevented from evolving to the giant branch. Since this is not observed, Mestel (1953) invoked the choking effect of the composition gradients that build up due to the nuclear burning.

However, these early studies overlooked the fact that the circulation carries angular momentum: starting from (unknown!) initial conditions, the star undergoes a transient phase which lasts indeed about an Eddington-Sweet time, after which it settles into a quasi-stationary regime where the circulation is governed mainly by the torque applied to the star. For instance, when the star loses angular momentum through a wind, the circulation adjusts such as to transport that momentum to the surface (Zahn 1992). The resulting rotation is then non-uniform, and a baroclinic state sets in, with the temperature varying with latitude along isobars. On the other hand, when the star does not exchange angular momentum, the circulation would die altogether as predicted by Busse (1982), if it had not to compensate a weak turbulent transport down the gradient of angular velocity, because there again the rotation is non-uniform. Thus, in the absence of other processes that will be discussed in §4.2, it is the loss (or the gain) of angular momentum which drives the circulation and determines its strength, and not the rotation as such, which intervenes only in the transient phases.

A similar circulation is also induced at the boundary of a radiation zone, when the adjacent convection zone is in differential rotation. This occurs in the Sun, where helioseismology has revealed that the rotation rate changes from latitude-dependent in the convection zone, to almost uniform in the radiative interior below, with the transition occurring in a thin boundary layer, the *tachocline*.

What causes the thinness of the tachocline (it extends less than 5% in radius) still remains to be elucidated. Indeed, the differential rotation which is imposed by the convection zone should penetrate deep into the radiation zone, through thermal diffusion. Such penetration could be prevented by strongly anisotropic turbulence, as was suggested initially by Spiegel and Zahn (1992). An alternative explanation was put forward by Gough and McIntyre (1998), involving a fossil magnetic field; however we shall see in §4.2.1 that this solution is ruled out by recent numerical simulations. But the tachocline spread could be prevented by the cyclic dynamo field, whose penetration is limited by (ohmic) turbulent diffusion – the classical skin effect; that field would act with the shear induced toroidal field to suppress the differential rotation, as was shown by Forgács-Dajka and Petrovay (2002; see also Forgács-Dajka 2004).

3.2. Shear turbulence caused by differential rotation

As we have seen, the rotation regime that results from the applied torques is not uniform, and this flow is prone to various instabilities, which generate turbulence and therefore mixing. For a comprehensive review of these instabilities, we refer the reader to Pinsonneault (1997) or Maeder & Meynet (2000). Here we shall consider only those which apparently play a major role. namely the shear instabilities.

3.2.1. Turbulence produced by the vertical shear

Let us first consider the instability that is produced by the shear in the vertical direction, $\Omega(r)$. This instability is very likely to occur, because the Reynolds number characterizing such shearing flows in stars is extremely high, due to the large sizes involved. Depending on the velocity profile, the instability may be linear, or of finite amplitude. In the absence of stratification, turbulence can be sustained whenever the Reynolds number $Re = w\ell/\nu$ is larger than about $Re_c = 40$ (Schatzman et al. 2000). (The Reynolds number Re is expressed here in terms of the velocity w and the size ℓ of the largest turbulent eddies, and ν is the kinematic viscosity.) However the stable entropy stratification in the radiation zone hinders the shear instability: in the absence of thermal dissipation, the instability occurs only if locally

$$\frac{N^2}{(\mathrm{d}V_\mathrm{h}/\mathrm{d}z)^2} \leqslant Ri_\mathrm{c}, \qquad (3.2)$$

where V_h is the horizontal velocity and z the vertical coordinate. The buoyancy frequency N is given by

$$N^2 = \frac{g\delta}{H_P}\left[\left(\frac{\partial \ln T}{\partial \ln P}\right)_\mathrm{ad} - \frac{\mathrm{d}\ln T}{\mathrm{d}\ln P}\right], \qquad (3.3)$$

where g is gravity, H_P the pressure scale height and $\delta = -(\partial \ln \rho/\partial \ln T)_P$. This condition (3.2) is known as the *Richardson criterion*; the critical Richardson number Ri_c is of order unity, and depends somewhat on the rotational profile.

In a stellar radiation zone, this criterion is modified because the perturbations are no longer adiabatic, due to thermal diffusion. When the radiative diffusivity K exceeds the turbulent diffusivity $D_\mathrm{v} = w\ell$, which is generally the case, the instability criterion takes the form (Townsend 1958; Dudis 1974; Zahn 1974; Lignières et al. 1999)

$$\frac{N^2}{(\mathrm{d}V_\mathrm{h}/\mathrm{d}z)^2}\left(\frac{w\ell}{K}\right) \leqslant Ri_\mathrm{c} \quad \text{when} \quad \left(\frac{w\ell}{K}\right) \leqslant 1. \qquad (3.4)$$

From the largest eddies that fulfill this condition, one can estimate the turbulent diffusivity acting in the vertical direction in the radiation zone of a star; it scales as the square of the local shear:

$$D_\mathrm{v} = w\ell = Ri_\mathrm{c} K \frac{\Omega^2}{N^2}\sin^2\theta\left(\frac{\mathrm{d}\ln\Omega}{\mathrm{d}\ln r}\right)^2, \qquad (3.5)$$

provided that $D_\mathrm{v} \geqslant Re_\mathrm{c}\nu \approx 40\nu$.

The instability criterion (3.4) holds as such in regions of uniform composition, where the stability is enforced only by the temperature gradient; when the molecular weight μ varies with depth, it seems at first sight that one should replace this criterion by the original one, expression (3.2), where now the buoyancy frequency is dominated by the gradient of molecular weight:

$$N^2 \to N_\mu^2 := \frac{g\varphi}{H_P}\frac{\mathrm{d}\ln\mu}{\mathrm{d}\ln P}, \qquad (3.6)$$

with $\varphi = (\partial \ln \rho/\partial \ln \mu)_{P,T}$. As Meynet & Maeder (1997) pointed out, this very severe condition would prevent any mixing in early-type main-sequence stars, contrary to what is observed. We shall see next how that stabilizing action of μ-gradients can be overcome.

3.2.2. Turbulence produced by the horizontal shear

Likewise, the horizontal shear $\Omega(\theta)$ will also generate turbulence, probably through a finite-amplitude instability, because most plausible rotation profiles in stars are linearly stable. What type of turbulence then occurs is still a matter of debate. Its vertical component will be constrained by the stratification and therefore it is likely that this turbulence will be highly anisotropic, with much stronger transport in the horizontal than in the vertical direction, i.e. $D_\mathrm{h} \gg D_\mathrm{v}$.

The postulated anisotropy seems plausible, given the strong stratification in stellar radiation zones and the fact that no restoring force opposes the horizontal displacements; such turbulence was conjectured already in Zahn (1975). We shall further assume that this turbulence acts to suppress its cause, namely the differential rotation in latitude. This will thus lead to a 'shellular' rotation state, where the angular velocity depends little on latitude, and where we can assume that $\Omega \sim \Omega(r)$.

An interesting property of such anisotropic turbulence is that it interferes with the advective transport due to the meridional circulation, and turns it into a vertical diffusion, as it was pointed out by Chaboyer & Zahn (1992). To first approximation, the vertical component of the circulation velocity is given by $u_r(r,\theta) = U_2(r)P_2(\cos\theta)$, where P_2 is the Legendre polynomial of degree 2; then the resulting effective diffusivity is

$$D_\mathrm{eff} = \frac{1}{30} \frac{(rU_2)^2}{D_\mathrm{h}}, \qquad (3.7)$$

provided that $rU_2 \geqslant D_\mathrm{h}$. Unfortunately, a reliable prescription for that horizontal diffusivity D_h is still lacking, in spite of recent developments. Maeder (2003) derived a new expression from energy considerations, whereas Mathis *et al.* (2004) adapted the prescription drawn from laboratory experiments by Richard & Zahn (1999); the two recipes give similar results (which does not mean that they are valid).

Another property of such anisotropic turbulence is that, by smoothing out chemical inhomogeneities on level surfaces, it reduces the stabilizing effect of the vertical μ-gradient. The Richardson criterion for the vertical shear instability then involves also the horizontal diffusivity D_h, and the vertical component of the turbulent viscosity can be derived as before (Talon & Zahn 1997):

$$D_\mathrm{v} = Ri_\mathrm{c} \left[\frac{N_T^2}{K + D_\mathrm{h}} + \frac{N_\mu^2}{D_\mathrm{h}} \right]^{-1} \sin^2\theta \left(\frac{\mathrm{d}\Omega}{\mathrm{d}\ln r} \right)^2, \qquad (3.8)$$

where N_T^2 stands for the thermal part of the buoyancy frequency given by expression (3.3). Thanks to this horizontal erosion, the μ-gradients are much less effective in hindering the mixing.

4. Rotational mixing

The two transport processes that have just been discussed (meridional circulation and shear-induced turbulence) are both linked with the (differential) rotation. Therefore, when modeling the evolution of a star including these mixing processes, it is necessary to calculate also the secular evolution of its rotation rate $\Omega(r,\theta)$. The problem is simple to handle, if one deals just with laminar flows. However, as already mentioned, the differential rotation which is generated by this advection generates turbulence, which in turn transports angular momentum. In addition, torques may be applied to the star, either externally due to mass loss or accretion, or to tides exerted by a companion or an accretion disk, or else internally, due to magnetic stresses and to internal gravity waves, as we shall see next. To take all these effects into account, one has thus to solve the complete transport equation for the angular momentum:

$$\frac{\partial}{\partial t}\left[\rho r^2 \sin^2\theta\, \Omega\right] + \nabla \cdot \left[\rho \mathbf{u}\, r^2 \sin^2\theta\, \Omega\right] = \text{applied torques}, \tag{4.1}$$

$\mathbf{u}$ being the circulation velocity.

The problem may be significantly simplified if the horizontal shear is nearly suppressed by the anisotropic turbulence mentioned above. To lowest order Ω is then a function of r only, and all perturbations from the non-rotating state separate in r and colatitude θ, as illustrated here for the vertical component of the meridional velocity: $u_r(r,\theta) = U_2(r)P_2(\cos\theta)$. For a detailed account of how this modelization is implemented, we refer to Zahn (1992), Maeder & Zahn (1998) and Mathis & Zahn (2004).

4.1. *Rotational mixing of type I*

We first examine the simplest case, that we call 'rotational mixing of type I', where the angular momentum is transported by the same processes that are responsible for the mixing, namely the meridional circulation and the turbulent diffusion. The angular velocity $\Omega(r)$ then obeys the following transport equation, obtained by averaging equation (4.1) over θ:

$$\frac{\partial}{\partial t}\left[\rho r^2 \Omega\right] = \frac{1}{5r^2}\frac{\partial}{\partial r}\left[\rho U_2 r^2 \Omega\right] + \frac{1}{r^2}\frac{\partial}{\partial r}\left[\rho \nu_v r^4 \frac{\partial \Omega}{\partial r}\right], \tag{4.2}$$

with $\nu_v \approx D_v$ given by equation (3.8). In spite of the fact that this equation is only one-dimensional, it captures the advective character of the angular momentum transport by the meridional circulation: depending on the sign of U_2, i.e. on the sense of the circulation in each hemisphere, angular momentum may be transported up the gradient of Ω, which is never the case when the effect of meridional circulation is modeled just as a diffusive process, as it is often done (cf. Pinsonneault et al. 1989 or Chaboyer et al. 1995).

We stated already that the circulation is governed mainly by the applied torques. When there is no or little loss of angular momentum from the star, the circulation settles into a regime of differential rotation where a weak inward flux of angular momentum compensates the turbulent diffusion directed outwards, down the gradient of $\Omega(r)$. On the other hand, when the star loses angular momentum (for instance through a wind), the circulation adjusts itself such as to transport the required amount of angular momentum towards the surface (Zahn 1992).

Massive main-sequence stars belong to the first category, and the implementation of rotational mixing has been quite successful in improving their models (Maeder & Meynet 2000). The theoretical isochrones fit well the observed ones, even without introducing convective overshoot from the core, and such rotational mixing accounts well for the observed enhancement of He and N at the surface of early-type stars. This was first illustrated with the evolution of a 9 $M_\odot$ star by Talon et al. (1997), and then generalized to all massive stars by Meynet and Maeder (2000). Combined with a suitable prescription for the mass loss, this type of mixing also yields the observed proportion of blue and red giants. Finally, such mixing accounts perfectly for the destruction of Li on the blue side of the Li gap, as was shown by Charbonnel and Talon (1999); in their calculations they took the same value for the adjustable parameter characterizing the anisotropic turbulence than that which was used to model the 9 $M_\odot$ star by Talon et al. (1997).

The second case applies to solar-type stars, whose modeling has been much less successful, until very recently. In those stars, which lose most of their angular momentum in their youth through a magnetized wind (Schatzman 1962), the meridional circulation adjusts so as to carry the required angular momentum towards the surface, at least in the absence of other processes (Zahn 1992). One would then expect that the amount of mixing, and hence the depletion of light elements, be proportional to the loss of angular momentum. This would have as consequence that tidally-locked binaries would destroy less Li than single stars, since the angular momentum carried away by their winds would not be drawn from their internal rotation, but from their orbital motion; hence there would be no need to transport angular momentum inside such stars, and their Li would be preserved (Zahn 1994). However, there is no sign that tidally-locked binaries are less Li-depleted than single stars, as was shown by Balachandran (2002); hence one must conclude that the transport of matter is not correlated with the transport of angular momentum.

Another hint is provided by the fact that lithium is seriously depleted in the Sun, and that the somewhat less fragile beryllium is not, as was shown by Balachandran and Bell (1997). This contrasted behavior may be explained if the mixing occurs only in the vicinity of the convection zone, i.e. in the tachocline. As a matter of fact, the meridional circulation which is induced in that layer by the differential rotation of the convection zone (see section 3.1), can produce such localized mixing, as was shown by Brun et al. (1999).

But the most severe observational test is that models built according to equation (4.2), i.e. including only turbulence and meridional circulation, conserve a rapidly spinning core (Pinsonneault et al. 1989; Matias & Zahn 1998), which is ruled out by helioseismology. We must therefore conclude that another, more powerful process is responsible for the transport of angular momentum in solar-type stars; it is that process which then shapes the rotation profile and therefore, though indirectly, determines the extent of mixing.

4.2. Rotational mixing of type II

In what we call rotational mixing of type II, the chemical elements are still transported by the meridional circulation and the turbulence caused by differential rotation, but the angular momentum is carried by another process, which is powerful enough to establish an almost uniform rotation profile in the solar radiative interior. Two candidates have been proposed for this extra transport: magnetic torquing and internal gravity waves.

4.2.1. Magnetic field

Magnetic fields, because they are almost 'frozen' in the highly conducting stellar material, are very powerful in reducing differential motions. This was already pointed out by Mestel (1953), who claimed that a fossil magnetic field of moderate strength would render stellar rotation nearly uniform. More precisely, in presence of an axisymmetric poloidal field, the angular velocity tends to become uniform along the field lines, a property referred to as Ferraro's law (1937). Thus if the poloidal field lines lie entirely in the radiation zone of a star, they impose there uniform rotation.

The situation changes however when the poloidal field connects with a differentially rotating convection zone, such as that in the Sun. Then this differential rotation is transmitted along the field lines, and the result is a non uniformly rotating radiation zone, as illustrated by the calculations made by Charbonneau and MacGregor (1993). Can this fate be prevented, for instance if the field is advected by a suitable flow?

Gough and McIntyre (1998) sketched a solution to this problem in which the ohmic diffusion of the fossil field into the solar convection zone was prevented by the down-flow associated with the thermally driven circulation in the tachocline. In turn, that field was invoked to oppose the penetration of the tachocline into the deeper interior. The authors were aware of course that the circulation had also up-flows, which would advect the field into the convection zone, but if these up-flows cover only a small fraction of the domain, the field would imprint a small amount of differential rotation in the radiation zone.

Clearly, more detailed calculations were required to settle the question. The first of these were performed by Garaud (2002), who built a set of stationary solutions for various parameters. These solutions already hinted that the field lines would penetrate into the convection zone over a broad band of high latitudes, and therefore imprint a substantial amount of differential rotation in the radiation zone. This behavior has been confirmed by Brun and Zahn (2006) through 3-dimensional time-dependent calculations: they found that even a deeply buried poloidal magnetic field will eventually connect with the convection zone, and induce differential rotation below. We are thus led to conclude that, in the Sun at least, it is not the magnetic field that is responsible for the uniform rotation of the radiative interior.

However in other stars, such as magnetic A-type stars, fossil fields presumably play a much more important role, as was recently shown by Braithwaite and Spruit (2004): they certainly interfere with the circulation to modify the mixing in the radiation zone (Mathis & Zahn 2005). Moreover, certain field configurations are unstable, and they could produce MHD turbulence, and possibly mixing, before they relax into a stable state.

This is now being explored intensively through numerical simulations; the first indication we have gleaned from them (Zahn *et al.* 2007) is that this MHD turbulence behaves rather as a superposition of Alfvén waves, and that it does not produce any mixing.

4.2.2. *Internal gravity waves*

Since magnetic fields seem unable to enforce uniform rotation in the radiative interior of solar-type stars, we must turn to the other possible mechanism, namely the transport of angular momentum by the internal gravity waves emitted by the turbulent motions at the base of the convection zone. The importance of that process was already anticipated by Press (1981) and Schatzman (1993).

The restoring force operating on gravity waves is the buoyancy force: therefore they travel only in stably stratified regions, i.e. in radiation zones. Presumably a whole spectrum of such waves is emitted at the base of the convection zone of late-type stars; they penetrate into the radiation zone, transporting angular momentum which they deposit wherever they are dissipated through radiative damping. It is by shaping the rotation profile that they indirectly participate in the mixing of chemicals.

Let us first examine the behavior of those waves which are of short wavelength, and are dissipated close to the convection zone. Prograde waves carry positive angular momentum, retrograde waves negative angular momentum. When they travel in a medium which is rotating faster then the region where they have been emitted, their frequency is Doppler-shifted, leading to higher dissipation for the prograde waves than for the retrograde waves. For this reason the angular velocity tends to increase where it was already high, and its slope with depth steepens until the shear becomes unstable. That turbulent layer then merges with the convection zone. But in the meanwhile the retrograde waves have deposited negative angular momentum somewhat further down, thus building there another shear layer of opposite direction, which now takes the place of the former one. And the cycle repeats. A similar phenomenon is observed in the Earth atmosphere, where it is called the quasi-biennial oscillation (cf. McIntyre 1994).

The question then arises how this very thin shear layer, located just below the convection zone, affects the waves of longer period and wavelength, which are much less damped, and are therefore able to reach the core of the star. If there is no slope in angular velocity, the shear layer oscillation is perfectly symmetrical in time, and its effect is the same on the prograde and on the retrograde waves. But if Ω increases even slightly with depth in that layer, the prograde waves will be more dissipated, which allows the retrograde waves to extract angular momentum from the deep interior. This scenario has been tested through numerical simulations performed by Talon et al. (2002), using a rather crude approach where the Coriolis force was neglected and where an arbitrary turbulent viscosity was imposed. Recently it has been confirmed through more detailed and more realistic calculations performed by Talon and Charbonnel; beside the internal gravity waves, they include also the meridional circulation and the shear induced turbulence (Talon & Charbonnel 2003, 2004, 2005). Their calculations are one-dimensional, with the flux of angular momentum averaged over level surfaces, which are assumed to rotate uniformly due to the anisotropic shear-induced turbulence discussed above.

The result is spectacular: internal gravity waves succeed in achieving nearly uniform rotation in solar-type stars at the solar age (see also Charbonnel & Talon 2005). Furthermore their models predict the observed Li abundances as a function of age, which is a crucial test to validate the mixing processes. They explain the Spite plateau for population II stars, and the Li dip in galactic clusters. For the first time, we now have a coherent and consistent picture of internal mixing in all main sequence stars!

5. Thermohaline mixing

Numerous spectroscopic observations provide compelling evidence for a non-canonical mixing process that modifies the surface abundances of Li, C and N of low-mass red giants when they reach the bump in the luminosity function. These abundance patterns cannot explained by the rotational mixing described above (see Charbonnel & Palacios 2004). Based on numerical simulations, Eggleton et al. (2006) proposed that a molecular weight inversion created by the ^{3}He$(^3$He,2p$)^4$He reaction may be at the origin of this mixing, and they ascribed it to the Rayleigh-Taylor instability.

In fact, while the inverse μ-gradient gradually builds up below the convective envelope, the first instability that appears is the thermohaline instability, as it was shown by Charbonnel and Zahn (2007). This instability is well known in oceanography, where it arises when salty warm water lies above cold fresh water (Stern 1960). In the laboratory, the instability takes the form of 'salt fingers'; since heat diffuses faster than salt, these fingers sink because they grow increasingly heavier than their environment, until they become turbulent and dissolve. In stellar interiors, the role of salt is played by a heavier species, such as helium in a hydrogen-rich medium.

In a star, this instability occurs in a stable stratification that satisfies the Ledoux criterion for convective stability, but where the molecular weight decreases with depth:

$$N_T^2 + N_\mu^2 > 0, \quad N_\mu^2 < 0. \tag{5.1}$$

(The buyoancy frequencies N_T^2 and N_μ^2 have been defined above in §3.2.1.)

In its fully developed regime, the thermohaline instability manifests itself as a complex phenomenon, displaying fingers, 'staircases', collective instabilities, etc. Numerical simulations reveal that its properties strongly depend on the governing parameters, and on the applied boundary conditions. As yet, there is no simple way to describe the mixing achieved by that instability in stars. Lacking of something better, one can follow Ulrich (1972), who assimilated this mixing to a diffusive process, and was the first to derive a prescription for the turbulent diffusivity; for a general equation of state, it is

$$D_t = C_t K \frac{-N_\mu^2}{N_T^2 + N_\mu^2} \quad \text{for } N_\mu^2 < 0, \tag{5.2}$$

K being the thermal diffusivity. Ulrich's non-dimensional coefficient C_t involves the aspect ratio α (length/width) of the fingers: $C_t = 8\pi^2\alpha^2/3$; for the value he advocates, $\alpha = 5$, this coefficient is rather large: $C_t = 658$.

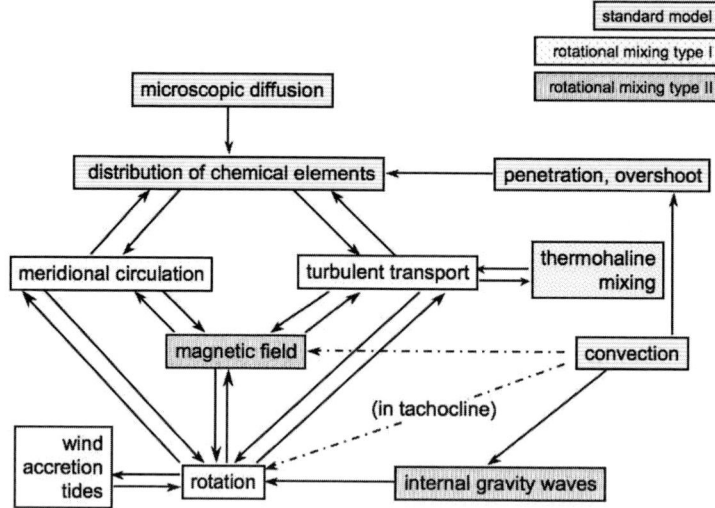

Figure 1. Mixing processes in a stellar radiation zone: a highly non-linear problem, with multiple interactions. In the standard model, the only mixing processes taken into account are convective penetration and microscopic diffusion. Rotational mixing involves the meridional circulation induced by external torques or structural changes and the shear turbulence due to differential rotation; in rotational mixing of type I, angular momentum is carried exclusively by the same processes. In rotational mixing of type II, the angular momentum is transported either by magnetic stresses or by internal gravity waves. (Adapted from Mathis & Zahn 2005, courtesy A&A).

When applying this prescription to a evolutionary sequence, the models predict the abundance ratios [^{7}Li/H], [^{12}C/^{13}C], [C/Fe] and [N/Fe] that are in excellent agreement with the observed values, as it was shown by Charbonnel and Zahn (2007).

Elsewhere in this volume, S. Vauclair describes another occurrence of thermohaline mixing, when a star accretes heavy material, such as telluric planets (see Vauclair 2004).

6. Conclusion

To conclude, we are entering a new era of modeling stellar interiors, where rotation will be explicitly taken into account, and where the transport processes discussed above will be implemented in the models. This is not an easy task, because it involves intricate couplings, as illustrated by the flow chart displayed in Fig. 1. The (differential) rotation is responsible for the meridional circulation and the shear-induced turbulence, which transport angular momentum and react back on the rotation profile. These mechanisms modify the distribution of chemical elements, which itself interferes with them. A magnetic field modifies the rotation and therefore also the circulation and the turbulent transport; it may become unstable and will then contribute to the mixing. Internal gravity waves emitted at the edge of convection zones will transport angular momentum, and modify the rotation profile. Furthermore some mixing occurs in the vicinity of convection zones, due to convective penetration, and to the circulation induced in the tachoclines through the differential rotation imposed by the adjacent convection zones. Finally, thermohaline mixing occurs in stably stratified regions, when the molecular weight increases with height.

Most pieces of this scheme are now based on sound physics and robust prescriptions, but some weaknesses remain to be remedied.

Above all, we need to improve the description of the turbulent transport due to shear instabilities, in particular that operating in the horizontal direction. Also, the way we handle convective penetration and the generation of internal gravity waves is far from satisfactory – but the same could be said about our present treatment of the convection zones. We must clarify whether the gravity waves are able to diffuse chemicals, beside transporting angular momentum, as was argued by Schatzman (1993). We still don't know for sure which physical process prevents the spread of the solar tachocline. We have to introduce magnetic fields in our models, at least where we think that they could play a role. Finally, we are in great need of a reliable prescription for the thermohaline mixing. In all these subjects, there is much to be expected from the high-resolution numerical simulations that are now undertaken by several teams.

In the meanwhile, the transport processes discussed above have, as they stand, already been introduced in several stellar evolution codes, in collaborative efforts such as described by Decressin et al. (2008). Their treatment will continue to greatly benefit from observational constraints, in particular from those we anticipate from asteroseismology. It is clear that in not so distant future those non-standard mixing processes will be an integrant part of the standard model.

References

Balachandran, S. C. 2002, Highlights of Astronomy 12, 276
Balachandran, S. C. & Bell, R. A. 1997, BAAS, 29, 1325
Braithwaite, J. & Spruit, H. C. 2004, Nature 431, 819
Brun, A. S., Turck-Chièze, S., & Zahn, J.-P. 1999, ApJ 525, 1032
Brun, A. S. & Zahn, J.-P. 2006, A&A 457, 665
Busse, F. H. 1982, ApJ 259, 759
Chaboyer, B., Demarque, P., & Pinsonneault, M. H. 1995, ApJ 441, 865
Chaboyer, B. & Zahn, J.-P. 1992, A&A 253, 173
Charbonneau, P. & MacGregor, K. B. 1993, ApJ 417, 762
Charbonnel, C. & Palacios, A. 2004, IAU Symp. 215, 440
Charbonnel, C. & Talon, S. 1999, A&A 351, 635
Charbonnel, C. & Talon, S. 2005, Science 309, 2189
Charbonnel, C. & Zahn, J.-P. 2007, A&A 467, L15
Decressin, T., Mathis, S., Palacios, A., Siess, L., Talon, S., Charbonnel, C., & Zahn, J.-P. 2008, A&A (submitted)
Deliyannis, C, Demarque, P., & Kawaler, S. 1990, ApJS 73, 21
Dudis, J. J. 1974, J. Fluid Mech. 64, 65
Eddington, A. S. 1926, Observatory 48,73
Eggleton, P. P., Dearborn, D. S. P., & Lattanzio, J. C. 2006, Science 314, 1580
Ferraro, V. C. A. 1937, MNRAS 97, 458
Forgács-Dajka, E. 2004, A&A 413, 1143
Forgács-Dajka, E. & Petrovay, K. 2002, A&A 389, 629
Garaud, P. 2002, MNRAS 329, 1
Gough, D. O. & McIntyre, M. E. 1998, Nat 394, 755
Herrero, A., Kudritski, R. P., Vilchez, J. M. et al. 1992, A&A 261, 209
Herrero, A., Puls, J., & Villamariz, L. R. 2000, A&A 354, 193
Korn, A. J., Grundahl, F., Rochard, O., Barklem, P. S., Mashonkina, L., Collet, R., Piskunov, N., & Gustavsson, B. 2006, Nat 442, 657

Lignières, F., Califano, F., & Mangeney, A. 1999, A&A 349, 1027
Maeder, A. 2003, A&A 399, 263
Maeder, A. & Meynet, G. 2000, ARA&A 38, 143
Maeder, A. & Zahn, J.-P. 1998, A&A 334, 1000
Mathis, S., Palacios, A., & Zahn, J.-P. 2004, A&A 425, 243
Mathis, S. & Zahn, J.-P. 2004, A&A 425, 229
Mathis, S. & Zahn, J.-P. 2005, A&A 440, 653
Matias, J. & Zahn, J.-P. 1998, *Sounding Solar and Stellar Interiors*, IAU Symp. 181, (ed. Provost, J. & Schmider, F.-X.) poster vol. p. 103
McIntyre, M. E. 1994, *The Solar Engine and its Influence on the Terrestrial Atmosphere and Climate (NATO ASI Subseries I, Global Environmental Change)* 25, 293 (Cambridge Univ. Press), p. 557
Mestel, L. 1953, MNRAS 113, 716
Meynet, G. & Maeder, A. 1997, A&A 321, 465
Meynet, G. & Maeder, A. 2000, A&A 361, 101
Michaud, G. 1970, ApJ 160, 641
Michaud, G., Charland, Y., Vauclair, S., & Vauclair, G. 1976, ApJ 210, 447
Pinsonneault, M. 1997, ARA&A 35, 557
Pinsonneault, M., Kawaler, S. D., Sofia, S., & Demarque, P. 1989, ApJ 338, 424
Press, W. H. 1981, ApJ 245, 286
Proffitt, C. R. & Michaud, G. 1991, ApJ 371, 584
Richard, D. & Zahn, J.-P. 1999, A&A 347, 734
Richard, O., Michaud, G., & Richer, J. 2002, ApJ 568, 979
Schatzman, E. 1962, Ann. Ap 25, 18
Schatzman, E. 1993, A&A 279, 431
Schatzman, E., Zahn, J.-P., & Morel, P. 2000, A&A 364, 876
Spiegel, E. A. & Zahn, J.-P. 1992, A&A 265, 106
Spite, F. & Spite, M. 1922, A&A 115, 357
Stern, M. E., 1960, Tellus, 12, 172
Sweet, P. A. 1950, MNRAS 110, 548
Talon, S. & Charbonnel, C. 2003, A&A 405, 1025
Talon, S. & Charbonnel, C. 2004, A&A 418, 1051
Talon, S. & Charbonnel, C. 2005, A&A 440, 981
Talon, S., Kumar, P., & Zahn, J.-P. 2002, ApJ 574L, 175
Talon, S. & Zahn, J.-P. 1997, A&A 317, 749
Talon, S., Zahn, J.-P., Maeder, A., & Meynet, G. 1997, A&A 322, 209
Théado, S. & Vauclair, S. 2001, A&A 375, 70
Townsend, A. A. 1958, J. Fluid Mech. 4, 361
Ulrich, R. K., 1972, ApJ, 172, 165
Vauclair, G., Vauclair, S., & Michaud, G. 1978, ApJ 223, 920
Vauclair, G., Vauclair, S., & Pamjatnikh, A. 1974, A&A 31, 63
Vauclair, S. 1999, A&A 351, 973
Vauclair, S. 2004, ApJ 605, 874
Vitense, E. 1953, Z. Astrophys. 32, 135
Vogt, H. 1925, Astron. Nachr. 223, 229
Von Zeipel, H. 1924, MNRAS 84, 665
Zahn, J.-P. 1974, *Stellar Instability and Evolution*, IAU Symp. 59, p. 185
Zahn, J.-P. 1975, Mém. Soc. Roy. Sci. Liège 6 série, 8, 31
Zahn, J.-P. 1991, A&A 252, 179
Zahn, J.-P. 1992, A&A 265, 115
Zahn, J.-P. 1994, A&A 288, 829
Zahn, J.-P., Brun, A. S., & Mathis, S. 2007, A&A 475, 145

Discussion

CANUTO: Salt-Finger existence is heavily dependent on the strength of the shear present whatever its origin – Lab data on SF do not include shear av therefore they must be used with care when applied to the real world – A theory of SF+Turbulence has recently been proposed.

ZAHN: I agree that shear will play a role, and we know that there is shear because of differential rotation.

The speaker, J.-P. Zahn (middle), is discussing with S. Mathis (left) and G. Meynet.

Lithium Depletions in Late-type Dwarfs

D. R. Xiong[1] and L. Deng[2]

[1] Purple Mountain Observatory, Chinese Academy of Sciences, Nanjing 210008
email: xiongdr@pmo.ac.cn

[2] National Astronomical Observatories, Chinese Academy of Sciences, Beijing 100012
email: licai@bao.ac.cn

Abstract. Using a self-consistant dynamic theory of non-local convection in chemically inhomogeneous stars, the lithium depletion in MS stars of masses 0.725–1.5 $M_\odot$ are calculated. Both of the overshooting and microdiffusion are included in a consistent way. The comparisons of theretical reasults with the observed Li depletions in open clusters show that the general characters of Li depletions can be reproduced by theory. The overshooting mixing and microdiffusion induced by gravitational setting and radiative accelerations may be two main mechanisms of Li depletion.

Keywords. Convection- star: abundance-star: late-type-stars:evolution- open clusters and associations: general

---

It is well known that the Li abundance is very significant for the theory of big-bang nucleosynthesis and tracking the extension of the surface convection zone in stars during the course of their evolution.

The mechanism of Li depletion is still not completely known. A number of theoretical approach has been presented so far, such as the mass loss (Weymann & Sears, 1965; Schramm et al., 1990), wave-driven mixing (Garcia López & Spruit, 1991; Montalban & Schatzman, 1996), rotationally induced mixing (Charboneau et al., 1992; Chaboyer et al., 1995; Pinsonneault, et al., 1999), microdiffusion and turbulent mixing (Michaud, 1986; Turcotte et al., 1998), overshooting mixing (Straus et al., 1976; Xiong & Deng, 2001) and so on. However, there is no any one single mechanism which can interpret all the known observed characters of Li depletion. The Li depletion, in our opinion, shoult be the result of several physical processes, rather than from a single mechanism. The overshooting mixing and microdiffusion induced by gravitation setting and ratiation acceleration seem to be a reasonable combination.

Using our non-local convection theory in chamically inhomogeous stars (Xiong, 1981), we calculated the Li depletions for 20 series of evolutionary models with mass of 0.725-1.5 $M_\odot$. The solar abundance (X = 0.70, Z = 0.02) is assumed. The overshooting mixing and microdiffusion induced by gravitational setting are included in a selfconsistant way. The fundamental equations and the numerical scheme will be described in our latter paper (Xiong & Deng, 2008). Figure 1 illustrates the Li abundances as function of stellar age for the evolutionary models with different masses as marked on the curves. In Fig. 1a the microdiffusion is neglected. In Fig. 1b both of the microdiffusion and overshooting mixing are included. It can be seen that the Li abundances decrease (approximately) exponentially with age of stars. The e-folding time of Li depletion are not a monotonic function of stellar mass. It achives the maximum at $M \approx 1.1 M_\odot$. This is due to the fact that the Li depletion is resulted dominantly from macrodiffusion for warm stars ($M \geqslant 1.1 M_\odot$ or $T_e \geqslant 6100K$), and it is resulted dominantly from the overshooting mixing for the cool stars ($M \leqslant 1.0 M_\odot$ or $T_e \leqslant 5800K$). The sign $\odot$ in Figs. 1 & 3d is

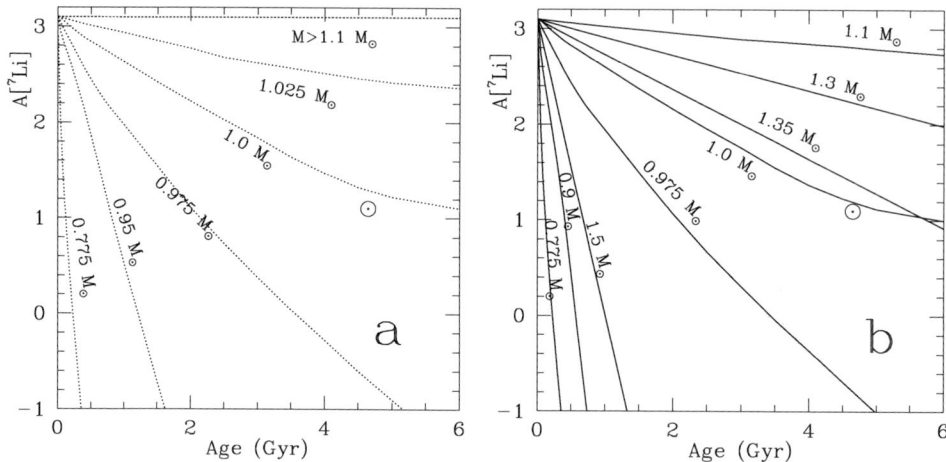

Figure 1. The Li abundances as functions of star's age for the different masses marked on the curves. (a) the microdiffusion is neglected, (b) both of the microdiffusion and overshooting mixing are included.

the observed lacation of the Sun. It is necessary to indicate that the depth of our solar model agrees with the requirement of helioseismology.

Figs. 2a – d show the lithium distribution in stellar interior at different ages. The short dashed lines mark the bottom boundary of convectively unstable zone. The sharply drop at bottom is due to violent burn of lithium. It can be seen that the convective overshooting (the shadowy regions) penetrates deeply into the convectively stable zone and the full mixiong zone is extended greatly. For the warm stars with mass greater than abut 1.1 $M_\odot$ (Figs. 2a & b), the convection zone is too shallow. The gravitational setting is not enough to bring the surface lithium to the deep burn region. It drive the surface lithium to the deeper interior of stars and the surface lithium will be stored in the lower radiative region under the overshooting zone, so the surface lithium abundance decreases. The convection zone becomes deeper with decrease of stellar temperature. overshooting brings the surface lithium to the deep burn region, therefore, the surface lithium abundance decreases (Figs 2c & d). We can know from the above analysis that the mechanisms of lithium depletion are different for the warm and cool stars. For the warm stars the microdiffusion is domanial, however the oversooting mixing becomes domanial for the cool stars.

Figure 3 give The Li abundance vs. the effect temperature for the clusters with different ages, and the corresponding theoretical isochne lines of Li abundance are also drawn here. It can be seen from Fig. 3 that within the observational uncertain the theoretical results reproduced roughly the general profiles of Li depletion except the following points: (1) The Li depletions seem to be underestimated for stars arroud $Te \approx 5900K$. It is possible there are depletion mechanisms ignored in this work; (2) In Fig. 3d the Li abundances of five stars on the right of theoretical isochne line seem to be too high. However, two of them are only given the upper limits of their Li abundance. The another two stars, named as I-2a and I-2b in the Table 1 of Jones et al. (1999), are the two components of a double-lined spectroscopic binary. Their Li abundance are quite uncertain. The samples of lower temperature star are too few for giving a confident conclusion.

It is the conclusion that the overshooting mixing and microdiffusions induced by gravitation setting and radiative acceleration may be two main mechanisms of Li depletion. Not only the convective overshooting interpret the lithium depletion of cool stars but

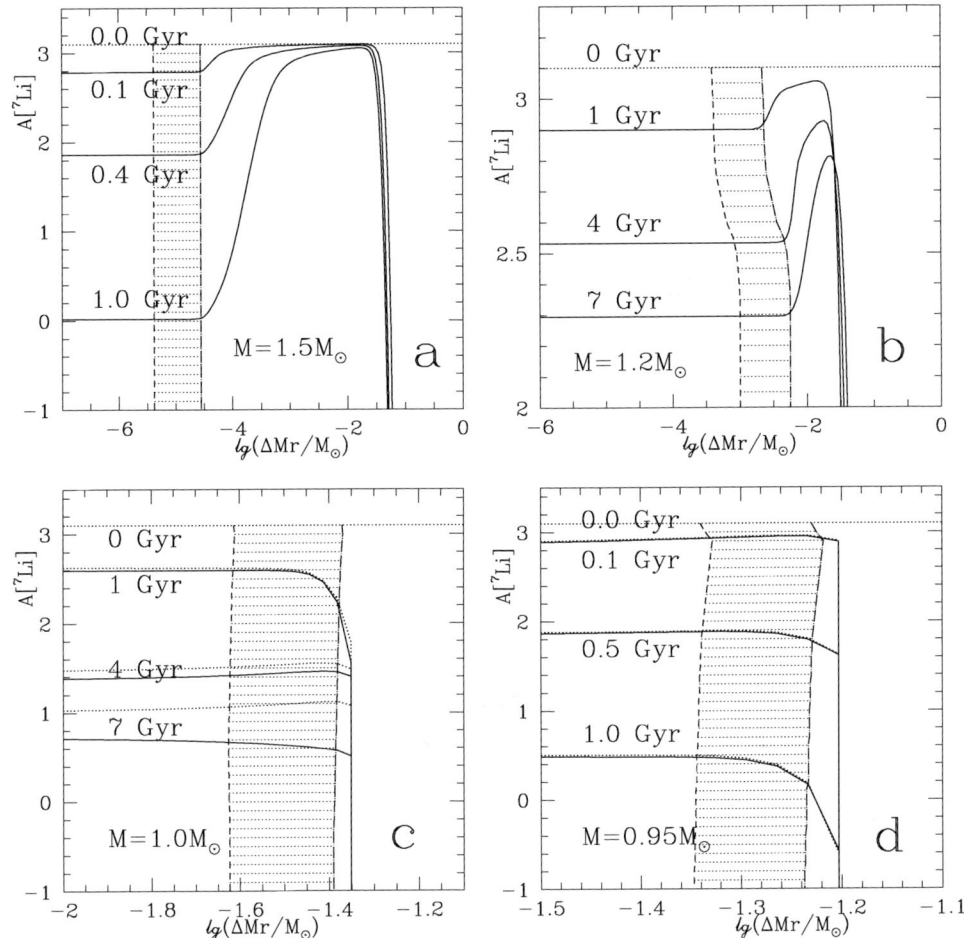

Figure 2. The lithium distribution in stellar interior at different ages, (a) $M = 1.5 M_\odot$ (b) $M = 1.2 M_\odot$ (c) $M = 1.0 M_\odot$ (d) $M = 0.95 M_\odot$

also it increase the mass of the surface full mixing region, therefore the too quiuck depletion of lithium in warm stars predicted by single microdiffusion theory is removed automatically. It is not necessary to induce a phenomenological turbulent diffusion such as induced by Proffitt & Michaut (1991) and Richard, Michaud & Richer (2005). In other words, convective overshootng is just the so called turbulet diffusion needed by theim.

Acknowledgements

This work is supported in part by NSFC through grants 10773029 and 10778719.

References

Chaboyer, B., Demarque, P., & Pinsonneault, M.H., 1995, *ApJ* 441, 865
Charboneau, P., Vauclair, S., & Zahn, J. P., 1992 *A&A*, 255, 191
Garcia López, R.J. & Spruit, H. C., 1991 *ApJ*, 377, 268
Jones, B. F., Fischer, D., & Soderblom, D. R., 1999 *AJ*, 117, 330
Michaud, G., 1986 *ApJ*, 302, 650
Montalban, J. & Schtzman, E., 1996 [5 *A&A*, 305,513
Pinsonneault, M. H., Walker, T. P., Steigman, G., & Narayanan, V. K., 1999 *ApJ*, 527, 180

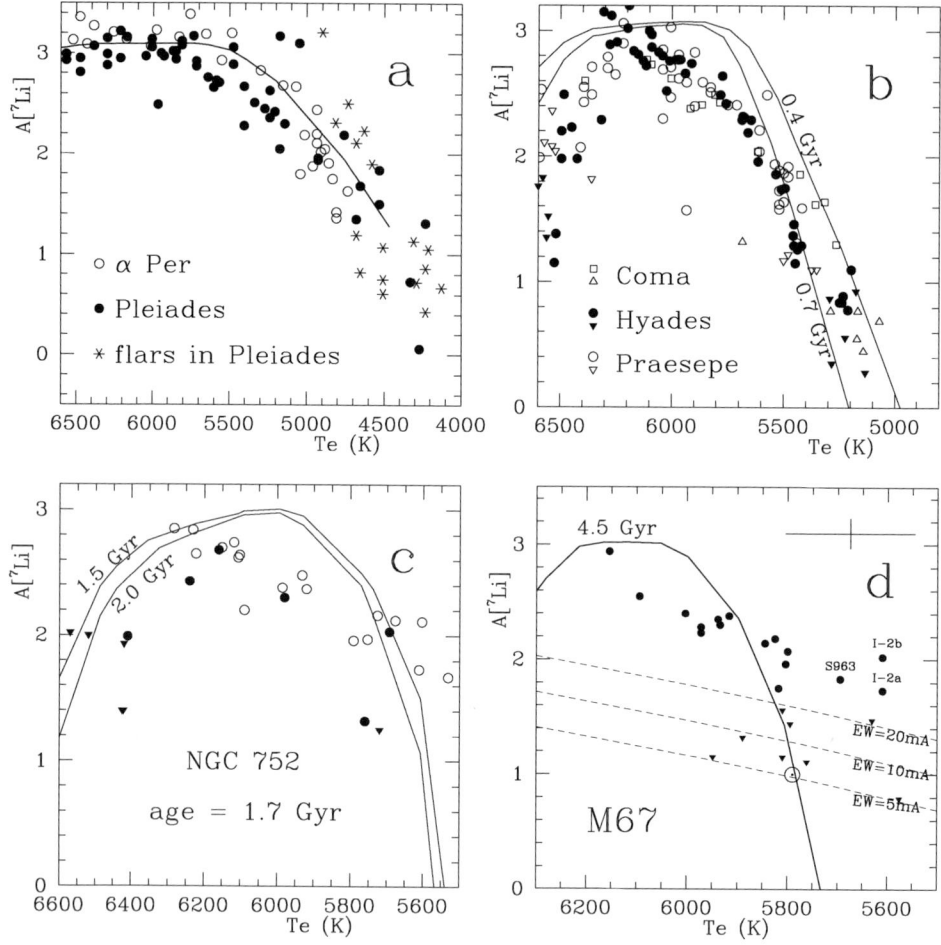

Figure 3. The Li abundances versus effective temperature of stars for (a) α Persei, Pleiades, (b) Coma Berenices, Hyades Prasepe, (c) NGC 752, and (d) M67 clusters. The triangles and inverse triangles give the upper limits for Li abundance. The estimated uncertainties of LI abundance and effect temperature are indicated in the upper-right corner of the plots.

Proffitt, C. R. & Michaud, G. 1991 *ApJ*, 380, 238
Richard, O., Michaud, G., & Richer, J., 2005 *ApJ*, 619, 538
Schramm, D. N., Steigman, G., & Dearborn, D. S. P., 1990 *ApJ*, 359, L55
Straus, J. M., Blake, J. B., & Schramm, D. N., 1976 *ApJ*, 204, 481
Turcotte, S., Richer, J., & Michaud, G., 1998 *ApJ*, 504, 559
Weymann, R. & Sears, R. L., 1965 *ApJ*, 142, 174
Xiong, D. R., 1981 *Scientia Sinica*, 24,1406
Xiong, D. R. & Deng, L. 2001 *MNRAS*, 327, 1137
Xiong, D. R. & Deng, L. 2008 *MNRAS*, prepared

Discussion

ASPLUND: Have you investigated how much ^{7}Li depletion your convective overshooting model predict for metal-poor turn-off stars?

XIONG: We have only considered the depletion of Lithium for solar type stars. We understand that Lithium depletion in metal-poor stars is extremely important in exploring the initial cosmological Lithium abundance and nucleosynthesis of Big-Bang. Due to the lack of reliable evolutionary models of metal-poor stars, such work can only be done in a future time when such models become available.

LANGER: Are there any free parameters in your model for overshooting? and: Are your models (for main sequence stars) in strict thermal equilibrium?

XIONG: 1). Indeed, there is also a free parameter in our statistical theory of non-local turbulent convection, which is similar to that of MLT representing the characteristic length (wave number) of the turbulent spectrum. However, such a parameter is not at all "free". Instead, it has to be regulated by observational constraints from stellar evolution, depletion of Lithium in stars, helioseismology and stellar pulsation, etc. As we have shown in our previous work, the theory stands against all these challenges.

2). The models computed so far are all located on the main sequence, therefore the assumption of perfect thermal equilibrium is reasonable.

KUPKA: Have you also made any computations for the Li abundance in hotter (main sequence) stars than the ones you have shown here? I mean early F type and late A type, and compared to some data? This could be useful to test the model for microscopic diffusion you have used in your computations in the OV zone of cool main sequence stars too!

XIONG: We have done Lithium depletion models only for late F type – K type stars so far. The late A type – early F type stars are not modelled, mainly because we could not good monochromatic opacity. For microdiffussion, only gravitational settling has been taken into account. Radiative acceleration has been excluded. Once monochromatic opacity is available, radiative acceleration can be included into microdiffusion easily. In that case, the combination of microdiffusion and overshooting mixing can be well applied to the studies of Li depletion and A_p, F_m phenomena in A-F type stars. Turbulent diffusion is very likely not needed, pretty the same way as in the present work that overshooting can naturally reproduce the effect of turbulent diffusion

The speaker, D.R. Xiong (right), is chatting with Jørgen Christensen-Dalsgaard during a coffee time.

The Art of Modelling Stars in the 21st Century
Proceedings IAU Symposium No. 252, 2008
L. Deng & K.L. Chan, eds.

Angular momentum and overshooting: two as yet unsolved problems in stellar mixing

V. M. Canuto[1,2]† and Y. Cheng[1,3]

[1] NASA, Goddard Institute for Space Studies, New York, NY 10025, USA
email: vcanuto@giss.nasa.gov, ycheng@giss.nasa.gov

[2] Dept. of Appl. Phys. and Appl. Math., Columbia University, New York, NY 10027, USA
[3] Ctr. Clim. Sys. Res., Columbia University, New York, NY 10025, USA

Abstract. Helioseismological data have given us two interesting results: *the differential-to-uniform solar rotation curve and the extent of the overshooting region (OV)*. As of today, no model (including numerical simulations) has been able to reproduce these findings. Here, we first present a new model for the angular momentum. It contains new terms representing *vorticity and buoyancy* that were left out in all previous formulations without a clear justification. It is shown that they extract angular momentum from the stellar core, a welcome feature since the standard angular momentum equation leads to a rotation curve that is considerably higher than what is observed. As for the overshooting extent, all models yield values that are an order of magnitude larger than the helio data of $0.07 H_p$. We propose a criterion whose main ingredient is a new flux conservation law that includes new terms, one of which increases the dissipation in the radiative zone and thus lowers the OV extent, a tendency in the desired direction. Since we have not coupled the new models to a solar structure-evolution code, we cannot at this stage carry out a comparison with the helio data. The purpose is to exhibit the fact that in both cases the missing ingredients are of such nature as to improve the previous model predictions. A proper quantification remains to be done.

Keywords. Stars: abundances, convection, turbulence

---

1. Introduction

Since mixing in stars is a complex interplay of processes as diverse as unstable stratification, stable stratification, differential rotation, gravity waves, double-diffusion, etc., the formalism employed to model mixing should be sufficiently general to account for such a large variety of processes. And yet, the literature shows that this is not the case, the two methodologies being employed being: a) large scale numerical simulations and b) heuristic models. In the first case (e.g., Brummell *et al.*, 2002), the values of several parameters, such as the Prandtl number, are widely different from those in stars. The authors of those studies have however stated that their primary goal was the elucidation of the intertwined physical processes and not to provide stellar studies with tools to model the processes of interest. The consequence is that mixing processes are still modeled using heuristic arguments that have severe limitations, as we discuss in section 5. Our assessment is that heuristic models should be abandoned for lack of physical completeness. As a substitute we suggest, work out and assess models of at most algebraic complexity that avoid the guessing work that heuristic models always entail. It is instructive to point out that an analogous situation existed in geophysics, specifically in modeling atmospheric and oceanic mixing. More than 25 years ago, it was decided to forgo the heuristic approach in favor of a more predictive and flexible tool known as RSM

† Present address: NASA-GISS, 2880 Broadway, New York, NY 10025, USA.

(Reynolds Stress Methodology) which is now commonly used. For reasons unclear to the present authors, stellar mixing studies are lagging behind geophysical studies and this paper will thus discuss a new model as well as the limitations of the heuristic models. The RSM is a set of equations for the turbulent correlations of the velocity and temperature fields that follow from the Navier-Stokes equations and the temperature equations. The RMS'main features can be summarized as follows: *mathematical structure*, the relevant equations are a set of linear algebraic equations and thus pose no particular numerical problems; *flexibility*, which is one of the key advantages, means that adding new processes such as rotation, vorticity, double-diffusion, etc, does not require guessing work since the RSM has a well defined set of procedural rules; *assessment*, the results of the RSM can be assessed before being used in a stellar context, an important feature that none of the heuristic model satisfies, raising the justified doubt that these models were constructed and tailored to a specific astrophysical setting, a feature that limits their predictive power. Since the RSM was discussed in detail elsewhere, we refer to that work for more details (Canuto, 2008).

2. The angular momentum problem

The angular momentum equation is given by ($\Gamma = sin\theta$):

$$\frac{\partial}{\partial t}\left(r^2\Omega\right) = -\Gamma^{-1}r^{-2}\frac{\partial}{\partial r}\left(r^3\tau_{r\phi}\right) - \Gamma^{-3}\frac{\partial}{\partial \theta}\left(\Gamma^2\tau_{\theta\phi}\right) + ... \tag{2.1}$$

where $\tau_{ij} = \overline{u_i u_j}$ are the Reynolds stresses. If one assumes that $\tau_{r\phi}$ is governed only by shear S_{ij}, one has:

$$\tau_{r\phi} = -2K_m S_{r\phi}, \quad S_{r\phi} = \frac{1}{2}r\Gamma\frac{\partial\Omega}{\partial r}, \quad 2S_{ij} = \overline{u}_{i,j} + \overline{u}_{j,i} \tag{2.2}$$

where K_m is a momentum diffusivity, (2.1) then becomes (Talon and Zahn, 1998; Talon and Charbonnel, 2003; Palacios *et al.*, 2003, 2006):

$$\frac{\partial}{\partial t}\left(r^2\Omega\right) = r^{-2}\frac{\partial}{\partial r}\left(r^4 K_m \frac{\partial\Omega}{\partial r}\right) + ... \tag{2.3}$$

Thompson *et al.* (2003) have written that (2.3) predicts rotation of the solar interior at a rate several times higher than the surface rate in stark disagreement with helio data of nearly uniform rotation. The first obvious conclusion is that since shear alone does not fully represent the mean flow, one must also include *vorticity*:

$$V_{ij} = \frac{1}{2}\left(\overline{u}_{i,j} - \overline{u}_{j,i}\right), \quad V_{r\phi} = -\frac{1}{2}r^{-1}\Gamma\frac{\partial}{\partial r}\left(r^2\Omega\right) \tag{2.4}$$

which leads to a real "diffuse nature" of the angular momentum. Furthermore, it seems natural that if one wants to explain the different behavior of the solar rotation curve in the *convective and radiative* zones, one must have an "ingredient" capable of differentiating between the two regimes. The obvious candidate is the buoyancy flux that is positive in the first case and negative in the second. On the basis of these qualitative considerations, one concludes that in addition to shear and vorticity, the Reynolds stresses must also depend on buoyancy and thus $\tau_{ij}(S,V,B)$. Finally, one must account for possible radiative losses and thus the formalism must include a Peclet number. We conclude that the final form of the Reynolds stresses must be:

$$\tau_{ij}(S,V,B,Pe) \tag{2.5}$$

3. Reynolds stresses and heat fluxes

In the presence of shear, vorticity, buoyancy and radiative losses, the general form of the traceless Reynolds stress tensor $b_{ij} = \tau_{ij} - \delta_{ij}2K/3$ is (Canuto and Minotti, 2001):

$$A\tau^{-1}b_{ij} = -\frac{8K}{15}S_{ij} - \frac{1}{2}Z_{ij} + \frac{1}{2}B_{ij} \tag{3.1}$$

where

$$Z_{ij} = b_{ik}V_{jk} + b_{jk}V_{ik}, \quad B_{ij} = \alpha\left(g_i J_j + g_j J_i\right) - \frac{2}{3}\delta_{ij}\alpha g_k J_k \tag{3.2}$$

Here, $A = 5$, $\alpha = -\rho_0^{-1}\partial\overline{\rho}/\partial T$ is the volume expansion coefficient and $J_i = \overline{u_i\theta}$ is the heat flux for which the RSM provides the following equations (Canuto and Minotti, 2001):

$$\tau^{-1}A_{ij}J_j = -\tau_{ij}\frac{\partial T}{\partial x_j}, \quad \tau = 2K\epsilon^{-1} \tag{3.3}$$

$$A_{ij} = \lambda_5\delta_{ij} + \lambda_6\tau S_{ij} + \lambda_7\tau V_{ij} + \lambda_8\tau^2\alpha g_i\frac{\partial T}{\partial x_j} + 2\epsilon_{ipj}\Omega_p^0 \tag{3.4}$$

with $\lambda_{6,7,8} = 0.786, 0.643, 0.547$. Canuto and Dubovikov (1998) and Canuto (2008) showed that the Peclet number dependence enters via the two remaining variables:

$$\lambda_5^{-1} = aPe(1 + bPe)^{-1}(1 + Ri)^{-1}, \quad \lambda_8 = cPe(1 + dPe)^{-1} \tag{3.5}$$

$$a = (4\pi^2)^{-1}, \quad b = 5a(1 + \sigma_t^{-1}), \quad c = \frac{8}{3}(7\pi^2)^{-1}, \quad d = 4(7\pi^2)^{-1}\sigma_t^{-1} \tag{3.6}$$

where $\sigma_t = 0.72$ and Ri is the Richardson number which is present only in the stable stratification case.

4. New results of the RSM model

We have numerically solved the set of linear algebraic Eqs.(3) and in figure 1 we present the heat diffusivity as a function of Ri and for different values of Pe; in figure 2 we plot the momentum diffusivity while in Fig.3 we plot the turbulent Prandtl number $\sigma_t(Ri, Pe) \equiv K_m/K_h$, which is the ratio between momentun and heat diffusivities. As one can see, the ratio in not constant but increases with Ri. For large Pe, we have superimposed a variety of LES, DNS, lab and direct measurements (Canuto et al., 2008) that the model reproduces quite well.

An example of the results that are obtained from our method is given in figure 1.

5. Previous heuristic mixing model

Here, we compare the results of Figs.1–3 with those of the heuristic relations used by various authors (Mathis et al., 2004; Palacios et al., 2003, 2006; Charbonnel and Talon, 2005, 2007; Maeder and Meynet, 2001):

$$Pe \gg 1: \quad \frac{K_{m,h}}{\chi} = 2\frac{Ri(cr)}{Ri} = \frac{1}{3Ri} \tag{5.1}$$

where $\chi(cm^2 s^{-1})$ is the radiative diffusivity. (5.1) is consistent with Fig. 1 if:

$$Pe \approx 10^2 \tag{5.2}$$

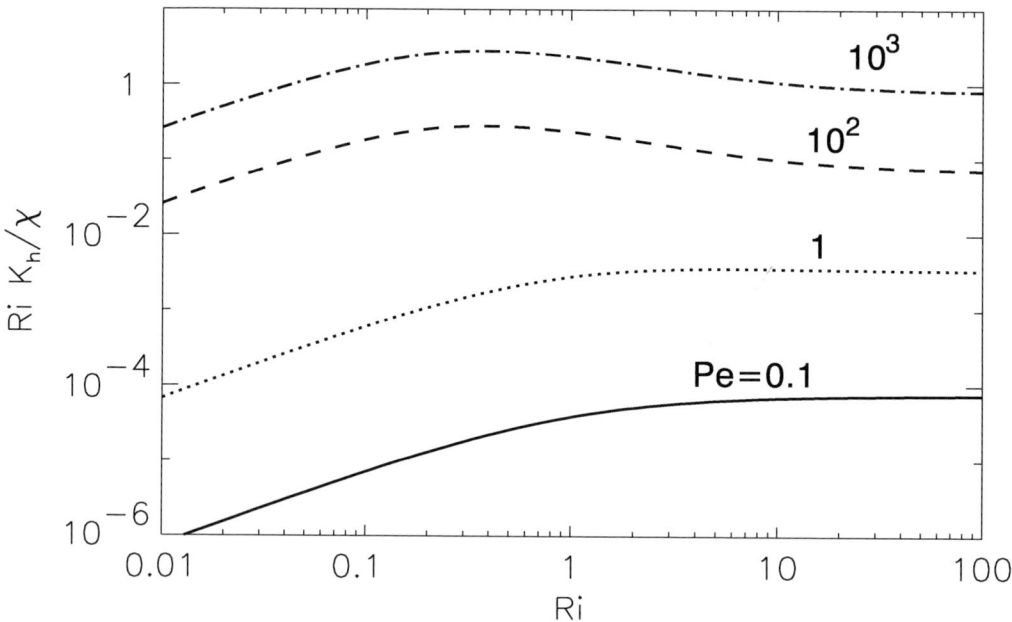

Figure 1. The heat diffusivity K_h in units of the radiative diffusivity vs. the Richardson number Ri for different values of the Peclet number. As expected, the stronger the effect of stable stratification, the larger is the value of Ri and the smaller is the resulting diffusivity. We plot K_h multiplied by Ri in order to allow a direct comparison with the heuristic model (5.1). See the text for details.

As for the momentum diffusivity in Fig. 2, (5.1) is not satisfied. The only possibility is to impose (5.1) but that in turn implies a unique value of Ri equal to:

$$Ri \approx 0.1 \qquad (5.3)$$

This means that the heuristic model is valid for only one combination of Pe, Ri given by (5.2) and (5.3) which is quite unusual for any model. Finally, (5.1) implies a turbulent Prandtl number of unity whereas Fig. 3 shows that it is a rather strong function of Ri unless one limits the validity of the model to $Ri \ll 1$.

6. New angular momentum equation

Using a method of symbolic algebra, we have solved Eqs. (3) without meridional currents. Introducing three dimensionless variables x, X and Z to characterize stratification, shear and vorticity in units of the dynamical time scale $\tau = 2K/\epsilon$:

$$x \equiv (\tau N)^2, \quad X \equiv \tau S_{r\phi}, \quad Z \equiv \tau V_{r\phi} \qquad (6.1)$$

the explicit form of the Reynolds stress that enters (2.1) turns out to be:

$$\tau_{r\phi} = -2K_m^{(1)} S_{r\phi} - 2x K_m^{(2)} V_{r\phi} \qquad (6.2)$$

which yields the new angular momentum equation to:

$$r^2 \frac{\partial}{\partial t}(r^2 \Omega) = \frac{\partial}{\partial r}\left(K_m^{(1)} r^4 \frac{\partial \Omega}{\partial r}\right) + \frac{\partial}{\partial r}\left(x K_m^{(2)} r^2 \frac{\partial r^2 \Omega}{\partial r}\right) \qquad (6.3)$$

Several considerations are in order:1) the presence of vorticity has now introduced a new term which has the character of a true diffusion of angular momentum whereas the

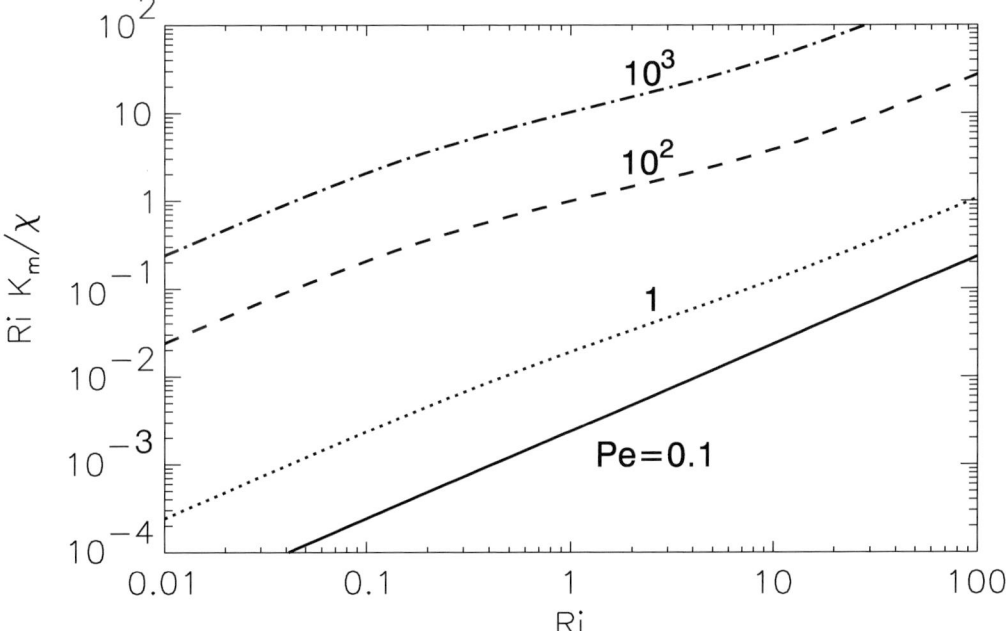

Figure 2. Same as in figure 1 but for the momentum diffusivity K_m.

first traditional term, in spite of being usually called "diffusion", is not, 2) the new term entails vorticity multiplied by buoyancy represented by the variable x which is negative in the convective zone and positive in the radiative, stably stratified, zone, 3) in the latter zone, where we can take $\Omega = constant$ as from the helio data, only the second term in (6.3) survives and since $x > 0$, this implies an *outward transport of angular momentum from the radiative interior*, an "extraction process" that in principle drives it toward a state of uniform rotation, 4) since in the radiative zone turbulence is much weaker than in the CZ, the eddies life time is correspondingly longer and the variable x is an increasing function of Ri making its largest contribution to the second term in (6.3), 5) the two momentum diffusivities are not the same since they themselves depend on x, X, Z but for the purposes of this paper their expressions are not relevant (they can be provided by request to the authors). 6) even if one assumes that the two diffusivities in (6.3) are the same and of the form (5.1), the first term in (6.3) decreases like Ri^{-1} while the combination xK_m decreases with Ri with a lower power.

In summary, *the inclusion of both buoyancy and vorticity leads to a new angular momentum equation which may provide a better model for the helio data since it contains a mechanism to extract angular momentum from the stellar core that is absent in the commonly used formula (2.3).*

7. New equation for the OV extent

As for the OV extent, numerical simulations (Brummel et al., 2002) yield a value of about $0.5H_p$ which is an order of magnitude larger than the helio data of $0.07H_p$. To reconcile model results with the data, we suggest a new criteria for the OV extent. Consider the equation for the turbulent kinetic energy $K(D/Dt \equiv \partial/\partial t + \bar{u}_i \partial/\partial x_i)$:

$$\frac{DK}{Dt} + \frac{\partial F_i^{ke}}{\partial x_i} = P_b + P_m - \epsilon \qquad (7.1)$$

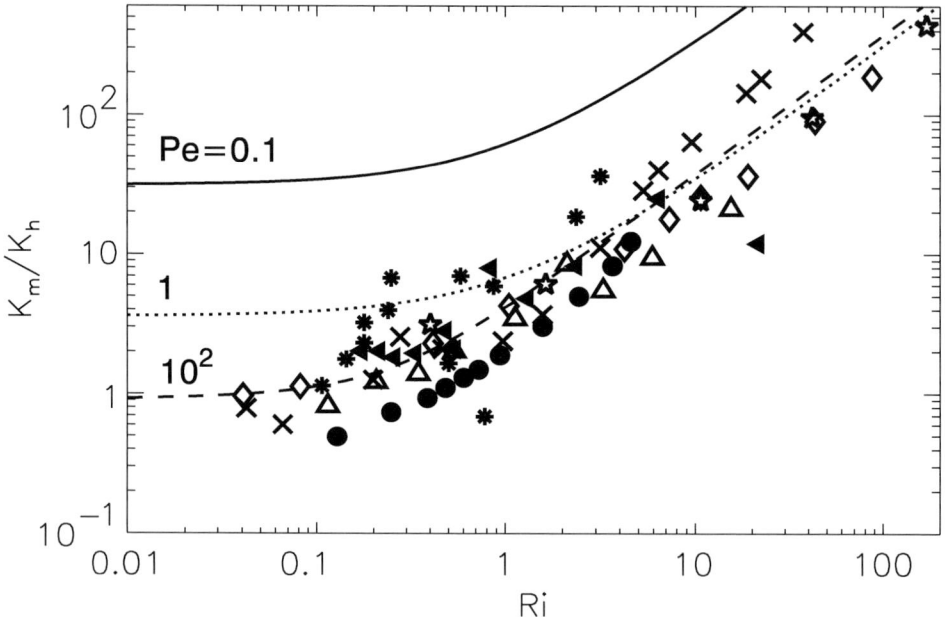

Figure 3. Turbulent Prandtl number $\sigma_t = K_m/K_h$ vs. Ri for different Pe. The data corresponding to the $Pe > 1$ case, are as follows: meteorological observations (Kondo et al., 1978, slanting black triangles; Bertin et al., 1997, snow-flakes), lab experiments (Strang and Fernando, 2001, black circles; Rehmann and Koseff, 2004, slanting crosses; Ohya, 2001, diamonds), LES (Zilitinkevich et al., 2007, 2008, triangles), DNS (Stretch et al., 2001, five-pointed stars).

where F_i^{ke} is the flux of K and $P_{b,m} = (\alpha g_i J_i, -\tau_{ij} S_{ij})$ are the productions of buoyancy and shear. Next, consider the flux conservation law (Canuto, 1997):

$$F_i^{rad} + F_i^{conv} + F_i^{ke} + \overline{u}_j(\mathbf{E}\delta_{ij} + \tau_{ij}) = constant = C \quad (7.2)$$

$$E = c_p T + K + \overline{K} + G, \quad \overline{K} = \frac{1}{2}\overline{\mathbf{u}}^2, \quad g_i \overline{u}_i = \frac{DG}{Dt} \quad (7.3)$$

where $F_i^{conv} = c_p J_i$. While the "traditional" flux conservation law used in stellar models contains only the first two terms in (7.2), one must also account for the flux of K represented by the third term; however, in the presence of mean currents, one has new terms, the first of which represents the flux $\overline{u}_i \mathbf{E}$, where $\mathbf{E}$ is the sum of enthalpy $c_p T$, turbulent kinetic energy K, mean field kinetic energy $\overline{K}$ and gravitational energy G while the other term is the flux of the Reynolds stresses. Eliminating the heat flux between (7.1) and 7.2), yields in the stationary case the equation for the flux of K:

$$\frac{\partial F_i^{ke}}{\partial x_i} + \frac{\alpha}{c_p} g_i F_i^{ke} = \Phi_{old} + \Phi_{new} \quad (7.4)$$

$$\Phi_{old} = C - \alpha c_p^{-1} g_i F_i^{rad}, \quad \Phi_{new} = -\tau_{ij} S_{ij} - \alpha c_p^{-1} g_i (\tau_{ij} \overline{u}_j + \mathbf{E}\overline{u}_i) - \epsilon \quad (7.5)$$

After some algebra, one obtains that:

$$\Phi_{new} = -\left(1 + T_2 \frac{r}{H_p}\right)\epsilon + (T_0 + T_1)\epsilon - H_p^{-1}(\mathbf{E} + \overline{u_r^2})\overline{u}_r \quad (7.6)$$

The key result is that the first term shows that the dissipation increases with depth thus reducing the extent of the OV compared to the standard criterion without Φ_{new}. This is

predicated on the fact that:
$$T_2 \sim (\tau N)^2 (\tau \Omega)^2 \tag{7.7}$$
is positive in the radiative zone since $N^2 > 0$. At the same time, the eddies life time is the largest since turbulence is weak and thus $(\tau N)^2 > 1$.

In conclusion, *the new term in the OV equation (7.4) contributes a term that reduces the OV extent which is a desired feature since no model has thus far been able to reproduce the helio data of* $OV \approx 0.07 H_p$.

8. Conclusions

The form of the standard angular momentum equation (2.3) yields results that are not in agreement with helio data since the extraction of angular momentum from the radiative zone is too inefficient. We show that (2.3) is based on a very restricted form of the Reynolds stresses. If one uses the full form of the Reynolds stresses that entails *shear, vorticity and buoyancy*, the combination of the last two ingredients gives rise to a new term that is larger than the canonical one that contains only shear and which leads to extraction of angular momentum from the core, a tendency in the right direction.

As for the OV extent, the key ingredient is the new flux conservation law that entails Reynolds stresses and mean flows. One of the new terms leads to an increased dissipation of the flux of turbulent kinetic energy which in turn entails a smaller OV extent, a welcome feature since thus far all models have yielded an OV extent about an order of magnitude larger than the helio data of $0.07 H_p$.

An interesting aspect of the new models is the relative simplicity of the equations determining Reynolds stresses and heat fluxes since they are given by linear algebraic equations. This is especially relevant if one considers the amount of information they contain: stable stratification, unstable stratification, rigid rotation, shear, and radiative losses (Peclet number). Having established the qualitative behavior of the two models, what is needed next is a specific computation in conjunction with a stellar code.

Acknowledgements

One of the authors (VMC) would like to thank to Dr. F. Kupka for very helpful suggestions and information on a variety of topics discussed in this paper.

References

Bertin, F., Barat, J., & Wilson, R. 1997, *Radio Science* 32, 791
Brummell, N. H., Clune, T. L., & Toomre, J. 2002, *ApJ* 570, 825
Brun, A. S. & Toomre, J. 2002, *ApJ* 570, 865
Canuto, V. M. 1997, *ApJ* 482, 827
Canuto, V. M., Minotti, F., & Shilling, O. 1994, *ApJ* 425, 303
Canuto, V. M. & Dubovikov, M. S. 1998, *ApJ* 493, 834
Canuto, V. M. & Minotti F. 2001, *Mon. Not R. Astron. Soc.* 328, 829
Canuto, V. M., Cheng, Y., Howard, A. M., & Esau, I. N. 2008, *J. Atmos. Sci.*, in press
Canuto, V. M. 2008, in: W. Hillebrandt & F. Kupka (eds.), *Interdisciplinary Aspects of Turbulence*, Lecture Notes in Physics (Berlin: Springer)
Charbonnel, C. & Talon, S. 2005, *Science* 309, 2189
Charbonnel, C. & Talon, S. 2007, *Science* 318, 922
Kondo, J., Kanechika, O., & Yasuda, N. 1978, *J. Atmos. Sci.* 35, 1012
Maeder, A. & Meynet, G. 2001, *A&A* 373, 555
Mathis, S., Palacios, A., & Zahn, J. P. 2004, *A&A* 425, 243
Ohya, Y. 2001, *Boundary-Layer Meteorol.* 98, 57

Palacios, A., Talon, S., Charbonnel, C., & Forestini, M. 2003, *A&A* 399, 603

Palacios, A., Charbonnel, C., Talon, S., & Seiss, L. 2006, *A&A* 453, 261

Rehmann, C. R. & Koseff, J. R. 2004, *Dynamics of Atmospheres and Ocean* 37, 271

Roxburgh, I. W. 1978, *A&A* 65, 281

Strang, E. J. & Fernando, H. J. S. 2001, *J. Phys. Ocean.* 31, 2026

Stretch, D. D., Rottman, J. W., Nomura, K. K., & Venayagamoorthy, S. K. 2001, in: B.B. Dally (eds.), *Proc. Fourteenth Australasian Fluid Mech. Conf.*, Adelaide University, South Australia, 612-628

Talon, S. & Zahn, J. P. 1998, *A&A* 329, 315

Talon, S. & Charbonnel, C. 2003, *A&A* 405, 1025

Thompson, M. J., Christensen-Dalsgaard, J., Miesh, M. S., & Toomre, J. 2003, *Annu. Rev. Astron. Astrophys.* 41, 599

Zilitinkevich, S. S., Elperin, T., Kleeorin, N., & Rogachevskii, I. 2007, *Boundary Layer Meteorol.* 125, 167

Zilitinkevich, S. S., Elperin, T., Kleeorin, N., Rogachevskii, I., Esau, I., Mauritsen, T., & Miles, M. W. 2008, *Quart. J. Roy. Meteor. Soc.*, in press

Discussion

LANGER: For obtaining angular momentum transport in radiative zones, do you assume turbulence to exist?

CANUTO: I do assume, as does everybody else, that in the radiative zone there is a variety of instabilities which ultimately will give give rise to a turbulent flow.

ZAHN: You mentioned the fact that numerical simulations predict an overshoot which is much too strong. But that is so because such simulations are run with a Peclet number which is too low, owing to the lack of resolution. For a given strength of the convection, the higher thermal diffusion, i.e. the lower the Peclet number, and the deeper is the overshoot, because the buoyancy force is lessened.

CANUTO: One thing is what simulations do and another is what the physics of the problem dictates – since the Peclet's number is directly proportional to the rsm turbulent velocity and since the latter is getting smaller as one approaches the bottom of the CZ, so does the Peclet's number – inside the radiative region, such rsm velocity is even smaller and so is Pe – Thus, small Pe, small rsm and small OV distance go together.

The Art of Modelling Stars in the 21st Century
Proceedings IAU Symposium No. 252, 2008
L. Deng & K.L. Chan, eds.

Radiation-hydrodynamics simulations of surface convection in low-mass stars: connections to stellar structure and asteroseismology

Hans-G. Ludwig[1,2], Elisabetta Caffau[1] and A. Kučinskas[3]

[1]Observatoire de Paris-Meudon, GEPI, 92195 Meudon Cedex, France,
email: [Hans.Ludwig,Elisabetta.Caffau]@obspm.fr

[2]CIFIST Marie Curie Excellence Team, Observatoire de Paris-Meudo

[3]Institute of Theoretical Physics and Astronomy, Goštauto 12, Vilnius 01108, Lithuania
email: ak@itpa.lt

Abstract. Radiation-hydrodynamical simulations of surface convection in low-mass stars can be exploited to derive estimates of i) the efficiency of the convective energy transport in the stellar surface layers; ii) the convection-related photometric micro-variability. We comment on the universality of the mixing-length parameter, and point out potential pitfalls in the process of its calibration which may be in part responsible for the contradictory findings about its variability across the Hertzsprung-Russell digramme. We further comment on the modelling of the photometric micro-variability in HD 49933 – one of the first main *COROT* targets.

Keywords. convection, hydrodynamics, stars: atmospheres, stars: evolution, stars: oscillations

1. Introduction

Radiation-hydrodynamical (RHD) simulations of surface convection in low-mass stars have reached a high level of maturity. Such simulations provide quantitative predictions about the spatial and temporal statistics of the flows taking place in the stellar surface layers. We used the 3D RHD code CO^{5}BOLD to study convective flows in late-type stars at different metalicity. This contribution deals with two distinct applications of CO^{5}BOLD models. The first one is related to the efficiency of the convective energy transport in convective envelopes, the second one to the low-level photometric variability related temporal evolution of the surface granulation pattern. We shall rather discuss problems than solutions with a focus on dwarfs.

2. Convective energy transport in the envelopes of late type-dwarfs

It is well known from the theory of stellar structure that convection is generally an efficient means of transporting energy, and it establishes a thermal structure close to adiabatic. Only in vicinity of the boundaries of convective regions noticeable deviations from adiabaticity occur. In convective envelopes of late-type stars the upper boundary of the convective envelope – usually located close to or even in the optically thin layers – constitutes the bottle-neck for the energy transport through the stellar envelope assigning a special role to it. Despite it is geometrically thin and contains little mass it largely determines the properties of the convective envelope as a whole. It is the value of the entropy of the adiabatically stratified bulk of the convective s_{env} which is most important from the point of view of stellar structure since it influences the resulting radius and effective temperature of a stellar model. s_{env} is controlled by the efficiency of convective and radiative energy transport in the thin, superadiabatically stratified surface layers. Detailed RHD simulations can be applied to model this region allowing to quantify the mutual efficiency of the convective and radiative energy transport and predict s_{env}. Comparing the RHD predictions to standard 1D models based on mixing-length theory (MLT) the value of s_{env} can be translated into a corresponding mixing-length parameter α_{MLT} (see Ludwig et al. 1999 for details).

In stellar evolution calculations the free mixing-length parameter is usually calibrated against the Sun. However, it is unclear whether mixing-length theory provides a suitable scaling of the convective efficiency at constant α_{MLT} across the Hertzsprung-Russell diagram (HRD), and a lot of work was invested to address this issue empirically. Unfortunately, hitherto, no coherent picture emerged. Here we report on an update of earlier work on the theoretical calibration of α_{MLT} based on RHD simulations. While not at all comprehensive it illustrates some of the pitfalls in the process which may in part be responsible of the blurred and sometimes even contradictory picture which emerged so far concerning the variability of the mixing-length parameter.

2.1. Comments on the functional dependence of α_{MLT} on stellar parameters

Discussions which took place during the symposium led the authors to add a comment about the question which stellar properties govern the value of the mixing-length parameter α_{MLT}. We argued above that RHD models of the surface layers are able to provide information about s_{env} and α_{MLT}. The models are characterised by the *atmospheric parameters* and consequently the functional dependence of α_{MLT} can be described in terms of them. Whether the standard atmospheric parameters T_{eff} and $\log g$ together with the chemical composition are the most suitable coordinates is not clear. One might speculate that, e.g., the surface opacity is a physically more relevant quantity. Nevertheless, for unevolved late-type stars the conditions at the stellar surface govern the global envelope structure and the standard atmospheric parameters are suitable coordinates to parameterise them. Global stellar parameters (mass, radius, or age) play merely an implicit role. The situation only changes when the size of the granular cells or the thickness of the superadiabatic layer become comparable to the stellar radius which might happen in giants. To our opinion one should usually avoid to express changes of α_{MLT} in terms of, e.g., stellar mass or age since it tends to obscure the underlying physics.

2.2. α_{MLT} from six 3D RHD models

Squares in Figs. 1 and 2 mark the parameter combinations in the T_{eff}-$\log g$-plane of six 3D RHD models we used for calibrating α_{MLT}. All atmospheres belong to dwarfs, three of solar metalicity, three of a metalicity of 1/100 solar. Our new 3D results are superimposed on earlier results obtained from a grid of 2D RHD models by Ludwig et al. (1999) and Freytag et al. (1999). There are obvious differences of up to ≈ 0.8 between the 3D and 2D calibrated values. Does this indicate that a theoretical calibration with the help of RHD models is hopelessly inaccurate?

The first question to answer is what is a big and what is a small difference when it comes to α_{MLT}. At present interferometry and analysis of eclipsing binaries provide stellar radii to an accuracy of about 1 %. Looking at stellar structure models (see, e.g., Lebreton et al. 2001, their Fig. 10) for late-type main-sequence stars the highest sensitivity of the stellar radius and effective temperature to changes of α_{MLT} is found at about one solar mass. Christensen-Dalsgaard (1997) finds for the Sun a sensitivity of $\delta \ln R \approx -0.24 \delta \ln \alpha_{MLT}$. Hence, considering the present observationally achievable accuracies we would like to

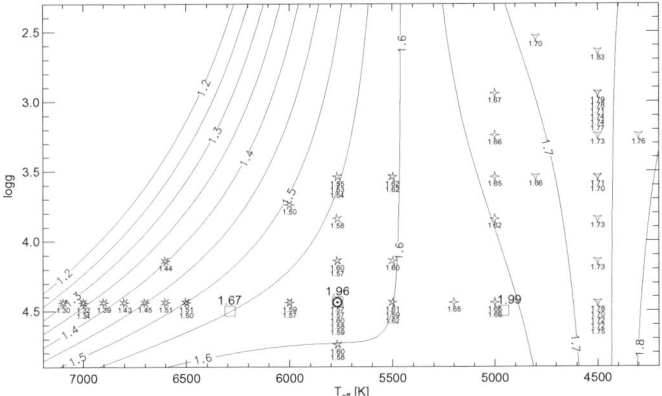

Figure 1. Theoretically calibrated mixing-length parameters at solar metalicity in the $T_{\rm eff}$-$\log g$-plane. Squares mark our new results based on 3D models, the other symbols mark earlier 2D results. The obtained values are given by the numbers, the isolines represent a smooth fit to the 2D data.

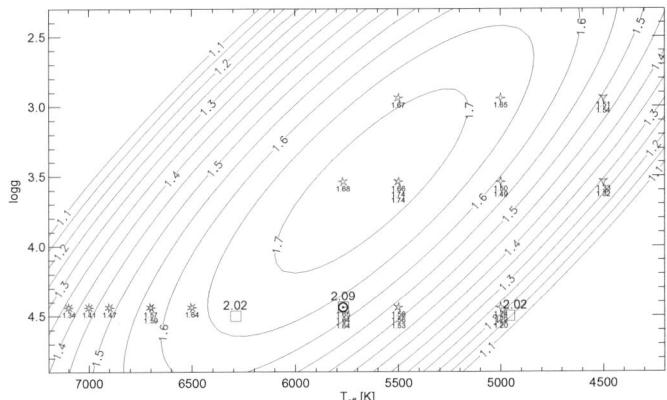

Figure 2. Same as Fig. 1 but for 1/100 of the solar metalicity.

know $\alpha_{\rm MLT}$ to better than 4%, and the differences between the 2D and 3D results are clearly relevant.

We did not mention yet the methodological changes we introduced when calibrating our 3D-based $\alpha_{\rm MLT}$'s in comparison to the earlier works: i) Our new values were calculated with a different mixing-length dialect. The earlier results assumed the formulation given by Böhm-Vitense (1958) while we now used the formulation given by Mihalas (1978). ii) Following the general trend in stellar evolution theory, our 1D comparison models are now full-fledged stellar atmosphere models instead of integrating a prescribed $T(\tau)$-relation to describe the atmospheric temperature run as was done in the earlier results. As we shall see in a moment this is a crucial point, in particular for the cool, metal-poor model at 5000 K. iii) It is now well established that the convective energy transport operates more efficiently in 3D than in restricting 2D symmetry. Hence, we would expect a systematic bias towards higher mixing-length parameters in 3D relative to 2D.

Having in mind that we expect larger $\alpha_{\rm MLT}$-values in 3D we think that the absolute differences are perhaps not surprising and now rather focus on differential trends with $T_{\rm eff}$. The statistical uncertainties in the 2D calibration amount to ±0.05. Taking together the different flavours of MLT, and the different 1D comparison models used in the 2D

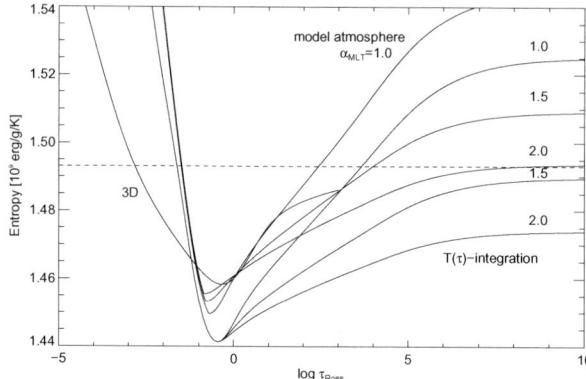

Figure 3. Entropy profiles for 1D stellar atmosphere and stellar structure models for various mixing-length parameters in comparison to the mean 3D stratification for the metal-poor model at $T_{\rm eff}$=5000 K, $\log g$=4.5, and [M/H]=-2.0. All 1D models employ the MLT formulation by Mihalas (1978). The dashed line depicts the entropy $s_{\rm env}$ predicted by the 3D model for the adiabatically stratified part of the convective envelope.

and 3D calibrations the trends start to look similar in 2D and 3D with the exception of the coolest metal-poor model which is far off. What is the reason?

Figure 3 shows a comparison between entropy profiles of 1D MLT based models and of the horizontally and time-wised (on surfaces of equal optical Rosseland depth) averaged 3D model. This kind of comparison provides the calibrated $\alpha_{\rm MLT}$. It is obvious that the result depends on the employed 1D models. Using stellar atmosphere models produces a mixing-length parameter which is almost by 0.5 larger than one based on models which use a prescribed $T(\tau)$-relation in the convectively stable part of the stellar atmosphere. The reason is to a large extend the difference in the entropy minimum $s_{\rm min}$ attained in the deep atmospheric layer in combination with the small overall entropy jump – the difference $s_{\rm env} - s_{\rm min}$ – in the star which amounts to only $\approx 1/6$ of the solar value. Stars of higher $T_{\rm eff}$ show larger entropy jumps which makes the exact level of the entropy minimum less critical. The quite different differential behaviour of $\alpha_{\rm MLT}$ in 3D relative to 2D is mainly a result of the different choice of comparison model.

While one might consider this as mere technicality we rather believe that part of the confusion about trends and absolute values of the mixing-length parameter have their origin – besides observational problems – in different and not clearly specified procedures of how the MLT-related quantities are computed and the atmospheric structure integration are actually performed in models. It is, e.g., still quite common to use simple grey $T(\tau)$-relations for describing the atmospheric temperature run. Figure 3 illustrates that the particular choice might generate mismatches corresponding to substantial changes in the value of the mixing-length parameter necessary to restore the actual scaling of the envelope entropy with changing stellar effective temperature.

Finally, Fig. 3 illustrates that even full-fledged model atmospheres are not always able to reproduce the entropy minimum in 3D models closely. The reason is that for metal-poor dwarfs 3D models predict large deviations from radiative equilibrium conditions in the formally convectively stable part of the atmosphere (in the figure apparent by the low entropy of the 3D model at low optical depth). In 1D stellar atmospheresradiative equilibrium conditions are assumed to fix the atmospheric temperatures. Trampedach (2007) puts forward the idea to extract the mean $T(\tau)$-relation from 3D models for performing the atmospheric structure integration. This obviously provides a better match to the 3D atmospheric entropy minimum, and hence a "cleaner" calibration of the

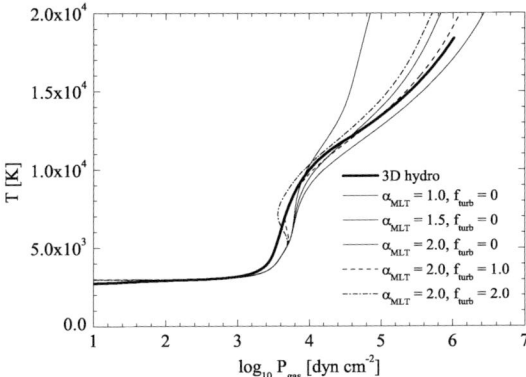

Figure 4. Temperature-pressure profiles of a 3D red giant model (thick solid line) in comparison to 1D stellar atmosphere models of different $\alpha_{\rm MLT}$ leaving out (thin solid lines) or including (dashed and dashed-dotted line) turbulent pressure (for details see text).

mixing-length parameter, however, also is more difficult to implement in existing stellar evolution codes. The 3D calibrated values of $\alpha_{\rm MLT}$ by Trampedach (for solar metalicity and MLT formulation by Böhm-Vitense) are somewhat lower than our values presented here but show a similar trend with $T_{\rm eff}$ on the main-sequence.

2.3. Turbulent pressure trouble

In the main-sequence models discussed above turbulent pressure plays generally only a small role but becomes relatively more important towards lower gravities – and causes extra trouble when one is interested in a well-defined calibration of the mixing-length parameter. Figure 4 shows the average temperature profile of a 3D red giant model ($T_{\rm eff}=3600\,{\rm K}$, $\log g=1.0$, [M/H]=0.0) in comparison to standard 1D model atmospheres of the same atmospheric parameters. While turbulent pressure $P_{\rm turb}$ is naturally included in the RHD simulation it is modelled in a ad-hoc fashion in 1D models assuming a paramterisation $P_{\rm turb} = f_{\rm turb}\rho v_c^2$, where $f_{\rm turb}$ is a free parameter of order unity, ρ the mass density and v_c the convective velocity according MLT.

Figure 4 shows that in the 3D models the turbulent pressure "lifts" the temperature-gas pressure profile towards lower pressures which is essentially impossible to reproduce in the 1D models – irrespective of the choice of $\alpha_{\rm MLT}$ and $f_{\rm turb}$. The failure is related to the local nature of MLT confining the action of the turbulent pressure gradients strictly to the convectively unstable regions. While formally one can still match the thermal profile of the 3D model in the deeper layers by 1D profile with suitably chosen $\alpha_{\rm MLT}$ and/or $f_{\rm turb}$ such a match becomes physically little motivated and is unlikely to provide a robust scaling with changing atmospheric parameters. An improved 1D convection description including non-local effects like overshooting is clearly desirable to handle this situation. An empirical calibration of $\alpha_{\rm MLT}$ using giants is likely to suffer from ambiguities related to the way turbulent pressure is treated in 1D models.

3. Modelling the photometric micro-variability in HD 49933

The stochastically changing granular flow pattern on the surface of late-type stars causes low-level residual brightness fluctuations in stellar disk-integrated light. Ludwig (2006) described a method to obtain predictions of the observable temporal power spectrum from local-box RHD simulations if – besides the atmospheric parameters – the radius of the target star is known. One of the first primary asteroseismic targets of

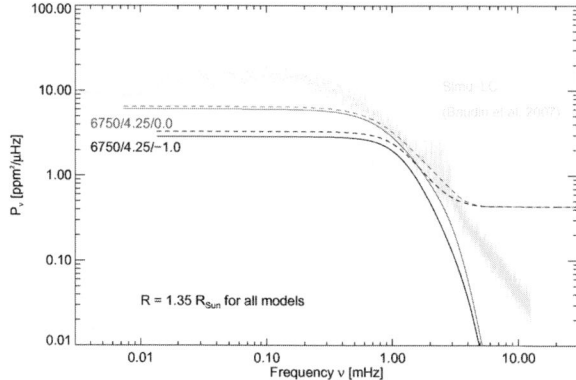

Figure 5. Predictions for the temporal power spectral density of the photometric brightness fluctuations exhibited by HD 49933. 3D RHD simulations are depicted by solid lines and labelled by $T_{\rm eff}/\log g/{\rm [M/H]}$. Dashed lines depict the result when adding the estimated photometric noise level. For further comparison a spectrum obtained with the *COROT* light curve simulator (Baudin *et al.* 2007) is shown.

the *COROT* satellite mission was the metal-depleted ($-0.3 \geqslant {\rm [M/H]} \geqslant -0.4$) F-dwarf HD 49933 for which a high-precision photometric light-curve was obtained. The data is not released yet so that we are only showing theoretical predictions obtained from two RHD simulation runs at $T_{\rm eff}=6750$ K, $\log g=4.25$ and two metallicities [M/H]=0.0 and -1.0 bracketing the observed metallicity of HD 49933 and having atmospheric parameters close to the spectroscopically measured values. Figure 5 depicts the power spectra obtained from the two RHD simulations. For smoothing, the raw RHD spectra were fitted by a simple analytical model, and oscillatory peaks from the acoustic eigenmodes of the computational box were removed. Hence, the pure convection-related signal is shown. The observed signal should fall between the two model predictions, at least at around 1 mHz where the photon noise and the signal due to magnetic activity are expected not to dominate. Unfortunately, at present it looks that this is not quite the case. An actual comparison will be presented in an upcoming paper by Ludwig & Samadi, so stay tuned.

References

Böhm-Vitense, E. 1958, *Zs. f. Astrophys.* 46, 108
Baudin, F., Samadi, R., Appourchaux, T., & Michel, E. 2007, arXiv:0710.3378
Christensen-Dalsgaard J. 1997, in: Pijpers, F. P. and Christensen-Dalsgaard, J. and Rosenthal, C. S. (eds.), *SCORe'96 : Solar Convection and Oscillations and their Relationship*, ASSL 225, 3
Freytag, B., Ludwig, H.-G., & Steffen, M. 1999, in: Gimenez, A. and Guinan, E. F. and Montesinos, B. (eds.), *Theory and Tests of Convection in Stellar Structure*, ASPC 173, 225
Lebreton Y., Fernandes J., & Lejeune T. 2001, *A&A* 374, 540
Ludwig, H.-G., Freytag, B., Steffen, M. 1999, *A&A* 346, 111
Ludwig, H.-G. 2006, *A&A* 445, 661
Mihalas, D. 1978, *Stellar Atmospheres*, Freeman and Company
Trampedach, R. 2007, in: *Unsolved Problems in Stellar Physics: A Conference in Honour of Douglas Gough*, AIPCS 948, 141

Discussion

CANUTO: Do you really believe that $\alpha > 1$ is physically realistic?

LUDWIG: Yes, I do not see a problem with $\alpha > 1$ in the framework of standard mixing-length theory (MLT). On the one hand, MLT is a simplistic approach so that its intrinsic parameters should not interpret overly strictly. On the other hand, the velocity field in stratified convection shows a correlation length which regularly supersedes the local pressure scale height (see, e.g., the works of Chan & Sofia 1987, 1989).

TURCK-CHIEZE: (1) Do you apply your 3D atmospheric model to predict the solar granulation in comparing with the GOLF/SoHO spectrum working in Doppler velocity? (2) Don't you think that the slope at low frequency of the CoRoT HD 49933 comes from an instrumental effect (naturally increase at low freq in 1/f).

LUDWIG: (1) I did not apply the 3D atmosphere model yet to obtain predictions for the granulation signal in Doppler signal. The scaling procedure from local box to disk – integrated signal has not been fully worked out yet. Moreover, if one intends to do a detailed modeling of the involved spectral line(s) the necessary computational effort becomes substantial. (2) No, not in the frequency range > 0.01 mHz. Calibration stars have been observed with CoRoT which do not show this increase which makes the interpretation as signature of magnetic activity much more likely.

CHRISTENSEN-DALSGAARD: Comment: In parallel region, Tranmpedach has developed determination of α_{MLT} from Nordlund-Stein simulations, going back to $\sim$ 1996. It would be very interesting to compare these two sets of results.

LUDWIG: Yes, Regner Trampedach presented values of the mixing-length parameter obtained from a small number ($\sim$ 6) 3D models at this time. The qualitative functional dependence of the ML parameter agreed with the dependence found in the 2D models, despite differences in the numerical approaches. It would be certainly worthwhile to take a fresh look at the predictions of the two sets of 3D models especially based on the analogous sets of 1D comparison models.

After meeting discussion: C. Charbonnel and the speaker, Hans-G. Ludwig (right).

On the boundaries of a convective zone and the extent of overshooting

L. Deng[1] and D.R. Xiong[2]

[1] National Astronomical Observatories, Chinese Academy of Sciences, Beijing 100012
licai@bao.ac.cn

[2] Purple Mountain Observatory, Chinese Academy of Sciences, Nanjing 210008
Xiongdr@pmo.ac.cn

Abstract. In this work, we will show that a proper definition of the boundary of a convective zone should be the place where the convective energy flux (i.e. the correlation of turbulent velocity and temperature) changes its sign. Therefore, it is convectively unstable region when the flux is positive, and it is convective overshooting zone when the flux becomes negative. In our nonlocal convection theory, convection is already sub-adiabatic ($\nabla < \nabla_{ad}$) far before reaching the unstable boundary; while in the overshooting zone below the convective zone, convection is sub-adiabatic and super-radiative ($\nabla_{rad} < \nabla < \nabla_{ad}$). The transition between the adiabatic temperature gradient and the radiative one is continuous and smooth instead of a sudden switch. In the unstable zone, the temperature gradient is approaching radiative rather than going to adiabatic. The distance of convective overshooting is different for different physical quantities. The overshooting distance in the context of stellar evolution, measured by the extent of mixing of stellar matter, should be more extended than that of other physical quantities. It is estimated as large as 0.25–1.7 H_p depending on the evolutionary timescale.

Keywords. convection, stars:evolution

1. Introduction

As the classical treatment of convection, the local theory has been used in modelling stellar structure and evolution. In the calculation of massive star evolution, Schwarzschild & Härm (1958) discovered that the hydrogen rich radiative envelope just outside the helium rich convective core cannot be convectively stable, and that leaded to the paradox of so called semi-convection. To solve that problem, the idea of semi-convection was initiated, i.e. the region outside the convective core is in a state of semi-convection. Stellar matter in this region is nearly in neutral stability ($\nabla \leqslant \nabla_{ad}$), therefore convective energy transport due to this mild convection can be neglected, while the mixing of chemical compositions should be important, which makes a gradient of molecular weight in this region (otherwise called semi-convection zone). There had been a great debate in the community for a long period since then on whether the Schwarzschild or Ledoux criteria should be applied for the neutral stability of convection, and whether the semi-convective zone should be very wide or rather narrow. Stothers (1970) commented on various establishments of semi-convection. Evolutionary scenarios for massive stars with or without semi-convection were also discussed (e.g. Chiosi & Summa 1970). It has been realized later that the problem of semi-convection is in fact due to the non-locality of stellar convection. Therefore various theories of non-local theory of stellar convection have been worked out (Spiegel 1963, Ulrich 1970, Xiong 1977, 1981, 1989, Kufuss 1986, Grossman et al. 1993, Canuto 1993, Canuto & Dubovikov 1998). Generally speaking, non-local theory of convection makes the results better match observations than the local ones.

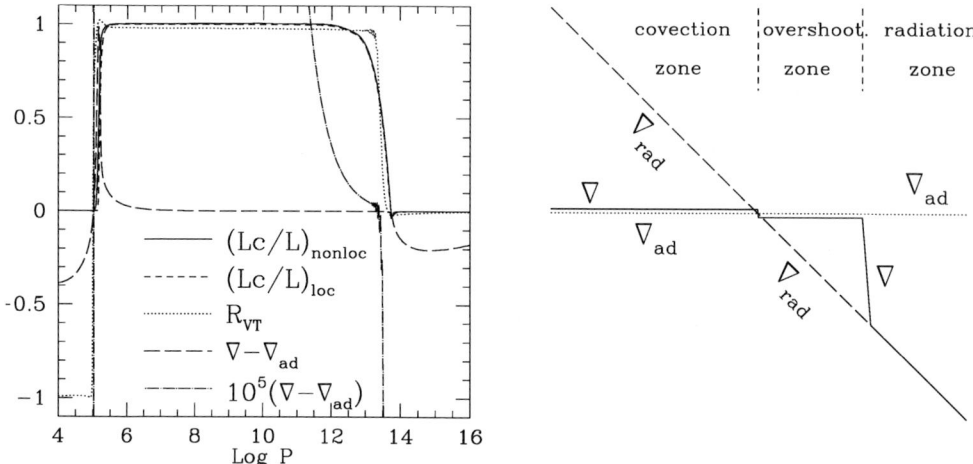

Figure 1. The left panel: the super-adiabatic temperature gradient $\nabla - \nabla_{ad}$, turbulent velocity-temperature correlation R_{VT}, and the fractional convective flux L_c/L versus the depth ($\log P$) for a non-local convection model of the Sun. The dashed line is the fractional convective flux L_c/L for a local model with the same depth of convective zone. The right panel: A sketch of the lower convective and overshooting zones in the usual phenomenological non-local mixing length theory (Monteiro et al. 2000).

However, the non-local theory of convection is rather complicated, which is much less straightforward to be understood, much more difficult to be applied and much more computing power demanding than the phenomenological local (Böhm-Vitense 1958) and non-local mixing length theories (Maeder 1975, Bressan et al. 1981). In this work, we will address the issues of the boundary of a convective zone and the related problems of convective overshoot.

2. How to define the boundaries of a convective zone?

Normally in a local theory of convection, the boundary of convective zone is given by the so called Schwarzschild's criterion, $\nabla = \nabla_{ad}$,

and for chemical inhomogeneous media, the Ledoux criterion applies, $\nabla = \nabla_{ad} + \nabla_\mu$, which is derived by analysis of the local convective stability.

Based on our nonlocal statistic theory of correlation functions, the boundary of the (non-local) convective zone is set at where the turbulent velocity-temperature correlation V vanishes (for definition of the quantities, see our previous work, e.g. Deng & Xiong [2008]),

$$V = 0; \qquad (2.1)$$

Passing through the boundary, V changes its sign: within the convective zone:

$$V > 0, \qquad (2.2)$$

and in the overshooting zone:

$$V < 0. \qquad (2.3)$$

It is clear from the left panel in Fig. 1 that if the boundary is defined as such, the structures of the local and non-local convection models with the same depth of convective zone should be similar. It can be understood by the fact that, within the stellar interior, the turbulent kinetic energy flux (L_t) is generally much less than that of thermal convection (L_c), and the turbulent pressure ($P_t = \rho x^2$) is much less than that of gas (P_g). The thermal structure derived from our non-local convection theory is different from that from usual non-local mixing length theory as sketched in the right panel of Fig. 1.

The left panel in Fig. 1 clearly demonstrates that convective motions near the upper and lower boundaries of the convective zone are very different. This is due to the fact that, in the atmosphere, the density is very low and $P_e = x/x_c < 1$, therefore convective energy transfer is inefficient. As a result, there exists a thin super-adiabatic layer atop of the convective zone. Passing through the upper boundary, the turbulent velocity-temperature correlation $R_{VT} = V/xZ^{1/2}$ drops quickly from ~ 1 to ~ -1. This theoretical prediction agrees the observations of the solar granular velocity field (Leighton et al. 1962; Salucci et al. 1994) and the results of hydrodynamic simulations (Kupka 2003). Contrary to the situations in the solar atmosphere, convection is highly efficient in terms of energy transfer ($P_e \gg 1$) in the deep interiors of the Sun. Towards the lower boundary of the convective zone, the turbulent velocity-temperature correlation R_{VT} decreases abruptly and approaches zero ($|R_{VT}| \ll 1$). What makes it so different at the two boundaries is the distinct the effective Peclet number (see Deng & Xiong 2008 for details).

3. How extended is convective overshooting?

Due to the complexity of non-local convection theory, almost all the modellings of stellar structure and evolution are still using the local convection theory. Convective overshooting is defined as the penetration of convective motion through the classical boundary of convectively unstable zone into the adjacent stable region. The extent of overshooting is not the same for different physical quantities following our dynamic theory of convection, and this leads to some troubles in understanding the true nature and estimation the extension of overshooting.

The overshooting distance defined by dropping off of Lithium abundance by a factor of e is shown in the 8th column of table 1, while that defined by the distance from the boundary of convective zone to where Lithium becomes zero is given in the last column of table 1. It is clearly from figs. 2a–d that Lithium abundance vanishes very quickly after the e-folding depletion, the distance between them is less than $0.1 \mathrm{H}_P$. Completely different from the Lithium abundance profile in fig. 2, turbulent velocity (x), temperature fluctuation ($Z^{1/2}$) and velocity–temperature correlation (V) decrease exponentially with depth in the overshooting zone. The e-folding distances determined from the curves are given in the 2nd–4th colums (for the upper part of overshooting zone), and 4th–7th columns (the lower part of overshooting zone). The e-folding distances given by turbulent velocity and temperature fields are very close to the analytic asymptote in our theory (Xiong 1989b). They hardly change with stellar mass, and are rather different from the e-folding distances defined by Lithium depletion.

The overshooting distance that is important to stellar evolution is the extension of the non-local convective mixing of chemical elements. Obviously, it is neither that of convective energy transfer nor those of turbulent velocity and temperature fluctuations. Calculations of massive star evolution under our complete non-local theory of convection shown that the non-local convective mixing overshoots a very extended distance (Xiong 1986).

Although the overshooting at the bottom of the solar convective zone cannot be observed directly, we fortunately have another excellent indicator for the extension of overshooting, which is the Lithium abundance of the Sun and solar type stars.

Table 1. The e-folding lengths measured in local pressure scale height.

M/M$_\odot$	upper oversh. zone			lower oversh. zone			Li	Dcut
	x	z	V	x	z	V		
0.800	0.47	0.36	0.20	0.25	0.25	0.080	0.26	0.30
0.850	0.48	0.36	0.21	0.25	0.25	0.081	0.36	0.38
0.900	0.50	0.35	0.20	0.25	0.25	0.082	0.42	0.54
0.925	0.50	0.36	0.21	0.25	0.25	0.069	0.50	0.58
0.950	0.50	0.33	0.20	0.26	0.25	0.076	0.60	0.69
0.975	0.49	0.33	0.20	0.31	0.31	0.104	0.67	0.74
1.000	0.50	0.32	0.20	0.29	0.31	0.093	0.85	0.91
1.025	0.63	0.30	0.21	0.29	0.29	0.096	1.05	1.11
1.050	0.52	0.30	0.19	0.30	0.36	0.092	1.25	1.33
1.075	0.60	0.30	0.21	0.38	0.38	0.114	1.64	1.69

Figs. 2a–2d show the Lithium abundance depletions due to overshooting for 4 given mass (indicated in the corresponding panels) stellar models as functions of depth (log P), when the effect of evolution is not considered. The dashed lines in the plots indicate the boundary of convective zone. As from fig. 2c, there is a gradually accelerating reduction of Lithium abundance in the overshooting zone. In the upper part of the overshooting zone, the mixing caused by non-local convection is very efficient, the abundance keeps the same as in the convectively unstable zone for about 0.4 H_P in length downwards. It is then followed by a partially mixed region where Lithium abundance is reduced quicker toward the center and vanishes suddenly, such a partial mixing region is about 0.5H_P in depth. Even deeper is the non-mixing zone. If the overshooting distance is taken as the e-folding length of the abundance from the bottom of the convective zone, it reads about 0.83H_P; otherwise if we count the deepest bottom of the mixing process, it reads about 0.9H_P for the solar model.

4. Summary and discussions

Detailed discussions on the definition of the boundary of convective zone, and the distance of convective overshooting are presented in this paper. The main results can be summarized in the following:

(a) Choosing the place where the convective flux (or equivalently the turbulent velocity–temperature correlation) changes it sign as the boundary is the most proper and convenient, where the convective flux is greater than zero is convectively unstable zone, while it is the overshooting zone when the convective flux is smaller than zero;

(b) It is not quite right to talk about a general overshooting distance for stellar convection. The distance of overshooting is different for different physical quantities. The overshooting distances of turbulent velocity and temperature fluctuations are quite extended, the e-folding lengths can reach 0.25–0.5H_P. Very extended and weak overshooting can still induce fairly efficient mixing in a very long timescale of evolution. Therefore we anticipate a very extended overshooting for mixing of matters, the e-folding lengths of which is generally larger than that of turbulent velocity and temperature. The one in massive stellar model, for instance, can reach 1 pressure scale height (Xiong 1986).

Using a similar non-local convection theory, we have calculated the early stages of the post-main-sequence evolution for massive stars. The overshooting distances from the cores are rather extended, being typically 1–2H_P, the fractional mass within the overshooting zone $M_{over}/M_0 \sim 0.1 - -0.5$ (Xiong 1986).

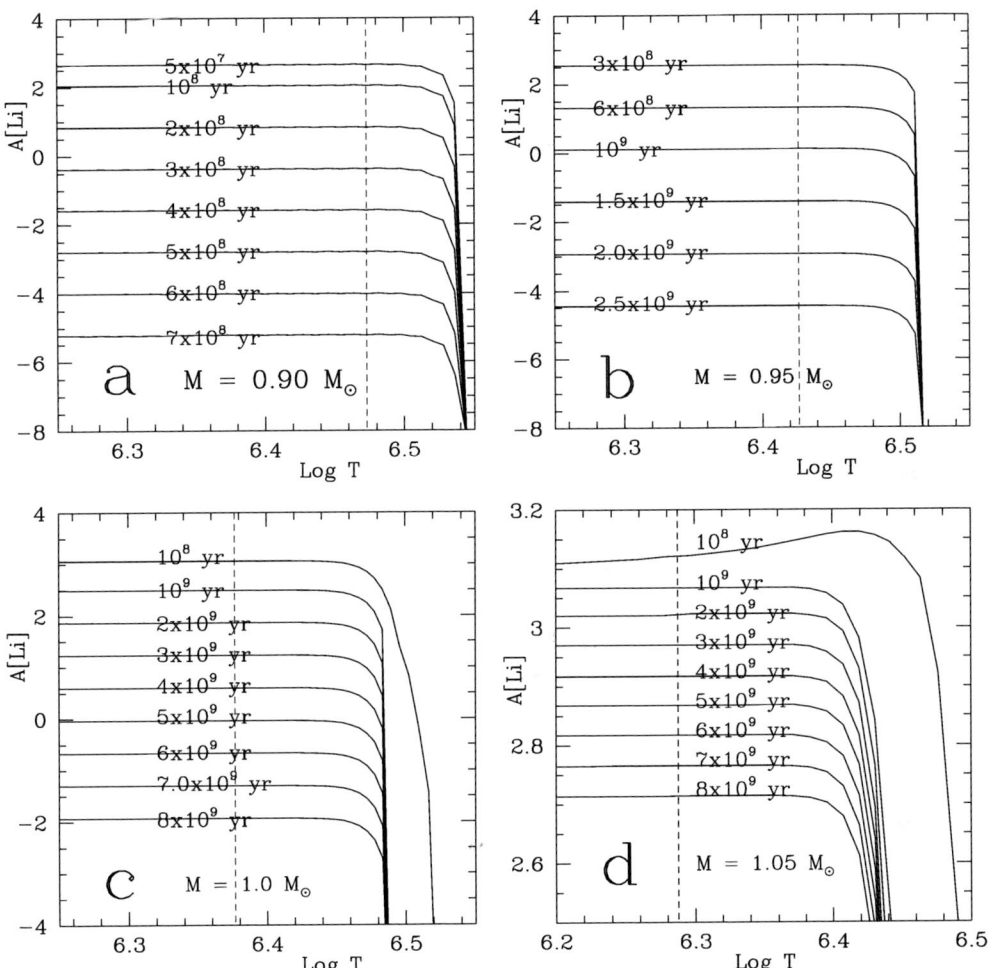

Figure 2. The Lithium abundance versus depth ($\log T$) and age (labeled on the curves) for main sequence stars. The dashed vertical line indicates the location of the lower convective boundary. Evolution is not considered in the calculations. a). M = $0.90 M_\odot$; b). M = $0.95 M_\odot$, c). M = $1.0 M_\odot$ and d). M = $1.05 M_\odot$.

Acknowledgements

The Chinese National Natural Science Foundation (CNNSF) is acknowledged for support through grants 10778719, 10773029, 10573022 and 10333060.

References

Böhm-Vitense, E., 1958, Astrophysik. 46, 108
Bressan, A., Bertelli, G. & Chiosi, C., 1981, *A&A*, 102, 25
Canuto, V. M., 1993, *ApJ*, 416, 331
Canuto, V. M. & Dubovikov, M., 1998, *ApJ*, 493, 834
Chiosi, C. & Summa, C., 1970, *ApSS*, 8, 478
Deng, L. & Xiong, D. R., 2006, *ApJ*, 643, 426
Deng, L. & Xiong, D. R., 2008, *MNRAS*, 386, 1979
Grossman, S. A., Narayan, R., & Arnett, D., 1993, *ApJ*, 407, 284
Kuhfuss, R., 1986, *A&A*, 160, 116

Kupka, F., 2003, in Modelling of Stellar Atmosphere, IAU Symp. 210, eds N. Piskunov, W. W. Weiss & D. G. Gray, p.143 (Pub. ASP)
Leighton, R. B., Neyes, R. W., & Simon, G. W., 1962, *ApJ*, 135, 474
Maeder, A., 1975, *A&A*, 40, 303
Monteiro, M. J. P. F. G., Christensen-Dalsgaard, J., & Thompson, M. J., 2000, *MNRAS*, 316, 165
Salucci, G., Bertello, L., Gavallini, F. Ceppatelli, G. *et al.*, 2004, *MNRAS*, 285, 322
Schwarzschild, M. & Harm, R., 1958, *ApJ*, 128,348
Spiegel, E. A., 1963, *ApJ*, 138, 216
Stothers, R., 1970, *MNRAS*, 151, 65
Ulrich, R. K., 1970, *ApSS*, 7, 183
Xiong, D. R., 1977, *AcASn*, 18, 86
Xiong, D. R., 1981, *SciSn*, 23, 1139
Xiong, D. R., 1986, *A&A*, 167, 239
Xiong, D. R., 1989, *A&A*, 213, 176

Discussion

LUDWIG: In case of overshooting at the upper boundary: in case of the Sun this means overshooting into optically thin regions. How do you handle the radiative transfer in this situation

DENG: For the case of the Sun, the overshooting under consideration does not enter the atmosphere (optically thin regions) therefore it is not an issue here.

CANUTO: How did you model the third-order moments?

DENG: In Xiong's theory of convection, the correlation are truncated at the second order, the 3rd order moments are dealt with by a gradient type approximation.

CHRISTENSEN-DALSGAARD: This is a comment: The helioseismic constraint limits the sharpness of the transition in the sound-speed gradient at the base of the convection zone as found by, e.g. in the model of Zahn. The observations in fact required a smooth transition than in models without overshoot. Your model, with subadiabtic gradient in the convection zone, may satisfy that. We should look at the temperature structure in your models. Also, compare with the model of Rempel.

DENG: Indeed, we should discuss this later, and we are really interested in doing some work together especially on inversion using our models.

LANGER: This is a question actually for Canuto: What is the key difference between Canuto's and Deng/Xiong's model for overshooting?

CANUTO: The Deng-Xiong's model assumes a specific form for the 3rd-order moment F_{KE} that represents the flux of T_{KE}. It was shown (many years ago) that the DGA (down-gradient-approximation) used by Deng-Xiong may be very far from the true form of F_{KE}. I have therefore decided to avoid entirely any modelling of F_{KE}. Rather I derived a differential equation for F_{KE} which contains only 2nd-order moments. The equation for F_{KE} is now being solved.

A new two-columns description for convective transport in stars

Alexander Stökl

CRAL, Université de Lyon, CNRS (UMR5574), École Normale Supérieure de Lyon, F-69007 Lyon, France
email: alexander.stoekl@ens-lyon.fr

Abstract. Assuming that the largest convective patterns generate the majority of convective transport, we devise a numerical scheme simplifying the convective velocity field using two parallel radial columns to represent up- and downstream flows. Horizontal exchange is described by fluid flow and radiation over the interface between those two columns. The main parameters of this convective description have a straightforward geometrical meaning, namely the diameter of the columns (representing the size of the convective cells) and the ratio of cross section between up- and downdrafts. For this geometrical setup, the equations of radiation hydrodynamics are solved time-dependently using an implicit scheme which has the advantage of being devoid of any time step limits. In order to demonstrate our approach, we present comparisons with detailed 2D hydrodynamics computations for the example of convection zones in Cepheids.

Keywords. hydrodynamics — Cepheids — convection — methods: numerical

1. Introduction

Current two- and three-dimensional hydrodynamics simulations of stellar photospheric convection are remarkably successful in reproducing the observed properties. In particular for the sun, comparisons with observational constraints from line profiles (e.g. Asplund et al., 2000) and the solar granulation pattern (e.g. Stein & Nordlund, 1998, Wedemeyer et al., 2004) show the high level of fidelity of multi-dimensional simulations. Indirectly, this also confirms various other (non-observed) properties of the numerical models such as temperature structure, convective flux, and convective velocities.

Motivated by this success of multidimensional hydrodynamics in modelling convective transport, we devised a numerical scheme which describes the circulating convective flow in the most simple manner with two parallel radial columns. Convective up- and downdraft motions are represented by radial fluid flow in these two columns while a horizontal component of the flow over the interface between these two columns closes the circulating motion. These two columns thereby do not stand for an individual convective cell but should be considered as a representation of all up- and downdraft motions.

This discretization scheme has three main parameters: First, the typical horizontal length scale D that can be interpreted as a diameter of the columns or as their distance from each other. Physically, this corresponds to the characteristic size of the flow patterns or eddies of the modelled convection.

Secondly, the parameter $cf1$ ('cf' for *column fraction*) describes the fraction of the sphere allocated to column 1. Accordingly, $cf2 = 1 - cf1$ is the relative cross section of column 2. This different size of the columns can be used to model convection zones with narrow down- and wide updrafts as observed, e.g., in the solar granulation.

Finally, a third constitutive parameter specifies details of the horizontal advection from one column to the other; but since this parameter not used in the examples below, we will not discuss it here further.

The big advantage of this two-columns discretization scheme is that its very simple setup allows for an efficient and straightforward application of the implicit solution method. This results in a code that is almost as fast as an implicit 1D-Code (as used, e.g., for stellar evolution or for stellar pulsations) while including the effects of convective transport by the simple, yet hydrodynamical consistent, circulating flow.

2. The discretization scheme

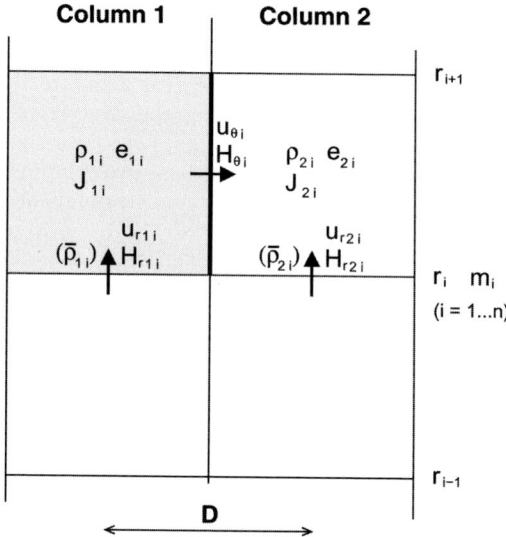

Figure 1. The two-columns discretization scheme. The primary variables (compare Table 1) are included in their staggered-mesh localization; D is the typical distance/diameter of the columns. Advection occurs, as indicated by arrows, in radial direction as well as over the interface between the two columns (thick line).

Figure 1 shows the setup of the two-columns discretization scheme. Note that, even though not drawn as such, all this takes place in spherical symmetry; in general, the two columns also do not have the same size. The area shaded in gray in Fig. 1 is the discretization volume for the scalar quantities ρ, e, J and the corresponding equations (see Table 1). Vector variables in radial (u_r, H_r, $\bar{\rho}$) and horizontal (u_θ, H_θ) direction are discretized in similar volumes located in staggered-mesh positions. For a detailed description of the discretization scheme see Stökl (2008).

The radial distribution of the grid points r_i is determined by an adaptive grid equation (Dorfi & Drury, 1987). This adaptive grid equation is solved implicitly together with the physical equations and continuously adjusts the grid resolution according to the evolving physical structures. The adaptive grid also allows the usage of a Lagrangian outer boundary condition so that the grid can follow radius variations of the star, e.g. due to stellar pulsations or structural resettling caused by the onset of convective transport.

3. Method of solution

The system of discrete equations compiled in Table 1 is solved by an implicit solver. This method has the advantage of being not affected by the CFL (after Courant, Friedrichs & Lewy 1928) time step limit and in principle allows arbitrarily large time steps. The

Table 1. The set of 16 primary variables and the corresponding discrete equations. A moment description for the radiation field (Mihalas & Mihalas, 1984) has been adopted for the equations of radiation hydrodynamics. The radially averaged densities $\overline{\rho}$ are used for assembling the radial momentum $\overline{\rho}u_r$; this formalism is necessary to implement the second order advection scheme for the momentum within the 5-point discretization stencil.

Variable	Description	Equation
r_i	Radius	Adaptive grid equation
m_i	Integrated mass	Poisson equation
$\rho_{1\,i}, \rho_{2\,i}$	Density	Equation of continuity
$\overline{\rho}_{1\,i}, \overline{\rho}_{2\,i}$	Averaged density	Radial averaging of ρ
$e_{1\,i}, e_{2\,i}$	Specific internal energy	Equation of energy
$u_{1\,i}, u_{2\,i}$	Radial velocity	Equation of motion, radial component
$u_{\theta\,i}$	Horizontal velocity	Equation of motion, horizontal component
$J_{1\,i}, J_{2\,i}$	0th moment of intensity	Radiation energy equation
$H_{1\,i}, H_{2\,i}$	1st moment of intensity, radial	Radiation flux Eq., radial component
$H_{\theta\,i}$	1st moment of intensity, horizontal	Radiation flux Eq., horizontal component

Closures: – tabulated equation of state (temperature, gas pressure), evaluated separately in each column: $T_1 = T(\rho_1, e_1)$, $T_2 = T(\rho_2, e_2)$, $P_1 = P(\rho_1, e_1)$, $P_2 = P(\rho_2, e_2)$
– tabulated opacities (Rosseland mean), evaluated separately in each column: $\kappa_1 = \kappa(\rho_1, e_1)$, $\kappa_2 = \kappa(\rho_2, e_2)$
– closure of radiation moments with an Eddington factor $f_{\mathrm{edd}} = \mathbf{K}/J = 1/3$

inclusion of elliptical parts into the system of physical equations (Poisson and grid equation) is also only possible with an implicit scheme.

As the system of equations contains nonlinear terms, a Newton-Raphson iteration is used where each iteration step involves the inversion of the Jacobi matrix. According to the number of equations, the Jacobian consists of 16×16 sub-matrices which form a pentadiagonal banded structure reflecting the discretization on a 5-point stencil. The inversion of the Jacobi matrix uses the customary approach of elimination of the two lower sub-diagonals and subsequent back-substitution of the resulting upper triangular matrix.

Usually, the models have 500 radial grid points. Due to the adaptive grid, the majority of them clusters around the steep gradients in the photosphere and in the convective region.

4. Stationary solutions

The stationary solution for the convective velocity field is obtained by computing the temporal evolution starting from a hydrostatic, purely radiative initial model. In order to make the convective circulation go in the intended sense of rotation (i.e. wide up-, and narrow downdrafts) small radial velocity perturbations ($u \leqslant 1\,\mathrm{m/sec}$) are applied to the hydrostatic model using the Schwarzschild convection criterion as a guide. Starting from these perturbations, the convective velocities develop quickly and after a dynamic phase of growth which lasts about a thermal time scale of the involved part of the envelope, the convection approaches the stationary solution. This evolution typically requires about 1000 time steps that increase from a few seconds at the start up to 10^{10} sec for the stationary solution.

The ultimate aim of our work is the investigation of the interaction of convection and pulsation in Cepheids. Hence, as a first step, we computed stationary solutions for Cepheid convection zones.

Figure 2 illustrates the results for a rather typical Cepheid with $T_{\text{eff}} = 5400\,K$, $L = 10^3\,L_\odot$, and $M = 4.75\,M_\odot$. The geometrical parameters are $D = 20\,H_{p0}$ and $cf1 = 0.8$. According to the (arbitrary) convention of using column 1 for up-, and column 2 for downdrafts, the latter corresponds to a downstream cross section of 20% of the sphere. The typical horizontal extension D of the columns is specified relative to the characteristic photospheric pressure scale height, $H_{p0} = \mathcal{R}\,T_{\text{eff}}/g$. This formalism is motivated by Freytag et al. (1997) where a horizontal scale of photospheric convection of about $10\,H_{p0}$ was found to be a good estimate for a broad range of stellar parameters.

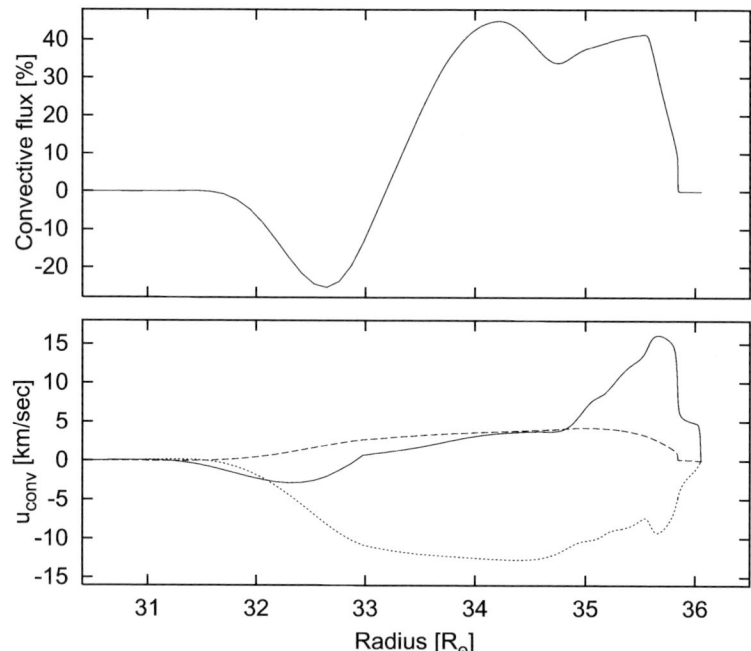

Figure 2. Stationary solution for a Cepheid convection zone: The upper panel shows the convective transport in units of the total luminosity. The convective velocities are given in the lower panel: updraft (dashed line), downdraft (dotted line) and horizontal (solid line). A positive sign of the horizontal velocity corresponds to a flow from column 1 to column 2, i.e. from updraft to downdraft. The figure focuses only on the outer convective region, the model actually extends down to about $3.6\,R_\odot$.

5. Comparison with 2D-computations

In order to verify the convection zones computed with the two-columns scheme, Fig. 3 compares them with results from 2D hydrodynamics carried out with CO^{5}BOLD (Freytag, 2008). The figure gives examples for convection zones in cool and hot Cepheids which are qualitatively quite different:

For lower effective temperatures, a deep convective region contains both the H and He II ionization zones. In that case, convection carries a substantial fraction of the energy flux and the extended downdrafts lead to a pronounced overshoot at the lower boundary.

For hotter stars, where radiative transport is more effective, convection becomes less vigorous and is thus no longer able to bridge the gap between the two ionization zones; hence, there only remains a thin convective shell at the H ionization zone. The He II ionization zone only appears as a very slight bump in the convective flux. However, even though not visible in Fig. 3, there is still a substantial, extended convective velocity field.

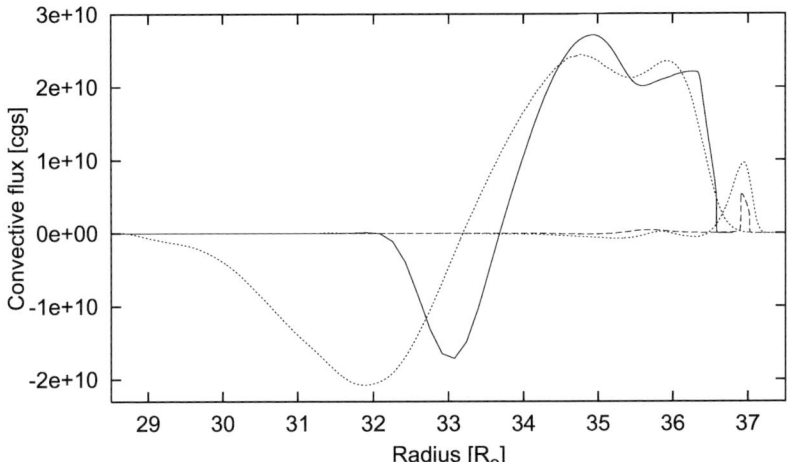

Figure 3. Comparison of the two-columns scheme with 2D hydrodynamics. Shown are convective fluxes computed with the two-columns code for Cepheids with effective temperatures of 5200 K (solid line) and 5800 K (dashed line) as well as the corresponding results obtained with CO^{5}BOLD (both dotted lines).

The comparison of the two-columns scheme with 2D hydrodynamics confirms that the two-columns scheme is able to reproduce many properties of the convection zone:
- An extended lower overshoot.
- The efficiency of convective transport, also in the overshoot region.
- Two distinctive bumps in the convective flux correlated to H and He II ionization.
- An inward flux of kinetic energy (not illustrated).
- The magnitude of convective velocities (not illustrated).
- The transition with increasing effective temperature from an extended convection zone embracing both the H and He II ionization zones to a thin, inefficient convective shell.

Most differences in the results between the two-columns scheme and the 2D simulations, in particular the different depth of the overshoot, are probably due to the simplified geometrical picture of the two columns scheme. An other prominent effect are the considerably smeared-out curves for the 2D results (see upper boundaries of the convective regions) that are caused by the spatial and temporal averaging necessary for extracting such integral quantities from the 2D simulations. Generally, the results for the convection zones in Cepheids agree well with those found for the similar A-type stars (Steffen et al., 2005).

The main advantage of the two-columns model over multi-dimensional hydrodynamics are the large time steps possible with the implicit solution method. The two-columns code requires about one minute CPU-time to follow the temporal evolution of the convection zone up to the stationary solution. In contrast, the 2D simulations of the Cepheid convection zones with CO^{5}BOLD took up to six months to reach sufficiently stationary states.

6. Summary

The two-columns scheme allows for a non-local, time-dependent description of convection and the main parameters of the scheme have a straightforward geometrical meaning. Moreover, the convective solutions are qualitatively quite robust and do not vary drastically within the reasonable parameter range (Stökl, 2008).

Due to the implicit solution method, very large time steps are possible which make the method in many applications much faster than multi-D hydrodynamics. The simplified convective circulation of the two-columns scheme has stationary solutions. Hence, this method can be applied to problems where long time series are required, e.g. stellar pulsation or stellar evolution, without the necessity of resolving the dynamical time scale of convective turbulence.

Comparisons with 2D hydrodynamics proved that the scheme is able to reproduce many important properties of convection such as convective flux, overshoot, kinetic energy flux, and the range of convective velocities. First time dependent Cepheid pulsations computed with the two-columns scheme have also already demonstrated the feasibility of our approach to simulate simultaneously convection and pulsations.

Despite these achievements, the two-columns model is basically still a parameter-depended description. The good quantitative agreement between two-columns and 2D is in part also a result of suitable values for the free-parameters. However, the adopted parameters $cf1 = 0.8$ and $D = 20\,H_{p0}$ are by no means the outcome of an extensive parameter tuning.

An other limit of the scheme is the very coarse spatial resolution in horizontal direction and the simplistic description of the (vertical and horizontal) spectrum of convective velocities. So while the method is able to reproduce the main macroscopic properties of convection through the simple circulation fluid flow, one should not expect a correct description of more subtle details or turbulence effects from such a simple scheme.

Acknowledgements

This work was funded by Agence Nationale de la Recherche under the ANR project number NT05-3 42319.

References

Asplund, M., Nordlund, Å., Trampedach, R., Allende Prieto, C., & Stein, R. F.: 2000, A&A, 359, 729
Courant, R., Friedrichs, K., & Lewy, H.: 1928, Math. Ann., 100, 32
Dorfi, E. A., & Drury, L.O'C.: 1987, J. Comp. Phys., 69, 175
Freytag, B., Holweger, H., Steffen, M., & Ludwig, H.-G.: 1997, in Science with the VLT Interferometer, ed. F. Paresce, Springer, Berlin, p. 316
Freytag, B.: 2008, in preparation
Mihalas, D. & Mihalas, B. W.: 1984, Foundations of Radiation Hydrodynamics, Oxford University Press, New York
Nordlund, Å., Spruit, H. C. ; Ludwig, H.-G., & Trampedach, R.: 1997, A&A, 328, 229
Steffen, M., Freytag, B., & Ludwig, H.-G.: 2005, in Proc. 13th Cool Stars Workshop, eds. F. Favata et al., ESA SP-560, p. 985
Stein, R. F. & Nordlund, Å.: 1998, ApJ, 499, 914
Stökl, A.: 2008, A&A, submitted
Wedemeyer, S., Freytag, B., Steffen, M., Ludwig, H.-G., & Holweger, H.: 2004, A&A, 414, 1121

Discussion

KUPKA: The type of model for convection you are using here is known in meteorology under the name of mass-flux model. If one ties both velocity and temperature to the same column (up≡hot, cold≡down), sign symmetries in some third (and fourth) order correlations are violated and one runs into troubles with that particularly in overshooting zones. Secondly, I wonder about your choice of the column fraction value. There are problems known to exist for A-stars where the simulations are known to not recover the observed, blue-wards curved line profiles. For line profiles of Cepheids this non-solar-like shape is even more prominent. I'd thus not rely on the simulations as a valid test of models until they themselves have been probed with observational data.

STOEKL: (1) I'm not familiar with these meteorology models – however, my implementation does not couple the sign of velocity and temperature variations. I can not argue about 3rd & 4rd order moments as I've never translated the 2-columns scheme into that formulism. (2) The value for the fraction in cross section for up- and downdrafts is left as a free parameter. Therefore, when observations – of line profiles or of whatever – can provide constrains, this is a very welcome input. Despite the mentioned problems of 2D (and 3D) hydro simulations, I think that the qualitatively agreement of 2 columns & 2D simulations is a promising result.

WOITKE: What is the difference between your "2 column" approach and a regular radiation-hydro code with an extremely coarse resolution like 500×2?

STOEKL: There is none. – in principle – the geometric interpretation of the 2-columns-scheme is not that of a usual 2D–grid.

Conference photograph: A. Stökl is presenting his work.

Thermohaline Convection in Main Sequence Stars

S. Vauclair

Laboratoire d'Astrophysique de Toulouse-Tarbes, Université de Toulouse, CNRS,
14 av. E. Belin, 31400 Toulouse, France

Abstract. Thermohaline convection is a well known subject in oceanography, which has long been put aside in stellar physics. In the ocean, it occurs when warm salted layers sit on top of cool and less salted ones. Then the salted water rapidly diffuses downwards even in the presence of stabilizing temperature gradients, due to double diffusion between the falling blobs and their surroundings. A similar process may occur in stars in case of inverse μ-gradients in a thermally stabilized medium. Here we describe this process and some of its stellar applications.

Keywords. stars:evolution, convection, stars: abundances

1. What is thermohaline convection?

Thermohaline convection is a wellknown process in oceanography : warm salted layers on the top of cool unsalted ones rapidly diffuse downwards even in the presence of stabilizing temperature gradients, due to the different diffusivities of heat and salt. When a warm salted blob falls down in cool fresh water, the heat diffuses out more quickly than the salt. The blob goes on falling due to its weight until it mixes with the surroundings. This leads to the so-called salt fingers (Figure 1). Thermohaline convection is a double diffusive convection. When the gradients are reversed, another kind of double diffusive convection occurs, which is generally referred to as semi convection.

The condition for the salt fingers to develop is related to the density variations induced by temperature and salinity perturbations. Two important characteristic numbers are defined :
• the density anomaly ratio
$$R_\rho = \alpha \nabla T / \beta \nabla S \qquad (1.1)$$
where $\alpha = -(\frac{1}{\rho}\frac{\partial \rho}{\partial T})_{S,P}$ and $\beta = (\frac{1}{\rho}\frac{\partial \rho}{\partial S})_{T,P}$ while ∇T and ∇S are the average temperature and salinity gradients in the considered zone
• the so-called "Lewis number"
$$\tau = \kappa_S / \kappa_T = \tau_T / \tau_S \qquad (1.2)$$
where κ_S and κ_T are the saline and thermal diffusivities while τ_S and τ_T are the saline and thermal diffusion time scales.

The density gradient is unstable and overturns into dynamical convection for $R_\rho < 1$ while the salt fingers grow for $R_\rho \geqslant 1$. On the other hand they cannot form if R_ρ is larger than the ratio of the thermal to saline diffusivities τ^{-1} as in this case the salinity difference between the blobs and the surroundings is not large enough to overcome buoyancy.

Salt fingers can grow if the following condition is satisfied :
$$1 \leqslant R_\rho \leqslant \tau^{-1} \qquad (1.3)$$

In the ocean, τ is typically 0.01.

2. The stellar case

Thermohaline convection may occur in stellar radiative zones in the presence of inverse μ-gradients. In this case $\nabla_\mu = \mathrm{d}\ln\mu/\mathrm{d}\ln P$ plays the role of the salinity gradient while the difference $\nabla_{ad} - \nabla$ (where ∇_{ad} and ∇ are the usual adiabatic and local (radiative) gradients $\mathrm{d}\ln T/\mathrm{d}\ln P$) plays the role of the temperature gradient. The medium can become dynamically unstable if (Ledoux criterion):

$$\nabla_{crit} = \frac{\phi}{\delta}\nabla_\mu + \nabla_{ad} - \nabla < 0 \qquad (2.1)$$

where $\phi = (\partial \ln \rho/\partial \ln \mu)$ and $\delta = (\partial \ln \rho/\partial \ln T)$. When ∇_{crit} vanishes, marginal stability is achieved and thermohaline convection may begin as a "secular process", namely on a thermal time scale (short compared to the stellar lifetime!).

As for the ocean case, salt fingers will form if the following condition is verified:

$$1 \leqslant |\frac{\delta(\nabla_{ad} - \nabla)}{\phi(\nabla_\mu)}| \leqslant \tau^{-1} \qquad (2.2)$$

with $\tau = D_\mu/D_T = \tau_T/\tau_\mu$ where D_T and D_μ are the thermal and molecular diffusion coefficients while τ_T and τ_μ are the corresponding time scales.

In stars the value of the τ ratio is typically 10^{-10} if D_μ is the molecular (or "microscopic") diffusion coefficient but it can go up by many orders of magnitude when the shear flow instabilities which induce mixing between the edges of the fingers and the surroundings are taken into account.

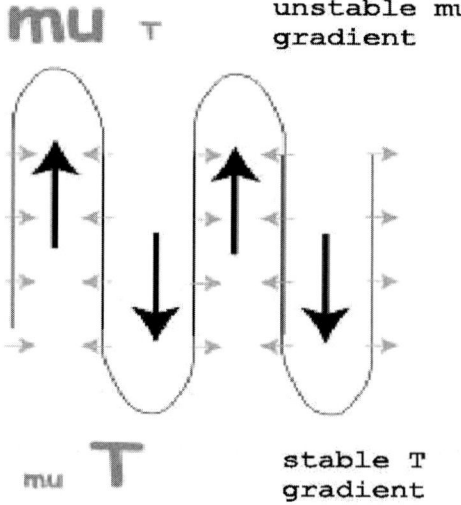

Figure 1. Schematic graph of thermohaline convection (salt or mu fingers). An important effect to take into account in the computations of thermohaline diffusion is the shear flow instabilities which occur at the edge of the fingers.

The effects of thermohaline convection as a mixing process in stars are far from trivial. Many detailed studies in the water case have been published, for example Gargett & Ruddick 2003) who gave precise comparisons between numerical simulations and laboratory experiments. However the stellar case is different as mixing then occurs in a compressible fluid.

As pointed out by Charbonnel & Zahn 2007, two different parametrisation recipes, as given by Ulrich 1972 or Kippenhahn *et al.* 1980 can differ by two orders of magnitude. This illustrates enough that treating thermohaline convection as a simple diffusion process, using a simple diffusion coefficient, may lead to wrong results.

Vauclair 2004 gave a discussion of these parameterisation procedures. The basic problem here concerns the vertical shear flow instability which occurs between the fingers and the inter-fingers medium, which is not taken into account in the Ulrich 1972 procedure. This instability leads to local turbulence which increases the mixing at the edge of the fingers. Consequently, a process of self destruction appears for the blobs, so that fingers eventually reach a regime where they cannot form anymore: this effect is taken into account in the Kippenhahn *et al.* 1980 procedure.

All this becomes much more complicate if one wants to take into account the competition between thermohaline convection and other macroscopic motions. This question has recently been addressed by Canuto *et al.* 2008

3. Where does thermohaline convection take place in stars?

Thermohaline convection may occur and has to be taken into account any time inverted μ-gradients are built. The first situation which was pointed out concerned ^{3}He burning regions (Abrams & Iben 1970, Ulrich 1971). It has recently been revisited by Eggleton *et al.* 2006 and Charbonnel & Zahn 2007 for the case of red giants: this is treated in other contributions to this meeting.

Situations with ^{4}He enhancement in stellar outer layers were also discussed several times, e.g. due to mass transfer (Stothers & Simon 1969). Another case occurs in main sequence helium rich stars, observed with effective temperatures around 20000 degrees, which are helium rich only in their outer layers. This may be explained by helium diffusion in a stellar wind (Vauclair 1975). The observations, which show that helium is enhanced by a factor two in average, prove that thermohaline mixing must be important in this case, otherwise the helium overabundance would be much larger. However thermohaline mixing must not be completely efficient, otherwise no helium enhancement would be left at all. The reason may be due to the presence of a magnetic field in this case. Extension of this kind of studies have been discussed for roAp stars by Balmforth *et al.* 2001.

In more recent work, detailed computations of iron accumulation in stars, due to atomic diffusion processes, have been performed in several frameworks. Charpinet *et al.* 1996 found that, in the case of sdB stars, iron accumulation could help triggering stellar oscillations due to the iron-related kappa-mechanism. Richard *et al.* 2001 found that such an iron accumulation can lead to a special iron-induced convective zone which may drastically change the internal stellar structure during main sequence evolution. As the iron accumulation leads to an inverted μ-gradient, the effects of thermohaline convection should be tested in all these computations, which has not been done yet.

Other cases where thermohaline convection must occur are related to accretion of metal-rich matter onto main sequence stars. Let us discuss two quite different cases. The first case is related to planetary formation and migration in the early times of planetary systems. The question arises whether the overmetallicity observed in exoplanets-host stars may be due to accretion of planetary material. One of the arguments against this was that the observed overabundances in exoplanets host stars is constant for main sequence stars, whatever their masses, while the depth of their convective zones were highly variable. Vauclair 2004 showed that this argument does not hold, due to thermohaline convection. Now the question arises whether some overmetallicity can remain in the outer layers of the stars or not. This question is not solved yet. It seems however, for other statistical reasons, that the overmetallicity must be primordial.

The second case is quite different. It concerns carbon enhanced main-sequence stars (CEMPs). These stars with abundance anomalies are supposed to have suffer some accretion of material coming from an AGB companion. Stancliffe *et al.* (2007) pointed out that in case of accretion of metal-rich matter, this material would subsequently fall down inside the star due to thermohaline convection. In a more recent paper (Thompson *et al.* 2008), we suggested that, between the stellar birth and the time when the AGB sends its processed material onto it, the main sequence star had time to suffer helium and heavy element diffusion below its convective zone, thereby creating a stabilizing μ-gradient. In the presence of this diffusion-induced μ-gradient, outside matter may accumulate in the convection zone until the overall μ-gradient becomes flat. This effect can save the whole process in this case (Figure 2).

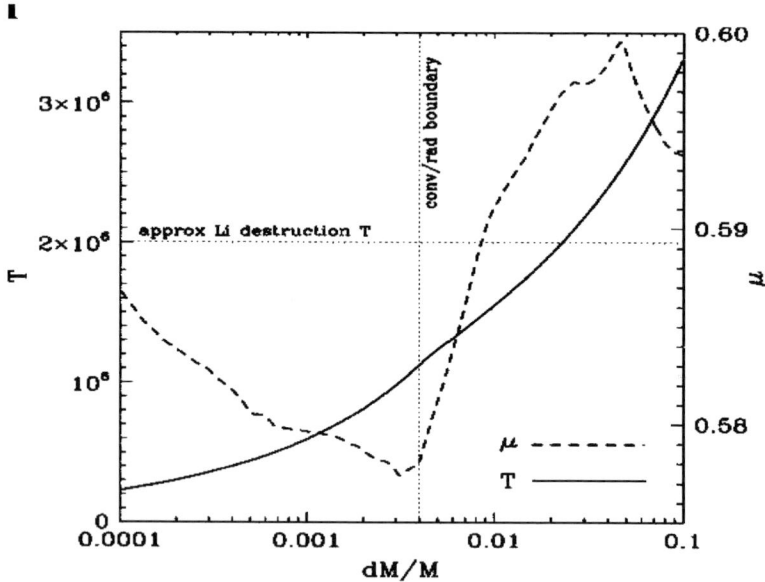

Figure 2. Mean molecular weight (including partial ionisation) and temperature profiles, as a function of the fractional mass, in a 0.78 Msun star with [Fe/H] = -2.3, at an age of 3.75 Gyr. This may correspond to the internal structure of a main sequence star just before the accretion of heavy material from an AGB companion. Helium depletion due to gravitational settling leads to a pre-existing stabilizing μ-gradient, which then allows a large part of the accreted matter to remain in the convective zone (after Thompson *et al.* 2008)

In summary, thermohaline convection, which has long been forgotten in astrophysics, has to be taken into account in several important cases during stellar evolution. Its competition with other processes like meridional circulation, magnetic fields, etc. will have to be computed and discussed thoroughly for a better understanding of stellar evolution.

References

Abrams, Z. & Iben, I., Jr. 1970, ApJ, 162, L125
Balmforth, N. J., Cunha, M. S., Dolez, N., Gough, D. O., & Vauclair, S. 2001, MNRAS, 23, 362
Canuto, V. M., Cheng, Y., & Howard, A. M. 2008, Geophys. Res. Let., 35, 2, L02613
Charbonnel, C. & Zahn, J.-P. 2007b, A&A, 476, L29
Charbonnel, C. & Zahn, J.-P. 2007a, A&A, 467, L15
charpinet, S., Fontaine, G., Brassard, P., & Dorman, B. 1996, ApJ, 471, L103

Eggleton, P. P., Dearborn, D. S. P., & Lattansio, J. C. 2006, Science, 314, 1580
Gargett, A. & Ruddick, B. 2003, ed., Progress in Oceanography, vol 56
Kippenhahn, R., Ruschenplatt, G., & Thomas, H. C. 1980, A&A, 91, 175
Richard, O., Michaud, G., & Richer, J. 2001, ApJ, 558, 377
Stancliffe, R. J., Glenneck, E., Izzard, R. G., & Pold, O. R. 2007, A&A, 464, L57
Stothers, R., & Simon, N. R. 1969, ApJ, 157, 673
Thompson, I. B., Ivans, I. I., Bisterzo, S., Sneden, C., Gallino, R., Vauclair, S., Burkey, G. S., Shectman, S. A., & Preston, G. W. 2008, ApJ, 667, 556
Ulrich, R. K. 1971, ApJ, 168, 57
Ulrich, R. K. 1972, ApJ, 172, 165
Vauclair, S. 1975, A&A, 45, 233
Vauclair, S., Dolez, N., & Gough, D. O. 1991, A&A, 252, 618
Vauclair, S. 2004, ApJ, 605, 874

Discussion

CHARBONNEL: It's certainly important to look at the effect of atomic diffusion on the inhibition of thermohaline in MS stars, and the case of carbon stars should bring important constraints on these interactions. However, it is not relevant in the case of red giants, because the first dredge-up erases completely the effects of atomic diffusion that occurred on the main sequence.

VAUCLAIR: Yes of course, I agree and I did not speak of giants, only main-sequence stars. Matteo Cantiello will speak about giants just after me.

STEPIEŃ: You mentioned that the diffusion rate determines the extent of fingers (or, equivalently their life time). Is friction peeling off the original bullet not more important?

VAUCLAIR: Yes indeed, this why I said that the particle transfer between the interior and exterior of the fingers was more rapid that original atomic diffusion and that this had to be taken into account in the computation.

Conference photograph

S. Vauclair is presenting her contribution.

| The Art of Modelling Stars in the 21st Century
Proceedings IAU Symposium No. 252, 2008
L. Deng & K.L. Chan, eds.

Thermohaline mixing in low-mass giants

M. Cantiello[1] and N. Langer[1]

[1]Astronomical Institute Utrecht, Princetonplein 5, 3584 CC Utrecht, The Netherlands
email: M.Cantiello@uu.nl, N.Langer@uu.nl

Abstract. Thermohaline mixing has recently been proposed to occur in low mass red giants, with large consequences for the chemical yields of low mass stars. We investigate the role of thermohaline mixing during the evolution of stars between $1\,\mathrm{M}_\odot$ and $3\,\mathrm{M}_\odot$, in comparison to other mixing processes acting in these stars. We confirm that thermohaline mixing has the potential to destroy most of the ^{3}He which is produced earlier on the main sequence during the red giant stage. In our models we find that this process is working only in stars with initial mass $M \lesssim 1.5\,\mathrm{M}_\odot$. Moreover, we report that thermohaline mixing can be present during core helium burning and beyond in stars which still have a ^{3}He reservoir. While rotational and magnetic mixing is negligible compared to the thermohaline mixing in the relevant layers, the interaction of thermohaline motions with differential rotation and magnetic fields may be essential to establish the time scale of thermohaline mixing in red giants.

Keywords. stars: evolution, stars: abundances, stars: AGB and post-AGB, stars: magnetic fields, stars: rotation, ISM: abundances

1. Introduction

Thermohaline mixing is not usually considered as an important mixing process in single stars, since the ashes of thermonuclear fusion consists of heavier nuclei than its fuel, and stars usually burn from the inside out. The condition for thermohaline mixing, however, is that the mean molecular weight (μ) decreases inward. Recently Charbonnel & Zahn (2007a, CZ07) identified thermohaline mixing as an important mixing process which significantly modifies the surface composition of red giants after the first dredge-up. The work by CZ07 was triggered by the paper of Eggleton *et al.* (2006, EDL06), who found a μ-inversion in their $1\,\mathrm{M}_\odot$ stellar evolution model, occurring after the so-called luminosity bump on the red giant branch (RGB), which is produced after the first dredge-up, when the hydrogen-burning shell reaches the chemically homogeneous part of the envelope. The μ-inversion is produced by the reaction ^{3}He(^{3}He,2p)^{4}He (as predicted by Ulrich (1972)). It does not show up earlier, since the magnitude of the μ-inversion is small and negligible compared to a stabilizing μ-stratification.

The mixing process below the convective envelope in models of low-mass stars turns out to be essential for the prediction of the chemical yield of ^{3}He (EDL06), and to understand the surface abundances of red giants, in particular the ^{12}C/^{13}C ratio, and the ^{7}Li, carbon and nitrogen abundances (CZ07). This may also be important for other occurrences of thermohaline mixing in stars, i.e., in single stars when a μ-inversion is produced by off-center ignition in semi-degenerate cores, or in stars which accrete chemically enriched matter from a companion in a close binary (e.g., Stancliffe *et al.* 2007). Accreted metal-rich matter during the phases of planetary formation also leads to thermohaline mixing. The host stars of exoplanets present a metallicity excess compared to stars in which no planets have been detected. This metallicity excess can be reconciled with the overabundances expected in cases of accretion if thermohaline mixing is included in the picture (Vauclair 2004).

2. Method

We use a 1-D hydrodynamic stellar evolution code (Yoon et al. 2006, and references therein). Mixing is treated as a diffusive process, the contributions to the diffusion coefficient are convection, semiconvection, thermohaline mixing, rotationally induced mixing and magnetic diffusion. The code includes the effect of centrifugal force on the stellar structure, and the transport of angular momentum is also treated as a diffusive process (Endal & Sofia 1978; Pinsonneault et al. 1989). The condition for the occurrence of thermohaline mixing is

$$\frac{\varphi}{\delta} \nabla_\mu \leqslant \nabla - \nabla_{\rm ad} \leqslant 0 \qquad (2.1)$$

i.e. the instability operates in regions that are stable against convection (according to the Ledoux criterion) and where an inversion in the mean molecular weight is present. Here $\varphi = (\partial \ln \rho / \partial \ln \mu)_{P,T}$, $\delta = -(\partial \ln \rho / \partial \ln T)_{P,\mu}$, $\nabla_\mu = d \ln \mu / d \ln P$, $\nabla_{\rm ad} = (\partial \ln T / \partial \ln P)_{\rm ad}$, and $\nabla = d \ln T / d \ln P$. Numerically, we treat thermohaline mixing through a diffusion scheme (Braun 1997; Wellstein et al. 2001). The corresponding diffusion coefficient is based on the work of Stern (1960), Ulrich (1972), and Kippenhahn et al. (1980):

$$D_{th} = -\alpha_{\rm th} \frac{3K}{2 \rho c_P} \frac{\frac{\varphi}{\delta} \nabla_\mu}{(\nabla_{\rm ad} - \nabla)} \qquad (2.2)$$

where ρ is the density, $K = 4acT^3/(3\kappa\rho)$ the thermal conductivity, and $c_P = (dq/dT)_P$ the specific heat capacity. The quantity $\alpha_{\rm th}$ is an efficiency parameter for the thermohaline mixing. The value of this parameter depends on the geometry of the fingers arising from the instability and is still a matter of debate (Ulrich 1972; Kippenhahn et al. 1980; Charbonnel & Zahn 2007a). Most of the calculations have been performed with $\alpha_{\rm th} = 2.0$, corresponding to the prescription of Kippenhahn et al. (1980), although we also investigated the effect of using different values of $\alpha_{\rm th}$.

In the code rotational mixing is included. Four different diffusion coefficients are calculated for dynamical shear, secular shear, Eddington-Sweet circulation and Goldreich-Schubert-Fricke instability. Details on the physics of these instabilities and their implementation in the code can be found in Heger et al. (2000).

Chemical mixing and transport of angular momentum due to magnetic fields (Spruit 2002) is included as in Heger et al. (2000). The contribution of magnetic fields to the mixing is also calculated and added to the total diffusion coefficient D entering the diffusion equation.

We compute evolutionary models of $1.0\,M_\odot$, $1.5\,M_\odot$, $2.0\,M_\odot$ and $3.0\,M_\odot$ at solar metallicity (Z=0.02). The initial equatorial velocities of these models were chosen to be 10, 45, 140 and 250 km s^{-1} (Tassoul 2000); we assume the stars are rigidly rotating at the zero-age main sequence. Throughout the evolution of all models, the mass-loss rate of Reimers (1975) was used.

3. Results

We confirm the presence of an inversion in the mean molecular weight in the outer wing of the H-burning shell, after the luminosity bump on the red giant branch. According to inequality (2.1) this inversion gives rise to thermohaline mixing in the radiative buffer layer, the radiative region between the H-burning shell and the convective envelope. Results of our calculations for the 1.0 and 2.0 $M_\odot$ models are shown in Fig. 1. In our $1\,M_\odot$ model, thermohaline mixing develops at the luminosity bump and transports chemical species in the radiative layer between the H-burning shell and the convective envelope.

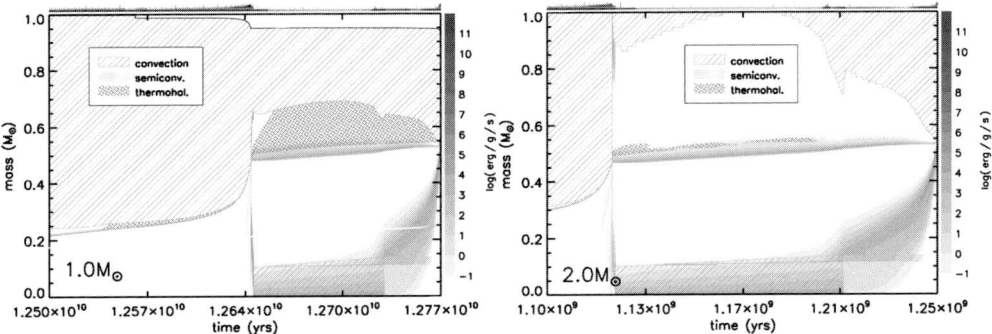

Figure 1. Left: evolution of the internal structure of a $1.0\,\mathrm{M}_\odot$ star from the onset of thermohaline mixing to the AGB phase. Green hatched regions indicate convection, yellow filled regions represent semiconvection and red cross hatched regions indicate thermohaline mixing, as displayed in the legend. Blue shading shows regions of nuclear energy generation. **Right:** same as left panel, but for a $2.0\,\mathrm{M}_\odot$ star.

This results in a change to the stellar surface abundances. The left panel of Fig. 2 shows the evolution of the ^{3}He surface abundance, and of the ratio ^{12}C/^{13}C at the surface as a function of time in the $1\,\mathrm{M}_\odot$ model, confirming the result of EDL06 and CZ07, namely that thermohaline mixing is efficiently depleting ^{3}He and lowering the ratio ^{12}C/^{13}C on the giant branch. Interestingly the $2\,\mathrm{M}_\odot$ model shows a different behaviour: thermohaline mixing is not able to connect the H-burning shell and the convective envelope, resulting in no change to the surface abundance of ^{3}He and to the ratio ^{12}C/^{13}C after the luminosity bump. Our models with $\alpha_\mathrm{th} = 2.0$ show thermohaline mixing to be important during the RGB only in stars with mass $M \lesssim 1.5\,\mathrm{M}_\odot$.

While CZ07 and EDL07 investigate thermohaline mixing only during the RGB phase, we followed the evolution of our models until the thermally pulsating AGB phase (TP-AGB). In fact a μ-inversion is always created if a H-burning shell is active in a chemically homogeneous layer; this happens not only during the RGB phase, but also during the horizontal branch (HB) and the AGB. The size of the μ-inversion depends on the local amount of ^{3}He. As a consequence, the efficiency of thermohaline mixing in different evolutionary phases depends on the amount of ^{3}He left by previous mixing episodes.

After the core He-flash, helium is burned in the core, while a H-burning shell is active below the convective envelope. We found that during this phase thermohaline mixing is present and can spread through the whole radiative buffer layer in our $1\,\mathrm{M}_\odot$ model (left panel in Fig. 1). In this model the surface abundances change also during this phase because the H-burning shell and the envelope are connected. This is shown in Fig. 2, left panel, where surface abundances change after the luminosity peak corresponding to the He-flash. We stress that using the prescription of Kippenhahn *et al.* (1980) for thermohaline mixing allows our model to reach this phase without completely burning the ^{3}He; models of CZ07 almost completely deplete ^{3}He in the envelope during the RGB phase because of their higher diffusion coefficient (Ulrich 1972). In this case thermohaline mixing would be much less efficient, during the subsequent evolutionary phases, due to the lower abundance of ^{3}He.

The last nuclear-burning phase of a low-mass star is the AGB, and is characterized by the presence of two burning shells and a degenerate core. The star burns H in a shell and the ashes of this process feed an underlying He shell. During the most luminous part of the AGB the He shell periodically experiences thermal pulses (TPs); in stars more massive than about $1.2\,\mathrm{M}_\odot$ these TPs are associated with a deep penetration of the

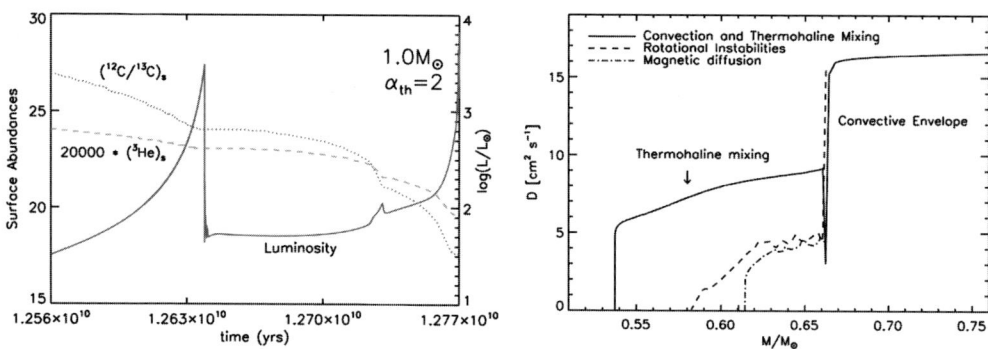

Figure 2. Left: evolution of the surface abundance of the ^{12}C/^{13}C ratio (dotted red line) and ^{3}He (dashed green line), and of the luminosity (solid blue line) from the onset of thermohaline mixing up to the AGB phase for a $1.0\,M_\odot$ star. **Right:** diffusion coefficients in the region between the H burning shell and the convective envelope for the $1.0\,M_\odot$ model during core He-burning (t = 1.267×10^{10} years). The black, continuous line shows convective and thermohaline mixing diffusion coefficients, the green, dashed line is the sum of the diffusion coefficients due to rotational instabilities while the blue, dot-dashed line shows the magnitude of magnetic diffusion coefficient.

convective envelope, the so-called third dredge-up (3DUP) (Wallerstein & Knapp 1998). We find thermohaline mixing to be present during the TP-AGB phase. Depending on the mass of the model the diffusion process is able to connect the H-burning shell with the convective envelope during the whole interpulse phase. In a $1\,M_\odot$ model thermohaline mixing connects the H-burning shell to the convective envelope (Fig. 3, left panel), confirming that this mixing process is more efficient at lower masses. The occurrence of thermohaline mixing in this late evolutionary phase is critically determined by the mixing history of the star. This is because the amount of ^{3}He present during the TP-AGB phase is controlled by the occurrence of thermohaline mixing and by its efficiency ($\alpha_{\rm th}$) in previous evolutionary phases.

Thermohaline mixing during the TP-AGB is a circulation of the type inferred by Cameron & Fowler (1971), since it carries ^{7}Be out of the high-temperature zone. ^{7}Be can then produce ^{7}Li by electron capture, which is transported in the convective envelope. This results in surface enrichment of ^{7}Li, as shown in the right panel of Fig. 3. Interestingly Uttenthaler et al. (2007) recently reported the detection of low-mass, Li-rich AGB stars in the galactic bulge. Given their low mass, these stars are not expected to experience any hot bottom burning. We argue that thermohaline mixing is a possible explanation for their abundance anomalies.

4. Rotation and magnetic fields

In our models we found that in the relevant layers thermohaline mixing has generally higher diffusion coefficients than rotational instabilities and magnetic diffusion. This result is not valid for magnetic stars, stars that possess anomalous surface fields of a few 10^2 to about 10^4 G, believed to be of fossil origin (Charbonnel & Zahn 2007b). The right panel of Fig. 2 clearly shows that rotational and magnetic mixing are negligible compared to the thermohaline mixing in our $1.0\,M_\odot$ model. The only rotational instability acting on a shorter timescale is the dynamical shear instability, visible in the right panel of Fig. 2 (dashed line) as a spike present at the lower boundary of the convective envelope. This instability works on the dynamical timescale in regions of a star characterized by a high degree of differential rotation. However, if present, this instability acts only in a very

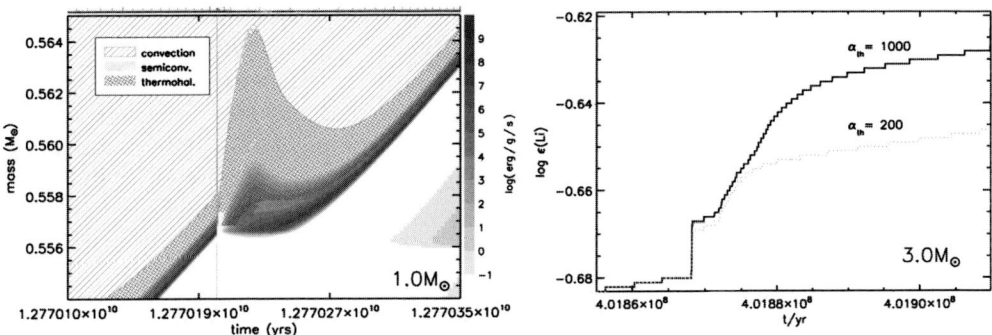

Figure 3. **Left:** evolution of the region between the H burning shell source and the convective envelope during a thermal pulse in a $1.0\,M_\odot$ star. Green hatched regions indicate convection and red crossed regions indicate thermohaline mixing. Blue shading shows regions of nuclear energy generation. **Right:** evolution of Lithium surface abundance during one thermal pulse in a $3\,M_\odot$ model. The black, continuous line shows a model evolved with $\alpha_{th}=1000$; the blue, dotted line refers to the same model evolved with a thermohaline mixing efficiency $\alpha_{th}=200$. In both cases the model is experiencing third dredge-up. The evolution of the star prior to the TP-AGB has been calculated with $\alpha_{th}=2$.

small region (in mass coordinate) at the bottom of the convective envelope. As a result thermohaline mixing is still setting the timescale for the diffusion of chemical species from the convective envelope to the hydrogen-burning shell.

The interaction between rotation and thermohaline mixing is more difficult to address, since it requires full hydrodynamic calculations. We expect the speed of thermohaline mixing to be affected by differential rotation and rotational instabilities, since these are able to change the geometry of the fingers, and to create turbulence, respectively. Canuto et al. (2008) recently suggested that turbulence must be taken into account to explain both laboratory and ocean data of double diffusive processes.

The interaction of magnetic fields with thermohaline mixing must also be considered. This depends on the geometry and the magnitude of the magnetic field: if the magnetic field is stronger or of the same order of the equipartition value the magnetic field allows plasma motions only along the direction of the field lines. This could result in the inhibition of thermohaline mixing, as discussed with detailed calculations by Zahn and Charbonnel (2008). However strong magnetic fields are observed in only a small fraction of low-mass stars (Wolff (1968); Power et al. (2007)). Concerning the interaction of weak fields with thermohaline motions, a change in the speed of the fingers is also expected and fully MHD calculations are needed to understand the process.

5. Discussion

We comfirm the results of EDL06 and CL07: thermohaline mixing in low-mass giants is capable of destroying large quantities of ^{3}He, as well as decreasing the ratio ^{12}C/^{13}C. Thermohaline mixing starts when the hydrogen burning shell moves into the chemically homogeneous layers established by the first dredge-up. Our models show further that thermohaline mixing remains important during core helium burning, and can still be relevant during the AGB phase – including the termally-pulsing AGB stage. This can result in important changes to the surface abundances of low-mass stars. The quantitative discussion is complicated by the fact that the capability of thermohaline mixing to change surface abundances depends on an uncertain efficiency parameter, as well as on the local ^{3}He abundance. The efficiency parameter α_{th} is still a matter of debate and is probably strongly affected by the interaction of thermohaline mixing with rotation and magnetic fields. The local ^{3}He abundance depends on the previous history of mixing.

Our calculations show that in the relevant layers, thermohaline mixing generally has a higher diffusion coefficient than rotational instabilities and magnetic diffusion. Still, the interaction of thermohaline mixing with magneto-rotational instabilities is important: we expect the speed of the mixing to be strongly affected by the presence of differential rotation and magnetic fields. To better understand the picture it would be desirable to have realistic MHD simulations of thermohaline mixing.

Acknowledgements

MC wishes to thank V. Canuto, J.-P. Zahn and C. Charbonnel for useful discussions. MC acknowledges LKBF and the IAU for financial support.

References

Braun, H. 1997, PhD thesis, Ludwig-Maximilians-Univ. München
Cameron, A. G. W. & Fowler, W. A. 1971, *ApJ*, 164, 111
Canuto, V. M., Cheng, Y., & Howard, A. M. 2008, *Geophysical Research Letters*, 35, 2
Charbonnel, C. & Zahn, J.-P. 2007, *A&A*, 467, L15
Charbonnel, C. & Zahn, J.-P. 2007, *A&A*, 476, L29
Eggleton, P. P., Dearborn, D. S. P., & Lattanzio, J. C. 2006, *Science*, 314, 1580
Endal, A. S. & Sofia, S. 1978, *ApJ*, 220, 279
Heger, A., Langer, N., & Woosley, S. E. 2000, *ApJ*, 528, 368
Kippenhahn, R., Ruschenplatt, G., & Thomas, H.-C. 1980, *A&A*, 91, 175
Pinsonneault, M. H., Kawaler, S. D., Sofia, S., & Demarque, P. 1989, *ApJ*, 338, 424
Power, J., Wade, G. A., Hanes, D. A *et al.* 2007, Physics of Magnetic Stars, 89
Reimers, D. 1975, Memoires of the Societe Royale des Sciences de Liege, 8, 369
Spruit, H. C. 2002, *A&A*, 381, 923
Stancliffe, R. J., Glebbeek, E., Izzard, R. G., & Pols, O. R. 2007, *A&A*, 464, L57
Tassoul, J.-L. 2000, Stellar Rotation, Cambridge University Press, 36
Ulrich, R. K. 1972, *ApJ*, 172, 165
Uttenthaler, S., Lebzelter, T., Palmerini, S. *et al.* 2007, *A&A*, 471, L41
Vauclair, S. 2004, *ApJ*, 605, 874
Wallerstein, G. & Knapp, G.R 1998, *ARAA*, 36, 369
Wellstein, S., Langer, N., & Braun, H. 2001, *A&A*, 369, 939
Wolff, S. C. 1968, *PASP*, 80, 281
Yoon, S.-C., Langer, N., & Norman, C. 2006, *A&A*, 460, 199

R. WALDMAN: Another possible case where thermohaline mixing can occur is off-center carbon burning. Has anybody done work on that?

Discussion

M. CANTIELLO: Yes, thermohaline mixing should occur during off-center carbon burning. I think this problem has been studied by Thomas (1965).

P. WOITKE: To what extent is thermohaline mixing included in current standard stellar evolution codes (Iben, Lattanzio...)? Are you the first one who has solved this consistently?

M. CANTIELLO: I think it is not standard in most of stellar evolution codes at the moment, but thermohaline mixing has been implemented in our code in 2001 (Wellstein *et al.*) following the prescription of Kippenhahn (1980). Charbonnel & Zahn (2007) studied the problem of thermohaline mixing driving the RGB consistently.

C. CHARBONNEL: Comment: Observations of $^{12}C/^{13}C$ give very strong constraints on the efficiency of thermohaline convection in RGB stars. Question: Thermohaline convection should have destroyed most of the ^3He in a $1M_\odot$ star on the RGB. So how can you have thermohaline convection on the early-AGB for such a star, if the ^3He reservoir is empty?

M. CANTIELLO: Yes. Thermohaline mixing can be present during core He burning and on the AGB and TP-AGB if ^3He reservoir is not empty. This is the case in our models since we use Kippenhahn's prescription. Also stars that avoid thermohaline mixing on RGB ($\sim 5\%$ of low mass stars) could show thermohaline mixing in subsequent phases. Interestingly there are observations of a few AGB stars which show surface abundances in agreement with thermohaline mixing occurring during this phase.

A snapshot of the meeting. Faces recognizable from left to right: Georges Meynet, Martin Asplund, Lee Anne Willson, Peter Woitke, Shiyang Jiang, Ofer Yaron, Y. C. Kim.

A snapshot of the meeting. Faces recognizable from left to right: Casey Meakin, L. Piau, Yi Hu, P. Kavella, Onno Pols, Georges Meynet.

Large-scale numerical simulations: Convection in an annular channel rotating about a vertical axis with side-walls

Yingli Chang[1] and Yu Liu[2,3]

[1]Shanghai fisheries university.334 Jungong Road, Shanghai 200090, China
[2]Shanghai Astronomical Observatory, 80 Nandan Road, Shanghai, China
[3]Graduate University of Chinese Academy of Science, China
email: yuliu@shao.ac.cn

Abstract. For the purpose of understanding the dynamics of planetary atmospheres, we use the annular convection model to simulate the dynamics of atmospheres of Jupiter and Saturn. The model (annular channel) rotates about a vertical axis with side-walls, and it is heated from below.

We use the software NaSt3DGP (a parallel software package to solve the 3D incompressible fluid dynamic problems in Cartesian coordinates by using Finite Difference Method) for the computation. It's reliability is tested by our application to simulate fully three-dimensional nonlinear convection in a box with lateral stress-free side-walls, uniformly heated from below. We found that, at moderately large Rayleigh numbers, the complex formation of multiple-jet flows can be maintained by the traveling convective eddies; we also found that the type of the sidewall velocity condition does not play an essential role in determining the primary properties of strongly nonlinear convection.

Keywords. Solver, rotation, annular channel, simulation convection

1. Introduction

Convection in a rotating Bénard layer has been also studied extensively in attempting to offer valuable insight into the general physics and dynamics in rotating convective systems because of its mathematical simplicity. It is also worth mentioning that the annular system with the gravity vector perpendicular to the axis of rotation, mimicking the equatorial region of a rotating spherical shell, has been extensively studied (for example, Busse, 1970)

Xhinhao Liao et al investigated linear convection and weakly nonlinear differential rotation in an annular channel with the no-slip side-walls, stress-free top and bottom, uniformly heated from below and rotating about a vertical axis, the exactly same model as that studied by Davies-Jones and Gilman (1971).

Our main work will report numerical simulations of nonlinear convection in a box with lateral stress-free side-walls,no-slip and stress-free top and bottom,uniformly heated from below.

2. Convection in a wide-gap channel

We first look at the results of numerical simulations of nonlinear convection in a wide-gap annular channel with stress-free sidewalls. Then we look at the results of three-dimensional numerical simulations of no-slip side walls.

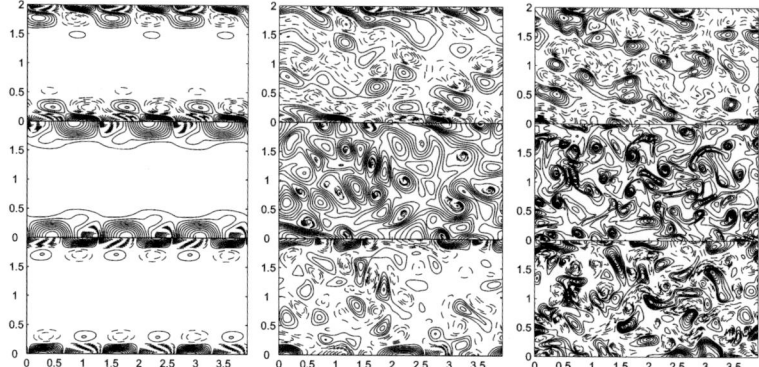

Figure 1. Contours of u_x (top), Θ (middle) and u_z (bottom) are shown in a horizontal xy-plane. $R = 7 \times 10^4$ (left), $R = 2.5 \times 10^5$ (middle), $R = 2.5 \times 10^6$ (right), $\tau = 10^3$, $\Gamma = 2$, $P_r = 7.0$, stress-free b.c.

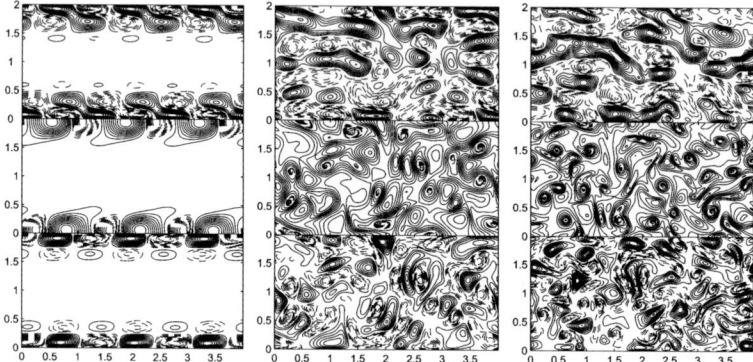

Figure 2. Contours of u_x (top), Θ (middle) and u_z (bottom) are shown in a horizontal xy-plane. $R = 5 \times 10^4$ (left), $R = 2.5 \times 10^5$ (middle), $R = 1 \times 10^6$ (right), $\tau = 10^3$, $P_r = 7.0$, non-slip side walls.

3. Discussion

A major conclusion drawn from the results of the numerical simulations is that at moderately large Rayleigh numbers, the complex formation of multiple-jet flows can be maintained by the travelling convective eddies; and the type of the sidewall velocity condition does not play an essential role in determining the primary properties of strongly nonlinear convection. This is significant with regard to possible geophysical and astrophysical applications because the well-known tangent cylindrical surface in rapidly rotating spherical shells may be regarded as an effective stress-free sidewall.

Acknowledgements

This work is supported by Ocean Common Weal Cause Research Project in 2008 (2008 05016), Shanghai fund for outstanding youth (10050 8101 06 4101) and Shanghai fisheries university fund for PHD (A 2401 06 0342).

References

Busse, F. H. 1970, *J. Fluid Mech.* 10, 441–460
Liao, X., Zhang, K., & Chang, Y. 2005, *Geophys. Astrophys. Fluid Dyn.* 99, 441–460
Davies-Jones, R. P. & Gilman P. A. 1971, *J. Fluid Mech.* 46, 65–81

Comparing the Nucleosynthesis Parameters of s+r Stars and Ba Stars

Wen-Yuan Cui, Dong-Nuan Cui and Bo Zhang

Department of Physics, Hebei Normal University, 113 Yuhua Dong Road, Shijiazhuang 050016, P.R. China
email: wenyuancui@126.com; zhangbo@hebtu.edu.cn

Abstract. In this paper, we use a parametric model of the asymptotic giant branch (AGB) stars, in which the ^{13}C neutron source is activated in radiative conditions during the interpulse periods, to calculate the nucleosynthesis in 29 very metal-poor double-enhanced stars (i.e. s+r stars) and 26 barium stars (i.e. Ba stars), respectively. Through a statistical analyzing on the corresponding parameters obtained for the above stars, we get the possible conditions which the s+r stars formed in. We find that the value of neutron exposures of most s+r stars is greater than that of Ba stars. In the very metal-poor stars, the Ba stars stars should belong to the binary systems with large initial orbital separation, by comparing the s-process-component coefficient (C_s) values with those of s+r stars. For s+r stars, there is strong correlation between their C_s and C_r (r-process-component coefficient) but no correlation for Ba stars. This strongly confirms the possibility that the s+r stars should form through the accretion-induced collapse (AIC) or type 1.5 supernova mechanism.

Keywords. nucleosynthesis, stars: abundances, AGB and post-AGB

1. Introduction

It is interesting to adopt the parametric model for metal-poor stars presented by Aoki, Reyan, Norris, *et al.* (2001) and developed by Zhang, Ma, & Zhou (2006) to study the physical conditions that could reproduce the observed abundance pattern. Using the parametric model, we investigate the characteristics of the nucleosynthesis pathway that produces the special abundance ratios of 29 s+r stars and 26 Ba stars, respectively. The calculated results are presented in Sect. 2 in which we also discuss the differences between the parameters of s+r stars and those of Ba stars. Conclusions are given in Sect. 3.

2. Results and Discussion

We explored the origin of the neutron-capture elements in s+r stars and Ba stars. In the parametric model, there are four parameters: the neutron exposure per thermal pulse $\Delta\tau$, the overlap factor r, C_s and C_r. Using the observed data from 29 s+r stars and 26 Ba stars, the model parameters can be obtained from the parametric approach, respectively.

Histograms given in figure 1 show the distribution of the number ratio of s+r stars and that of Ba stars via the nucleosynthesis parameters. From figure 1, We can see that the $\Delta\tau$ values of Ba stars are almost in the range $0 \sim 0.5$ mb^{-1}, but those of s+r stars are almost in the range $0.5 \sim 1.0$ mb^{-1}. The $\Delta\tau$ range of s+r stars ([Fe/H]$\leqslant -2.0$) is obviously greater than that of Ba stars ([Fe/H]> -1.0). This confirms the fact that the neutron density in the nucleosynthesis region increases with declining metallicity (Gallino, Arlandini, Busso, *et al.* 1998; Cui, & Zhang 2006). The right panel in figure 1

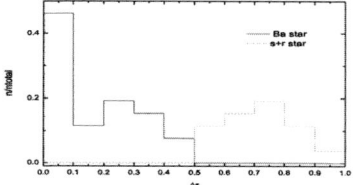
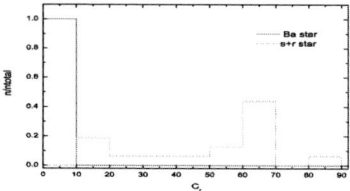

Figure 1. Histograms showing the distribution of the number ratio of s+r stars and that of Ba stars via the nucleosynthesis parameters. Left: via $\Delta\tau$; Right: via C_r.

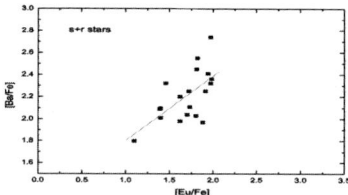
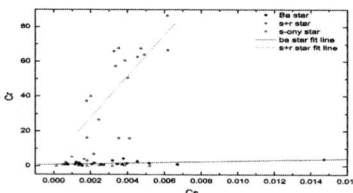

Figure 2. Left: Correlation between [Ba/Fe] and [Eu/Fe]; Right: between C_r and C_s.

shows that the C_r values of Ba stars distribute in a small range $0 \sim 10$, but in a wide range $10 \sim 90$ for s+r stars. Obviously, the C_r values of s+r stars are greater than those of Ba stars which is similar to s-only stars. This implies that there are different origins for the r-process elements between s+r stars and Ba stars. The Ba stars should formed in a normal homogeneous molecular cloud for their low r-process element abundances. From figure 2, we can see that [Ba/Fe] increases with [Eu/Fe], and C_r also increases with C_s for s+r stars. This strongly implies that the s+r stars form in binary systems which should have suffered the AIC or Type 1.5 supernova event. Because if they come through pre-enrichment event, i.e. they form in a nearby molecular cloud polluted by Type II supernova, the abundances of r-process elements will exhibit large spread. For Ba stars and s-only stars, C_r does not increase with C_s, furthermore there is not large spread. This implies that there is not rapid neutron-capture event taking place in the binary systems which Ba stars and s-only stars belong to.

3. Conclusions

In this paper, we carry a statistical analyzing on the nucleosynthesis parameters of 29 s+r stars and 26 Ba stars. We give the possible mechanisms of s+r stars, Ba stars and s-only stars, respectively.

Acknowledgements

This work is supported by the National Natural Science Foundation of China under grant no. 10673002, and the Doctoral grant of Hebei Normal University.

References

Aoki, W., Reyan, S. G., Norris, J. E., Beers, T. C., Ando, H., et al. 2001, *ApJ* 561, 346
Cui, W. Y. & Zhang, B. 2006, *MNRAS* 368, 305
Gallino, R., Arlandini, C., Busso, M., Lugaro, M., Travaglio, C., et al. 1998, *ApJ* 497, 388
Zhang, B., Ma, K., & Zhou, G. D. 2006, *ApJ* 642, 1075

The Art of Modelling Stars in the 21st Century
Proceedings IAU Symposium No. 252, 2008
L. Deng & K.L. Chan, eds.

© 2008 International Astronomical Union
doi:10.1017/S1743921308022588

Asteroseismic study of solar-like stars: A method of estimating stellar age

Y. K. Tang[1,3] S. L. Bi[2,1] and N. Gai[1,3]

[1] National Astronomical Observatories/Yunnan Observatory, Chinese Academy of Sciences,
Kunming 650011, P. R. China
email: bisl@bnu.edu.cn; tangyanke@ynao.ac.cn

[2] Department of Astronomy Beijing Normal University, Beijing 100875, P. R. China

[3] Graduate School of the Chinese Academy of Sciences, Beijing 100039, P. R. China

Abstract. Asteroseismology, as a tool to use the indirect information contained in stellar oscillations to probe the stellar interiors, is an active field of research presently. Stellar age, as a fundamental property of star apart from its mass, is most difficult to estimate. In addition, the estimating of stellar age can provide the chance to study the time evolution of astronomical phenomena. In our poster, we summarize our previous work and further present a method to determine age of low-mass main-sequence star.

Keywords. Stars: evolution, stars: oscillations

1. Introduction

Due to the frequencies of these oscillations depend on density, temperature, gas motion, and other properties of the stellar interior, it can take the window to "see" the interior of stars and help us to know the stellar internal structure and understand the stellar evolution. With the advance of observational technique, several stars have been detected the solar-like oscillations. Using the latest asteroseismic data, we reconstruct the model of α Cen B and 70 Ophiuchi A (Tang *et al.* 2008a, 2008b). In additional, Bi *et al.* (2008) have performed preliminary seismological analysis of two MOST targets.

2. Using asteroseismic diagram to estimate stellar age

Following the asymptotic formula for the frequency $\nu_{n,l}$ of a stellar p-mode of order n and degree l was given by Tassoul(1980):

$$\nu_{n,l} \simeq (n + \frac{l}{2} + \epsilon)\nu_0 - [Al(l+1) - B]\nu_0^2 \nu_{n,l}^{-1}, \qquad (2.1)$$

where, ν_0 and A are related to the run of sound speed. Based on some quantities as diagnostic purpose to probe the stellar interior proposed by some authors (Christensen-Dalsgaard 1988; Gough 2003; *et al.*), like $\delta\nu_{n,l} = \nu_{n,l} - \nu_{n-1,l+2}$, we definite another quantity (Tang *et al.* 2008)

$$r_{01} = \frac{\langle \delta\nu_{0,2} \rangle}{\langle \delta\nu_{1,3} \rangle}. \qquad (2.2)$$

The r_{01} comes from the perturbation to the gravitational potential, neglected in equation (2.1), which affects modes of the lowest degrees most strongly and which probably increases with evolution due to the increasing central density for modes of the lowest degrees which penetrate most deeply and hence affect $\delta\nu_{0,2}$ more than $\delta\nu_{1,3}$, leading to the dependence of r_{01} on age. We compute some models with initial heavy metal

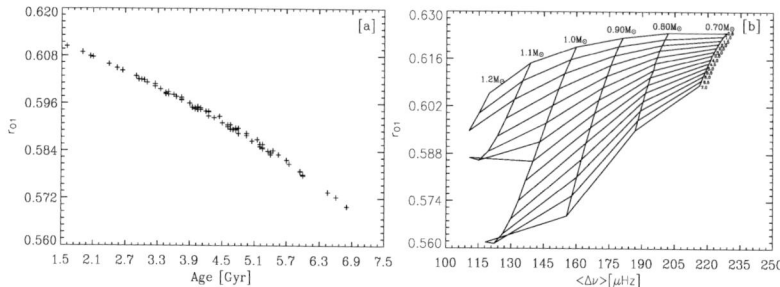

Figure 1. [a]: The ratio of small separations adjacent in l vs. age for each of 129 stellar models described in Tang et al. (2008a). [b]: ($\langle\Delta\nu\rangle$, r_{01}) diagram for stellar models. The vertical lines are evolutionary tracks, labeled by the mass in the top, whereas the transverse lines are isopleths with constant age, labeled by the age from 0.5 Gyr to 7.0 Gyr increasing with 0.5 (unit is Gyr).

abundance $Z_i = 0.02$, initial helium abundance $Y_i = 0.28$ and mixing-length parameter $\alpha = 1.7$ using the Yale stellar evolution code (YREC; Guenther et al. 1992) and analyze the pulsation of low degree p-modes ($l = 0 - 3$) for selected models in each given mass is implemented using the Guenther's pulsation code under the adiabatic approximation (Guenther 1994). Considering the r_{01} is tightly correlated with age, we construct another asteroseismic diagram shown in Fig. 1b, based on the values obtained from the above computation. Interestingly, the age of star can be marked in this asteroseismic diagram Fig. 1b. It is convenient to obtain the stellar important parameters: the mass and the age.

3. Discussion

1. The ($\langle\Delta\nu\rangle$, r_{01}) diagram as a new asteroseismic diagnostic tool can estimate the mass and the age of solar-like stars. The virtue is that the age of stars can be marked in the diagram, so we can obtain the mass and age directly.

2. We will discuss the effects of the assumed initial abundance of helium and the mixing length parameter on the asteroseismic diagram in future work.

Acknowledgements

This work was supported by The Ministry of Science and Technology of the People's republic of China through grant 2007CB815406, and by NSFC grants 10173021, 10433030, 10773003, and 10778601.

References

Bi, S. L., Basu, Sarbani & Li, L. H. 2008, ApJ 673, 1093
Christensen-Dalsgaard, J. 1988, In Proc. IAU Symposium No 123, Advances in helio- and asteroseismology, p. 295–298, eds Christensen-Dalsgaard, J. & Frandsen, S., Reidel, Dordrecht
Gough, D. O. 2003, AP&SS, 284, 165
Guenther, D. B., et al. 1992, ApJ 387, 372
Guenther, D. B. 1994, ApJ 422, 400
Tang, Y. K., Bi, S. L., & Gai, N. 2008a, New Astronomy 13, 541
Tang, Y. K., Bi, S. L., & Gai, N. 2008b, ChJAA, in press
Tassoul, M. 1980, ApJS 43, 469

The evolution of 'the moment of inertia' of stars

Y.-C. Kim[1]† and S. Barnes[2]

[1] Astronomy Department, Yonsei University, Seoul, 120-749, Korea
email: yckim@yonsei.ac.kr
[2] Lowell Observatory, Flagstaff, AZ 86001, USA
email: barnes@lowell.edu

Abstract. Observations of the rotation periods of cool open cluster stars display a distinctive dichotomy when plotted against stellar mass/color. Other measures of stellar activity are also known to be dependent on stellar mass and structure, especially the onset and characteristics of convection zones. One proposal for understanding the observed rotation period dichotomy suggested dependencies on the moment of inertia of either the whole star or that of only the outer convection zone (Barnes 2003).

The moment of inertia of stars with the mass between 0.1Msun and 3.0Msun have been calculated using a version of Yale Stellar evolution code (aka YREC). Each star has been evolved from stellar birthline to the onset of the core He burning. For easy comparison to observations, we have calculated the isochrones of these quantities as well as the convective turnover time, of interest to the activity community.

Keywords. stars: interiors stars: evolution

1. Introduction

The stellar rotation rate is the prime parameter determining the level of stellar magnetic activity. The activity level responds to changes in the stellar rotation rate that are the result of the loss of angular momentum through the magnetized stellar wind, and of evolutionary changes in the moment of inertia; both of these effects may also affect the internal differential rotation. Because of the evolutionary changes in the stellar interior, reflected in radial expansion or contraction and associated changes in density, the moment of inertia of the star changes with time. The magnetic brake applies directly to the top of the convective envelope. Therefore, it is of interest to study the moment of inertia of radiative interior and of the convective envelope separately.

The historical impetus for the belief in a connection between stellar structure and rotation-activity goes back to the work of Schatzman (1962) and Kraft (1967). This work suggested that the presence (absence) of a surface convection zone is the determinant of whether (or not) stars can spin down over time. Using rotation periods rather than the $v \sin i$ measurements, we can test whether this point in the theoretical models does indeed coincide exactly with the change in rotation. Consequently, we have been working on how to relate the best models of cool stars we can generate to the best rotation-activity datasets we can assemble. This paper is about the modeling component of this project. A comparison of the observation with these models will be presented elsewhere (Barnes & Kim 2008 in preparation).

† Visiting Astronomer, Lowell Observatory, Flagstaff, AZ 86001, USA

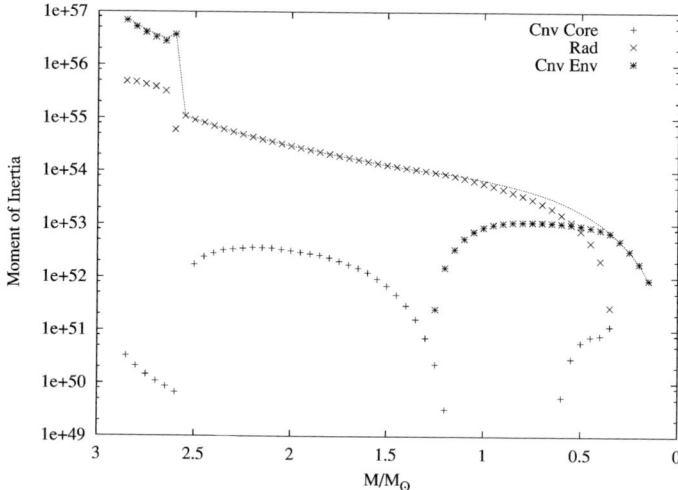

Figure 1. The isochrone of the moment of inertia for $500Myr$

2. Computation

All models used in this paper have been constructed using a version of the Yale Rotating Stellar Evolution Code (YREC; Demarque *et al.* 2007). A series of stellar models with masses ranging from 0.1 to $3.0M_\odot$ have been evolved from the stellar birth line (Palla & Stahler 1993), to the onset of the core Helium burning. Along the evolution, the moments of inertia are computed for the radiative interior and for the convective envelope and for the convective core separately, as well as the total value. Also, the convective turnover timescale has been calculated within the mixing length scheme. And the 500Myr isochrone is shown in Figure 1.

3. Summary

The moment of inertia of radiative interior and of the convective envelope as well as the total moment of inertia of a star have been computed along evolutionary tracks. Other characteristic parameters of the convection, for example the convective turnover time scale, have been calculated. This series of computations form the database, to be compared with observation to study a connection between stellar structure and rotation-activity.

Acknowledgements

YCK is supported by the Astrophysical Center for the Structure and Evolution of the Cosmos (ARCSEC) of Korea Science and Engineering Foundation (KOSEF) through the Science Research Center (SRC) Program.

References

Barnes, S. A. 2003, *ApJ* 586, 464
Demarque, P., Guenther, D. B., Li, L. H., Mazumdar, A. & Straka, C. W. 2007, *Ap&SS* e, 447
Kraft, R. P. 1967, *ApJ* 150, 551
Palla, F. & Stahler, S. W. 1993, *ApJ* 418, 414
Schatzman, E. 1962, *Annales d'Astrophysique* 25, 18

The spectral properties of a large sample of low surface brightness disk galaxies

Y. C. Liang[1]†, G. H. Zhong[1,2], L. C. Deng[1] and B. Zhang[2,1]

[1] National Astronomical Observatories, Chinese Academy of Sciences, Beijing, 100012, China

[2] Department of Physics, Hebei Normal University, Shijiazhuang 050016, China

Abstract. We present the spectral properties of a large sample of nearly face-on low surface brightness (LSB) disk galaxies selected from the SDSS-DR4 main galaxy sample. About 12,282 LSB galaxies have been selected from the photometry database with their B-band central surface brightness $\mu_0(B)$ ranging from 22 to 24.5 mag arcsec^{-2}. About 7000 of such LSBGs have measured emission lines ([OII]3727, [OIII]5007, Hβ, Hα, [NII]6583) with the S/N ratio greater than 5σ. Their spectral diagnostic diagram of [NII]/Hα vs. [OIII]/Hβ shows that $\sim$89% of them are star-forming galaxies, and $\sim$11% could be classified as AGNs. The relations of $\mu_0(B)$ vs. 12+log(O/H) and $\mu_0(B)$ vs. stellar masses M_* of these star-forming LSB galaxies show that their O/H and M_* increase following the increasing $\mu_0(B)$. The majority of these LSBGs are on the higher branch of metallicity.

Keywords. Galaxies: abundances, galaxies: evolution, galaxies: fundamental parameters, galaxies: spiral, galaxies: starburst

1. Introduction

The first quantitative suggestion about low surface brightness galaxies (LSBGs) is the so-called Freeman's law. It was noticed that the central surface brightness of 28 out of 36 disk galaxies of Freeman (1970) fell in the range of $\mu_0(B)$=21.65$\pm$0.3 mag arcsec^{-2}. Some consequent surveys were searching for more LSBGs. However, the detected numbers were only several hundreds up to Impey *et al.* (1996) since the LSBGs emit much less light per area than the normal galaxies which make them to be found difficultly. The modern digital sky survey, such as the Sloan Digital Sky Survey (SDSS), must greatly improve the detected numbers of LSBGs, and could provide much more interests of them by taking advantage of the high quality photometric and spectral observations. We have selected a large sample of LSBGs from the main galaxy sample of SDSS-DR4 (12,282, named as Sample-L, see Zhong *et al.* in this proceedings). Some of their property parameters and correlations from photometry have been presented, as well as the hints for stellar populations from color-color diagrams. In this work, we will present the spectral properties of these LSBGs.

2. The diagnostic diagrams

About 7000 galaxies among the Sample-L were obtained their optical spectra by the SDSS with high quality, i.e., their [OII]3727, [OIII]5007, Hβ, Hα and [NII]6583 were detected with a signal-to-noise (S/N) ratio greater than 5σ. The distribution of them in the BPT diagram (Baldwin, Phillips, Terlevich 1981) was given in the first panel of Fig.1. The diagnostic diagram of [NII]/Hα vs. [OIII]/Hβ show that $\sim$89% of them are star- forming galaxies, and $\sim$11% could be classified as AGNs by using the diagnostic

† email: ycliang@bao.ac.cn

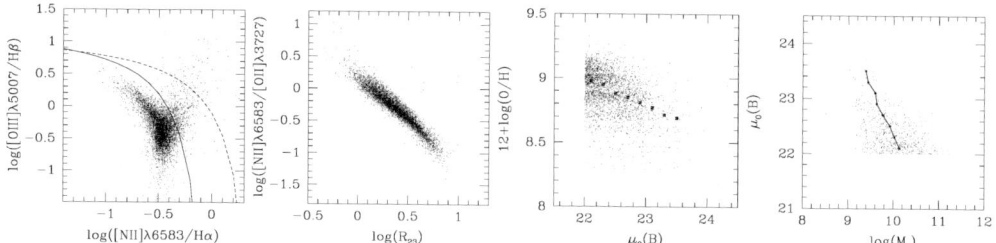

Figure 1. The properties of the large sample of LSBGs with high quality spectral observations selected from the main galaxy sample of SDSS-DR4: (a) [NII]/Hα vs. [OIII]/Hβ diagnostic diagram; (b) [NII]6583/[OII]3727 was plotted against the abundance indicating line ratio R_{23}; (c) The relations of $\mu_0(B)$ versus 12+log(O/H); (d) The relations of $\mu_0(B)$ versus stellar masses.

line from Kauffmann et al. (2003, the solid). The dashed line was taken from Kewley et al. (2001) to give the upper limit of the diagnostic relation.

In the second panel of Fig.1, [NII]6583/[OII]3727 was plotted against the abundance indicating line ratio R_{23} (=([OII]+[OIII])/Hβ) (McGaugh 1994). The transition through R_{23} turnover region for O/H occurs around log([NII]/[OII])~-1, so the majority of these LSBGs are on the higher branch of metallicity.

3. The relations of $\mu_0(B)$ versus O/H and $\mu_0(B)$ versus M_*

In the third panel of Fig.1, we present the relations of $\mu_0(B)$ vs. 12+log(O/H) of the star-forming LSBGs. It shows that the galaxies with larger $\mu_0(B)$ generally have higher metallicities. In the last panel of Fig.1, the relations of $\mu_0(B)$ vs. stellar masses M_* of the star-forming LSBGs are given, which show that their M_* increase following the increasing $\mu_0(B)$ generally. The lines in the two panels refer to the median values in the bins of 0.2 dex in $\mu_0(B)$.

4. Conclusions

We select a large sample of nearly face-on disk galaxies from the SDSS-DR4 main galaxy sample. Most of them are star-forming galaxies with strong emission lines. The majority of them are on the upper branch of metallicity with 12+log(O/H)>8.4 (logR_{23} < 0.8), i.e. not much metal-poor LSB galaxies were detected and obtained their spectra even in the SDSS survey. The LSBGs with higher $\mu_0(B)$ have higher metallicities and are more massive generally.

Acknowledgements

We thank the NSFC grant support under Nos. 10403006, 10433010, 10673002, 10573022, 10333060, 10521001, and the National Basic Research Program of China (973 Program) Nos. 2007CB815404, 06.

References

Baldwin, J. A., Phillips, M. M., & Terlevich, R. 1981, *PASP* 93, 5
Impey, C., Sprayberry, D., Irwin, M. K. et al. 1996, *ApJS* 105, 209
Kauffmann, G., Heckman, T. M., Tremonti, C. A., et al. 2003, *MNRAS* 346, 1055
Kewley, L. J., Heisler, C. A., Dopita, M. A. et al. 2001, *ApJS* 132, 37
McGaugh, S. S. 1994, *ApJ* 426, 135
Zhong, G. H., Liang, Y. C., Deng, L. C., & Zhang, B. 2008, *this proceedings*

Observations of clusters with ages from 0.01 to 1.0 Gyr in the Large Magellanic Cloud

Q. Liu[1,2]†, R. de Grijs[3,1], L. C. Deng[1], Y. Hu[1,2] and I. Baraffe[4]

[1]National Astronomical Observatories, Chinese Academy of Sciences, Beijing 100012, P. R. China
email:liuq@bao.ac.cn
[2]Graduate University of Chinese Academy of Sciences, Beijing, 100049, P. R. China
[3]Department of Physics & Astronomy,The University of Sheffield, Sheffield S3 7RH,UK
[4]CRAL, École Normale Supérieure, 46 allée d'Italie, 69007 Lyon, France

Abstract. The stellar initial mass function (IMF) is a very important question in modern astrophysics. Globular clusters (GCs) are good samples for studying the IMF, but the Galactic GCs can provide only one time-scale evolutionary stage. The Large Magellanic Cloud (LMC) is an ideal environment for studying the IMF because it contains compact clusters at different evolutionary stages. By studying the IMF at different evolutionary stages, we can see how the mass function evolves with time.

Keywords. Stars: mass function, Large Magellanic Cloud, galaxy: star clusters

1. Introduction

The stellar IMF plays an important role in many astronomical questions. GCs are perfect targets for the study of the IMF, because all stars in a GC have the same age, metallicity and distance from us. Unfortunately, Galactic GCs can only provide evolutionary information on one time scale due to their old ages. LMC is an ideal environment for studying the IMF of star clusters, because the LMC contains a large population of rich compact star clusters covering ages from 0.001 to 10 Gyr, which makes it possible to study star clusters at various evolutionary stages.

2. Observations and data reduction

The supreme spatial resolution of the Hubble Space Telescope (HST) makes it possible to observe stellar systems in extragalactic environments. HST GO program 7307 obtained WFPC2 and STIS imaging observations of 3 pairs of compact star clusters in the LMC. Two clusters in each pair have similar ages, metallicities and distances from the center of the LMC. We used the IRAF/APPHOT package to do aperture photometry and adopted 2-pixel aperture radii on both the WFPC2 and STIS images, then we adopted the method of Holtzman *et al.* (1995) to convert the aperture-corrected WFPC2 photometry to the standard Johnson-Cousins V, I photometry. We take the cluster NGC1818 in the youngest pair as an example in this paper.

3. Model and Results

Many faint stars are pre-main sequence (PMS) stars, and their ages are obtained by fitting the lower part of the clusters MS with Padova isochrones (Girardi *et al.* 2000)

† Present address: No. 20A Datun Road, Chaoyang District Beijing 100012, China

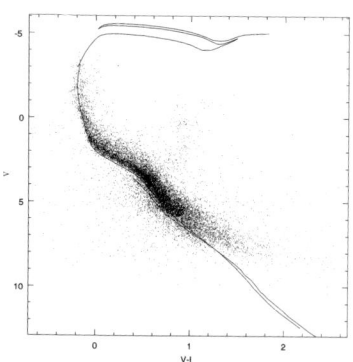
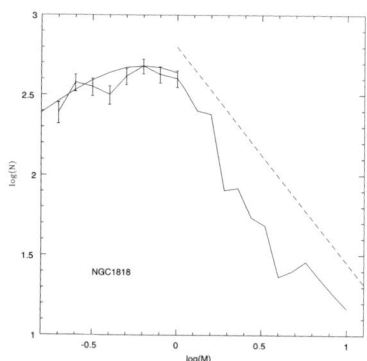

Figure 1. Age and metallicity of NGC1818.   Figure 2. The complete MF of NGC1818.

and the evolutionary models of Baraffe *et al.* (1998). The best fit is shown in Fig1. The red line is the Padova isochrone of $\log(\text{age/yr}) = 7.65$ and $Z = 0.008$. There is no corresponding model of the same metallicity in Baraffe *et al.* (1998), so we did some tests to prove that the differences are negligible. Hence, we adopt the model of $Z = 0.006$ to calculate the masses of stars. The green line in Fig1 is the improved Baraffe isochrone of $\log(\text{age/yr}) = 7.65$ and $Z = 0.006$. Fig. 2 shows the mass function of NGC1818 above 0.2 $M_\odot$.

4. Conclusions

The IMF is usually assumed to be universal. For NGC 1818, the MF above 1.0 $M_\odot$ follows the power-law distribution with slope -1.35 (de Grijs *et al.* 2002, black line in Fig. 2), and the green dashed line shows the Salpeter IMF (1955) within this mass range. We tried to use a log-normal distribution function (green solid curve in Fig. 2) to fit the MF between 0.2–1.0 $M_\odot$ (red line in Fig2). The result is identical to the IMF of Galactic GCs (Figure 3 in Paresce *et al.* 2000), so we think that the IMF below 1.0 $M_\odot$ in NGC 1818 may follow a log-normal distribution.

Acknowledgements

We would like to acknowledge the useful comments of a referee concerning the solution procedure used in §4. A.N.O. is supported by SERC under grant number GR/F/12345.

References

Baraffe, I., Chabrier, G., Allard, F., & Hauschildt, P. H. 1998, *A&A* 337, 403
de Grijs, R., Gilmore, G. F., Johnson, R. A., & Mackey, A. D 2002, *MNRAS* 331, 245
Girardi, L., Bressan, A., Bertelli, G., & Chiosi, C. 2000, *A&As* 141, 371
Holtzman, J. A., Burrows, C. J., Casertano, S., Hester, J. J., Trauger, J. T., Watson, A. M., & Worthey, G. 1995, *PASP* 107, 1065
Paresce, F. & De Marchi, G. 2000, *ApJ* 534, 870
Salpeter, E. E. 1955, *AJ* 101, 1865

Asteroseismic constraints on the OPAL opacity interpolation

W. M. Yang and M. Li

Department of Physics and Chemistry, Henan Polytechnic University, Jiaozuo 454000, China
email: yang.wuming@yahoo.com.cn

Abstract. There is a difference of a few Kelvins in the effective temperature between a model used only two-point interpolation of opacity and a model used piecewise linear interpolation of opacity. However the frequency difference between the models is of the order of several microHertz at a certain stage, which is almost 10 times worse than the observational precision of p-modes of solar-like stars. Therefore, the two-point interpolation of opacity is unsuitable in modelling of solar-like stars with element diffusion.

Keywords. stars: evolution, diffusion, stars: oscillations

Timescales of evolution and element diffusion are similar in solar-like stars with mass 1.0–1.5 $M_\odot$ (Turcotte *et al.* 1998). Thus element diffusion should be calculated in modelling of solar-like stars. Furthermore, helioseismology has confirmed the importance of including element diffusion and settling in solar modelling: models including effects of element diffusion and settling are better agreement with the seismically inferred sound-speed and density profiles than are models that ignored the effects (Christensen-Dalsgaard *et al.* 1993; Basu *et al.* 2000).

Heavy-element abundance, Z, is constant in main-sequence (MS) stage of solar-like stars without considering metal settling, thus evolution of these stars need only one set of opacity tables at a fixed Z_1. However the Z is a variable in the models with metal settling and then the second set of opacity tables must be obtained at a fixed Z_2. Opacity at the desired Z, X, T, and ρ can be obtained by two-point interpolation, i.e.,

$$\kappa(Z, X, T, \rho) = \kappa_2(Z_2, X, T, \rho) + \frac{\kappa_1(Z_1, X, T, \rho) - \kappa_2(Z_2, X, T, \rho)}{Z_1 - Z_2} \cdot (Z - Z_2), \quad (0.1)$$

The value of Z_1 and Z_2 is chosen at a suitable value, respectively.

In order to study the impact of opacity interpolation, using the Yale Rotating Evolution Code (YREC7) in its nonrotating configuration, we construct four models listed in Table 1. All parameters of the models are same except the parameters of opacity interpolation. All models evolve from pre-main sequence (PMS) to somewhere near the end of the MS. The OPAL eos (Rogers & Nayfonov 2002), OPAL opacity (Iglesias & Rogers 1996), and the Alexander & Ferguson(1994) opacity for low temperature are used. Element diffusion is implemented following the prescription of Thoul *et al.* (1994).

The changes in the effective temperature between our models at the same age are several Kelvin, which is within the error of observation of stellar effective temperature. The frequency differences between MT1 and MT2 are zero. In Fig. 1A, we represent the frequency differences between MT2 and MT3 at the same age. The difference increases, however, from about 1 μHz at the age of 1 Gyr to around 4 μHz at the age of 6 Gyr. But the difference at the age of 7 Gyr is less than that at the age of 6 Gyr. The frequency differences between MM1 and MT3 are shown in Fig. 1B. The frequency differences arrive at a maximum at the age of about 3 Gyr. Then with increase in age,

Table 1. Model parameters.

Model	Mass $M_\odot$	Z_0	Two-point interpolation $Z_1 \,---\, Z_2$	Piecewise linear interpolation $\delta z = Z_{i+1} - Z_i$
MT1	1.10	0.022	0.022 — 0.021	....
MT2	1.10	0.022	0.022 — 0.020	....
MT3	1.10	0.022	0.022 — 0.019	....
MM1	1.10	0.022	....	0.001

Notes: The Z_0 is the initial metal abundance. Piecewise linear interpolation: if $Z_i \leqslant Z < Z_{i+1}$, YREC will interpolate between $\kappa(Z_i)$ and $\kappa(Z_{i+1})$ to obtain the opacity at the required Z.

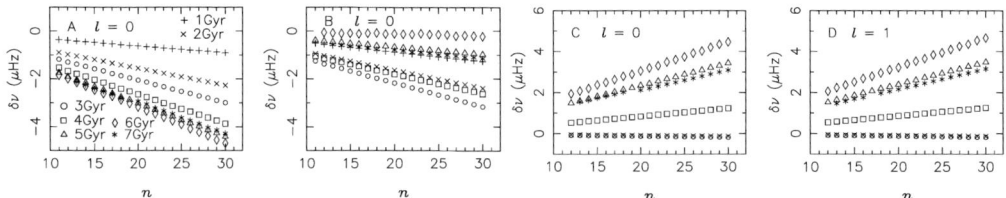

Figure 1. Frequency differences at different ages labeled by the different symbols. A: ν_{MT2}-ν_{MT3}; B: ν_{MM1}-ν_{MT3}; C and D: ν_{MM1}-ν_{MT2}.

the differences decrease. At the age of around 6 Gyr, the differences are almost zero. Thus the discrepancy between MM1 and MT3 mainly occurs between the ages of 2 and 4 Gyr. In Fig. 1C and D, we represent the frequency differences between MM1 and MT2. The differences are almost zero when the age of the models is less than 3 Gyr. Then with increase in age, the frequency differences increase. At the age of around 6 Gyr, the differences arrive at a maximum. The differences at the age of 4–7 Gyr are of the order of several microHertz, which is almost 10 times larger than the uncertainty of observation of stellar p-modes that is expected to reach 0.1–0.4 μHz (Thèado *et al.* 2005; Bedding *et al.* 2004).

Heavy-element abundance is a constant in the models of PMS, and the Z_s and Z_c are close to the Z_0 in the early evolutionary stage of MS. Thus Z_1 is specified to be Z_0 in models MT1, MT2, and MT3, and one of Z_i should be equal to Z_0 in MM1. The differences of the opacity interpolation in MT2, MT3, and MM1 should result in the difference between MT2, MT3, and MM1 and then lead the frequency difference between MT2, MT3, and MM1. The frequency difference between the models MT2, MT3, and MM1 is of the order of several microHertz at a certain phase, which is almost 10 times worse than the observational precision of p-modes of solar-like stars. Consequently, in modelling of solar-like star with metal settling, the two-point interpolation of opacity is unsuitable, and at least piecewise linear interpolation is required.

References

Alexander, D. R. & Ferguson, J. W. 1994, *ApJ*, 437, 846
Basu, S., Pinsonneault, M. H., & Bahcall, J. N. 2000, *ApJ*, 529, 1084
Bedding, T. R., Kjeldsen, H., Butler, P. R., *et al.* 2004, *ApJ*, 614, 380
Christensen-Dalsgaard, J., Proffitt, C. R. & Thompson, M. J. 1993, *ApJ*, 403, L75
Iglesias, C. & Rogers, F. J., 1996, *ApJ*, 464, 943
Rogers, F. J. & Nayfonov, A, 2002, *ApJ*, 576,1064
Thado, S., Vauclair, S., Castro, M., Charpinet, S., & Dolez, N. 2005, *A&A*, 437, 553
Thoul, A. A., Bahcall, J. N., & Loeb, A. 1994, *ApJ*, 421, 828
Turcotte, S., Richer, J., & Michard, G. 1998, *ApJ*, 504, 559

The electric currents from viscosity in differentially moved plasma

Zhiliang Yang[1] and Hongbin Wang[1]

[1] Astronomy Department, Beijing Normal University, Beijing 100875, China
email: zlyang@bnu.edu.cn

Abstract. From the two-fluid plasma theory, we derived a kind of current in differentially moved plasma. The current is from the different viscosities between electrons and ions. The higher temperature and lighter mass of the electrons make the viscosity of electrons much stronger than ions. In this way, the electrons will have smaller velocity than ions in the differentially moved layer and contribute a net current in the plasma. The value of the current depends on the temperature and density of electrons in the plasma and the differential velocity $\nabla^2 v$.

Keywords. Sun, Plasma, Electric Current.

1. Two-fluid Models of plasma

The two-fluid plasma equations with viscosity included are given by :

$$\frac{\partial n_\alpha}{\partial t} + \nabla \cdot n_\alpha \mathbf{v}_\alpha = 0 \tag{1.1}$$

$$\frac{d\mathbf{v}_\alpha}{dt} = \frac{q_\alpha}{m_\alpha}(\mathbf{E} + \frac{1}{c}\mathbf{v}_\alpha \times \mathbf{B}) - \frac{\nabla P_\alpha}{n_\alpha m_\alpha} + \frac{\mathbf{M}_{\alpha\beta}}{n_\alpha m_\alpha} - \frac{\mu_\alpha \nabla^2 \mathbf{v}_\alpha}{n_\alpha m_\alpha} \tag{1.2}$$

where $\mathbf{v}_\alpha$, $\mathbf{v}_\beta$ refers to electrons and/or protons (e,i), $\mu_\alpha \nabla^2 \mathbf{v}_\alpha$ is the viscous force acted on protons or electrons, $\mathbf{M}_{\alpha\beta}$ is the collision between electrons and ions, and the electron charge is $q_e = -e$ in the proton and electron plasma.

In the steady case, $\frac{\partial B}{\partial t} = 0$, that is the same as $E = 0$ in plasma, with the Hall current and Bierman's battery ignored, we can have:

$$\mathbf{j} = \frac{c^2}{4\pi\eta}(\frac{1}{c}\mathbf{v}_i \times \mathbf{B} - \frac{m_e}{nem_i}\mu_i \nabla^2 \mathbf{v}_i + \frac{1}{ne}\mu_e \nabla^2 \mathbf{v}_e) \tag{1.3}$$

The term $\frac{cm_e}{nem_i}\mu_i \nabla^2 \mathbf{v}_i - \frac{c}{ne}\mu_e \nabla^2 \mathbf{v}_e$ is the current from the viscosity effect. In the differential moved plasma, the ions and electrons are supposed to have the same shear velocity gradients. This may cause a net current in the shearing layer and generate magnetic field.

The viscosity for electron and proton in plasma have the form (Braginskii 1957, Spitzer, 1962):

$$\mu_i = 2.21 \times 10^{-15} \frac{T_i^{5/2} A_i^{1/2}}{Z^4 ln\Lambda} (\frac{gm}{cmsec}) \tag{1.4}$$

$$\mu_e = 2.21 \times 10^{-15} \frac{T_e^{5/2} A_e^{1/2}}{Z^4 ln\Lambda} (\frac{gm}{cmsec}) \tag{1.5}$$

where A_i, A_e is the atomic weight of electrons and positive ions, $A_i = 1$ and $A_e = 1/1836$, Z is the ionic charge, Te and T_i are the temperature of electron and ion in K.

The currents which are perpendicular to and parallel to the plasma velocity are,

$$\mathbf{j}_\perp = \frac{c}{4\pi\eta} \mathbf{v} \times \mathbf{B} \tag{1.6}$$

and

$$\mathbf{j}_\parallel = \frac{c^2 m_e}{4\pi e} P_{me} \nabla^2 \mathbf{v} \tag{1.7}$$

The parameter P_{me} is the magnetic Prandtl number of electrons,

$$P_{me} = \frac{\nu_e}{\eta} = 5.66 \times 10^{-3} T_e^{5/2} T^{3/2}/(n \ln \Lambda) \tag{1.8}$$

where T_e is the temperature of electrons in K and n is the electron/ion density in cm^{-3}. $\ln \Lambda$ is usually between $5 - -20$.

2. The Current in Coronal Loops

The current parallel to the velocity of plasma depends on the sheared velocity $\nabla^2 \mathbf{v}$, and the magnetic Prandtl number of electrons.

$$\mathbf{j}_\parallel = 0.136 P_{me} \nabla^2 \mathbf{v} (A/m^2) \tag{2.1}$$

with the velocity in unit of m/s. And the magnetic Prandtl number of electrons depends strongly on the temperature and density of electrons.

There is a lot of theoretical models and observations for the density and temperature of electrons in the coronal loops (Robb & Cally 1992; Spadaro et al. 2003). The electron temperature in the loops is about $10^6 K$ and the electron density is about $10^{10} cm^{-3}$. Supposing the plasma temperature in the loops is $10^6 K$, we can get the magnetic Prandtl number in the coronal loops is $P_{me} = 10^{11}$.

In a recent observation, Lin (Lin et al, 2006) shows that the velocity difference in a field of $13''$ is from $230 km/s$ to $320 km/s$. We can estimate $\nabla^2 \mathbf{v}$ as $\delta v/L^2$. δv is the velocity fluctuation in a field with scale L. It is about $2.25 \times 10^{-13} m^{-1} s^{-1}$. The total current in this area will be $2.7 \times 10^2 P_{me} A$. If we choose P_{me} to be 10^{10}, the current in a coronal loop will be $10^{12} A$.

Acknowledgements

The research is supported by The National Fund of Sciences of China under grant number 10778614.

References

Braginskii, S.I. 1957, J. Exptl. Theoret. Phys. (U.S.S.R) 33 ,459
Lin, C., Banerjee, D., O'Shea, E., & Doyle, J.G. 2006, AA 450, 1181
Robb T. D. & Cally, P. S. 1992, ApJ 397, 329
Spadaro, D., Lanza, A. F., Lanzafame, A. C., et al. 2003, ApJ 582, 486
Spitzer, L. Jr. 1962, Physics of Fully Ionized Gases, Interscience Publishers, John Wiley & Sons, Inc.

Solar scandium abundance

H. W. Zhang[1,2], T. Gehren[2] and G. Zhao[3]

[1]Department of Astronomy, School of Physics, Peking University, Beijing 100871, P.R. China
email: zhw@bac.pku.edu.cn

[2]Institut für Astronomie und Astrophysik der Universität München, Scheinerstrasse 1, D-81679 München, Germany

[3]National Astronomical Observatories, Chinese Academy of Sciences, Beijing 100012, P.R. China

Abstract. We investigate the formation of neutral and singly ionized scandium lines in the solar photospheres. Extensive statistical equilibrium calculations were carried out for a model atom, which comprises 92 terms for Sc I and 79 for Sc II. Synthetic line profiles calculated from the level populations according to the NLTE departure coefficients were compared with the observed solar spectral atlas. Abundance determinations using the ODF model lead to a solar Sc abundance of between $\log \varepsilon_\odot = 3.07$ and 3.13, depending on the choice of f values.

Keywords. Sun: abundances, line: formation, line: profiles

1. Introduction

The solar photospheric abundances serve as a reference for abundance determinations in metal-poor stars, so a reliable set of photospheric abundances is important. Ever since Anders & Grevesse (1989) published their widely used solar elemental abundance table, many revisions and updates to photospheric and meteoritic abundances of the elements have become available, although the solar photospheric scandium abundance has not been updated for quite a long time. The photospheric abundance value of $\log \varepsilon_\odot(\mathrm{Sc}) = 3.10 \pm 0.09$ adopted by Grevesse (1984) was changed to 3.05 ± 0.08 by Youssef & Amer (1989). Neuforge (1993) obtained 3.14 ± 0.12 from the Sc I lines and 3.20 ± 0.07 from Sc II lines. The average value of 3.17 ± 0.10 was adopted by Grevesse & Noels (1993) and was kept in the newest tabular version of Grevesse et al. (2007), which is somewhat higher than the meteoritic value of 3.04 ± 0.04.

It should be noted that local thermodynamic equilibrium (LTE) has been assumed in previous papers about scandium abundance determinations, and NLTE investigation of the scandium element has never been published. In general, departures from LTE are commonplace and often quite important, in particular for low surface gravities or metallicities, with minority ions and low-excitation transitions the most vulnerable.

In this paper we investigate the statistical equilibrium and formation of neutral and singly-ionized scandium lines in the solar photosphere.

2. NLTE line formation calculations

Our atomic reference model is constructed from 92 and 79 terms for neutral and singly-ionized scandium, respectively. The number of bound-bound transitions treated in the NLTE calculations is 1104 for Sc I and 1034 for Sc II. For bound-free radiative transitions in the Sc atom, hydrogen-like photoionization cross-sections are adopted.

In our calculations for Sc, we take into account inelastic collisions with electrons and hydrogen atoms leading to both excitation and ionization. Because laboratory measurements and detailed quantum mechanical calculations for collision cross-sections are absent, approximate formulae are applied.

3. Analysis of scandium lines in the solar spectrum

We investigate the formation of Sc I and Sc II lines in the solar atmosphere and derive the scandium abundance based on spectrum synthesis. Lines in the solar spectrum are calculated using the plane-parallel hydrostatic MAFAGS-ODF solar model atmosphere with $T_{\rm eff} = 5780$ K, $\log g = 4.44$, [Fe/H] $= 0.00$, $\xi_t = 0.90$ km s^{-1}.

The observed solar flux spectrum was taken from the Kitt Peak Atlas (Kurucz et al. 1984). Spectrum synthesis was employed to determine the abundance of scandium in the solar atmosphere. For the solar abundance analysis we selected 4 Sc I and 17 Sc II lines, which ideally should satisfy the following conditions: they are relatively free from blends, and oscillator strengths and hyperfine splitting parameters are available. Line profiles are computed under both LTE and NLTE assumptions: fitted to the observed profiles by means of scandium abundance variations.

4. The solar scandium abundance

Two sets of oscillator strengths are applied and compared in our abundance determinations: (i) theoretical values taken from Kurucz' database (http://kurucz.harvard.edu), and (ii) experimental data of Lawler & Dakin (1989). Using the values obtained for $\log gf\varepsilon_\odot$ and the $\log gf$ values from different data sets, we computed Sc abundances for the individual lines.

Under the LTE assumption, absolute solar abundances determined from Sc I lines are significantly lower than the values obtained for the Sc II lines. The mean LTE abundances for 4 Sc I and 17 Sc II lines are 2.90 ± 0.09 dex and 3.10 ± 0.05 dex, respectively. This discrepancy of the ionization equilibrium is resolved in NLTE calculations. Under NLTE assumption, Sc I and Sc II lines give very consistent abundance results, i.e. 3.08 ± 0.05 dex and 3.07 ± 0.04 dex, respectively. Using Kurucz' gf values, the mean abundance for all 21 Sc lines under NLTE is 3.07 ± 0.04 dex. Using instead the laboratory $\log gf$ values of Lawler & Dakin (1989), the mean abundance of 17 Sc lines under NLTE is 3.13 ± 0.05 dex.

Our abundance result by using Kurucz' f-values lead to a solar Sc abundance well in agreement with the meteoritic value, whereas the experimental data of Lawler & Dakin deviate from that reference by nearly 2σ.

References

Anders, E. & Grevesse, N. 1989, *Geochim. Cosmochim. Acta*, 53, 197
Grevesse, N. 1984, *Phys.Scr.*, 8, 49
Grevesse, N. & Noels, A. 1993, in: N. Prantzos, E. Vangioni-Flam, & M. Casse (eds.), *Origin and Evolution of the Elements* (Cambridge: Cambridge Univ. Press), p. 15
Grevesse, N., Asplund, M., & Sauval, A. J. 2007, *Space Sci. Rev.*, 130, 105
Kurucz, R. L., Furenlid, I., Brault, J., et al. 1984, *Solar Flux Atlas from 296 to 1300nm*, Kitt Peak National Solar Observatory
Lawler, J. E. & Dakin, J. T. 1989, *J. Opt. Soc. Am. B*, 6, 1457
Neuforge, C. 1993, in: N. Prantzos, E. Vangioni-Flam, & M. Casse (eds.), *Origin and Evolution of the Elements* (Cambridge: Cambridge Univ. Press), p. 63
Youssef, N. H. & Amer, M. A. 1989, *A&A*, 220, 281

Radio Star Candidates from FIRST and 2MASS Databases

Yanxia Zhang and Yongheng Zhao

[1] National Astronomical Observatories, Chinese Academy of Sciences, Beijing 100012, China
email: zyx@bao.ac.cn; yzhao@lamost.org

Abstract. We positionally cross-identified FIRST (the Faint Images of the Radio Sky at Twenty centimeters) catalogue with 2MASS (the Two Micron All Sky Survey) pointed-source database and collected the data from radio band and near infrared band. Then the data were cross-matched with the Véron-Cetty & Véron 2006 catalog and the Tycho-2 catalog, respectively. Therefore the known samples of quasars and stars are obtained. We applied principal component analysis (PCA) on the known sample. The overall sample may be projected in the principal component space. From the space, we can easily locate the area that radio stars occupy, and select out radio star candidates. With the follow-up observation of these candidates, the properties of radio stars may be studied.

Keywords. techniques: miscellaneous; methods: statistical; methods: data analysis; astronomical data bases: miscellaneous

1. Data Sample and Chosen Attributes

We cross-matched the Two Micron All Sky Survey (2MASS) catalogue with the Faint Images of the Radio Sky at Twenty centimeters (FIRST) catalogue within 5 arcsec radius, and obtained 153135 entries with one to one matching between the FIRST and 2MASS catalogues. The entries were then cross-identified with the Véron-Cetty & Véron 2006 catalog and the Tycho-2 catalog within 5 arcsec radius, respectively. Similarly, we obtained 2389 quasars and 1353 stars from the 2MASS and FIRST catalogues. The chosen attributes from different bands are $logF$peak (Fpeak: peak flux density at 1.4 GHz), $logF$int (Fint: integrated flux density at 1.4 GHz), $f$maj (fitted major axis before deconvolution), $f$min (fitted minor axis before deconvolution), $f$pa (fitted position angle before deconvolution), $j - h$ (near infrared index), $h - k$ (near infrared index), $k + 2.5logF$int, $k + 2.5logF$peak, $j + 2.5logF$peak, $j + 2.5logF$int.

2. Method

Principal component analysis (PCA) is a statistical method that permits the determination of the minimum number of independent or uncorrelated variables underlying a larger number of observed variables. Thus, PCA is used as a technique for both data compression and analysis (Zhang & Zhao, 2003 and reference therein).

3. Result and Discussion

We applied PCA on the known sample, and plotted them in the principal component space shown in Fig. 1. In order to clearly to see the difference between stars and quasars, we gave the distribution of the first principal component (PC1) and the second principal

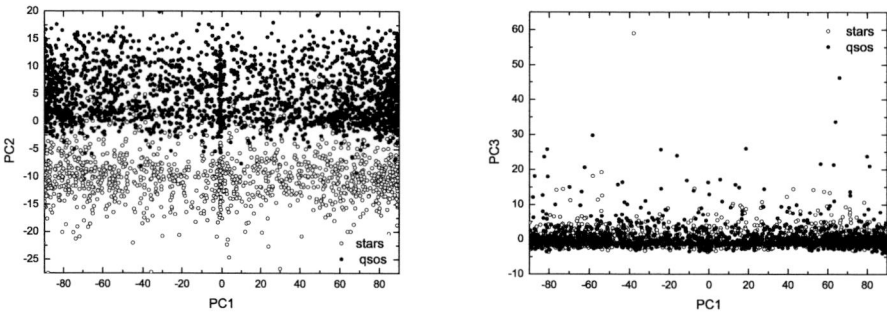

Figure 1. The distribution of the known stars (open circles) and quasars (filled circles) in the principal component space.

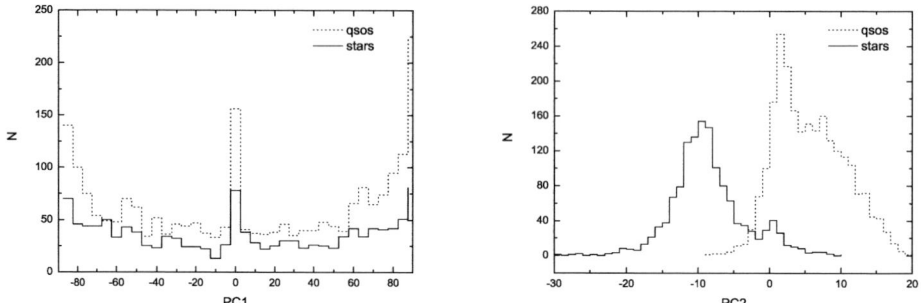

Figure 2. The distribution of the known stars and quasars as a function of PC1 (left panel) and PC2 (right panel), respectively.

component (PC2), as shown in Fig. 2. It is obvious from Fig. 1 and Fig. 2 that PC2 is the important factor to separate quasars from stars and has a good cutoff value, i.e. PC2= -2.6. As a result, the objects are identified as stars when PC2< -2.6, and as quasars when PC2> -2.6. The classification result is indicated in Table 1. The number of misclassified stars and quasars is 162 and 58, separately. The accuracy of stars and quasars adds up to 88.0% and 97.6%, respectively. The overall accuracy amounts to 94.1%. So PCA as unsupervised method is an effective approach used for the classification problem in astronomy. If we want to obtain radio star candidates from 2MASS and FIRST catalogues, we can project the overall sample into the principal component space, and select the sources whose PC2< -2.6 as radio star candidates, similarly choose those whose PC2> -2.6 as quasar candidates.

Acknowledgements

This paper is funded by National Natural Science Foundation of China under grants No.10473013, No.10778724 and No.90412016.

References

Zhang, Yanxia & Zhao, Yongheng 2003, *PASP* 115, 1006

A large sample of low surface brightness disk galaxies from SDSS

G. H. Zhong,[1,2]† Y. C. Liang,[1]‡ L. C. Deng[1] and B. Zhang[2,1]

[1]National Astronomical Observatories, Chinese Academy of Sciences, 20A Datun Road, Chaoyang District, Beijing 100012, China
[2]Department of Physicals, Hebei Normal University, Shijiazhuang 050016, China

Abstract. We present the properties of a large sample (12,282) of nearly face-on low surface brightness disk galaxies selected from the main galaxy sample of SDSS-DR4. Those properties includes B-band central surface brightness $\mu_0(B)$, scale lengths h, distances D, integrated magnitudes, colors and some resulted relations. This sample has $\mu_0(B)$ from 22 to 24.5 mag arcsec^{-2} with a median value of 22.44 mag arcsec^{-2}. They are quite bright with M_B taking values from -18 to -23 mag with a median value of -20.08 mag. The disk scale lengths h are from 2 kpc to 19 kpc. There exist clear correlations between $\log h$ and M_B, $\log h$ and $\log D$. Both the optical-optical and optical-NIR color-color relations show most of them have a mix of young and old stellar populations.

Keywords. Galaxies: distances and redshifts, galaxies: fundamental parameters, galaxies: photometry, galaxies: spiral, galaxies: stellar content

1. Introduction

Low surface brightness galaxies (LSBGs) are galaxies that emit much less light per area than normal galaxies. Yet, owing to their faintness compared with the night sky, they are hard to find. Hence their contribution to the local galaxy population has been underestimated for a long time. With the improvements of modern digital sky survey, such as the Sloan Digital Sky Survey (SDSS), the numbers of the detected LSBGs could be greatly improved. Therefore, we propose to search for a large sample of LSBGs from the SDSS database, which could help to understand the properties of LSBGs in details.

2. The Sample

The sample used in this work is selected from the main galaxy sample (Strauss *et al.* 2002) of SDSS-DR4 following the next steps:

(a) $fracDev_r < 0.25$: this requires the galaxy having an exponential light profile since the parameter $fracDev_r$ indicates the fraction of luminosity contributed by the de Vaucouleurs profile to exponential profile in the r-band.

(b) $b/a > 0.75$: this is selecting the nearly face-on galaxies, and a and b are the semi-major and semi-minor axes of the fitted exponential disk, respectively.

(c) $M_B < -18$ mag: keeping $M_B < -18$ in mind is to exclude few dwarf galaxies contained in our sample.

(d) $\mu_0(B) \geqslant 22.0$ mag arcsec^{-2}, which is following Impey *et al.* (2001) and is a bit lower than the first suggestion of 21.65 mag arcsec^{-2} by Freeman (1970) for LSBGs.

† E-mail: ghzhong@bao.ac.cn
‡ E-mail: ycliang@bao.ac.cn

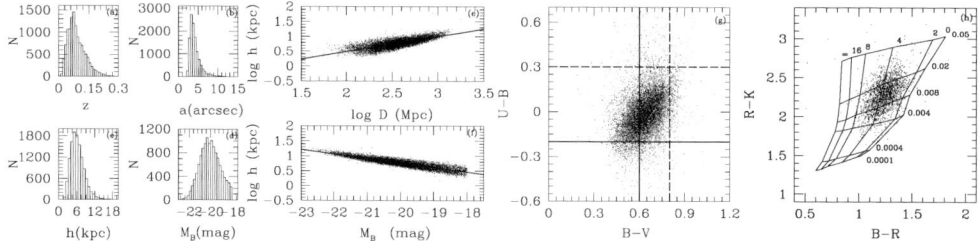

Figure 1. (a–d): histogram distributions for **Sample-L**: (a). the redshift, (b). the disk semi–major axis, (c). the disk radius, (d). the M_B; (e–f): correlations for **Sample-L**: (e). between $\log h$ and $\log D$, (f). between $\log h$ and M_B; (g). optical colur-color diagram of **Sample-L**; (h). optical-NIR color-color diagram of **Sample-L2**.

(e) Cross-correlation with 2MASS NIR photometry: after doing cross-correlation with 2MASS photometry, 1,878 galaxies are selected.

With step (a–d), we have selected a large sample of nearly face-on low surface brightness disk galaxies (12,282 galaxies, named **Sample-L**). Figure 1(a–d) show the histogram distributions of some property parameters of **Sample-L**. The sample selected by cross-correlation with 2MASS was named as **Sample-L2**.

3. Results and conclusions

In Fig. 1e, we show the correlation between $\log h$ and $\log D$ for **Sample-L**, which can be fitted by a least-square fit as: $\log h = 0.511(\pm 0.004) log D - 0.536(\pm 0.010)$, with standard deviation of 0.10 dex. This relation obviously show the selection effect of observations. Fig. 1f shows the correlation between M_B and $\log h$ for **Sample-L**. It can be fitted by a least-square fit as: $\log h = -0.150(\pm 0.0007) M_B - 2.245(\pm 0.014)$, with small standard deviation of 0.070 dex. This correlation suggests there are limits on both low luminosity and large scale length galaxies, otherwise such LSBGs are very rare (Impey *et al.* 2001).

Figure 1g shows $U - B$ vs. $B - V$ diagram of **Sample-L**, the solid and dashed lines are suggested by O'Neil *et al.* (1997a, b) to define the "very blue" and "very red" LSBGs. Figure 1f shows the trends in $B - R$ and $R - K$ optical-NIR diagram for **Sample-L2**. The overplots are the stellar population synthesis grids obtained by Bell *et al.* (2000) using the models of Bruzual & Charlot (2003), where the horizontal lines refer to the different metallicities and the vertical lines refer to the different characterized e-folding time scale of their star formation. Both the optical-optical and optical-NIR color-color relations show that most of the sample LSBGs have a mix of young and old stellar populations.

Acknowledgements

We thank the NSFC grant support under Nos. 10403006, 10673002,10573022 and the National Basic Research Program of China (973 Program) No.2007CB815404, 06.

References

Bell, E. F. & de Jong, R. S. 2000, *MNRAS* 312, 497
Bruzual, G. & Charlot, S. 2003, *MNRAS* 344, 1000
Freeman, K. C. 1970, *ApJ* 160, 811
Impey, C., Burkholdert, V., & Sprayberry, D. 2001, *AJ* 122, 2341
ONeil, K., Bothun, G. D., & Cornell, M. 1997a, *AJ* 113, 1212
ONeil, K., Bothun, G. D., Schombert, J., Cornell M. E., & Impey, C. D. 1997b, *AJ* 114, 2448
Strauss, M., *et al.* 2002, *AJ* 124, 1810

Researching turbulent convection models and the density gradient reversing

Q. S. Zhang[1,2] and Y. Li[1]

[1] National Astronomical Observatories/Yunnan Observatory, Chinese Academy of Sciences, PO Box 110, Kunming 650011, China
[2] Graduation School of Chinese Academy of Sciences, Beijing 100039, China
Email: mail_aoe_163@.com

Abstract. Turbulent convection models (TCM) provide a better way to study convection in stars than the MLT. Improving numerical method, we adopted larger diffusion parameters and smaller dissipation parameters in order to further correct the p-mode oscillation frequencies of TCM models. The density gradient reversing is discussed.

Keywords. convection, hydrodynamics, stars: interiors

1. TCM Solar models

Li and Yang (2007) tested the TCM, and found the p-mode oscillation frequencies of their TCM solar models are in better agreement with the observation than those of MLT models. It is expected that the p-mode oscillation frequencies of such models can be further improved by increasing diffusion parameters and decreasing dissipation parameters of the TCM they have adopted(Yang & Li, 2007). However, numerical problems precluded them from getting better model. After improving numerical method, we got better models. Table 1 shows solar models we calculated, they will be discussed in this paper.

2. The p-mode oscillation frequencies

Figures 1 and 2 show the frequency differences of p-mode oscillations with the spherical harmonic index $l = 3, 150$. It is obvious that differences of p-mode oscillations frequency of NLSM solar models are reduced more than 30% compared with those of SSM almost in all frequency range and for different l. Our results are better than those obtained by Yang and Li (2007) and so confirm the judgment that the p-mode oscillation frequencies of solar models with the TCM can be further improved by increasing diffusion parameters and decreasing dissipation parameters. Those results show the advantage of TCM.

Another important result is the appearance of density gradient reversing in our models, which is not found in the SSM. This result makes the relation of TCM's parameter and the density gradient reversing to be a more important and more interesting topic.

Table 1. The information of model parameters

Model	C_t	C_e	C_k	C_{t1}	C_{e1}	C_s	α	Y_0
NLSM(1)	0.25	0.10	2.50	0.11	0.20	0.05	0.05852	0.27509
NLSM(2)	0.20	0.10	2.50	0.15	0.25	0.10	0.05162	0.27509
NLSM(3)	3.00	1.25	2.50	0.0313	0.0313	0.0313	0.8966	0.27506
NLSM(4)	2.00	1.00	2.50	0.0277	0.0277	0.0277	0.6305	0.27506
NLSM(5)	1.70	0.85	2.50	0.0205	0.0205	0.0205	0.5339	0.27506
SSM	-	-	-	-	-	-	1.642	0.27506

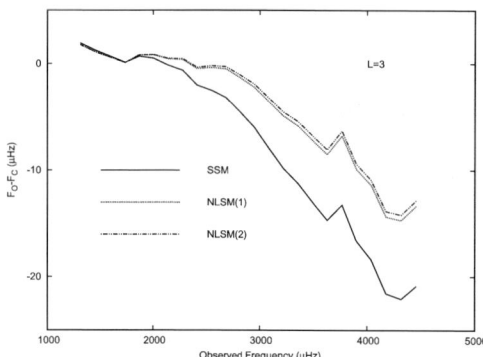

Figure 1. O-C of p-mode oscillation frequencies with $l = 3$.

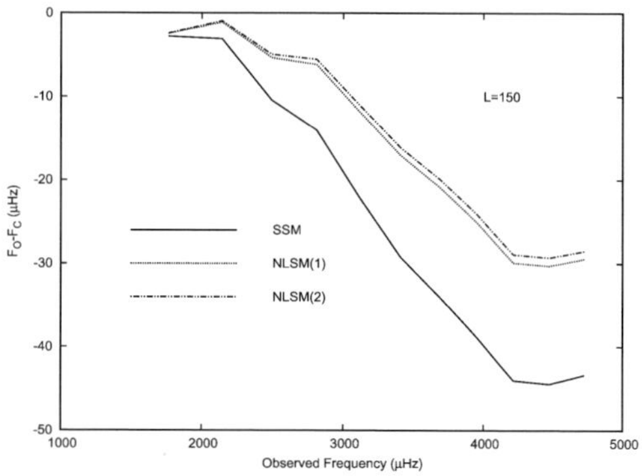

Figure 2. O-C of p-mode oscillation frequencies with $l = 150$.

3. Density gradient reversing

The density gradient reversing is impossible in radiative equilibrium zone for the maximum of gravitational potential. We found that density gradient reversed in the convection zone if $\nabla > \nabla_\rho = (\frac{\partial \ln T}{\partial \ln P})_\rho$ which is the criterion of it. Because $\nabla_\rho > \nabla_{ad}$, the reversing does in superadiabatic convection zone only if it does.

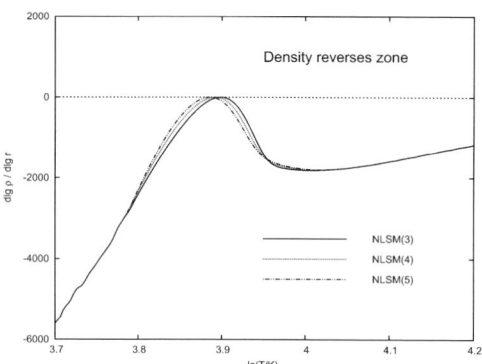
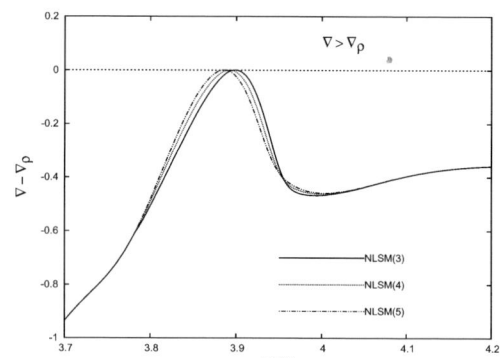

Figure 3. The density gradient of critical models in convection zone.

Figure 4. $\nabla - \nabla_\rho$ of critical models in convection zone.

We got some critical solar models(NLSM(3), NLSM(4) and NLSM(5), see table 1) of the density gradient reversing, Figure 4 shows their reversing density gradient, Figure 5 shows $\nabla_\rho > \nabla_{ad}$ of each model and validates the criterion. It can be found from table 1 that the density gradient reversing is dominated by the diffusion parameters C_{t1}, C_{e1} and C_s, because the ∇ increase if we enlarged those parameters(Li & Yang, 2007).

References

Li, Y. & Yang, J. Y., 2007, *MNRAS*, 375, 388
Yang, J. Y. & Li, Y., 2007, *MNRAS*, 375, 403

Helio- and asteroseismology

Jørgen Christensen-Dalsgaard[1]

[1]Danish AsteroSeismology Centre, and Department of Physics and Astronomy,
University of Aarhus, DK-8000 Aarhus C, Denmark
email: jcd@phys.au.dk

Abstract. Observations of solar and stellar oscillations are providing detailed information about stellar interiors. In the case of the Sun the set of observed frequencies is sufficiently detailed and accurate that the properties of the solar interior, such as sound speed, density and internal rotation, can be inferred with substantial precision and resolution. This allows detailed tests of solar modelling, with interesting and to some extent controversial results. Observations of solar-like oscillations in distant stars have started only recently, owing their very small amplitudes. However, developments in ground-based equipment and observations from space are revolutionizing this field, promising greatly increased insight into the structure and evolution of the stars.

Keywords. Sun: helioseismology – Sun: abundances – stars: oscillations – stars: evolution

1. Introduction

Most observations of stars provide only indirect information about the properties of stellar interiors. This is an obvious consequence of the fact, as noted by Eddington (1926), that the opaque nature of stellar material constitutes an 'impenetrable barrier', such that light from the star comes only from the stellar atmosphere. Thus our knowledge about the internal properties of stars are, to a large extent, based on theoretical modelling. However, the star can be penetrated by sound and gravity waves which therefore, if observed on the surface, carry information about the properties of stellar interiors. This forms the basis for helio- and asteroseismology. To study distant stars, and the more profound internal properties of the Sun, the most relevant observations are of frequencies of global modes; since these maintain phase coherence over several days, and in many cases much longer, the frequencies can be determined with very high accuracy. Also, they bear a relatively simply relationship to the properties of the stellar interior and hence provide rather direct constraints on those properties. In the solar case, in particular, the determination of a large number of accurate frequencies for a wide variety of modes allows inferences to be made of the solar internal structure and rotation with rather high resolution, in most of the Sun.

Here I provide a very brief overview of the properties of solar and stellar oscillations. More extensive presentations can be found, for example, in the book by Unno *et al.* (1989) and the review by Gough (1993). I then discuss some of the recent result on solar structure from helioseismology, including the consequences for solar modelling of the revised solar abundances. In addition, I provide an overview of the present and expected rapid development of asteroseismology, concentrating on the application to solar-like oscillations. Extensive reviews of helio- and asteroseismology were provided by Christensen-Dalsgaard (2002), Basu & Antia (2008), Cunha *et al.* (2007) and Aerts *et al.* (2008), while Christensen-Dalsgaard (2008a) discussed other aspects of asteroseismology.

2. Some properties of stellar oscillations

Stellar oscillations are characterized by the degree l and the azimuthal order m of the spherical harmonics which describe their dependence on position on the stellar surface; here l measures the total number of nodal lines on the surface and m, with $|m| \leqslant l$, the number of nodal lines in longitude. For each (l, m) there is a sequence of modes, characterized by the radial order n which, approximately, determines the number of nodes in the radial direction. Observations of distant stars, in light averaged over the stellar disk, are essentially sensitive only to modes of the lowest degree; for solar-like oscillations we only expect to detect modes of degrees up to 2–3. Solar observations with spatial resolution show oscillations with degrees exceeding 1000.

In the Sun and solar-like stars most observed modes are acoustic modes. An example of such a low-degree mode is illustrated in figure 1, showing that the eigenfunction extends to the core of the star. A simple analysis indicates that such acoustic modes are trapped between the near-surface region and an *inner turning point* at a distance r_t from the centre determined by

$$\frac{c(r_t)}{r_t} = \frac{\omega}{\sqrt{l(l+1)}}, \qquad (2.1)$$

where c is the adiabatic sound speed and ω is the angular frequency of the mode. The properties of this region largely determine the frequency of the mode. For $l = 0$ the modes extend essentially to the centre. Low-degree modes at the frequencies typically excited in solar-like oscillations have turning points in or near the core of the star and hence carry information about the structure of the core. With increasing degree r_t increases, and the high-degree modes observed in the Sun are trapped in the outer fraction of a per cent of the solar radius. This variation of penetration, and hence sensitivity, with degree allows the resolution of the solar internal properties through inverse analyses, given that acoustic modes over a broad range of degrees have been observed.

Low-degree acoustic modes satisfy an asymptotic relation that is very important for the interpretation of solar-like oscillations. This is normally written

$$\nu_{nl} \simeq \Delta\nu(n + l/2 + \alpha), \qquad (2.2)$$

where $\nu_{nl} = \omega_{nl}/(2\pi)$ is the cyclic frequency of a mode with radial order n and degree l, α is a quantity determined by the near-surface region of the star and

$$\Delta\nu = \left(2\int_0^R \frac{\mathrm{d}r}{c}\right)^{-1} \qquad (2.3)$$

is the inverse sound travel time across a stellar diameter. The resulting almost equally spaced pattern of frequencies is an important signature of solar-like oscillations (see figure 2 below), characterized by the *large frequency separation* $\Delta\nu_{nl} = \nu_{n-1\,l} - \nu_{nl}$. According to (2.2) the frequencies satisfy $\nu_{nl} \simeq \nu_{n-1\,l+2}$. Taking the asymptotic expansion to higher order lifts this degeneracy, yielding the *small frequency separation*

$$\delta\nu_{nl} = \nu_{nl} - \nu_{n-1\,l+2} \simeq -(4l+6)\frac{\Delta\nu}{4\pi^2\nu_{nl}}\int_0^R \frac{\mathrm{d}c}{\mathrm{d}r}\frac{\mathrm{d}r}{r}. \qquad (2.4)$$

This is predominantly sensitive to the properties of the sound speed in the core of the star, reflecting the different depth of the lower turning point in the two nearly degenerate modes.

For spherically symmetric stars the frequencies are independent of m. This degeneracy is lifted by rotation (or other departures from spherical symmetry). For acoustic modes

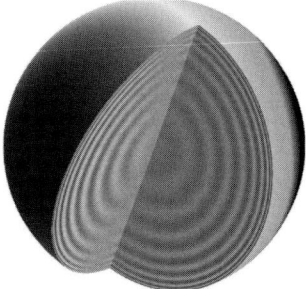

Figure 1. Properties of an eigenfunction in a solar model, for a mode of degree $l = 1$ and frequency $2700\,\mu$Hz. In the cut-out the greyscale indicates schematically the energy density of the mode.

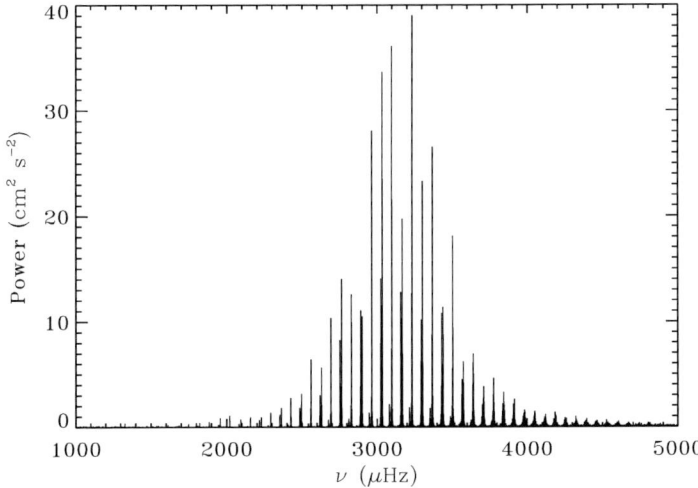

Figure 2. Power spectrum of integrated-disk velocity observations with the GOLF instrument on the SOHO spacecraft. The observations covered 805 days, starting 11 April 1996.

in a slowly rotating star the resulting frequencies are approximately given by

$$\omega_{nlm} \simeq \omega_{nl0} + m\langle\Omega\rangle \,, \tag{2.5}$$

where $\langle\Omega\rangle$ is a suitable average of the angular velocity Ω, approximately weighted by the energy density of the mode. In the solar case observations of a very large number of such rotational splittings have allowed detailed inferences of the internal rotation as a function of position (see Thompson et al. 2003, for a review).

Unlike the oscillations seen, for example, in Cepheids solar-like oscillations are characterized by being intrinsically damped and excited by the near-surface convection where motion at near-sonic speed provides an efficient source of acoustic noise. The resulting amplitudes are very small, typically well below $1\,\mathrm{m\,s^{-1}}$ in velocity and a few parts per million in intensity. Thus it is only through the development of very stable measurement techniques that the observation of these oscillations have become possible; in the case of stellar observations the push towards stable spectrographs to detect extra-solar planetary systems has been a major contribution to the recent successes in studying solar-like oscillations.

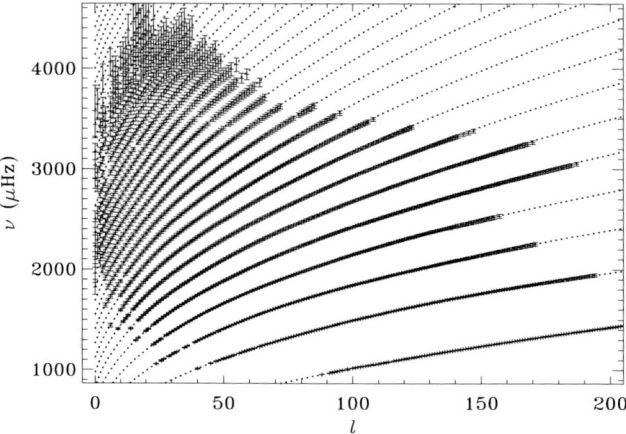

Figure 3. Inferred frequencies, as functions of degree, from 144 days of observations with the MDI instrument on the SOHO spacecraft; 1000σ error bars are indicated. The dotted curves show corresponding computed frequencies for Model S of Christensen-Dalsgaard et al. (1996).

3. Helioseismic results on solar structure

Very extensive data on solar oscillations have been obtained through several major projects, on the ground and on the SOHO spacecraft. Data on low-degree modes have been obtained by observing the Sun as a star, in particular in radial velocity with observations spanning more than three decades from the BiSON network (Chaplin et al. 2007a) and data from the GOLF instrument on SOHO (Gabriel et al. 1997). Spatially resolved velocity data have been obtained from the GONG network (Harvey et al. 1996) and the MDI instrument on SOHO (Scherrer et al. 1995). Figure 2 shows a power spectrum of observations from GOLF. This clearly reflects the asymptotic structure discussed in (2.2)–(2.4), with a dominant nearly uniform separation of the peaks and closely spaced pairs reflecting the small separation. Also, it is evident that the peaks are quite narrow, particularly at low frequency where the mode lifetime is at least several weeks.

The quality of the frequencies is illustrated in figure 3 which shows results from the MDI observations. Even though 1000σ error bars are shown, they are barely visible in large parts of the diagram, illustrating the extremely high accuracy of the frequencies. For a substantial fraction of the modes the relative frequency error is less than 5×10^{-6}.

A first step in the analysis of the observed frequencies is to compare with frequencies of solar models. As a typical, if slightly outdated, example I consider the so-called Model S of Christensen-Dalsgaard et al. (1996) which has seen widespread use in helioseismic analyses; further details of the computation were provided by Christensen-Dalsgaard (2008b). This uses detailed equation of state and opacity tables, and includes diffusion and settling of helium and heavy elements. The present surface ratio $Z_\mathrm{s}/X_\mathrm{s}$ between the abundances by mass of heavy elements and hydrogen was taken to be 0.0245 (Grevesse & Noels 1993). As is common in computations of solar models the model was adjusted to the correct surface radius, luminosity and composition by adjusting the initial composition and the mixing-length parameter describing the properties of near-surface convection. In figure 3 the dotted curves show the computed frequencies for this model. It is clear that the model reproduces the observed frequency structure and on this scale the differences between observations and model are barely discernible. In fact, close inspection shows a tendency for the computed frequencies to be slightly higher than the observations, particularly at high frequency. As discussed, for example, by Christensen-Dalsgaard et al. (1996) a detailed investigation of these frequency differences shows that they arise predominantly

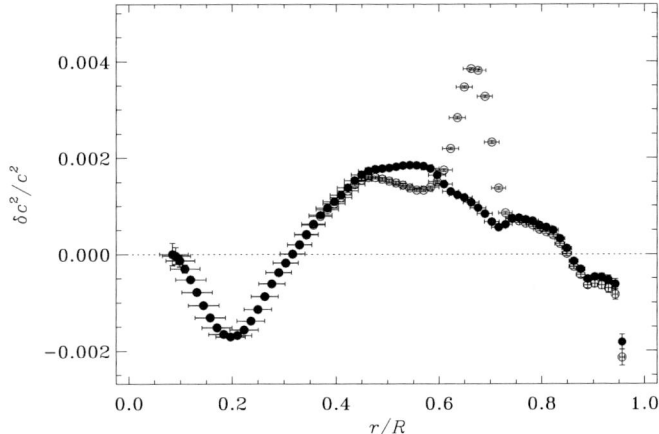

Figure 4. The open symbols show inferred relative differences in squared sound speed between the Sun and Model S, in the sense (Sun) − (model), based on inversion of the 'Best set' of observed frequencies described by Basu *et al.* (1997). One-sigma error bars are indicated but are barely visible. The horizontal bars provide an indication of the resolution of the inversion. The filled symbols show results for a corresponding model including weak turbulent diffusion beneath the convection zone. (From Christensen-Dalsgaard & Di Mauro 2007).

in the superficial layers of the Sun where the modelling of the energy transport and dynamical effects of convection, as well as of the excitation and damping of the oscillations, is inadequate. This must be kept in mind in the analysis of solar oscillations as well as solar-like oscillations in other stars. In the solar case, however, the properties of the effects can be used to suppress them in analyses to infer the structure of the solar interior (e.g., Dziembowski *et al.* 1990; Rabello-Soares *et al.* 1999).

Given the richness of the solar data much more detailed analyses than the simple comparison are possible, using procedures originally developed in geoseismology (see, for example Gough & Thompson 1991, for a review). Specific techniques for inferring solar structure were discussed in some detail by Rabello-Soares *et al.* (1999), while results of helioseismic inversion were discussed by Gough *et al.* (1996). As an example, in figure 4 the open symbols show the result of an analysis of Model S. It is evident that the differences, while highly significant, are comparatively small. A characteristic feature of the differences is the bump at $r \simeq 0.7R$, where the model sound speed is too low. This is in the region just beneath the convective envelope where a strong composition gradient is established by helium settling. An improved match to the observations can be obtained through partial mixing of this region, softening the composition gradient and hence increasing the hydrogen abundance and thus the sound speed (e.g., Brun *et al.* 1999; Elliott & Gough 1999). In figure 4 this is illustrated by the model shown by the filled symbols where mixing was introduced through the inclusion of a suitably parametrized turbulent diffusion (Christensen-Dalsgaard & Di Mauro 2007).

It is important to emphasize that these sound-speed differences, and the corresponding models, provide a well-defined estimate of the sound speed in the solar interior which depends little on the details of the model used as a reference (e.g., Basu *et al.* 2000). Thus in this sense the helioseismic determination is robust.

As discussed by Asplund (these proceedings) a careful analysis, including three-dimensional modelling of the solar atmosphere and effects of departures from local thermodynamic equilibrium, has resulted in a dramatic revision of the determination of the solar surface abundances of oxygen, nitrogen and carbon (for a review, see also Asplund 2005).

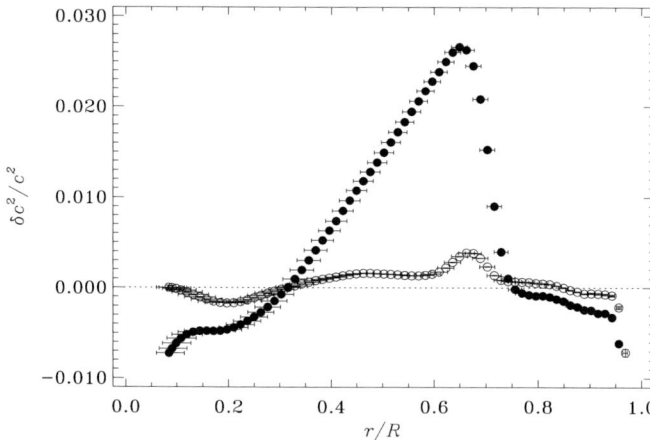

Figure 5. Inferred differences in squared sound speed between the Sun and two models, in the sense (Sun) − (model). The open symbols show results for Model S while the filled symbols show results for a corresponding model, but computed with the Asplund (2005) composition. For details, see the caption to figure 4. (From Christensen-Dalsgaard & Di Mauro 2007).

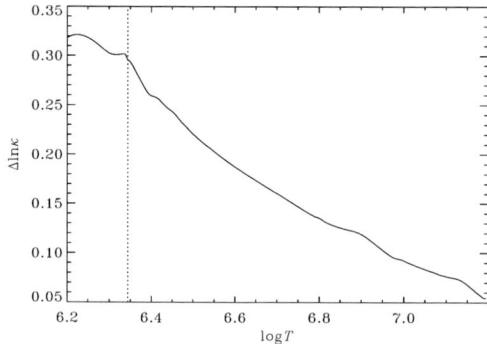

Figure 6. Intrinsic change in $\ln \kappa$, ln being natural logarithm, required to bring a model computed with the revised composition into agreement with Model S; $\Delta \ln \kappa$ has been assumed to depend only on temperature T. The vertical dotted line marks the base of the convective envelope.

As a result, $Z_{\rm s}/X_{\rm s}$ has been reduced from the value of 0.0245 assumed in Model S to 0.0165. This has dramatic consequences on the structure of the resulting models: the reduction in the heavy-element abundances leads to a comparable change in the opacity and hence a substantial change in the structure of the radiative interior. The result is a model deviating much more strongly from the helioseismic results. In figure 5 the open symbols again show results for Model S while the filled symbols are for a corresponding model but with the new composition. It is obvious that the revised abundances lead to a model in far worse agreement with the solar sound speed than Model S. Similar results were obtained, for example, by Turck-Chièze et al. (2004) and Bahcall et al. (2005). Also, Chaplin et al. (2007b) found that the small frequency separation (cf. (2.4)) in the model with the revised composition deviated strongly from the observed value, unlike Model S. An extensive review of such comparisons was given by Basu & Antia (2008).

There have been extensive efforts to modify the models in such a way as to accommodate the revised composition. As reviewed by Guzik (2006) these have met with little success. A simple, if possibly physically unrealistic, solution is to acknowledge that the dominant effect on the models of the change in composition arises from the opacity

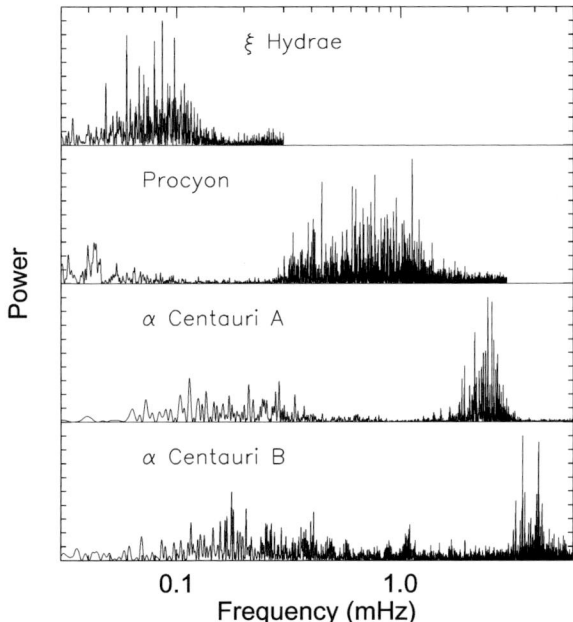

Figure 7. Power spectra as functions of cyclic frequency for four examples of solar-like oscillators. The maximum power corresponds to the following amplitudes: ξ Hydrae: 190 cm s^{-1}; Procyon: 42 cm s^{-1}; α Centauri A: 38 cm s^{-1}; α Centauri B: 12 cm s^{-1}.

change, and to compensate for this through a corresponding *intrinsic* change in the opacity (e.g., Montalbán *et al.* 2004; Bahcall *et al.* 2005). Christensen-Dalsgaard *et al.* (submitted) made a simple estimate of the correction to opacity κ, regarded as a function of temperature T, which would be needed to obtain a model structure similar to Model S, but using the new composition. The resulting change in $\ln \kappa$, ln being the natural logarithm, is shown in figure 6. This results in a model with only very small deviations from Model S, and hence providing a similarly good fit to the helioseismically inferred sound speed. However, it remains to be seen whether the required opacity change, of around 30% at the base of the convection zone and with this temperature dependence, is physically realistic.

It is evident that independent confirmation of the revision of the composition would be highly desirable. Thus it is very encouraging that Caffau & Ludwig (these proceedings) present an analysis in many way parallelling the one by Asplund and his collaborators, but on an independent basis (see also Caffau *et al.* 2008). Interestingly, the resulting oxygen abundance is intermediate between the old and new values. It seems likely that the solution to this serious problem for solar modelling may involve a number of factors, both in the composition determination and possibly in the calculation of interior opacities.

4. Asteroseismology of solar-like stars

Oscillations excited by near-surface convection, as observed in the Sun, are expected in all stars with substantial convective envelopes. Thus attempts to detect such oscillations started as soon as the global nature of the solar oscillations had been realized (e.g., Noyes *et al.* 1984; Gelly *et al.* 1986). However, the very low amplitudes clearly present a major challenge to such observations. Perhaps the first detection of the expected power envelope of solar-like oscillations in another star was made by Brown *et al.* (1991), observing

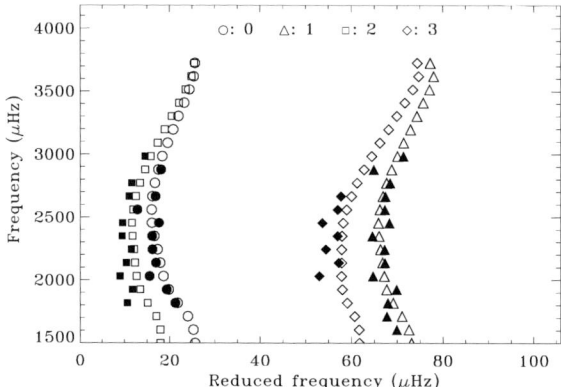

Figure 8. Echelle diagram for α Cen A. The abscissa is reduced frequency, corresponding to reducing $\nu_{nl} - \nu_0$ modulo the large frequency separation, taken to be $\Delta\nu = 106\,\mu$Hz; here ν_0 is a suitable reference frequency. The ordinate is frequency. The filled symbols are from the observations of Bedding et al. (2004), while the open symbols correspond to a model fitted to the data by Teixeira et al. (in preparation). The symbol type indicates the degree, as shown.

Procyon, while the first determination of solar-like oscillation frequencies, in the subgiant η Bootis, was made by Kjeldsen et al. (1995). In the last decade the development of very stable spectrographs, and extensive observational campaigns, has revolutionized the study of solar-like oscillations, resulting in the detection of such oscillations in a substantial number of stars (see Aerts et al. 2008, for a recent brief overview). A few examples of the resulting spectra are shown in figure 7. It should also be noted that evidence for solar-like oscillations has been found in very non-solar-like stars, such as red giants (Frandsen et al. 2002; Kiss & Bedding 2003; De Ridder et al. 2006) and long-period semiregular variables (Christensen-Dalsgaard et al. 2001).

Solar-like oscillations observed in distant stars are typically low-degree high-order acoustic modes and hence satisfy the asymptotic relations (2.2) and (2.4). This frequency pattern is an important diagnostic for the nature of the oscillations and the determination of average values of the large and small frequency separations is a likely early result of the analysis of the observations. The large frequency separation $\Delta\nu$ reflects the global properties of the star and hence, as do the frequencies, essentially scale as the square root of the mean density, $\Delta\nu \propto (M/R^3)^{1/2}$, where M and R are the mass and radius of the star. On the other hand, the small separation $\delta\nu$ (cf. (2.4)) is sensitive to the sound speed, and hence the composition structure, of the stellar core and hence changes as a result of the fusion of hydrogen to helium during stellar evolution, providing a measure of stellar age (e.g., Christensen-Dalsgaard 1984; Ulrich 1986; Gough 1987).

In the analysis of solar-like data the near-surface effects on the frequencies, discussed above, must be taken into account. Given their strong frequency dependence they directly affect $\Delta\nu$ but they also have a significant effect on $\delta\nu$. It was demonstrated by Roxburgh & Vorontsov (2003) that this effect can be suppressed by considering instead frequency-separation ratios, such as

$$r_{02} \equiv \frac{\nu_{n+1\,0} - \nu_{n\,2}}{\nu_{n+1\,1} - \nu_{n\,1}} \qquad (4.1)$$

(see also Otí Floranes et al. 2005). This similarly provides a measure of stellar age. As discussed by Houdek & Gough (these proceedings) a more careful analysis is required to take into account other features, such as acoustical 'glitches', in the structure of the star which may affect the age determination.

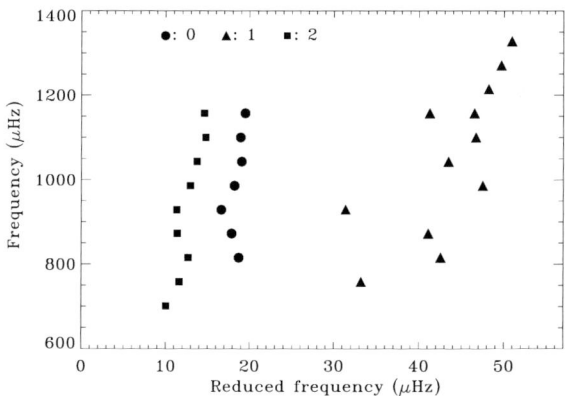

Figure 9. Echelle diagram, with $\Delta \nu = 57\,\mu\mathrm{Hz}$, based on observed frequencies for β Hydri obtained by Bedding *et al.* (2007). See caption to figure 8.

An asteroseismically very interesting case is the well-studied binary system α Centauri. The masses of the A and B components have been determined precisely, as $1.105\,\mathrm{M}_\odot$ and $0.934\,\mathrm{M}_\odot$, respectively, where $\mathrm{M}_\odot$ is the mass of the Sun (Pourbaix *et al.* 2002); thus the components nicely span the Sun. Both components show solar-like oscillations whose frequencies and other properties have been determined in substantial detail, from several observations of α Cen A (Bouchy & Carrier 2001, 2002; Bedding *et al.* 2004; Bazot *et al.* 2007) and α Cen B (Carrier & Bourban 2003; Kjeldsen *et al.* 2005). By fitting the frequencies as well as other observations the properties of the system, including its age and initial composition, have been determined with substantial precision (e.g., Eggenberger *et al.* 2004; Miglio & Montalbán 2005). An interesting issue is obviously whether the resulting fitted models are consistent with the observed frequencies or whether the observations indicate further significant problems in the modelling of the star. Indications that the latter may be the case are provided in figure 8. Here the frequencies are illustrated in a so-called *echelle diagram*, essentially corresponding to dividing the spectrum in segments of length $\Delta \nu$ and stacking them. The observed frequencies are compared with frequencies for a model resulting from a fit to the stellar parameters, suitably corrected for the effects of the near-surface errors. Although the agreement is generally good, there is a suggestion of a systematic shift in the position of the modes with $l = 3$. Given the present data, with a substantial scatter reflecting the stochastic nature of the excitation, this difference is barely significant; further observations, ideally extending much longer in time than the 5 days in the data illustrated, are highly desirable.

Another interesting case is the subgiant β Hydri which approximately represents a later evolutionary state of the Sun. Solar-like oscillations were discovered in this star, as an early example of the clear detection of power resulting from solar-like oscillations, by Bedding *et al.* (2001), and detailed results were obtained from a two-site campaign by Bedding *et al.* (2007). Figure 9 shows an echelle diagram for the frequencies obtained from the latter observations. Unlike the corresponding diagram for α Cen A this shows some irregularity for the modes identified as having $l = 1$. This probably arises as a result of the relatively late evolutionary stage of this star: this causes high values of the buoyancy frequency in the deep interior of the star, giving rise to so-called *mixed modes*, behaving like internal gravity waves, in the frequency range of the observed modes. Such modes do not satisfy the asymptotic relations (2.2) and (2.4) and hence do not follow the expected behaviour in the echelle diagram. A similar behaviour has been found in the case of the subgiant η Boo (Christensen-Dalsgaard *et al.* 1995; Guenther & Demarque 1996).

If this identification can be confirmed the observations potentially provide a sensitive probe of conditions in the stellar core.

5. Concluding remarks

Asteroseismology, of both solar-like and other types of stars, has seen a remarkable development over the past decade, and further major progress is expected in the coming years. Space asteroseismology was started by the WIRE (Buzasi 2004) and MOST (Walker *et al.* 2003) missions which have provided important results on a substantial number of stars, although lacking the sensitivity to provide detailed observations of solar-like oscillations in main-sequence stars. The French/European CoRoT mission (Baglin *et al.* 2006), launched in December 2006, has produced early very promising results on a number of stars, including a detailed study of the solar-like star HD 49933 (Appourchaux *et al.*, submitted); this mission will undoubtedly provide a breakthrough in space-based asteroseismology. Excellent data on a very large number of stars are expected from the NASA Kepler mission with scheduled launch in December 2009 (Borucki *et al.* 2003; Christensen-Dalsgaard *et al.* 2007). An almost overwhelming amount of asteroseismic data can result from the PLATO mission (Catala 2008), now under study for possible selection by ESA, which will observe of order 10^5 relatively bright stars.

These extremely promising photometric space missions do not eliminate the need for ground-based observations. To reach the highest precision in observations of solar-like oscillations, given the inevitable background from other processes in stellar atmospheres, Doppler-velocity observations are required (Harvey 1988). Nearly continuous observations of sufficient duration can only be obtained from dedicated facilities; on the other hand, for bright stars these can be of relatively modest size. A very interesting proposal is the SIAMOIS project to carry out such observations from Dome C in Antarctica (Mosser 2006). Observations from mid-latitudes require a dedicated network, similar to the helioseismic networks discussed above. One such project is the Stellar Observations Network Group (SONG) network (Grundahl *et al.* 2007, see also Grundahl *et al.*, these proceedings) which aims at setting up eight 1-meter class telescopes at suitably distributed sites.

Given these projects, and the further development of techniques for the analysis of the resulting data, the prospects seem excellent for using asteroseismology to address at least some of the many issues involved in modelling stars in the 21st century.

Acknowledgements

I am very grateful to the organizers for the invitation to attend this extremely interesting and enjoyable conference. H. Kjeldsen and P.-O. Quirion are thanked for help with figures.

References

Aerts, C., Christensen-Dalsgaard, J., Cunha, M., & Kurtz, D. W. 2008, *Solar Phys.*, in the press
Asplund, M. 2005, *Annu. Rev. Astron. Astrophys.* 43, 481
Baglin, A., Michel, E., Auvergne, M. and the CoRoT team 2006, in: K. Fletcher (ed.), *Proc. SOHO 18 / GONG 2006 / HELAS I Conf. Beyond the spherical Sun*, ESA SP-624 (Noordwijk, The Netherlands: ESA Publications Division)
Bahcall, J. N., Basu, S., Pinsonneault, M., & Serenelli, A. M. 2005, *Astrophys. J.* 618, 1049
Basu, S. & Antia, H. M. 2008, *Phys. Rep.* 457, 217

Basu, S., Chaplin, W. J., Christensen-Dalsgaard, J., Elsworth, Y., Isaak, G. R., New, R., Schou, J., Thompson, M. J., & Tomczyk, S. 1997, *Mon. Not. R. astr. Soc.* 292, 243

Basu, S., Pinsonneault, M. H., & Bahcall, J. N. 2000, *Astrophys. J.* 529, 1084

Bazot, M., Bouchy, F., Kjeldsen, H., Charpinet, S., Laymand, M., & Vauclair, S. 2007, *Astron. Astrophys.* 470, 295

Bedding, T. R., Butler, R. P., Kjeldsen, H., Baldry, I. K., O'Toole, S. J., Tinney, C. G., Marcey, G. W., Kienzle, F., & Carrier, F., 2001. *Astrophys. J.* 549, L105

Bedding, T. R., Kjeldsen, H., Arentoft, T., et al. 2007, *Astrophys. J.* 663, 1315

Bedding, T. R., Kjeldsen, H., Butler, R. P., McCarthy, C., Marcy, G. W., O'Toole, S. J., Tinney, C. G., & Wright, J. T. 2004, *Astrophys. J.* 614, 380

Borucki, W. J., Koch, D. G., Lissauer, J. J., et al. 2003, in: J. C. Blades & O. H. W. Siegmund (eds), *Future EUV/UV and Visible Space Astrophysics Missions and Instrumentation, Proceedings of the SPIE* Volume 4854, p. 129

Bouchy, F. & Carrier, F. 2001, *Astron. Astrophys.* 374, L5

Bouchy, F. & Carrier, F. 2002, *Astron. Astrophys.* 390, 205

Brown, T. M., Gilliland, R. L., Noyes, R. W., & Ramsey, L. W. 1991, *Astrophys. J.* 368, 599

Brun, A. S., Turck-Chièze, S., & Zahn, J. P. 1999, *Astrophys. J.* 525, 1032 (Erratum: *Astrophys. J.* 536, 1005).

Buzasi, D. L. 2004, in: F. Favata & S. Aigrain (eds), *Proc. 2nd Eddington workshop, "Stellar structure and habitable planet finding"*, ESA SP-538 (Noordwijk, The Netherlands: ESA Publications Division), p. 205

Caffau, E., Ludwig, H.-G., Steffen, M., Ayres, T. R., Bonifacio, P., Cayrel, R., Freytag, B., & Plez, B. 2008, *Astron. Astrophys.*, in the press [arXiv:0805.4398v1]

Carrier, F. & Bourban, G. 2003, *Astron. Astrophys.* 406, L23

Catala, C., and the PLATO consortium 2008, in: L. Gizon & M. Roth (eds), *Proc. HELAS II International Conference: Helioseismology, Asteroseismology and the MHD Connections, Göttingen, August 2007, J. Phys.: Conf. Ser.*, in the press

Chaplin, W. J., Elsworth, Y., Miller, B. A., & Verner, G. A. 2007a, *Astrophys. J.* 659, 1749

Chaplin, W. J., Serenelli, A. M., Basu, S., Elsworth, Y., New, R., & Verner, G. A. 2007b, *Astrophys. J.* 670, 872

Christensen-Dalsgaard, J. 1984, in: A. Mangeney, & F. Praderie (eds), *Space Research Prospects in Stellar Activity and Variability* (Paris Observatory Press), p. 11

Christensen-Dalsgaard, J. 2002, *Rev. Mod. Phys.* 74, 1073

Christensen-Dalsgaard, J. 2008a, *Mem. Soc. Astron. Ital.*, in the press.

Christensen-Dalsgaard, J. 2008b, *Astrophys. Space Sci.*, in the press [arXiv:0710.3114v1]

Christensen-Dalsgaard, J. & Di Mauro, M. P. 2007, in: C. W. Straka, Y. Lebreton & M. J. P. F. G. Monteiro (eds), *Stellar Evolution and Seismic Tools for Asteroseismology: Diffusive Processes in Stars and Seismic Analysis*, EAS Publ. Ser. 26 (Les Ulis, France: EDP Sciences), p. 3

Christensen-Dalsgaard, J., Bedding, T. R., & Kjeldsen, H. 1995, *Astrophys. J.* 443, L29

Christensen-Dalsgaard, J., Däppen, W., Ajukov, S. V., et al. 1996, *Science* 272, 1286

Christensen-Dalsgaard, J., Kjeldsen, H., & Mattei, J. A. 2001, *Astrophys. J.* 562, L141

Christensen-Dalsgaard, J., Arentoft, T., Brown, T. M., Gilliland, R. L., Kjeldsen, H., Borucki, W. J., & Koch, D. 2007, *Comm. in Asteroseismology* 150, 350

Cunha, M. S., Aerts, C., Christensen-Dalsgaard, J., et al. 2007, *Astron. Astrophys. Rev.* 14, 217

De Ridder, J., Barban, C., Carrier, F., Mazumdar, A., Eggenberger, P., Aerts, C., Deruyter, S., & Vanautgaerden, J. 2006, *Astron. Astrophys.* 448, 689

Dziembowski, W. A., Pamyatnykh, A. A., & Sienkiewicz, R. 1990, *Mon. Not. R. astr. Soc.* 244, 542

Eddington, A. S. 1926, *The internal constitution of the stars*, (Cambridge: Cambridge University Press)

Eggenberger, P., Charbonnel, C., Talon, S., Meynet, G., Maeder, A., Carrier, F., & Bourban, G. 2004, *Astron. Astrophys.* 417, 235

Elliott, J. R. & Gough, D. O. 1999, *Astrophys. J.* 516, 475

Frandsen, S., Carrier, F., Aerts, C., et al., 2002, *Astron. Astrophys.* 394, L5

Gabriel, A. H., Charra, J., Grec, G., et al. 1997, *Solar Phys.* 175, 207

Gelly, B., Grec, G., & Fossat, E. 1986, *Astron. Astrophys.* 164, 383

Gough, D. O. 1987, *Nature* 326, 257

Gough, D. O. 1993, in: Zahn, J.-P. & Zinn-Justin, J. (eds), *Astrophysical fluid dynamics, Les Houches Session XLVII* (Amsterdam: Elsevier), p. 399

Gough, D. O. & Thompson, M. J. 1991, in: A. N. Cox, W. C. Livingston & M. Matthews (eds), *Solar interior and atmosphere*, Space Science Series (University of Arizona Press), p. 519

Gough, D. O., Kosovichev, A. G., Toomre, J., *et al.*, 1996, *Science* 272, 1296

Grevesse, N. & Noels, A. 1993, in: N. Prantzos, E. Vangioni-Flam & M. Cassé (eds), *Origin and evolution of the Elements* (Cambridge: Cambridge Univ. Press), p. 15

Grundahl, F., Kjeldsen, H., Christensen-Dalsgaard, J., Arentoft, T., & Frandsen, S. 2007, *Comm. in Asteroseismology* 150, 300

Guenther, D. B. & Demarque, P. 1996, *Astrophys. J.* 456, 798

Guzik, J. A. 2006, in: K. Fletcher (ed.), *Proc. SOHO 18 / GONG 2006 / HELAS I Conf. Beyond the spherical Sun*, ESA SP-624 (Noordwijk, The Netherlands: ESA Publications Division)

Harvey, J. W. 1988, in: J. Christensen-Dalsgaard & S. Frandsen (eds), *Proc. IAU Symposium No 123, Advances in helio- and asteroseismology* (Dordrecht: Reidel), p. 497

Harvey, J. W., Hill, F., Hubbard, R. P., *et al.* 1996, *Science* 272, 1284

Kiss, L. L. & Bedding, T. R. 2003, *Mon. Not. R. astr. Soc.* 343, L79

Kjeldsen, H., Bedding, T. R., Butler, R. P., Christensen-Dalsgaard, J., Kiss, L. L., McCarthy, C., Marcy, G. W., Tinney, C. G., & Wright, J. T. 2005, *Astrophys. J.* 635, 1281

Kjeldsen, H., Bedding, T. R., Viskum, M. & Frandsen, S. 1995, *Astron. J.* 109, 1313

Miglio, A. & Montalbán, J. 2005, *Astron. Astrophys.* 441, 615

Montalbán, J., Miglio, A., Noels, A., Grevesse, N., & Di Mauro, M. P. 2004, in: D. Danesy, D. (ed.), *Proc. SOHO 14 - GONG 2004: "Helio- and Asteroseismology: Towards a golden future"* ESA SP-559 (Noordwijk, The Netherlands: ESA Publication Division), p. 574

Mosser, B. 2006, in: K. Fletcher (ed.), *Proc. SOHO 18 / GONG 2006 / HELAS I Conf. Beyond the spherical Sun*, ESA SP-624 (Noordwijk, The Netherlands: ESA Publications Division)

Noyes, R. W., Baliunas, S. L., Belserene, E., Duncan, D. K., Horne, J., & Widrow, L. 1984, *Astrophys. J.* 285, L23

Otí Floranes, H., Christensen-Dalsgaard, J., & Thompson, M. J. 2005, *Mon. Not. R. astr. Soc.* 356, 671

Pourbaix, D., Nidever, D., McCarthy, C., *et al.* 2002, *Astron. Astrophys.* 386, 280

Rabello-Soares, M. C., Basu, S., & Christensen-Dalsgaard, J. 1999, *Mon. Not. R. astr. Soc.* 309, 35

Roxburgh, I. W. & Vorontsov, S. V. 2003, *Astron. Astrophys.* 411, 215

Scherrer, P. H., Bogart, R. S., Bush, R. I., Hoeksema, J. T., Kosovichev, A. G., Schou, J., Rosenberg, W., Springer, L., Tarbell, T. D., Title, A., Wolfson, C. J., Zayer, I., and the MDI engineering team 1995, *Solar Phys.* 162, 129

Thompson, M. J., Christensen-Dalsgaard, J., Miesch, M. S., & Toomre, J. 2003, *Annu. Rev. Astron. Astrophys.* 41, 599

Turck-Chièze, S., Couvidat, S., Piau, L., Ferguson, J., Lambert, P., Ballot, J., García, R. A., & Nghiem, P. 2004, *Phys. Rev. Lett.* 93, 211102

Ulrich, R. K. 1986, *Astrophys. J.* 306, L37

Unno, W., Osaki, Y., Ando, H., Saio, H., & Shibahashi, H. 1989, *Nonradial Oscillations of Stars, 2nd Edition* (University of Tokyo Press).

Walker, G., Matthews, J., Kuschnig, R., *et al.* 2003, *Publ. Astron. Soc. Pacific* 115, 1023

Discussion

STEPIEŃ: Could the problem of lower metal abundance reported for the solar atmosphere be solved by assuming more efficient diffusion of heavy elements, so that the chemical composition below the convection zone is close to adopted earlier?

CHRISTENSEN-DALSGAARD: That is not likely. Since settling only changes Z by about 10% the change in settling rates required would probably be unphysical, and in fact attempts to solve the problem in this manner have in any case been unsuccessful.

DENG: You mentioned observations of solar like oscillation in Dome A, I have 2 questions: (1) What is your expectation from the current project at Dome A? (2) If a permanent base will be made in Dome A, is it a good replacement of SONG?

CHRISTENSEN-DALSGAARD: (1) The data obtained so far are likely not sufficiently precise to allow study of solar like oscillations, but they should be great for large-amplitude 'classical' pulsators. They will be very interesting to analyze. (2) Apart from the difficulty of operating delicate equipment on Dome A, that site only covers the southern sky and continuous dark sky is restricted to ~ 4 months. However, it could be a great complement to SONG. I should also mention the French SIAMOIS proposal to put an asteroseismic instrument on Dome C.

TURCK-CHIEZE: (1) In complement to the presentation I mention the existence of the successor of GOLF/SoHO (which discovers the first g modes) – the name is GOLF-NG. A prototype is now ready to observe in Tenerife. The modes at different heights in the atmosphere of the Sun – The objective will be to detect more g-modes first in Dome C. (2) Could you give us more on the whole program of SONG?

CHRISTENSEN-DALSGAARD: In my talk I concentrated on studying solar-like oscillations with Kepler and SONG, but both projects will certainly be open to other types of pulsating stars. For SONG a careful selection of targets will be required, given the limited number that can be observed. For Kepler procedures for target selection are being established to be discussed at the June KASC meeting.

Conference photograph: the speaker J. Christensen Dalsgaard (right) is discussing posters with G. Meynet (face not visible) and J. Vink.

Progress report on solar age calibration

G. Houdek[1] and D. O. Gough[1,2]

[1]Institute of Astronomy, Madingley Road, Cambridge CB3 0HA, UK
email: hg@ast.cam.ac.uk

[2]Department of Applied Mathematics and Theoretical Physics, Wilberforce Road,
Cambridge CB3 0WA, UK
email: douglas@ast.cam.ac.uk

Abstract. We report on an ongoing investigation into a seismic calibration of solar models designed for estimating the main-sequence age and a measure of the chemical abundances of the Sun. Only modes of low degree are employed, so that with appropriate modification the procedure could be applied to other stars. We have found that, as has been anticipated, a separation of the contributions to the seismic frequencies arising from the relatively smooth, glitch-free, background structure of the star and from glitches produced by helium ionization and the abrupt gradient change at the base of the convection zone renders the procedure more robust than earlier calibrations that fitted only raw frequencies to glitch-free asymptotics. As in the past, we use asymptotic analysis to design seismic signatures that are, to the best of our ability, contaminated as little as possible by those uncertain properties of the star that are not directly associated with age and chemical composition. The calibration itself, however, employs only numerically computed eigenfrequencies. It is based on a linear perturbation from a reference model. Two reference models have been used, one somewhat younger, the other somewhat older than the Sun. The two calibrations, which use BiSON data, are more-or-less consistent, and yield a main-sequence age $t_\odot = 4.68 \pm 0.02$ Gy, coupled with a formal initial heavy-element abundance $Z = 0.0169 \pm 0.0005$. The error analysis has not yet been completed, so the estimated precision must be taken with a pinch of salt.

Keywords. Sun: helioseismology, Sun: abundances, stars: abundances, stars: oscillations, stars: fundamental parameters.

---

1. Introduction

The only way by which the age of the Sun can be estimated to a useful degree of precision is by accepting the basic tenets of solar-evolution theory and measuring those aspects of the structure of the Sun that are predicted by the theory to be indicators of age. The structure measurements must be carried out seismologically, and evidently one expects greatest reliability of the results when all the available helioseismic data are employed. However, the most relevant modes are those of lowest degree, because it is they that penetrate most deeply into the energy-generating core where the relic helium-abundance variation records the integrated history of nuclear transmutation. Moreover, it is also only they that can be measured in other stars. Therefore, there has been some interest in calibrating theoretical stellar models using only low-degree modes. The prospect was first discussed in detail by Christensen-Dalsgaard (1984, 88), Ulrich (1986) and Gough (1987), although prior to that it had already been pointed out that the helioseismic frequency data that were available at the time indicated that either the initial helium abundance Y_0 or the age $t_\odot$, or both, are somewhat greater than the generally accepted values (Gough 1983), an inference which is consistent with our present findings. Subsequent, more careful, calibrations were carried out by Guenther (1989), Gough & Novotny (1990), Guenther & Demarque (1997), Weiss & Schlattl (1998), Dziembowski *et al.* (1999), Gough (2001) and Bonanno, Schlattl & Paternò (2002). Not all of them addressed the influence of uncertainties in Y_0 on the determination of $t_\odot$.

As a main-sequence star ages, helium is produced in the core, increasing the mean molecular mass μ, preferentially at the centre, and thereby reducing the sound speed. The resulting functional form of the sound speed $c(r)$ depends not only on age $t_\odot$ but also on the relative augmentation of $\mu(r)$, which itself depends on the initial absolute value of μ, and hence on Y_0. Gough (2001) tried to separate these two effects using the degree dependence of the small separation $d_{n,l} = 3(2l+3)^{-1}(\nu_{n,l} - \nu_{n-1,l+2})$ of cyclic frequencies $\nu_{n,l}$, where n is order and l is degree. This is possible, in principle, because modes of different degree and similar frequency sample the core differently. However, that difference is subtle, and the sensitivity to the relatively fine distinction between the effects of $t_\odot$ and Y_0 on the functional form of $c(r)$ in the core is low. Consequently the error in the calibration produced by errors in the observed frequency data is uncomfortably high.

This lack of sensitivity can be overcome by using, in addition to core-sensitive seismic signatures, the relatively small oscillatory component of the eigenfrequencies induced by the sound-speed glitch associated with helium ionization (Gough 2002), whose amplitude is close to being proportional to helium abundance Y (Houdek & Gough 2007a). The neglect of that component in the previously employed asymptotic signature $d_{n,l}$ had not only omitted an important diagnostic of Y but had also imprinted an oscillatory contamination in the calibration as the limits (k_1, k_2), where $k = n + \frac{1}{2}l$, of the adopted mode range was varied (Gough 2001). It therefore behoves us to decontaminate the core signature from glitch contributions produced in the outer layers of the star (from both helium ionization and the abrupt variation at the base of the convection zone, and also from hydrogen ionization and the superadiabatic convective boundary layer immediately beneath the photosphere). To this end a helioseismic glitch signature has been developed by Houdek & Gough (2007a), from which frequency contributions $\delta\nu_{n,l}$ can be computed and subtracted from the raw frequencies $\nu_{n,l}$ to produce effective glitch-free frequencies $\nu_{\text{sn},l}$ to which a glitch-free asymptotic formula (2.10) can be fitted. The solar calibration is then accomplished by interpolating the theoretical seismic signatures computed on a grid of solar models to the observations, using a standard grid to compute derivatives with respect to $t_\odot$ and Y_0, and a carefully computed reference solar model designed to be close to the Sun. The result of the first preliminary calibration by this method, using BiSON data, has been reported by Houdek & Gough (2007b). Here we enlarge on our discussion of the analysis, and we augment our results with a calibration based on a second reference solar model.

2. The seismic diagnostic and calibration method

Any abrupt variation in the stratification of a star (relative to the scale of the inverse radial wavenumber of a seismic mode of oscillation), which here we call an acoustic glitch, induces an oscillatory component in the spacing of the cyclic eigenfrequencies of seismic modes. Our interest is principally in the glitch caused by the depression in the first adiabatic exponent $\gamma_1 = (\partial \ln p/\partial \ln \rho)_s$ (where p, ρ and s are pressure, density and specific entropy) caused by helium ionization, which imparts a glitch in the sound speed $c(r)$. The deviation

$$\delta\nu_i := \nu_i - \nu_{\text{s}i}, \qquad (2.1)$$

where $i := (n, l)$, of the eigenfrequency ν_i from the corresponding frequency $\nu_{\text{s}i}$ of a similar smoothly stratified star is the indicator of Y that we use in conjunction with the indicators of core structure to determine the main-sequence age.

Approximate expressions for the frequency contributions $\delta\nu_i$ arising from acoustic glitches in solar-type stars were recently presented by Houdek & Gough (2007a). Here

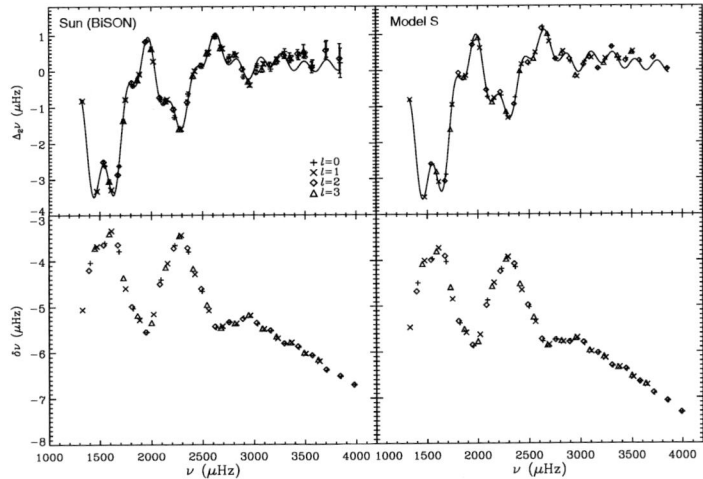

Figure 1. Top left: The symbols (with error bars obtained under the assumption that the raw frequency errors are independent) represent second differences, $\Delta_2\nu$, of low-degree solar frequencies from BiSON. Top right: The symbols are second differences $\Delta_2\nu$ of adiabatic pulsation eigenfrequencies of solar Model S of Christensen-Dalsgaard et al. (1996). The solid curves in both panels are the diagnostics (2.2) – (2.8), whose eleven parameters have been adjusted to fit the data optimally. Bottom: The symbols denote contributions $\delta\nu$ to the frequencies produced by the acoustic glitches of the Sun (left panel) and Model S (right panel).

we improve them by adopting the appropriate Airy functions $\mathrm{Ai}(-x)$ that are used as comparison functions in the JWKB approximations to the oscillation eigenfunctions, as did Houdek & Gough (2007b). The complete expression for $\delta\nu_i$ is then given by

$$\delta\nu_i = \delta_\gamma \nu_i + \delta_c \nu_i \,, \tag{2.2}$$

where

$$\begin{aligned}\delta_\gamma \nu = &-\sqrt{2\pi} A_{\mathrm{II}} \Delta_{\mathrm{II}}^{-1} \left[\nu + \tfrac{1}{2}(m+1)\nu_0\right] \\ &\times \left[\tilde\mu\tilde\beta \int_0^T \kappa_{\mathrm{I}}^{-1} e^{-(\tau-\tilde\eta\tau_{\mathrm{II}})^2/2\tilde\mu^2\Delta_{\mathrm{II}}^2}|x|^{1/2}|\mathrm{Ai}(-x)|^2\,\mathrm{d}\tau\right.\\ &\left.+\int_0^T \kappa_{\mathrm{II}}^{-1} e^{-(\tau-\tau_{\mathrm{II}})^2/2\Delta_{\mathrm{II}}^2}|x|^{1/2}|\mathrm{Ai}(-x)|^2\,\mathrm{d}\tau\right]\end{aligned} \tag{2.3}$$

arises from the variation in γ_1 induced by helium ionization, and

$$\begin{aligned}\delta_c \nu \simeq &\, A_c \nu_0^3 \nu^{-2}\left(1 + 1/16\pi^2\tau_0^2\nu^2\right)^{-1/2}\\ &\times \left\{\cos[2\psi_c + \tan^{-1}(4\pi\tau_0\nu)] - (16\pi^2\tilde\tau_c^2\nu^2 + 1)^{1/2}\right\}\end{aligned} \tag{2.4}$$

arises from the acoustic glitch at the base of the convection zone resulting from a near discontinuity (a true discontinuity in theoretical models using local mixing-length theory with a non-zero mixing length at the lower boundary of the convection zone) in the second derivative of density. Here, $m = 3.5$ is a constant, being a representative polytropic index in the expression for the approximate effective phase ψ appearing in the argument of the Airy function, and $\tilde\beta$, $\tilde\eta$ and $\tilde\mu$ are constants of order unity which account for the relation between the acoustic glitches caused by the first and second stages of ionization of helium (Houdek & Gough 2007a); τ is acoustic depth beneath the seismic surface of the star, and $T \simeq 1/2\nu_0$ is the total acoustic radius of the star; Δ_{II} and τ_{II} are respectively the acoustic width of the glitch and its acoustic depth beneath the seismic surface. The argument of the Airy function is $x = \mathrm{sgn}(\psi)|3\psi/2|^{2/3}$, where

$$\psi(\tau) = \kappa\omega\tilde{\tau} - (m+1)\cos^{-1}[(m+1)/\omega\tilde{\tau}] \qquad \text{if } \tilde{\tau} > \tau_\mathrm{t}, \tag{2.5}$$

and

$$\psi(\tau) = |\kappa|\omega\tilde{\tau} - (m+1)\ln[(m+1)/\omega\tilde{\tau} + |\kappa|] \qquad \text{if } \tilde{\tau} \leqslant \tau_\mathrm{t}, \tag{2.6}$$

in which $\tilde{\tau} = \tau + \omega^{-1}\epsilon_\mathrm{II}$, with $\omega = 2\pi\nu$, and τ_t is the location of the upper turning point of the mode; also $\kappa(\tau) = [1 - (m+1)^2/\omega^2\tilde{\tau}^2]^{1/2}$, and $\kappa_\mathrm{I} = \kappa(\tilde{\eta}\tau_\mathrm{II})$ and $\kappa_\mathrm{II} = \kappa(\tau_\mathrm{II})$. In addition

$$\psi_\mathrm{c} = \kappa_\mathrm{c}\omega\tilde{\tau}_\mathrm{c} - (m+1)\cos^{-1}[(m+1)/\tilde{\tau}_\mathrm{c}\omega] + \pi/4, \tag{2.7}$$

where $\kappa_\mathrm{c} = \kappa(\tau_\mathrm{c})$ and $\tilde{\tau}_\mathrm{c} = \tau_\mathrm{c} + \omega^{-1}\epsilon_\mathrm{c}$.

The seven coefficients $\eta_\alpha = (A_\mathrm{II}, \Delta_\mathrm{II}, \tau_\mathrm{II}, \epsilon_\mathrm{II}, A_\mathrm{c}, \tau_\mathrm{c}, \epsilon_\mathrm{c})$, $\alpha = 1, \ldots, 7$, are found by fitting the second difference

$$\Delta_{2i}\nu \equiv \nu_{n-1,l} - 2\nu_{n,l} + \nu_{n+1,l} \simeq \Delta_{2i}(\delta_\gamma\nu + \delta_\mathrm{c}\nu) + \sum_{k=0}^{3} a_k \nu_i^{-k} \equiv g_i(\nu_j; \eta_\alpha) \tag{2.8}$$

to the corresponding observations by minimizing

$$E_g = (\Delta_{2i}\nu - g_i)C^{-1}_{\Delta ij}(\Delta_{2j}\nu - g_j) \tag{2.9}$$

using the value of ν_0 obtained by fitting to (2.10), where $C^{-1}_{\Delta ij}$ is the (i,j) element of the inverse of the covariance matrix C_Δ of the observational errors in $\Delta_{2i}\nu$, computed, perforce, under the assumption that the errors in the frequency data ν_i are independent. The last term in equation (2.8) approximates smooth contributions arising, in part, from wave refraction in the stellar core, from hydrogen ionization and from the superadiabaticity of the upper boundary layer of the convection zone, introducing four more fitting coefficients $a_k = \eta_\alpha$, $k = 0, \ldots, 3$, $\alpha = 8, \ldots, 11$. The covariance matrix $C_{\eta\alpha\gamma}$ of the errors in η_α were established by Monte Carlo simulation.

The outcome of the fitting to the BiSON data (Basu et al. 2007) and to the adiabatically computed eigenfrequencies of solar Model S (Christensen-Dalsgaard et al. 1996) is displayed in Figure 1: the upper panels display the second differences, together with the fitted formula (2.8), the lower panels display the corresponding contributions $\delta\nu_i$ to the frequencies of oscillation from the acoustic glitches. All the frequencies displayed in the figure have been used in equation (2.9) for fitting (2.8).

To the resulting glitch-free frequencies ν_{si}, derived from equation (2.1), of both the solar observations and the eigenfrequencies of the reference solar model, is fitted the asymptotic expression

$$\nu_{si} \sim (n + \tfrac{1}{2}l + \hat{\epsilon})\nu_0 - \frac{AL^2 - B}{\nu_{si}}\nu_0^2 - \frac{CL^4 - DL^2 + E}{\nu_{si}^3}\nu_0^4 - \frac{FL^6 - GL^4 + HL^2 - I}{\nu_{si}^5}\nu_0^6 \equiv s(\nu_{si}; \xi_\beta), \tag{2.10}$$

by minimizing $(\nu_{si} - s_i)C^{-1}_{sij}(\nu_j - s_j)$, where $L^2 = l(l+1)$ and C_s is the covariance matrix of the observational errors in ν_{si}, from which we obtain both the coefficients $\xi_\beta = (\nu_0, \hat{\epsilon}, A, B, C, D, E, F, G, H, I)$, $\beta = 1, \ldots, 11$, and the covariance matrix $C_{\xi\beta\delta}$ of the errors. Following Gough (2001), we carry out this fitting in the frequency range given by $k_1 \leqslant k \leqslant k_2$, where $k = n + \tfrac{1}{2}l$ and $0 \leqslant l \leqslant 3$, and we vary k_1 and k_2. Each of the parameters ξ_β represents an integral of a function of the equilibrium stratification. The integrals A, C and F are of particular importance to our analysis, because C and F are dominated by conditions in the core, and, although the contributions to A from the core and the rest of the star are roughly equal in magnitude (and potentially have opposite signs), the latter is relatively insensitive to $t_\odot$ and Y_0. The integrands in the remaining integrals are either more evenly distributed throughout the Sun or are concentrated near the surface.

Table 1. Partial derivatives $H_{\alpha j}$ obtained from two sets of calibrated evolutionary models for the Sun. Values with respect to age $t_\odot$ are in units of Gy^{-1}.

$(\partial A/\partial t_\odot)_Z$	$(\partial A/\partial Z)_{t_\odot}$	$(\partial C/\partial t_\odot)_Z$	$(\partial C/\partial Z)_{t_\odot}$	$[\partial(-\delta\gamma_1/\gamma_1)/\partial t_\odot]_Z$	$[\partial(-\delta\gamma_1/\gamma_1)/\partial Z]_{t_\odot}$
-0.0469	-0.584	0.677	36.8	-0.00656	0.442

We have carried out age calibrations using combinations of the parameters

$$\zeta_\alpha = (A, C, -\delta\gamma_1/\gamma_1), \qquad \alpha = 1, 2, 3, \tag{2.11}$$

where $-\delta\gamma_1/\gamma_1 = A_{\mathrm{II}}/\sqrt{2\pi}\nu_0 \Delta_{\mathrm{II}}$ is a measure of the maximum depression in γ_1 in the second helium ionization zone. Presuming, as is normal, that the reference model is parametrically close to the Sun, we consider the reference value $\zeta_\alpha^{\mathrm{r}}$ to be approximated by a two-term Taylor expansion of ζ_α about the value $\zeta_\alpha^\odot$ of the Sun:

$$\zeta_\alpha^{\mathrm{r}} = \zeta_\alpha^\odot - \left(\frac{\partial \zeta_\alpha}{\partial t_\odot}\right)_Z \Delta t_\odot - \left(\frac{\partial \zeta_\alpha}{\partial Z}\right)_{t_\odot} \Delta Z + \epsilon_{\zeta\alpha}, \tag{2.12}$$

where $\Delta t_\odot$ and ΔZ are the deviations of age $t_\odot$ and initial heavy-element abundance Z from the reference model, and $\epsilon_{\zeta\alpha}$ are the formal errors in the calibration parameters whose covariance matrix $C_{\zeta\alpha\beta}$ can be derived from $C_{\xi\beta\delta}$ and $C_{\eta\alpha\gamma}$. A (parametrically local) maximum-likelihood fit then leads to the following set of linear equations:

$$H_{\alpha j} C_{\zeta\alpha\beta}^{-1} H_{\beta k} \Theta_{0k} = H_{\alpha j} C_{\zeta\alpha\beta}^{-1} \Delta_{0\beta}, \tag{2.13}$$

in which $\Theta_k = (\Delta t_\odot, \Delta Z) + \epsilon_{\Theta k} = \Theta_{0k} + \epsilon_{\Theta k}$, $k=1,2$, is the solution vector subject to (correlated) errors $\epsilon_{\Theta k}$, $\Delta_\beta = \zeta_\beta^\odot - \zeta_\beta^{\mathrm{r}} + \epsilon_{\zeta\beta} = \Delta_{0\beta} + \epsilon_{\zeta\beta}$, and the partial derivatives $H_{\alpha j} = [(\partial \zeta_\alpha/\partial t_\odot)_Z, (\partial \zeta_\alpha/\partial Z)_{t_\odot}]$, $j = 1, 2$.

A similar set of equations is obtained for the formal errors $\epsilon_{\Theta k}$:

$$H_{\alpha j} C_{\zeta\alpha\beta}^{-1} H_{\beta k} \epsilon_{\Theta k} = H_{\alpha j} C_{\zeta\alpha\beta}^{-1} \epsilon_{\zeta\beta}, \tag{2.14}$$

from which the error covariance matrix $C_{\Theta kq} = \overline{\epsilon_{\Theta k}\epsilon_{\Theta q}}$ can be computed from $C_{\zeta\alpha\beta}$.

The partial derivatives $H_{\alpha j}$ were obtained from the two sets of five calibrated evolutionary models for the Sun that were used in a similar calibration by Houdek & Gough (2007b), computed with the evolutionary programme by Christensen-Dalsgaard (1982), and adopting the Livermoore equation of state and the OPAL92 opacities. One set of models has a constant value for the heavy-element abundance $Z = 0.02$ but varying age; the other has constant age but varying Z. Note that, for prescribed relative abundances of heavy elements, the condition that the luminosity and radius of the Sun agree with observation defines a functional relation between Y_0, Z and $t_\odot$. The values of the partial derivatives $H_{\alpha j}$ are listed in Table 1.

3. Results

To illustrate the effect of taking $\delta\nu_i$ into account, we compare in Figure 2 a first assessment of $A(k_1, k_2)$ using the glitch-free frequencies $\nu_{\mathrm{s}i}$ (left panel) with that obtained from the raw frequencies ν_i (right panel). Recall that A represents a functional of the equilibrium structure of the star, and should not vary with k_1 and k_2. The range of values for A is the lower for $\nu_{\mathrm{s}i}$, as we had anticipated. We believe that the upturn of A for low values of k_1 in the left panel of the figure is a result of the failure of the asymptotic formula (2.2)–(2.4) when ν_i is low. The dipping of A at high values of k_1 and low values of k_2 occurs because the frequency range is too small for a reliable determination of the fitting coefficients ξ_β. We therefore adopt intermediate values for k_1 and high values for k_2, for which A is insensitive to the selected frequency range.

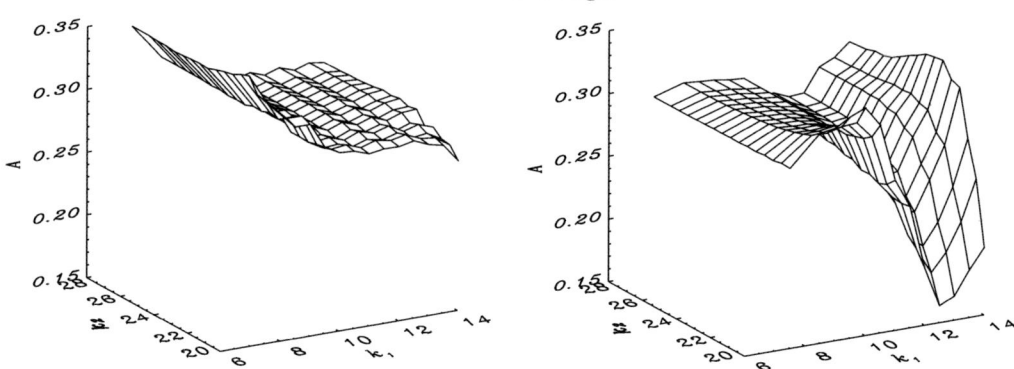

Figure 2. Asymptotic fitting coefficient A (see equation 2.10) as functions of k_1 and k_2 ($k = n + \frac{1}{2}l$). Results are shown for fitting (2.10) to the glitch-free frequencies ν_{si} (left panel) and to the raw frequencies ν_i (right panel).

Age calibrations using different combinations of the parameters ζ_α and two different reference models are summarized in Table 2. The younger reference model is 'Model S' (Christensen-Dalsgaard et al. 1996) which has age $t_\odot = 4.6$ Gy; the second is 'Model T', which has age $t_\odot = 4.7$ Gy. The same physics was adopted in the evolutionary calculations of both models. We notice in Table 2 that the calibration for the combination (A, C), i.e. without $\delta\gamma_1/\gamma_1$, is less stable to a change in the reference model than are the calibrations with combinations in which $\delta\gamma_1/\gamma_1$ is included, and therefore is less reliable, as we have explained in the introduction. If we ignore in Table 2 the results for (A, C) and combine the others, we obtain

$$t_\odot = 4.68 \pm 0.02\,\mathrm{Gy}\,, \quad Z = 0.0169 \pm 0.0005\,.$$

Including the calibrations with (A, C) does not change the outcome. Error contours corresponding to the calibration from Model S in the first row of Table 2 are plotted in Figure 3. Corresponding contours for Model T are the same, except that their centres are displaced to (4.677 Gy, 0.0170). One can adduce from our description of the analysis in Section 2 that our current treatment of the errors is not completely unbiassed; however, the potential bias is of the order of only $|\delta\nu_i/\nu_i|$, which is small.

The age we have found is greater than currently accepted values. The values of Z are somewhat smaller than those of Models S and T (0.01963), but we hasten to point out that they should not be regarded strictly as statements about the initial heavy-element abundance, but rather as measures of the opacity in the radiative interior. Asplund et al. (2004) have argued that the photospheric abundances of C, N and O had previously been overestimated, suggesting that the actual total heavy-element abundance is even lower than had previously been believed. However, that cannot imply that the opacity in the solar interior is necessarily comparably lower because it has been implicitly calibrated here (by accepting the tenets of solar-evolution theory, and the OPAL opacity calculations upon which the models are based), and indeed the opacity has already been determined seismologically from a broader spectrum of modes than has been adopted here (Gough 2004). The matter raised by Asplund et al. therefore challenges either the opacity calculations, the nuclear reaction rates, or the basic physics of stellar evolution, not helioseismology as some spectators have surmised. As we know already from seismological structure inversions, the solar models are not accurate by helioseismological standards. Therefore the properties inferred from these calibrations could be more contaminated by systematic error than by errors in the observed frequencies.

Table 2. Age calibrations with different combinations of ζ_α and for the two reference models: Model S with an age $t_\odot = 4.6$ Gy and Model T with an age $t_\odot = 4.7$ Gy. The first two columns show the results adopting Model S as the reference model, the third and fourth columns display the results for Model T.

ζ_α	$t_\odot$ (Gy)	Z	$t_\odot$ (Gy)	Z	$C^{1/2}_{\Theta 11}$	$-(-C_{\Theta 12})^{1/2}$	$C^{1/2}_{\Theta 22}$
$A, C, -\delta\gamma_1/\gamma_1$	4.679	0.0169	4.677	0.0170	0.017	-0.0023	0.0005
A, C	4.658	0.0177	4.673	0.0171	0.023	-0.0037	0.0007
$A, -\delta\gamma_1/\gamma_1$	4.673	0.0165	4.676	0.0169	0.017	-0.0019	0.0007
$C, -\delta\gamma_1/\gamma_1$	4.700	0.0169	4.680	0.0170	0.028	-0.0029	0.0005

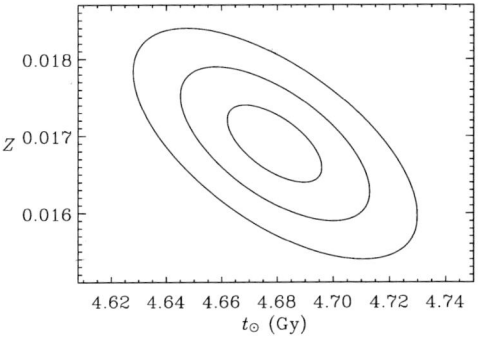

Figure 3. Error ellipses for the calibration using all three parameters ζ_α and Models S as the reference model: solutions $(t_\odot, Z)$ satisfying the frequency data within 1, 2 and 3 standard errors in those data reside in the inner, intermediate and outer ellipses, respectively.

Acknowledgements

We thank Jørgen Christensen-Dalsgaard for providing us with his stellar-evolutionary programme. GH acknowledges support by the STFC of the UK.

References

Asplund, M., Grevesse, N., Sauval, A. J., Allende Prieto, C., & Kiselman, D. 2004, *A&A* 417, 751
Basu, S., Chaplin, W. J., Elsworth, Y., New, A. M., Serenelli, G., & Verner, G. A. 2007, *ApJ* 655, 660
Bonanno, A., Schlattl, H., & Paternò, L. 2002, *A&A* 390, 1115
Christensen-Dalsgaard, J. 1982, *MNRAS* 199, 735
Christensen-Dalsgaard, J. 1984, in: Mangeney, A., Praderie, F., (eds), *Space Research Prospects in Stellar Activity and Variability*, Paris Observatory Press, Paris, p. 11
Christensen-Dalsgaard, J. 1988, in: Christensen-Dalsgaard, J., Frandsen, S., (eds), *Proc. IAU Symp. 123, Advances in helio- and asteroseismology*, Reidel, Dordrecht, p. 295
Christensen-Dalsgaard, J. et al. 1996, *Sci* 272, 1286
Dziembowski, W. A., Fiorentini, G., Ricci, B., & Sienkiewicz, R. 1999, *A&A* 343, 990
Gough, D. O. 1983, in: Shaver, P. A., Kunth, D., Kjär, K., (eds), *Primordial helium*, Southern Observatory, p. 117
Gough, D. O. 1987, *Nat.* 326, 257
Gough, D. O. 2001, in: von Hippel, T., Simpson, C., Manset, N., (eds), *ASP Conf. Ser. Vol. 245, Astrophysical ages and timescales*, Gough D. O. 1987, *Nat.* 326, 257
Gough, D. O. 2002, in: Favata, F., Roxburgh, I. W., Gadalí-Enríquez, D., (eds), *Proc 1st Eddington Workshop: Stellar structure and habitable planet finding*, ESA SP-485, Noordwijk, p. 65
Gough, D. O. 2004, in: Čelebonović, V., Däppen, W., Gough, D. O., (eds), *AIP Conf. Proc. Vol. 731, Equation-of-state and phase-transition issues in models of ordinary astrophysical matter*, Am. Inst. Phys., Melville, p. 119
Gough, D. O. & Novotny, E. 1990, *Solar Phys.*, 128, 143
Guenther, D. B. 1989, *ApJ* 339, 1156

Guenther, D. B. & Demarque, P. 1997, *ApJ* 484, 937
Houdek, G. & Gough, D. O. 2007a, *MNRAS* 375, 861
Houdek, G. & Gough, D. O. 2007b, in: Stancliffe, R. J., Dewi, J., Houdek, G., Martin, R. G., Tout, C. A., (eds), *AIP Conf. Proc.: Unsolved Problems in Stellar Physics*, American Institute of Physics, New York, p. 219
Ulrich, R. K. 1986, *ApJ* 306, L37
Weiss, A. & Schlattl, H. 1998, *A&A* 332, 215

Discussion

CHRISTENSEN-DALSGAARD: A comment: with SONG we expect to be able to carry out a similar analysis of distant stars, on which we of course know much less a priori.

HOUDEK: This seismic diagnostic has been developed with the aim to be able to use it also for distant stars. The accuracy of the observed frequency data required for such a diagnostic analysis is one part in 10^4 or better.

S. VAUCLAIR: Two small comments which are actually more relevant for stars that are slightly more massive than the Sun; First, I would like to point out that in case of helium settling below the convective zone the effect of the helium gradient in the second differences may become more important than the convective border, and than the effect of helium ionization. Second, the so-called asymptotic theory, which is very useful, may become quite wrong in some cases, at the end of the main sequence or the beginning of the sub-giant branch. The small frequency separation, which is always positive in the asymptotic theory, can become negative.

GOUGH & HOUDEK: You are certainly correct in implying that the amplitude of the oscillatory contribution to the second differences arising from helium settling beneath the convection zone can be greater in stars more massive than the Sun, which have shallower convection zones, although whether or not it is more important than the ionization signature depends upon the issue in question. The cumulative amount of settling increases with time, and therefore is a potential indicator of age. But the sound-speed profile that it produces depends on uncertain fluid-dynamical issues associated with the tachocline, the recession of the convection zone, and possible overshooting, so we would be wary of attempting to use its seismic signature in an age calibration. In the current state of our understanding we would instead prefer to separate it from the ionization signature and then ignore it, as we have for the Sun; that course is possible provided that the helium ionization zone is acoustically far from the base of the convection zone. We would use it separately to investigate tachocline structure, however, as indeed we are in the process of doing for the Sun.

One cannot deny that conditions in some stars might be such as to render it impossible to develop an adequate asymptotic theory of low-degree acoustic modes, although we do not share your apparently implied pessimism. It is perhaps worth pointing out that there have been instances when an asymptotic formula developed in one set of circumstances has been misused by applying it without modification in another, in which the conditions for the validity of the theory are not satisfied; if that is what you mean by "so-called asymptotic theory" we must surely agree. We must point out, however, that it is not true that even the simple asymptotic glitch-free formula (2.10) precludes a negative so-called small frequency separation.

The Art of Modelling Stars in the 21st Century
Proceedings IAU Symposium No. 252, 2008
L. Deng & K.L. Chan, eds.

© 2008 International Astronomical Union
doi:10.1017/S1743921308022709

Rate of change of the pulsation periods in the PG 1159 star PG 0122+200

G. Vauclair[1], J.-N. Fu[2], J.-E. Solheim[3], S.-L. Kim[4], M. Chevreton[5], N. Dolez[1], L. Chen[2], M. A. Wood[6], and I. M. Silver[6]

[1]Laboratoire d'Astrophysique de Toulouse-Tarbes, Université de Toulouse, CNRS, 14 av. E. Belin, 31400 Toulouse, France

[2]Department of Astronomy, Beijing Normal University, Beijing 100875 China

[3]Institute of Theoretical Astrophysics, University of Oslo, Oslo, Norway

[4] Korea Astronomy and Space Science Institute, Daejeon, Korea

[5] LESIA, Observatoire de Paris-Meudon, Meudon, France

[6] Department of Physics and Space Sciences & SARA Observatory, Florida Institute of Technology, Florida 32935, USA

Abstract. The pre-white dwarf pulsators of PG 1159 type, or GW Virginis variable stars, are in a phase of rapid evolution towards the white dwarf cooling sequence. The rate of change of their nonradial g-mode frequencies can be measured on a reasonably short time scale. From a theoretical point of view, it was expected that one could derive the rate of cooling of the stellar core from such measurements. At the cool end of the GW Virginis instability strip, it is predicted that the neutrinos flux dominates the cooling. PG 0122+200 which defines the red edge of the instability strip is in principle a good candidate to check this prediction. It has been followed-up through multisite photometric campaigns for about fifteen years. We report here the first determination of the rate of change of its 7 largest amplitude frequencies. We find that the amplitudes of the frequency variations are one to two orders of magnitude larger than predicted by theoretical models based on the assumption that these variations are uniquely caused by cooling. The time scales of the variations are much shorter than the ones expected from a neutrino dominated core cooling. These results point to the existence of other mechanisms responsible for the frequency variability. We discuss the role of nonlinearities as one possible mechanism.

Keywords. neutrinos, stars:evolution, stars: individual (PG 0122+200), stars: oscillations, stars: white dwarfs

1. Introduction

The PG 1159 stars form an evolutionary link between the central stars of planetary nebulae and the white dwarf cooling sequence. A fraction of the PG 1159 stars does pulsate and constitutes the subgroup of the GW Vir type of variable stars. The asteroseismology of these variable stars provides invaluable insight on their internal structure and evolutionary status. Among the four known GW Vir stars, which have all been studied with the Whole Earth Telescope network (WET; Nather *et al.* 1990), PG 0122+200 is the coolest one. Its atmospheric parameters T_{eff}= 80 000 K ± 4000 K, log g = 7.5 ± 0.5, and abundances (C/He = 0.3, C/O = 3, N/He = 10^{-2} by numbers) are typical of the GW Vir stars (Dreizler & Heber 1998). At this effective temperature, PG 0122+200 presently defines the red edge of the GW Vir instability strip. O'Brien *et al.* (1998, 2000) have shown that at this phase of the pre-white dwarf evolution, the neutrino losses become important in the cooling process. For a 0.60 $M_\odot$ PG 1159 star at the effective temperature of PG 0122+200, they estimate that the ratio of the neutrino luminosity over the photon luminosity could be $1 \leqslant L_\nu/L_\gamma \leqslant 2$. This makes PG 0122+200 of particular interest to study neutrino physics if one could measure its cooling rate from asteroseismology.

The predicted neutrino luminosity depends on the stellar parameters, mainly the total mass and the effective temperature, which can be determined precisely in principle from asteroseismology. Determining the stellar parameters of a non-radial g-mode pulsator relies on the capacity to determine the period spacing between a large enough number of pulsation modes and to correctly identify their degree ℓ. In the best case, a detailed comparaison with realistic models provides an even more precise estimate of the stellar parameters. After the discovery of its pulsations (Bond & Grauer 1987), PG 0122+200 was observed in 1986 (O'Brien et al. 1996), and then through a number of multisite campaigns in 1990 (Vauclair et al. 1995), in 1996 when it was the priority target of a WET campaign (O'Brien et al. 1998), in 2001 and 2002 (Fu et al. 2007) and more recently in 2005. From the cumulative pulsation periods observed during these campaigns, it was possible to identify 23 periods as $\ell = 1$ modes, composed of 7 triplets and 2 single modes, and to derive a precise value of the average period spacing, $\Delta P = 22.9$ s (Fu et al. 2007). These results were used by Córsico et al. (2007) to derive precise constraints on the PG 0122+200 parameters, the most important ones for estimating the neutrino luminosity being the total mass ($M_* = 0.556$ (+0.009, −0.014) $M_\odot$) and the effective temperature ($T_{eff} = 81540$ (+800, −1400) K) or alternatively the luminosity ($\log(L_*/L_\odot) = 1.14$ (+0.02, −0.04)).

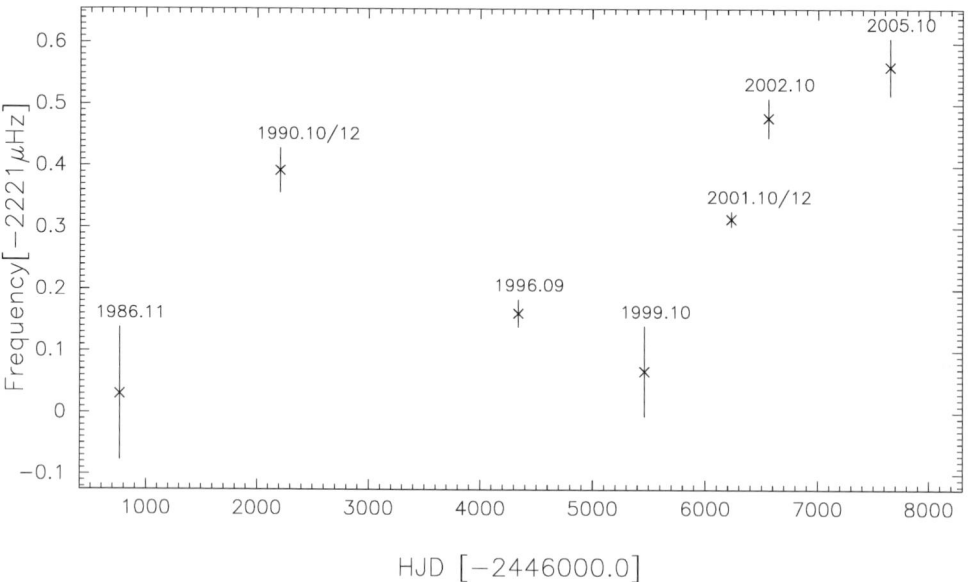

Figure 1. Evolution of the 2221 μHz frequency with time in PG 0122+200. The residual frequency, after subtraction of 2221 μHz is plotted as a function of time, expressed in Heliocentric Julian Days referred to a zero point at HJD = 2446000.0. Each frequency value is marked with the date of the corresponding campaign (year.month) and with the appropriate uncertainty. The value for 1986 is added for completness but this was a single site campaign. All other values correspond to multisite campaigns.

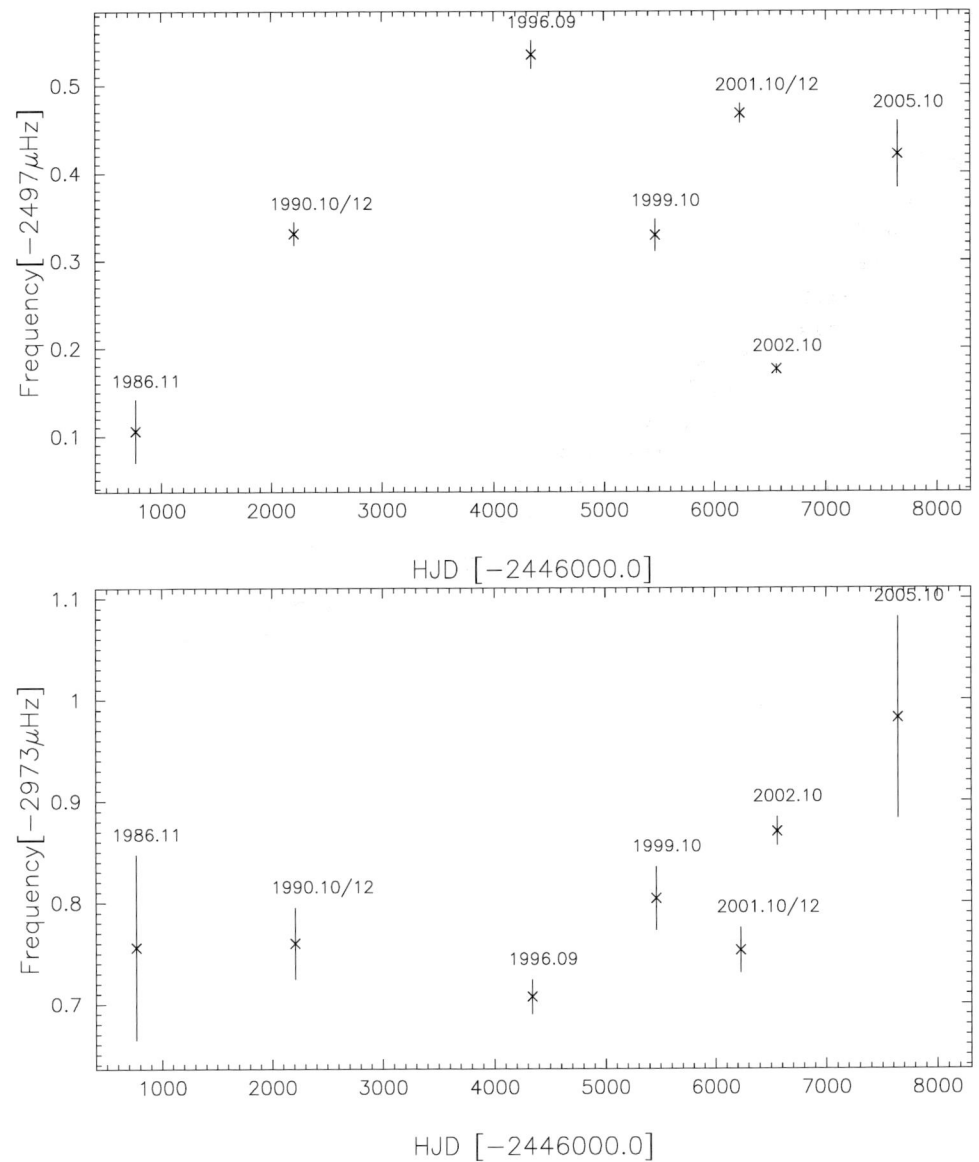

Figure 2. Same as Figure 1 for the three components of the triplet at 2490-2493-2497 μHz.

2. The rate of change of the pulsation periods

2.1. *The predictions of the model*

The evolutionary model by Córsico *et al.* (2007) relies on the neutrino production rate by Ito *et al.* (1989, 1992). According to the predictions of the best model, the rate of change of the pulsation periods ($\dot{P} = \mathrm{d}P/\mathrm{d}t$) varies between $1.22\times10^{-12} ss^{-1}$ for the $\ell = 1$, k = 12 mode of period 336.68 s and $3.26\times10^{-12} ss^{-1}$ for the $\ell = 1$, k = 24 mode of period 611.15 s. These values translate into evolutionary time scales between 1.2×10^{7} and 6×10^{6} years. On the fifteen years time scale covered by the multisites campaigns reported here, one should not have detected frequency variations of the pulsation modes larger than $-5\times10^{-3}\mu$Hz $\leqslant \delta$f $\leqslant -4\times10^{-3}\mu$Hz (i.e. 6×10^{-4} s $\leqslant \delta$P $\leqslant 1.5\times10^{-3}$ s in the period variations). Such small variations are not reachable within 15 years given the present uncertainties on the frequency determination. We estimate that it would take ≈50 years to get a significant $3\times\sigma$ signature of a frequency change induced by the cooling.

2.2. *The measured frequency variations*

We have reanalysed the data obtained since 1986 and redetermined the frequencies and the amplitudes in a homogenous way using the Period04 software (Lenz & Breger 2005). Since the uncertainties derived by Period04 only estimate the internal consistency of the solution, they are underestimated. We derived more realistic uncertainties through Monte-Carlo simulations. We were able to follow the frequency variations by the direct method for the 7 largest amplitude modes. This includes the three components of the triplets at 2221, 2224, 2228 μHz and at 2490, 2493, 2497 μHz and the prograde component of a third triplet at 2973 μHz. The other modes either do not show up in a large enough number of observations or have too low amplitude for a significant detection of frequency variations. The frequency variation of the retrograde component of the first triplet at 2221 μHz is shown in Fig. 1. The Fig. 2 shows the frequency variation for the three components of the second triplet and Fig. 3 the variation for the prograde component of the third triplet. Among these 5 modes, only the largest amplitude ones, at 2221 μHz, 2497 μHz and 2973 μHz, are present in all the observing campaigns. The other frequencies show similar behaviour. Since the variations are clearly nonlinear with time, the (O-C) method does not apply.

As can be seen on the various figures, the behaviour of the variations varies for different frequencies. The amplitude of the frequency variations are much larger than expected from the model predictions which are based on the assumption that the cooling, possibly dominated by neutrino luminosity, is the unique cause of the frequency changes. We observe frequency variations of the order of a fraction of μHz instead of the predicted 10^{-3} μHz. The frequencies vary on time scales of the order of a few years, considerably shorter than the cooling time scale.

3. Conclusions and interpretation

The multisite campaigns of observations of PG 0122+200 covering a period of 15 years (extending to 19 years if one includes a first single site campaign) show that the frequencies vary with time. The amplitude of the variations are one to two orders of magnitude larger than the ones derived from the best fit model of Córsico *et al.* (2007) and they do not follow a simple function of time. Their typical time scales are much shorter than the expected cooling time, for any reasonable assumption on the rate of neutrino production. Similar results have been observed in the prototype of the GW Vir pulsators PG 1159-035 (Costa & Kepler 2008). This result indicates that the observed frequency variations are not dominated by cooling. They are not either induced by an orbiting companion, i.e. a planet or a brown dwarf (which is the case of the sdB pulsator V391 Peg: Silvotti *et al.* 2007) since the variations of different frequencies are not correlated in phase and do not have the same amplitudes and time scales. We have to think about other mechanism(s). One potential mechanism producing such frequency and amplitude variations is the resonant coupling induced by rotation within triplets, as discussed by Goupil *et al.* (1998), who showed that, depending on the value of the ratio of the frequency mismatch induced in a triplet by second order effect of rotation and the mode growth-rate, one may encounter three different regimes. If this ratio is small, the frequencies are forced to be equally spaced within the triplet; this is the frequency lock regime. If the ratio is large, the resonant coupling is inefficient and the frequencies keep their nonresonant properties, i.e. keep their linear values. In the intermediate case, both the amplitudes and the frequencies undergo time modulation. The variations observed in PG 0122+200 indicate that the star could be in this intermediate regime. PG 0122+200 is clearly a good target to further explore this type of nonlinearities.

References

Bond, H. & Grauer, A. 1987, ApJ, 321, L123 AJ, 112, 2699
Corsico, A. H., Miller Bertolami, M. M., Althaus, L.G., Vauclair, G., & Werner, K. 2007, A&A, 475, 619
Costa, J. E. S. & Kepler, S. O. 2008, A&A, submitted
Dreizler, S. & Heber, U. 1998, A&A, 334, 618
Fu, J.-N., Vauclair, G., Solheim, J.-E. et al. 2007, A&A, 467, 237
Goupil, M. J., Dziembowski, W. A., & Fontaine, G. 1998, Baltic Astronomy, 7, 21 MNRAS 286, 303
Ito, N., Adachi, T., Nakagawa, M., Kohyama, Y., & Munakata, H. 1989, ApJ, 339, 354
Ito, N., Mutoh, H., Hikita, A., & Kohyama, Y. 1992, ApJ, 395, 622 Kawaler, S. D. 1989, ApJ, 336, 403 Asteroseismology, ed. J. Christensen-Dalsgaard & S. Frandsen (Dordrecht: Reidel), 329 al. 1995, ApJ, 450, 350 2004, A&A, 428, 969 330, 1041
Lenz, P. & Breger, M. 2005, Comm. in Asteroseismology, 146, 53 454, 845
Nather, R. E., Winget, D. E., Clemens, J. C., Hansen, C. J., & Hine, B. P. 1990, ApJ, 361, 309
O'Brien, M. S., Clemens, J. C., Kawaler, S. D., & Dehner, B. T. 1996, ApJ, 467, 397
O'Brien, M. S. & Kawaler, S. D. 2000, ApJ, 539, 372
O' Brien, M. S., Vauclair, G., & Kawaler, S. D. et al. 1998, ApJ, 495, 458 on Faint Blue Stars, eds. A. G. D. Philip, J. Liebert & R. A. Saffer ApJ, 545, 429
Silvotti, R. et al. 2007, Nature, 449, 189
Vauclair, G., Pfeiffer, B., Grauer, A., Belmonte, J.-A. et al. 1995, A&A, 299, 707 in 12th European Workshop on White Dwarf Stars, ASP Conference Series, Vol. 226, 2002, A&A, 381, 122 437 1991, ApJ, 378, 326

Discussion

TURCK-CHIEZE: Could we imagine variation of the outer layers or activity of these stars?

VAUCLAIR: The hottest PG1159 stars still show mass-loss going-on. In the case of the hottest one (RXJ2117+3412) the mass-loss rate is of the order of 10^{-7} $M_\odot$/yr. Any change of period oscillation with time would certainly depends on the mass loss process in this case, through the change in the surface layers structure and boundary conditions. However, in PG0122+200, there is no mass loss detected. The observed changes of period oscillations are not probably related on a stellar activity processes.

LUDWIG: What sets observationally the precision to which the oscillation frequencies can be determined?

VAUCLAIR: The precision to which the oscillation frequencies can be determined depends mainly on the length of the time series. The results I have shown some from multisite campaigns (except for the first 1986 data). It also depends on the amplitude of the mode, the frequency of a large amplitude mode is more precisely determined than the one of a smaller amplitude one. – Note also that it is essential to resolve the rotational splitting. In most case, the rotational splitting is of the order of a few μHz. Multisite campaigns resulting in a frequency resolution of 1 to 2 μHz are necessary, i.e., campaigns of at least 5 days. Most of our campaigns are 7 to 15 days long.

CHRISTENSEN-DALSGAARD: What is the magnitude of PG 0122? It would be very interesting to observe such star with Kepler; this could provide accurate determination of period changes over several years.

VAUCLAIR: PG 0122+200 is a B = 16.3, V = 16.8 faint star. Many of the newly discovered PG 1159 stars (from the SDSS) are fainter than 17. Of course, it would be extremely interesting to observe such rapid pulsators with Kepler for period change measurements. We carefully searched for white dwarf pulsators in the Kepler field. Unfortunately there is no known white dwarf pulsators in the Kepler field. We started a search for white dwarfs in the Kepler field, form Strömgren photometry. The following step will be to test the candidates with the right colors (to be inside the Z Ceti instability strip) for variability.

The Art of Modelling Stars in the 21st Century
Proceedings IAU Symposium No. 252, 2008
L.Deng & K.L.Chan, eds.
© 2008 International Astronomical Union
doi:10.1017/S1743921308022710

Deep inside low-mass stars

Corinne Charbonnel[1,2] and Suzanne Talon[3]

[1]Geneva Observatory, University of Geneva, ch. des Maillettes 51, 1290 Versoix, Switzerland
email: Corinne.Charbonnel@obs.unige.ch

[2]Laboratoire d'Astrophysique de Toulouse-Tarbes, Université de Toulouse, CNRS UMR 5572, 14 Av. E.Belin, 31400 Toulouse, France

[3]RQCHP, Département de physique, Université de Montréal, C. P. 6128, succ. centre-ville, Montréal, Québec, H3C 3J7
email: talonsuz@rqchp.qc.ca

Abstract. Low-mass stars exhibit, at all stages of their evolution, the signatures of complex physical processes that require challenging modeling beyond standard stellar theory. In this review, we recall the most striking observational evidences that probe the interaction and interdependence of various transport processes of chemicals and angular momentum in these objects. We then focus on the impact of atomic diffusion, large scale mixing due to rotation, and internal gravity waves on stellar properties on the main sequence and slightly beyond.

Keywords. hydrodynamics - instabilities - turbulence - waves - stars: abundances- stars: evolution - stars: interiors - stars: rotation

1. Clues on transport processes in low-mass stars

During the last couple of decades, it became obvious that "the art of modeling stars in the 21st century" will actually strongly rely on the art of modeling transport processes in stars. Observational evidences now give precise clues on the various processes that transport angular momentum and chemical elements in the radiative regions of low-mass stars, at various phases of their evolution. Here are a few examples of observations that require modeling beyond standard stellar theory: The abundance patterns of lithium in the Sun, in field and cluster main sequence and subgiant stars, as well as in Population II dwarfs; the rotation profile in the Sun inferred from helioseismology; the abundance patterns of lithium, beryllium, carbon and nitrogen on the red giant branch (see Charbonnel & Zahn 2007a, b, and this volume); the intrinsic s-process elements observed in AGB stars; the abundance of helium 3 and heavier elements in planetary nebulae; the observed spin of white dwarfs.

One of the most striking signatures of transport processes in low-mass stars is the so-called Li dip (see Fig. 1). This drop-off in the Li content of main-sequence F-stars in a range of ∼300 K centered around 6700 K was discovered in the Hyades by Wallerstein *et al.* (1965); its existence was latter confirmed by Boesgaard & Tripicco (1986). This feature appears in all open clusters older than ∼200 Myr, as well as in field stars (Balachandran 1995), an indication that it is a phenomenon occurring on the main sequence.

It was first suggested by Michaud (1986) that the Li dip could be due to element separation below the stellar convective envelope. He showed that Li is supported by radiative acceleration at the bottom of the convective envelope in stars with $T_{\rm eff} \geqslant 7000$ K, while gravitational settling dominates and leads to Li underabundances in stars with $T_{\rm eff}$ between 6800 and 6400 K. In cooler stars, at the age of the Hyades, atomic diffusion did not have enough time to modify the Li abundance in the deep surface convection zone. These predictions were obtained from first principles, the only free parameter being the mixing length parameter, which controls the effective temperature of the Li dip. In addition, a mass loss rate of the order of $10^{-15} M_\odot \, \mathrm{yr}^{-1}$ was required to reduce the expected Li overabundance on the hot side of the dip. More sophisticated models based solely on atomic diffusion were constructed by Richer & Michaud (1993), but this explanation suffers from two serious drawbacks:

▶ The expected concomitant underabundances of heavier elements (C, N, O, Mg, Si) are not observed in cluster stars (Takeda *et al.* 1998; Varenne & Monier 1999; Gebran *et al.* 2008).

▶ In this framework Li is not destroyed; it rather settles out of the convective envelope and accumulates in a buffer zone below. As a consequence, Li should be dredged-up as a star enters the Hertzsprung gap. This is not seen in the Li data, neither in the field nor in open cluster stars (Pilachowski *et al.* 1988; Deliyannis *et al.* 1997).

This suggests that another process is responsible for the existence of the Li dip. Boesgaard (1987) noticed that the effective temperature of the dip is also associated with a sharp drop in rotation velocities as can be seen in Fig. 1. Rotation was then suggested to play a dominant role in this mass range.

2. Rotation-induced transport in low-mass stars

In order to properly model rotation-induced mixing, one must follow the time evolution of the angular momentum distribution within a star, taking into account *all* relevant physical processes: contraction/expansion caused by the stellar evolution, mass loss or accretion, tidal effects, as well as internal redistribution of angular momentum through meridional circulation, turbulence, magnetic torques, waves, etc.

As discussed by Zahn in these proceedings, the description of the internal physical processes related to stellar rotation has been greatly improved during the last two decades. Talon & Charbonnel (1998; see also Charbonnel & Talon 1999, Palacios *et al.* 2003, Pasquini *et al.* 2004, and Decressin *et al.* in preparation) showed that the hot side of the Li dip (down to $T_{\mathrm{eff}} \sim 6500$ K) is very well reproduced using Zahn's (1992) model of rotation-induced mixing and the same free parameters as those required to explain abundance anomalies in more massive stars (see Zahn, and Meynet, this volume). In this framework, transport of angular momentum is dominated by the Eddington-Sweet meridional circulation and shear instabilities. The blue side of the Li dip is then attributed to enhanced mixing (and thus, Li burning) caused by the large angular momentum gradients created by surface braking. However these rotating models fail to reproduce the Li rise on the cool side of the dip, i.e., for stars with effective temperature lower than $\sim$6600 K. Why is it so?

This effective temperature appears to be a "pivotal" value in the rotation history of stars (see Fig. 1). Indeed, the physical processes responsible for the evolution of the surface velocity are different on each side of the Li dip, and the plot may be split into three temperature ranges associated with different dominant physical processes:

▶ Stars with T_{eff} higher than $\sim$6900 K have a very shallow convective envelope, which is not an efficient site for magnetic generation via a dynamo process†. Thus, contrary to Sun-like stars, they are not slowed down by a magnetic torque. As a result, these stars soon reach a stationary regime where there is no net flux of angular momentum‡, in which

---

† Here, we discuss the presence of an "external" magnetic field, which could interact with mass loss.

‡ In fact, there remains a small flux of angular momentum, just sufficient to counteract the effect of stellar contraction/expansion.

meridional circulation and turbulence counterbalance each other. The associated weak mixing is just sufficient to counteract atomic diffusion. These rotating models account nicely for the observed constancy of Li and CNO in these stars, and they also explain the Li behaviour in subgiant stars (Palacios et al. 2003; Pasquini et al. 2004).

In this region Am stars, which are known to be slow rotators, offer very good constraints to probe the processes that compete with atomic diffusion (e.g., Richer et al. 2000). Talon, Richard, & Michaud (2006) showed that the transport coefficients related to rotation-induced mixing lead to normal A stars for rotation velocities above ~ 100 km s^{-1}, and permit Am anomalies below, with a mild correlation with rotation, provided a reduction of turbulent mixing by horizontal turbulence is taken into account. Fossati et al. (2008) have confirmed observationally this correlation with rotation: in a large sample of stars in Praesepe, they find indeed a strong correlation between chemical pecularities of Am stars and their $v \sin i$.

▶ Between ~ 6900 and 6600 K, the convective envelope deepens and a weak magnetic torque, associated with the appearing dynamo, spins down the outer layers of the star. In this case, the transport of angular momentum by meridional circulation and shear turbulence increases, leading to a larger destruction of Li, in agreement with the data. The rotating model thus perfectly fits the blue side of the Li dip.

▶ Stars on the cool side of the Li dip ($T_{\rm eff} < 6600$ K) have an even deeper convective envelope sustaining a very efficient dynamo, which produces a strong magnetic torque that spins down the outer layers very efficiently. If we assume that all the angular momentum transport is assured by the wind-driven circulation in these stars†, we obtain too much Li depletion compared to the observations (see Fig. 2). On the basis of these results, Talon & Charbonnel (1998) suggested that the Li dip corresponds to a transition region where another internal process starts to efficiently transport angular momentum.

This proposition is directly linked to another observation that fails to be reproduced by the pure hydrodynamic models, namely the flat solar rotation profile revealed by helioseismology (Brown et al. 1989; Kosovichev et al. 1997; Couvidat et al. 2003; Garcia et al. 2007; see also Christensen-Daalsgard, this volume). At the solar age indeed, models relying only on turbulence and meridional circulation for momentum transport predict large angular velocity gradients that are not present in the Sun (Pinsonneault et al. 1989; Chaboyer et al. 1995; Talon 1997; Matias & Zahn 1998). This is an additional clue that another process participates to the transport of angular momentum in solar-type stars.

Two mechanisms have been proposed to explain the near uniformity of the solar rotation profile. The first rests on the possible existence of a magnetic field within the radiation zone (Charbonneau & MacGregor 1993; Eggenberger et al. 2005). The second invokes traveling internal gravity waves (hereafter IGWs) generated at the base of the convection envelope (Schatzman 1993; Zahn et al. 1997; Kumar & Quataert 1997; Talon et al. 2002). For either of these solutions to be convincing, they must be tested with numerical models coupling these processes with rotational instabilities and should explain all the aspects of the problem, including the lithium evolution with time.

In the case of the magnetic field, two different models have been suggested. First, in a series of 2–D numerical calculations based on a static poloidal field fully contained in the Sun's radiative zone, it has been shown that, at low latitudes, radial differential rotation can be severely limited (Charbonneau & MacGregor 1993; MacGregor & Charbonneau 1999). Shear is also reduced in these models, and, for the solar case, lithium burning could be reconciled with observations (Barnes et al. 1999). The temperature

---

† Let us mention that in the case of large shears meridional circulation is far more efficient than shear for angular momentum transport.

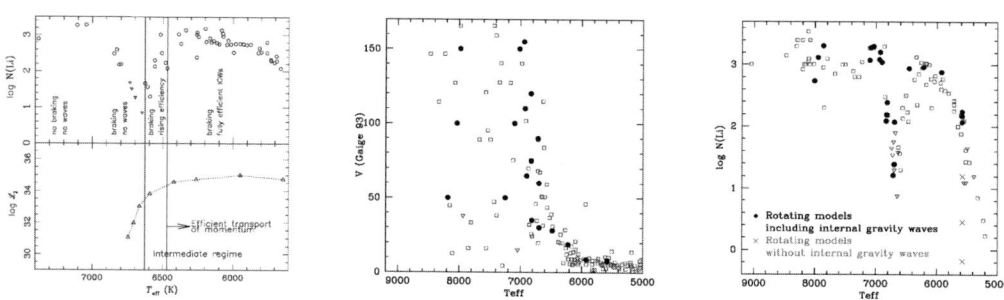

Figure 1. (**Left**) (*top*) Lithium abundance in the Hyades versus effective temperature. Also indicated are the approximate dependence of surface braking on effective temperature and the requirements for angular momentum transport so that rotational mixing can be expected to lead to the formation of the lithium dip (Talon & Charbonnel 1998). (*bottom*) Filtered angular momentum luminosity of waves below the SLO measuring the efficiency of wave induced angular momentum extraction. Adapted from Talon & Charbonnel (2003). (**Middle**) Projected rotational velocity in the Hyades (Gaigé 1993; open squares and triangles). (**Right**) Li abundances versus effective temperature in the Hyades, Coma B and Praesepe (Burkhart & Coupry 1998, 2000; open squares and triangles). Comparison with our models including IGWs (black dots; Charbonnel *et al.* in preparation); the crosses indicate the predictions for the 1 M$_\odot$ computed without IGW

dependence has not been examined in the framework of that model, and no natural relation between the existence of a surface convection zone and an internal poloidal field is expected. Furthermore, recent numerical simulations by Brun & Zahn (2006) have cast doubt on the possibility that such a field would remain confined to the radiative zone for over 4 Gyr (see however the recent results by Wood & McIntyre 2007). Second, Eggenberger *et al.* (2005) studied the impact of the Tayler-Spruit instability in solar models. The impact on light element abundances has not been quantified and once more, no correlation is expected with surface temperature.

The Li data suggest that the efficiency of the additional process is linked to the growth of the convective envelope in stars with effective temperatures around $T_{\rm eff} \simeq 6600$ K. As we shall see below, this is a characteristic of IGWs.

3. IGWs generation and momentum extraction in low-mass stars

In the Earth's atmosphere, wave-induced momentum transport is a key process in the understanding of several phenomena, the best known being the quasi-biennial oscillation of the stratosphere. In astrophysics, IGWs have initially been invoked as a source of mixing for chemicals (Press 1981; Garcia Lopez & Spruit 1991; Schatzman 1993; Montalban 1994; Montalban & Schatzman 1996, 2000; Young *et al.* 2003). Ando (1986) studied the transport of angular momentum associated with standing gravity waves in Be stars. He was the first to clearly state, in the stellar context, that IGWs carry angular momentum from the region where they are excited to the region where they are dissipated. Traveling IGWs have since been invoked as an important source of angular momentum redistribution in single stars (Schatzman 1993; Kumar & Quataert 1997; Talon *et al.* 1997, 2002; Charbonnel & Talon 2005).

3.1. *IGWs generation and wave spectrum*

The existence of IGWs in stars is expected from numerical simulations of penetrative convection both in 2– and 3–D (Hurlburt *et al.* 1986, 1994; Andersen 1994; Nordlund *et al.* 1996; Kiraga *et al.* 2000; Dintrans *et al.* 2005; Rogers & Glatzmaier 2005a, b). Ultimately, one may wish to obtain realistic wave fluxes from such simulations; however, conditions for these simulations are still too far from realistic to be used at this time (for more details on these aspects see Charbonnel & Talon 2007 and the discussion at the end of the present paper). From a theoretical viewpoint, two different and complementary processes excite IGWs: convective overshooting in a stable region (García López & Spruit 1991; Kiraga et al. 2003; Rogers & Glatzmaier 2004), and Reynolds stresses in the convection zone (Goldreich & Kumar 1990; Balmforth 1992; Goldreich et al. 1994). Here, we shall use the second mechanism, which has been calibrated on solar p-modes and, thus, seems more reliable at this time. We are fully aware that, as of now, this is the weakest point of wave modeling. Work in underway to evaluate analytically the contribution of penetrative plumes to excitation (Belkacem et al. 2008).

For the evolution of angular momentum deep inside the star, the relevant parameter is the net angular momentum luminosity slightly below the convective envelope. To estimate that luminosity, we first need to estimate the spectrum of excited waves. In the Goldreich et al. (1994) model that we use, driving is dominated by entropy fluctuations. In our calculations (Talon & Charbonnel 2003, 2004, 2005, 2007, 2008; Charbonnel & Talon 2005), we neglect wave generation by overshooting, although it is expected to be very efficient; our present estimate is thus a lower limit to the correct/total wave flux. As far as numerical simulations are concerned, some authors agree as to the order of magnitude we obtain in our wave-flux (Kiraga et al. 2000; Dintrans et al. 2005) while Rogers & Glatzmaier (2005) state that we over-estimate this flux. The reason for this discrepancy has been discussed in Charbonnel & Talon (2007), and we are confident that, as 3-D numerical simulations evolve to more realistic regimes, the simulated wave flux will rise with turbulence.

In Talon & Charbonnel (2005), we developed a formalism to incorporate the contribution of IGWs to the transport of angular momentum and chemical elements in stellar models. We showed that the development of a double-peaked shear layer (SLO, for Shear Layer Oscillation), acts as a filter for waves and discussed how the asymmetry of this filter produces momentum extraction from the core when it is rotating faster than the surface. Using only this filtered flux, it is possible to follow the contribution of internal waves over long (evolutionary) time-scales. Let us recall the main features of our formalism. The energy flux per unit frequency $\mathcal{F}_E$ is

$$\mathcal{F}_E(\ell,\omega) = \frac{\omega^2}{4\pi} \int dr \, \frac{\rho^2}{r^2} \left[\left(\frac{\partial \xi_r}{\partial r}\right)^2 + \ell(\ell+1) \left(\frac{\partial \xi_h}{\partial r}\right)^2 \right]$$
$$\times \exp\left[-h_\omega^2 \ell(\ell+1)/2r^2\right] \frac{v_c^3 L^4}{1+(\omega\tau_L)^{15/2}}, \qquad (3.1)$$

where ξ_r and $[\ell(\ell+1)]^{1/2}\xi_h$ are the radial and horizontal displacement wave-functions, which are normalized to unit energy flux just below the convection zone, v_c is the convective velocity, $L = \alpha_{\mathrm{MLT}} H_P$ the radial size of an energy bearing turbulent eddy, $\tau_L \approx L/v_c$ the characteristic convective time, and h_ω is the radial size of the largest eddy at depth r with characteristic frequency of ω or greater ($h_\omega = L \min\{1, (2\omega\tau_L)^{-3/2}\}$). The radial wave number k_r is related to the horizontal wave number k_h by

$$k_r^2 = \left(\frac{N^2}{\sigma^2} - 1\right) k_h^2 = \left(\frac{N^2}{\sigma^2} - 1\right) \frac{\ell(\ell+1)}{r^2} \qquad (3.2)$$

where N^2 is the Brunt-Väisälä frequency. In the convection zone, the mode is evanescent and the penetration depth varies as $\sqrt{\ell(\ell+1)}$. The momentum flux per unit frequency

$\mathcal{F}_J$ is then related to the energy flux by

$$\mathcal{F}_J(m,\ell,\omega) = \frac{2m}{\omega}\mathcal{F}_E(\ell,\omega) \tag{3.3}$$

(Goldreich & Nicholson 1989; Zahn, Talon & Matias 1997). We integrate this quantity horizontally to get an angular momentum luminosity

$$\mathcal{L}_J = 4\pi r^2 \mathcal{F}_J \tag{3.4}$$

which, in the absence of dissipation, is conserved (Bretherton 1969; Zahn *et al.* 1997). Each wave then travels inward and is damped by thermal diffusivity and by viscosity. The local momentum luminosity of waves is given by

$$\mathcal{L}_J(r) = \sum_{\sigma,\ell,m} \mathcal{L}_{J\ell,m}(r_{cz})\exp\left[-\tau(r,\sigma,\ell)\right] \tag{3.5}$$

where 'cz' refers to the base of the convection zone. τ corresponds to the integration of the local damping rate, and takes into account the mean molecular weight stratification

$$\tau(r,\sigma,\ell) = [\ell(\ell+1)]^{\frac{3}{2}} \int_r^{r_c} (K_T + \nu_v)\frac{NN_T^2}{\sigma^4}\left(\frac{N^2}{N^2-\sigma^2}\right)^{\frac{1}{2}}\frac{dr}{r^3} \tag{3.6}$$

(Zahn *et al.* 1997). In this expression, N_T^2 is the thermal part of the Brunt-Väisälä frequency, K_T is the thermal diffusivity and ν_v the (vertical) turbulent viscosity. σ is the local, Doppler-shifted frequency

$$\sigma(r) = \omega - m\left[\Omega(r) - \Omega_{cz}\right] \tag{3.7}$$

and ω is the wave frequency in the reference frame of the convection zone. Let us mention that, in this expression for damping, only the radial velocity gradients are taken into account. This is because angular momentum transport is dominated by the low frequency waves ($\sigma \ll N$), which implies that horizontal gradients are much smaller than vertical ones (*cf.* Eq. 3.2).

When meridional circulation, turbulence, and waves are all taken into account, the evolution of angular momentum follows

$$\rho\frac{d}{dt}\left[r^2\Omega\right] = \frac{1}{5r^2}\frac{\partial}{\partial r}\left[\rho r^4 \Omega U\right] + \frac{1}{r^2}\frac{\partial}{\partial r}\left[\rho\nu_v r^4\frac{\partial\Omega}{\partial r}\right] - \frac{3}{8\pi}\frac{1}{r^2}\frac{\partial}{\partial r}\mathcal{L}_J(r), \tag{3.8}$$

(Talon & Zahn 1998) where U is the radial meridional circulation velocity. This equation takes into account the advective nature of meridional circulation rather than modeling it as a diffusive process and assumes a "shellular" rotation (see Zahn 1992 for details). Horizontal averaging was performed, and meridional circulation is considered only at first order. When we calculate the fast SLO's dynamics, U is neglected in this equation. This is justified by the fact that, when shears are large such as in the SLO angular momentum redistribution is dominated by the (turbulent) diffusivity rather than by meridional circulation. However the complete equation is used when secular time-scales are involved as required when we compute full evolution models as in Charbonnel & Talon (2005).

3.2. *Shear layer oscillation (SLO) and filtered angular momentum luminosity*

One key feature when looking at the wave-mean flow interaction is that the dissipation of IGWs produces an increase in the local differential rotation: this is caused by the increased dissipation of waves that travel in the direction of the shear (see Eqs. 3.6 and 3.7). In conjunction with viscosity, this leads to the formation of an oscillating doubled-peak shear layer that oscillates on a short time-scale (Gough & McIntyre 1998; Ringot 1998; Kumar, Talon & Zahn 1999). This oscillation is similar to the Earth quasi-biennial oscillation that is also caused by the differential damping of internal waves in a shear region.

This SLO occurs if the deposition of angular momentum by IGWs is large enough when compared with (turbulent) viscosity (Kim & MacGregor 2001)†. To calculate the turbulence associated with this oscillation, we rely on a standard prescription for shear turbulence away from regions with mean molecular weight gradients

$$\nu_v = \frac{8}{5} Ri_{\rm crit} K \frac{(r\,\mathrm{d}\Omega/\mathrm{d}r)^2}{N_T^2} \qquad (3.9)$$

which take radiative losses into account (Townsend 1958; Maeder 1995). This coefficient is time-averaged over a complete oscillation cycle (for details, see TC05).

In the presence of differential rotation, the dissipation of prograde and retrograde waves in the SLO is not symmetric, and this leads to a finite amount of angular momentum being deposited in the interior beyond the SLO. This is the filtered angular momentum luminosity $\mathcal{L}_J^{\rm fil}$. Let us mention that in fact, the existence of a SLO is not even required to obtain this differential damping between prograde and retrograde waves, and thus, as long as differential rotation exists at the base of the convection zone, waves will have a net impact of the rotation rate of the interior.

The SLO's dynamics is studied by solving Eq. (3.8) with small time-steps and using the whole wave spectrum while for the secular evolution of the star, one has to use instead the filtered angular momentum luminosity. Let us stress that in the case of the secular evolution we do not follow the SLO dynamics, because of its very short time scale. Rather, we only consider the net angular momentum luminosity beyond the SLO, and its effect on chemicals is given by a local turbulence calculated from a study of the SLO's dynamic over very short time-scales. Let us also mention here that, for both the SLO and the filtered angular momentum luminosity, differential damping is required. Since this relies on the Doppler shift of the frequency (see Eqs. 3.6 and 3.7), angular momentum redistribution will be dominated by the low frequency waves that experience a larger Doppler shift, but that is not so low that they will be immediately damped. Numerical tests indicate that this occurs around $\omega \simeq 1\ \mu$Hz.

4. The case of Pop I low-mass stars

A very important property of IGWs is that their generation and efficiency in extracting angular momentum from stellar interiors depend on the structure of their convective envelope, which varies very strongly with the effective temperature of the star. Figure 1 shows the $T_{\rm eff}$-dependence of the filtered angular momentum luminosity of waves below the SLO, which directly measures the efficiency of wave-induced angular momentum extraction, in zero-age main sequence stars around the Li dip. It appears that the net momentum luminosity slightly increases with increasing $T_{\rm eff}$, presents a plateau, and suddenly drops at the $T_{\rm eff}$ of the dip. This clearly indicates that the momentum transport by IGWs has the proper $T_{\rm eff}$-dependence to be the required process to explain the cool side of the Li dip (Talon & Charbonnel 2003).

Talon et al. (2002) have shown, in a static model, that waves can efficiently extract angular momentum from a star that has a surface convection zone rotating slower than the interior. Charbonnel & Talon (2005) then calculated the evolution of the internal

† If viscosity is large, a stationary state can be reached.

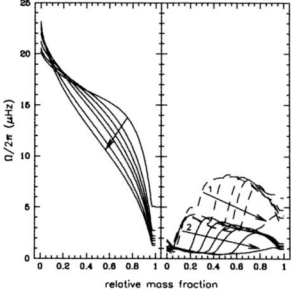 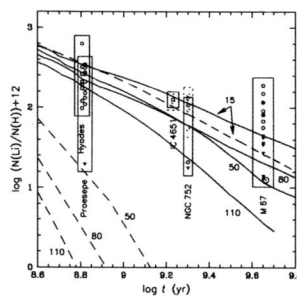

Figure 2. (**Left**) Evolution of the rotation profile in a solar-mass model with and without IGWs. The initial equatorial rotation velocity is 50 km s^{-1}, and identical surface magnetic braking is applied. (*left*) Model without IGWs. Curves correspond to ages of 0.2, 0.5, 0.7, 1.0, 1.5, 3.0 and 4.6 Gy that increase in the direction of the arrow. Differential rotation remains large at all times. (*right*) When IGWs are included, low-degree waves penetrate all the way to the core and deposit their negative angular momentum in the whole radiative region. Because the core's angular momentum is minute, it is spun down very efficiently. In the so-created "slow" region, damping of retrograde waves increases, leading to the formation of a front, which propagates from the core to the surface. Curves showing propagation of the first front (labeled 1) correspond to ages of 0.2, 0.21, 0.22, 0.23, 0.25 and 0.27 Gy. Further spin down leads to the formation of a second front (ages 0.5, 0.7, 1.0, 1.5, 3.0 and 4.6 Gy). The first front propagates faster than the second one due to stronger braking early in evolution. At the age of the Sun, the radiative region is rotating almost uniformly. From Charbonnel & Talon (2005). (**Right**) Evolution of surface lithium abundance with time for solar-mass stars. The vertical extent of boxes shows the range of lithium values as observed in various galactic clusters for stars with an effective temperature corresponding to that of the model ± 100 K at the cluster age, plus a typical error in abundance determination. The horizontal extent corresponds to the age uncertainty. Circles indicate abundance determinations, and triangles denote upper limits for individual stars. The solar value is shown with the usual symbol $\odot$. Solid lines correspond to models including IGWs and dashed lines to models without IGWs. Initial velocities are shown on the figure (in km s^{-1}). In the cases without IGWs, except for the slowest rotator, lithium depletion is too strong, by orders of magnitude, at all ages. When included, IGWs, by changing the shape of the internal velocity gradients, lead to a decrease of the associated transport of chemicals. Lithium is then much less depleted and predictions account very well for the data. At all considered ages, the observed dispersion in atmospheric lithium is well explained in terms of a spread in initial velocities. From Charbonnel & Talon (2005).

rotation profile for a solar-mass star with surface spin-down. We showed that, in that case, waves tend to slow down the core, creating "slow" fronts that propagate from the core to the surface (Fig. 2). These calculations confirmed, for the first time in a complete evolution of solar-mass models evolved from the pre-main sequence to 4.6 Gy, that IGWs play a major role in braking the solar core (Charbonnel & Talon 2005). This momentum transport reduces rotational mixing in low-mass stars, leading to a theoretical surface lithium abundance in agreement with observations made in solar mass stars in open clusters of various ages (Fig. 2).

Figure 1 shows our predictions for rotation velocities and Li surface abundances together with the observed data at the age of the Hyades. On the left side of the dip, IGWs play no role and the predictions are taken from Charbonnel & Talon (1999). On the cool side of the dip IGWs are at act and lead to the rise of the surface Li. The model at $\sim$5800 K corresponds to a 1.0 $M_\odot$ star. It was computed for 3 initial rotation velocities of 50, 80 and 110 km s^{-1} both in the case with (black dots) and without IGWs (crosses). Models with IGWs are in perfect agreement with the observations, both regarding the amplitude of the Li depletion and the dispersion at a given effective temperature. More details on these models will be given in Charbonnel *et al.* (in preparation).

5. The case of Pop II low-mass stars

In the context of primordial nucleosynthesis, it has long been debated whether Pop II stars could have depleted their surface Li abundance, just as their metal-rich counterpart did. Recent results on cosmic microwave background anisotropies, and especially those of the WMAP experiment, have firmly established that the primordial Li abundance is ~2.5 to 3 times higher than the measured Li value in dwarf stars along the so-called Spite plateau (Charbonnel & Primas 2005). The main theoretical difficulty to reproduce these data is that the Li abundance is remarkably constant in halo dwarfs, while it seems at first sight that depletion would lead to a larger dispersion.

A re-examination of Li data in halo stars available in the literature (Charbonnel & Primas 2005) has led for the first time to a very surprising result: the mean Li value as well as its dispersion appear to be lower for the dwarfs than for the subgiant stars. In addition, all the deviant stars, i.e., the stars with strong Li deficiency and those with abnormally high Li content, lie on or originate from the hot side of the Li plateau. These results indicate that halo stars that have now just passed the turnoff have experienced a Li history sightly different from that of their less massive counterparts.

We suggested that such a behaviour is the signature of a transport process for angular momentum whose efficiency changes on the extreme blue edge of the plateau. Such behaviour corresponds to that of the generation and filtering of IGWs in Pop II stars (Talon & Charbonnel 2004), just as it does in the case of Pop I stars. Indeed and as discussed previously, the generation of IGWs and, consequently, their efficiency in transporting angular momentum, depend on the structure of the stellar convective envelope, which in turn depends on the effective temperature of the star (Fig. 1). As in the case of Pop I stars on the red side of the Li dip, the net angular luminosity of IGWs is very high and constant in Pop II stars along the plateau up to $T_{\rm eff} \sim 6300\,{\rm K}$. There, IGWs should dominate the transport of angular momentum and enforce quasi solid-body rotation of the stellar interior on very short timescale. As a result, the surface Li depletion is expected to be independent of the initial angular momentum distribution, implying a very low dispersion of the Li abundance from star to star. In more massive stars however the efficiency of IGWs decreases and internal differential rotation is expected to be maintained under the effect of meridional circulation and turbulence. Consistently, variations of the initial angular momentum from star to star would lead to more Li dispersion and to more frequent abnormalities in the case of the most massive stars where IGWs are not fully efficient, as required by the observations. We note that the mass-dependence of the IGWs efficiency also leads to a natural explanation of fast horizontal branch rotators. The proper treatment of the effects of IGWs together with those of atomic diffusion, meridional circulation and shear turbulence has now to be undertaken in fully consistent models of halo stars.

6. IGWs in intermediate-mass stars

Although waves produced by the surface convection zone can be ignored safely for more massive stars (i.e., with $T_{\rm eff} \geq 6700$ K) while on the main sequence, it is not the case for later evolutionary stages. In particular, Talon & Charbonnel (2008) showed that angular momentum transport by IGWs emited by the convective envelope could be quite important in intermediate-mass stars on the pre-main sequence, at the end of the subgiant branch, and during the early-AGB phase. This implies that possible differential rotation, which could be a relic of the star's main sequence history and subsequent contraction, could be strongly reduced when the star reaches the AGB-phase. This could have profound impact on the subsequent evolution. In particular, this could help solving the long-standing problem of the production of the s-process elements (Herwig *et al.* 2003; Siess *et al.* 2004), as well as explaining the observed white dwarf spins (Suijs *et al.* 2008). Work is in progress to quantify these effects.

Let us also mention that waves can be excited at the boundary of the convective core. These waves travel towards the surface, and are fully damped there, as thermal diffusivity is quite large at the star's surface, and could have a large impact on rotation profiles there (Pantillon, Talon, & Charbonnel 2007). As these stars are fast rotators, the Coriolis force must be accounted for, and much work remains to be done in that direction (Pantillon *et al.* 2007; Mathis *et al.* 2008).

7. Conclusions

IGWs have a large impact on the evolution of low-mass stars, especially through their effect on the rotation profile, which then modifies meridional circulation and shear turbulence. Within this framework, the hydrodynamical models including the combined effects of meridional circulation, shear turbulence and internal gravity waves (using an excitation model that reproduces the solar p-modes) successfully shape both the rotation profile and the time evolution of the surface lithium abundance in low-mass stars. Up to now, no other theoretical model has achieved such a goal. Our comprehensive picture should have implications for other difficult unsolved problems related to the transport of chemicals and angular momentum in stars. We think in particular to the stars on the horizontal and asymptotic giant branches that exhibit unexplained abundance anomalies. No doubt that all these so-called "non-standard" physical processes must be part of the art of modelling stars in the 21st century.

References

Ando, H. 1986, *A&A*, 163, 97
Andersen, B. N. 1994, *Solar Phys.*, 152, 241
Balmforth, N. J. 1992, *MNRAS* 255, 639
Barnes, G., Charbonneau, P., & Mac Gregor, K. B. 1999, *ApJ* 511, 466
Belkacem, K., Talon, S., Goupil, M.-J., & Samadi, R. 2008, *A&A*, in preparation
Boesgaard, A. M. & Tripicco, M. J. 1986, *ApJ*, 302, L49
Boesgaard, A. M. 1987, *PASP* 99, 1067
Bretherton, F. P. 1969, *Quart. J. R. Met. Soc.*, 95, 213
Brown, T. M., *et al.* 1989, *ApJ* 343, 526
Brun, A. S. & Zahn, J. P. 2006, *A&A*, 457, 665
Burkhart, C. & Coupry, M. F. 1998, *A&A* 338, 1073
Burkhart, C. & Coupry, M. F. 2000, *A&A* 354, 216
Chaboyer, N., Demarque, P., Guenther, D. B., & Pinsonneault, M. H. 1995, *ApJ* 446, 435
Charbonneau, P. & Mac Gregor, K. B. 1993, *ApJ* 417, 762
Charbonnel, C. & Primas, F. 2005, *A&A* 442, 961
Charbonnel, C. & Talon, S. 1999, *A&A* 351, 635
Charbonnel, C. & Talon, S. 2005, *Science* 309, 2189
Charbonnel, C. & Talon, S. 2007, AIP Conference Proceedings, Volume 948, pp. 15-26
Charbonnel, C. & Zahn, J. P. 2007a, *A&A* 467, L15
Charbonnel, C. & Zahn, J. P. 2007b, *A&A* 476, L29
Couvidat, S., *et al.* 2003, *ApJ* 597, L77
Deliyannis, C. P., King, J. R., & Boesgaard, A. M. 1997, Kontikas E., *et al.* (eds), "Wide-field spectroscopy", p. 201
Dintrans, B., Brandenburg, A., & Nordlund, A. 2005, *A&A*, 438, 365

Eggenberger, P., Maeder, A., & Meynet, G. 2005, *A&A*, 440, L9
Fossati, L., et al. 2008, arXiv0803.3540F
Gaigé, Y. 1993, *A&A* 269, 267
García López, R. J. & Spruit, H. C. 1991, *ApJ* 377, 268
Gebran, M., Monier, R., Richard, O. 2008, arXiv0803.0947G
Goldreich, P. & Kumar, P. 1990, *ApJ* 363, 694
Goldreich, P., Murray, N., Kumar, P. 1994, *ApJ* 424, 466
Goldreich, P. & Nicholson, P. D. 1989, *ApJ* 332, 1079
Gough, D. O. & McIntyre, M. E. 1998, *Nature* 394, 755
Herwig, F., Langer, N., & Lugaro, M. 2003, *ApJ* 593, 1056
Hurlburt, N. E., Toomre, J., & Massaguer, J. M. 1986, *ApJ*, 311, 563
Hurlburt, N. E., Toomre, J., Massaguer, J. M., & Zahn, J. P. 1994, *ApJ*, 421, 245
Kim, E. & MacGregor, K. B. 2001, *ApJ*, 556, L117
Kiraga, M., Jahn, K., Stepien, K., & Zahn, J.-P. 2003, *Acta Astronomica* 53, 321
Kiraga, M., Rózyczka, M., Stepien, K., Jahn, K., Muthsam, H. 2000, *Acta Astronomica* 50, 93
Kosovichev, A., et al. 1997, *Sol. Phys.* 170, 43
Kumar, P. & Quataert, E. J. 1997, *ApJ* 575, L143
Kumar, P., Talon, S., Zahn, J.-P. 1999, *ApJ* 520, 859
MacGregor, K. & Charbonneau, P. 1999 *ApJ*, 519, 911
Maeder, A. 2005, *A&A*, 299, 84
Maeder, A. & Meynet, G. 2000, *ARAA* 38, 143
Mathis, S., Talon, S., Pantillon, F. P., Zahn, J.-P. 2008, *Solar Phys.*, available online
Matias, J. & Zahn, J.-P. 1998, Provost & Schmider (eds), "Sounding solar and stellar interiors", IAU Symp. 181
Michaud, G. 1986, *ApJ* 302, 650
Montalban, J. 1994 *A&A*, 281, 421
Montalban, J. & Schatzman, E. 1996 *A&A*, 305, 513
Montalban, J. & Schatzman, E. 2000 *A&A*, 354, 943
Nordlund, A., Stein, R. F., & Brandenburg, A. 1996 *Bull. Astron. Soc. of India*, 24, 261
Palacios, A., Talon, S., Charbonnel, C., Forestini, M. 2003, *A&A* 399, 603
Pantillon, F. P., Talon, S., Charbonnel, C. 2007, *A&A* 474, 155
Pasquini, L., Randich, S., Zoccali, M., Hill, V., Charbonnel, & C., Nordström, B. 2004, *A&A* 424, 951
Pilachowski, C. A., Saha, A., & Hobbs, L. M. 1988, *PASP*, 100, 474
Press, W. H. 1981 *ApJ*, 245, 286
Richer, J. & Michaud, G. 1993, *ApJ* 416, 312
Richer, J., Michaud, G., & Turcotte, S. 2000, *ApJ* 529, 338
Ringot, O. 1998, *A&A* 335, 89
Rogers, T. M. & Glatzmaier, G. A. 2005a, *ApJ*, 620, 432
Rogers, T. M. & Glatzmaier, G. A. 2005b, *MNRAS*, 364, 1135
Schatzman, E. 1993 *A&A* 279, 431
Siess, L., Goriely, S., Langer, N. 2004 *A&A* 415, 1089
Suijs, M. P. L., Langer, N., Poelarends, A. J., Yoon, S. C., Heger, A., & Herwig, F. 2008, *A&A* 481, L87
Takeda, Y., Kawanomoto, S., Takada-Hidai, M., & Sadakane, K. 1998, *PASJ* 50, 509
Talon S. 1997 *PhD Thesis*, Université Paris VII
Talon, S. & Zahn J.-P. 1997, *A&A* 317, 749
Talon, S., Kumar, P., & Zahn, J.-P. 2002, *ApJL* 574, 175
Talon, S. & Charbonnel, C. 1998, *A&A* 335, 959
Talon, S. & Charbonnel, C. 2003, *A&A* 405, 1025
Talon, S. & Charbonnel, C. 2004, *A&A* 418, 1051
Talon, S. & Charbonnel, C. 2005, *A&A* 440, 981
Talon, S. & Charbonnel, C. 2008, *A&A* in press, arXiv:0801.4643
Talon, S., Richard, O., Michaud, G. 2006, *ApJ* 645, 634

Talon, S. & Zahn, J. P. 1998, *A&A*, 329, 315
Townsend, 1958
Varenne, O. & Monier, R. 1999 *A&A* 351, 247
Wallerstein, G., Herbig., G. H., & Conti, P.S. 1965, *ApJ*, 141, 610
Wood, T. S., & McIntyre, M. E. 2007, AIP Conference Proceedings, Volume 948, pp. 303-308
Young, P. A., Knierman, K. A., Rigby, J. R., Arnett, D. 2003, *ApJ*, 585, 1114
Zahn, J. P. 1992, *A&A* 265, 115
Zahn, J. P., Talon, S., & Matias, J. 1997, *A&A* 322, 320

Discussion

WOITKE: Why do the simulations of Glatzmaier *et al.* for the excitation of gravity waves by convection fail? Do they yield too strong or too weak convection?

CHARBONNEL: The convective motions in this simulation are "too lazy". The numerical resolution is too small to resolve the hammering of plumes properly. Prandtl-numbers are too low by orders of magnitudes.

KUPKA: I would like to comment on the question of the resolution and turbulence in 3–D global solar simulations. If you take the case of 512^3 grid points, at the bottom of the solar convection zone you have a horizontal resolution of ~ 6000 km. The local pressure scale height there is ~ 50000 km. So if you consider the energy carrying scales to be of that size, horizontally $R_{eff} \sim (L/R)^{4/3} \sim 20$. Vertically the resolution is perhaps some 600 km, hence $R_{eff} \sim 370$. Also, for comparison, the so-called "extent of overshooting", according to helioseismology, is some 2500 km, if you take it to be $\sim 0.05 H_p$. Thus, the simulations cannot resolve shear-driven turbulence created by the flow on such scales. It is numerically simply too expensive to do that on computers currently available.

LANGER: You prefer g-modes as mechanism to slow down the core of MS stars, since this gives Li-depletion on the cool side of the dip but not on the hot side. Would you not expect a similarly different effect of internal magnetic transport for both sides of the dip, since the cool stars suffer magnetic braking but the hot stars don't?

CHARBONNEL: The question is more related to the transport of angular momentum inside the star, and not to the braking at the surface. For the moment there is no argument to suspect a T_{eff}-dependence of the angular momentum transport by magnetic field.

LUDWIG: The Li abundance on the Spite plateau is very homogeneous over almost 1000 K in T_{eff} and 2 orders of magnitude in metallicity. The production rate of gravity waves, and I presume its mixing is dependent on the mass of the convective envelope. Can you comment how the homogeneity of the Li abundance is consistent with the substantial dependence of the stellar structure.

CHARBONNEL: There is in fact a threshold value for the momentum luminosity of waves above which the waves are extremely efficient in transporting angular momentum. The dwarf stars that lie on the Li plateau, despite their slightly different internal structure, all managed to be above that threshold in their infancy. We thus expect that they all managed to become solid-body rotators. As a consequence, they must have depleted Li in a very homogeneous way.

Hydrodynamical Simulations of Turbulent Convection in a Rotating Red Giant Star

A. Palacios[1,2] and A. S. Brun[2]

[1] Université Montpellier II – GRAAL, CNRS – UMR 5024, place Eugène Bataillon, F-34095 Montpellier, France
email: palacios@graal.univ-montp2.fr

[2] CEA/DSM/IRFU/SAp – L'Orme des Merisiers bât 709 – F-91191 Gif-sur-Yvette (France)
email: asbrun@cea.fr

Abstract. We present 3-D hydrodynamical simulations of the extended turbulent convective envelope of a low-mass red giant star. These simulations, computed with the ASH code, aim at understanding the redistribution of angular momentum and heat in extended turbulent convection zones of these giant stars. We focus our study on the effects of turbulence and of the rotation rate on the convective patterns and on the distribution of angular momentum within the inner 50% of the convective envelope of such stars.

Keywords. Hydrodynamics, convection, stars: rotation, stars: evolution

1. Introduction

Turbulent convection is still poorly known and understood in the extended envelopes of giant stars, and its interaction with rotation also needs to be more deeply understood. As of today, our understanding of stellar evolution is based on 1-D stellar evolution codes that compute the late evolution of stars on the giant branch by assuming mixing length theory (Bohm-Vitense 1958). The mixing length parameter α used for red giant models is generally calibrated on the main sequence so as to fit the Sun (Brun *et al.* 2002). In more sophisticated rotating evolutionary models, the transport of angular momentum in radiative regions is ensured by meridional circulation and shear-induced turbulence (Zahn 1992), while solid-body rotation is assumed in convective regions (e.g. strong turbulent diffusion of the angular momentum) in the absence of a reliable formalism to describe the transport of angular momentum in these regions. These assumptions, both on the mixing length parameter and on the angular momentum distribution inside convective regions are based on simple arguments developed many decades ago to model the Sun (and even for our closest star this does not necessarily hold true today, Brun & Toomre 2002, Miesch *et al.* 2008). Further the structure and the properties of the convective envelope of red giant stars are completely different from that of the Sun. The combination of determinations of $v \sin i$ for stars in different evolutionary phases indicates that the assumption of solid-body rotation in extended low-density convective envelopes is most likely erroneous. In particular, the analysis of the rotation of horizontal branch stars in globular clusters led Sills & Pinsonneault (2000) to suggest that in order to reproduce the rotation velocity of HB stars considering that the progenitors of these stars on the red giant branch had almost zero surface rotation, a large amount of angular momentum should be retained in the inner part of the stars during the RGB ascension. This can be achieved if RGB convective envelopes rotate differentially. In the following work, we investigate more deeply the interaction of turbulent convection and rotation in such an extended convective zone using 3-D simulations in an attempt to address these important scientific topics. In section 2 we briefly discuss the numerical model, in §3 the patterns realized in the convective envelope, in §4 the angular momentum redistribution within the shell and conclude in §5.

2. Physical Inputs and Numerical Simulations with the ASH code

We have used the Anelastic Spherical Harmonic (ASH) code in its purely hydrodynamic mode to study the convection in extended envelope of a low-mass red giant (RGB) star (Palacios & Brun 2007). The ASH code has been successfully applied to study convection in different types of stars such as the Sun or massive stars (Brun & Toomre 2002, Miesch et al. 2006, Browning et al. 2004). These recent successes give credit to the present simulations. The reader is referred to Brun et al. (2004) for details on the code. Let us here briefly summarize the main characteristics of our simulations. To build the 3-D nonlinear simulation we used a 1-D red giant stellar model as initial state, whose characteristics are: $M_{ini} = 0.8\ M_\odot$, $L_* = 425\ L_\odot$ and $R_* = 39\ R_\odot$. In order use a density contrast that can be numerically handled by ASH at a reasonable computational cost (e.g. $\rho_{bottom}/\rho_{top} = 100$), we only consider the inner 50% of the convective envelope, e.g. $r \in [0.05\ R_*; 0.5\ R_*]$. This choice is also driven by the fact that getting too close to the surface would imply large Mach numbers that are not compatible with the anelastic assumption. We assume that the convective envelope initially rotates as a solid-body, and we adopt two different initial rotation rates : a tenth solar (case A), corresponding to $v_{eq,surf} \simeq 7$ km/s and a fiftieth solar (case B) corresponding to $v_{eq,surf} \simeq 1.4$ km/s. We assume rigid stress free boundary conditions at the edges of the computational domain, and impose a flux of radiative energy at the base of our domain that is extracted at the surface. The model parameters for cases A and B are given in Table 1.

Table 1. The Prandtl number is defined by ν_{top}/κ_{top}. The Rayleigh number $R_a = (-\partial\tilde{\rho}/\partial\tilde{S})\Delta S g L^3/\rho\nu\kappa$ (with ΔS the entropy contrast across the domain) is evaluated at mid-layer depth. The rms Reynolds number is given by $\tilde{R}_e = \tilde{v}L/\nu$, where $\tilde{v}$ is a representative rms convective velocity. In both cases $L = 1.22 \times 10^{22}$ cm. Time averages over the full computational domain of the total kinetic energy (KE), and that associated with the (axisymmetric) differential rotation (DRKE), the (axisymmetric) meridional circulation (MCKE) and the non-axisymmetric convection itself (CKE), all in units of ergs.cm^{-3}.

Simulation Parameters								
Case	$\Omega_0/\Omega_\odot$	N_r, N_θ, N_ϕ	P_{rot}	ν_{top}	κ_{top}	P_r	R_e	R_a
A	0.1	256, 256, 512	273 d ($\times$ 12)	1.2×10^{15}	1.2×10^{15}	1	400	7.5×10^5
B	0.02	256, 256, 512	1398 d ($\times$ 6)	1.2×10^{15}	1.2×10^{15}	1	375	4.5×10^5

Time and Volume Averaged Energy Densities			
KE	DRKE	CKE	MCKE
A	1.36×10^6	4.69×10^5 (34.4 %)	7.86×10^5 (57.7 %)	1.07×10^5 (7.9 %)
B	2.18×10^6	1.36×10^5 (6.2 %)	1.80×10^6 (82.4 %)	2.47×10^5 (11.4 %)

3. Convective patterns

In both cases, the convective instability sets in during the first hundred days of the simulation and then undergoes non-linear saturation. It subsequently reaches a statistical

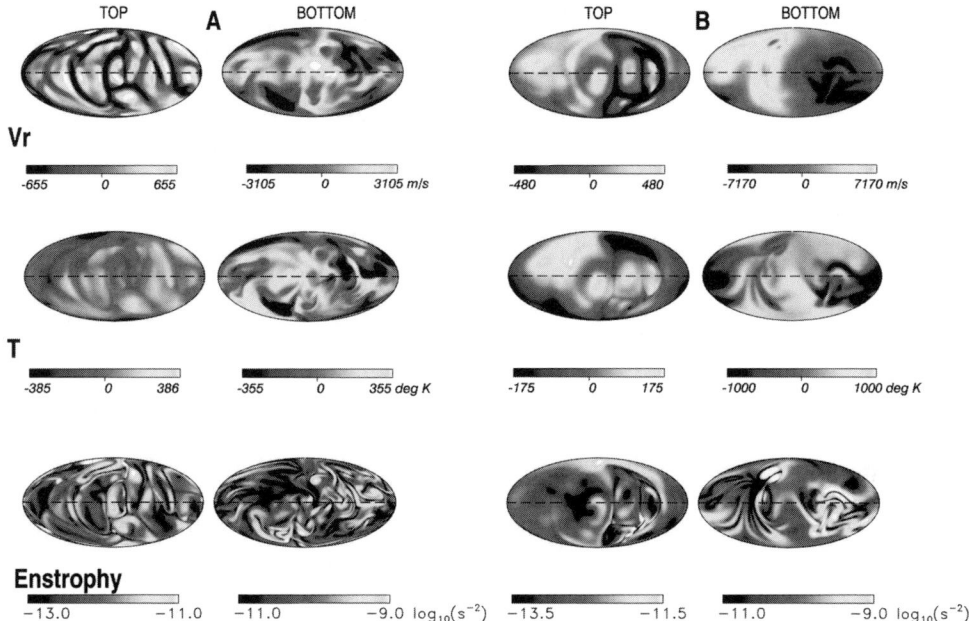

Figure 1. Convective patterns represented by the radial velocity V_r (first row) and the temperature T (second row) variations at the top and bottom edges of the computational domain for cases A (first two columns) and B (last two columns).

equilibrium that is maintained over the later evolution of the simulation. The simulations are thus considered to be relaxed (even though the model is still marginally evolving toward complete thermal equilibrium). The convective patterns that develop are shown in Fig. 1 after respectively 12 and 6 rotation periods of overall evolution for cases A and B respectively. Two different depths are shown : near the top (19 $R_\odot$), and at the bottom (3 $R_\odot$) of the computational domain.

The turbulent convection achieved in these simulations is characterized by large warm up-flows surrounded by cool narrower downflows at the top edge of the computational domain. For case A, the convective pattern consists of small number of cells uniformly distributed over the sphere, this pattern being toned down at larger depths. For case B, which rotates slowlier than case A, the number of cells is considerably reduced, and convection is strongly bipolar, with a warm and a cool side that are clearly distinct. This pattern resembles that obtained in non-rotating simulations (Woodward *et al.* 2003, Kuhlen *et al.* 2006), where the $\ell = 1$ mode is predicted to be dominant (case of incompressible fluid sphere at low R_a numbers, Chandrasekhar 1961).

In both cases, the strong correlation between radial velocity and temperature variations appearing in the maps results in an outward transport of heat. In order to actually transport the large luminosity of the star ($L \sim 400 L_\odot$), the velocity and temperature fluctuations are very large, up to 3000 m.s^{-1} and 400 K in case A, 7000 m.s^{-1} and 1000 K in case B. These variations are at least one order of magnitude larger than those found in the simulations of the solar convective envelope, where the radial velocity and temperature fluctuations do not exceed a few 100 m.s^{-1} and 10 K respectively. The even larger fluctuations that appear in case B indicate the stabilizing role of rotation (when significant) that leads to a less vigorous convective flow.

The convective luminosity is found to be significantly larger than the star's luminosity, in the bulk of the domain for case A and at the bottom of the domain for case B. This is to compensate a negative kinetic energy luminosity that can represent up to 70% of the total luminosity in case A and up to 120% in case B. This negative kinetic energy flux results from the strong asymmetry between up-flows and downflows in the bulk of the domain. It is thus even stronger in case B where the bipolarity of convection is strongly marked. This is an important result that contradicts the assumption of the mixing length theory, that assumes that the total and convective luminosity are equal.

The convection is found to possess a significant amount of vorticity that mainly concentrates around the strongest downflows as illustrated in Fig. 1 by the enstrophy map. It is obvious that case A possesses a more intricate and small scale enstrophy field which is the direct consequence of its larger rotation rate. Table 1 also summarizes the energetics of the convection for cases A and B.

In the bulk of our simulations, the kinetic energy density (KE) is dominated by convection, and the meridional circulation contributes significantly to the total energy (about 10% in both cases). This is clearly different than for the solar case, where the contribution of meridional circulation to the averaged total KE is less than 1%. For case B, the energy associated with meridional circulation is larger than that associated with differential rotation. This is actually not the case during the whole simulation : an alternation between phases with MCKE > DRKE and with MCKE < DRKE of variable length is found. This affects the meridional circulation pattern achieved in that simulation. When MCKE < DRKE, which is always true for case A, the meridional circulation pattern consists of one poleward cell per hemisphere. For case A, this pattern, presented in panel (a) of Fig. 3, is very stable over the simulation, in particular, no inversions of the meridional circulation are observed, and the total number of cells is always two. For case B, this pattern changes when MCKE > DRKE, and turns into a unique cell crossing the equator, and being equatorward in the northern hemisphere. Associated with this circulation is a strong bipolar structure, found both in the temperature and radial velocity fluctuations. Note however that the slow rotation rate of case B leads to a reduced rotational constrain on the flow. Forming axisymmetric averages around the axis of rotation may, in this case, lead to large fluctuation in the resulting averaged quantity by the simple misalignment of the large scale flow with respect to that axis (i.e the meridional flow may drift in latitude and longitude making it difficult to project into the meridional plane). The typical amplitudes of these large-scale circulations varies between 100 and 700 m.s^{-1} for case A and between 600 and 1600 m.s^{-1}, depending on the location in the computational domain. This is much larger than what is found in the Sun (about 20 m.s^{-1} at the solar surface). The turnover time for the meridional circulation in both cases is much longer that the typical convection turnover time (3.5 years vs. 150 days for case A).

4. Angular momentum redistribution in the shell

In this extended convective envelope, our choice of stress-free velocity boundaries ensures that no torque is applied to the system, and that angular momentum is conserved in the convective shell. In turbulent convection zones, the angular momentum is redistributed by various physical processes, which are in turn, viscous diffusion, Reynolds stresses, and mean large scale circulations. We may identify the contribution to the transport of angular momentum of each process by considering the mean radial ($\mathcal{F}_r$) and latitudinal ($\mathcal{F}_\theta$) angular momentum fluxes. The expressions of these fluxes are given in Brun & Toomre 2002. Figure 3 presents these integrated averaged fluxes along co-latitude and radius for case A and case B respectively. We may first note the good quality of the angular momentum balance characterized by the solid lines, that is very close to zero (e.g. no net flux) in both cases A and B when averaging over the entire length of the simulations.

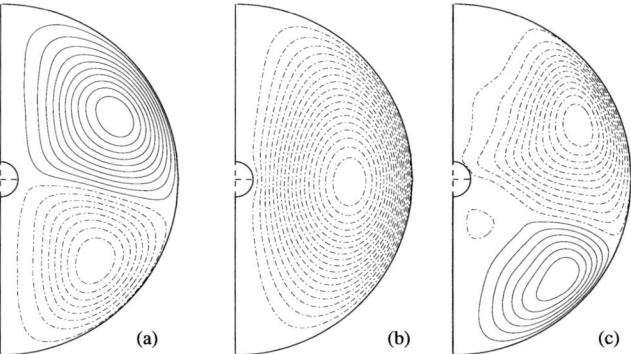

Figure 2. Streamlines of the meridional circulation currents. Dashed and solid lines indicate clockwise and counterclockwise circulation respectively. (a) corresponds to the mean pattern obtained in case A; (b) and (c) are associated with phases MCKE > DRKE and MCKE < DRKE respectively in case B.

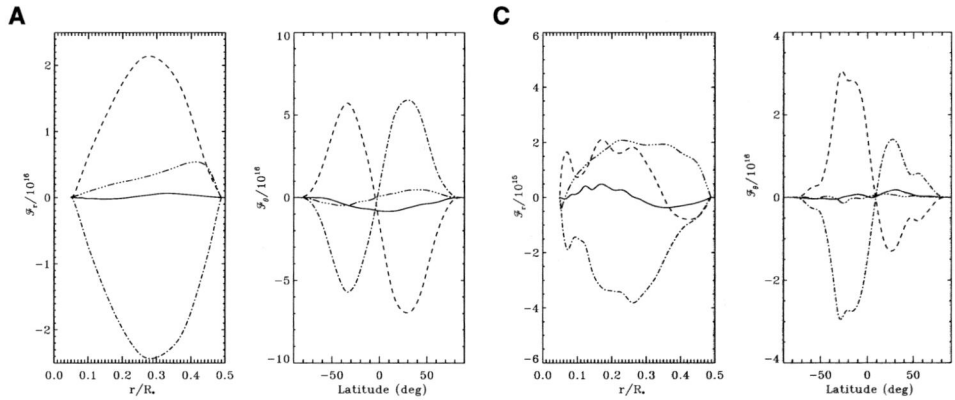

Figure 3. Time average of latitude line integral of the angular momentum flux $\mathcal{F}_r$ and of the radial line integral of the angular momentum $\mathcal{F}_\theta$ for cases A (left panels) and B(right panels). Viscous (dashed-triple-dotted), Reynolds stresses (dashed-dotted) and meridional circulation (dashed) contributions are represented together with the total fluxes (solid). Positive values indicate outward radial flux and northward latitudinal flux. Averages are over 12 and 6 rotation periods for cases A and B respectively.

In the latitudinal direction, the meridional circulation flux compensates that of the Reynolds stresses, with a poleward transport in the northern hemisphere for both cases A and B. The viscous transport is negligible. In the radial direction, the pictures differs in both cases. In the slowly rotating case (B) meridional circulation and Reynolds stresses are of opposite sign, but the viscous flux has the same amplitude and sign as meridional circulation flux, and plays an important role in the flux balance. As we will see in the forthcoming paragraph, this is associated with a shellular rotation achieved in this case. In the more rapidly rotating case (A), the balance in the radial directions resembles that in latitudinal direction: meridional circulation acts to transport the angular momentum outward, and is compensated by the Reynolds stresses (associated with $<v'_r v'_\phi>$). The viscous flux is negligible in this case.

The redistribution of angular momentum in the convective shell results in a strong departure from the initial solid-body rotation regime. The rotation achieved is substantially different according to the bulk angular velocity adopted as can be seen in Fig. 4.

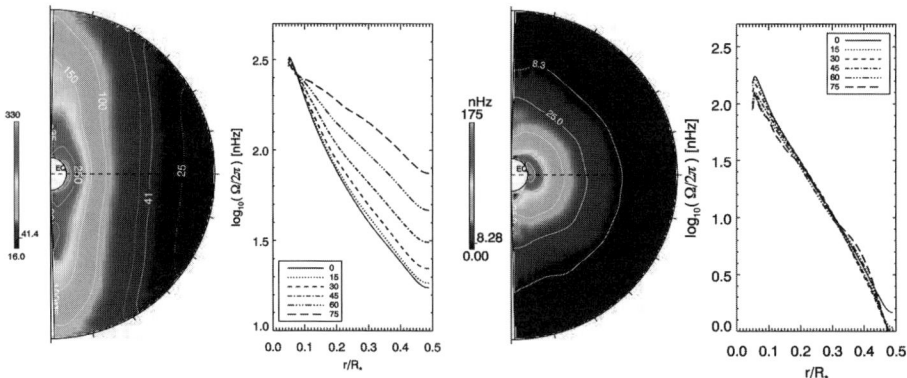

Figure 4. *2D plots* Temporal and longitudinal average of the angular velocity profile in cases A (left) and B (right). *1D plots* Radial profiles for selected latitudes (0,15,30,45,60,75 degrees) for cases A and B.

For case A, Ω_0 is larger than what would represent that of a real RGB star. A strong differential rotation is found in both radial and latitudinal directions ($\Delta\Omega/\Omega_0$ between 0° and 60° near the top of the domain close to 60%). The poles rotate faster than the equator (anti-solar rotation) and the inner and outer regions rotate in opposite directions. Ω contours are aligned with the rotation axis, resulting in an almost cylindrical pattern that might be attributed to the mild degree of turbulence of the simulation or to the absence of a strong latitudinal entropy contrast (Miesch *et al.* 2006).

For case B, which is more realistic compared to the actual rotation velocities determined in RGB stars, a strong shellular rotation is found : shells rotating as solid-bodies pile-up with increasing radial velocity towards the center. This leads to the large viscous transport found in the radial direction and to a more subtle balance of angular momentum (see §3). Independently of their latitudinal angular velocity profiles both models possess a strong radial differential rotation. This behaviour has also been reported by Steffen & Freytag (2007) in their rotating "star-in-a-box" experiments.

5. Conclusion

We have undertaken a series of hydrodynamical simulations of the turbulent convection in the rotating extended envelope of a RGB star, of which we have exposed here two cases with different bulk rotation rates.

The rotation rate appears to be crucial in determining both the convective pattern and the angular momentum redistribution. For more realistic values (case B) associated with $v_{eq,surf} \simeq 2$ km.s^{-1}, the convection is bipolar, as already observed in non-rotating simulations of the same kind. The enthalpy flux associated with the convection can be much larger that the total flux at the base of the domain, in order to compensate for the large negative kinetic fluxes. This is in contradiction with one of the basic assumptions of the mixing-length theory used in 1-D stellar evolution models to describe convection. The angular momentum redistribution leads to shellular rotation in the spherical shell, with a strong differential rotation in the radial direction. When rotation rate is increased (as in case A), the $\ell = 1$ mode does not dominate turbulent convection anymore, and the convective pattern becomes more complex. As in case B, the enthalpy flux again becomes larger that the total flux in the bulk of the domain, which confirms the contradiction with MLT. Concerning the redistribution of angular momentum, we observe the development and maintenance of a large cylindrical differential rotation within the inner part of the convective envelope of the RGB star. This regime contradicts the assumption of solid body rotation assumed in stellar evolution modeling. It could be of great importance for the global redistribution of angular momentum and chemical species in the underlying radiative region. We shall thus explore in more details the parameter space of our simulation in order to be able to propose a more realistic prescription for the angular momentum distribution within the convective envelope of red giant stars for 1-D stellar evolution.

Acknowledgement

The simulation has been carried out using the supercomputers of CEA-CCRT.

References

Bohm-Vitense, E. 1958, *Z. Astrophys.*, 54, 114
Browning, M. K., Brun, A. S., & Toomre, J. 2004, *ApJ*, 601, 512
Brun, A. S., Antia, H. M., Chitre, S. M., & Zahn, J.-P. 2002, *A&A*, 391, 725
Brun, A. S., Miesch, M. S., & Toomre, J. 2004, *ApJ*, 614, 1073
Brun, A. S., & Toomre, J. 2002, *ApJ*, 570, 865
Chandrasekhar, S. 1961, *Hydrodynamic and hydromagnetic stability - International Series of Monographs on Physics*, Oxford: Clarendon
Kuhlen, M., Woosley, S. E., & Glatzmaier, G. A. 2006, *ApJ*, 640, 407
Miesch, M. S., Brun, A. S., & Toomre, J. 2006, *ApJ* 641, 618
Miesch, M. S., Brun, A. S., Derosa, M., Toomre, J. 2008, *ApJ*, 673, 557
Palacios, A., & Brun, A. S., *Astronomical Notes*, 328, 1114
Palacios, A., Charbonnel, C., Talon, S., & Siess, L. 2006, *A&A* 453, 261
Sills, A., & Pinsonneault, M. 2000, *ApJ* 540, 489
Steffen, M. & Freytag, B, 2007, *Astronomical Notes*, 328, 1054
Sweigart, A. V. & Mengel, J. G. 1979, *ApJ* 229, 624
Woodward, P. R., Porter, D. H., & Jacobs, M. 2003, *3D Stellar Evolution*, 293, 45
Zahn, J.-P. 1992, *A&A* 265, 115

Discussion

LUDWIG: Remark: I found it very encouraging to see that two codes (your ASH code, and COSBOLD by Freytag & Steffen) give so similar results for rotating giants, considering the very different numerical approaches which are taken in the modeling.

PALACIOS: Thank you.

CHAN: (1) What flux did you use in the model? (2) What is the thermal relaxation time for that flux?

PALACIOS: (1) The flux used in the model comes from the 1D stellar evolution model, that is realistic as far as stellar parameters are concerned. (2) The simulations have not reached the thermal relaxation, but as we start from a realistic structure that is thermally relaxed, we assume that it is not so crucial for our purpose.

MEYNET: (1) How would you translate your 3D results into a rule for 1D models? (2) Do you plan to study the effects of a magnetic field?

PALACIOS: (1) We first want to explain the parameters of the simulations, but it is already clear that we should avoid considering uniform angular velocity in the extended envelopes of giants ($\Omega(r) = cte$). The assumption of uniform specific angular momentum

($r^2\Omega \propto cte$) is closer to what we have. (2) We have obtained European funding to do MHD simulations and we plan to investigate the role of magnetic fields in these envelopes in the near future.

KUPKA: How do you compute the length scale L in your estimates of Re ? – I think that using the total depth of the convective zone is not a very useful number, because for turbulence what matters is the energy carrying scale which is probably closer to the local pressure scale height or the width of the up- and downflows in your simulations. Thus, the effective Reynolds number is probably a lot smaller than 1000 in the red giant case and 4000 in the solar composition model.

PALACIOS: I have used the *?side, wide?* length of the computational domain.

The Art of Modelling Stars in the 21st Century
Proceedings IAU Symposium No. 252, 2008
L. Deng & K.L. Chan, eds.

© 2008 International Astronomical Union
doi:10.1017/S1743921308022734

Mixing-length parameter from binaries and clusters

Mutlu Yıldız

Ege University, Dept. of Astronomy and Space Sciences, Bornova, 35100 Izmir, Turkey
email: mutlu.yildiz@ege.edu.tr

Abstract. Mixing length parameter of the mixing length theory is a free parameter and used to calibrate radii of late-type stars. From the Hyades cluster and the α Centauri system, we find that the parameter depends on the stellar mass and some other properties of stars. Considering some other systems, we describe how the parameter depends on stellar parameters.

Keywords. convection, stars: interiors, stars: evolution, stars: late-type, binaries: eclipsing, stars: individual (α Cen, FL Lyr, V442 Cyg)

1. Introduction

Brightness and size of a star essentially depend on physical conditions in its nuclear core. For radius, another important issue is energy transport mechanism in outer regions. For the early-type stars, whose envelope is in radiative equilibrium, at least near the zero-age main-sequence (ZAMS), radius is determined by the heavy element abundance. For the late type stars, however, the energy transport mechanism is convection (Bohm-Vitense, 1958) and value of the mixing-length parameter plays an essential role for the size of their models. The value of mixing-length parameter is customarily taken as the value derived from the solar calibration, despite no strict clues for the constancy.

The disagreement between observed and model radii of late-type stars in eclipsing binaries is a long-standing problem. Popper (1997), for example, discusses many of possible reasons. In many studies on stellar interior variable mixing-length parameter is also considered as a solution for such a disagreement (see for exapmle, Lebreton *et al.* 2001).

In this talk, we discuss how the mixing-length parameter should depend on the stellar variables in order to fit the model values of physical properties to the observed values.

2. The mixing-length parameter from the Hyades open cluster

Recently, stellar mass dependence of the mixing-length parameter is derived for the stars in the Hyades open cluster by Yıldız *et al.* (20006). In order to fit the radii (or colors) of models to the observed values, the mixing-length parameter should be an increasing function of stellar mass:

$$\alpha = 9.19(M/M_\odot - 0.74)^{0.053} - 6.65. \quad (2.1)$$

The age of Hyades is about 700 Myr and it is very short time in comparison with the MS life time. Therefore this relation can be considered as the stellar mass dependence of the mixing length parameter near the ZAMS.

In deriving the expression given in Eq. 2.1, together with fundamental properties of binary stars (V818 Tau, 70 Tau, HD 27149 and θ^2 Tau) in Hyades, slopes of lower and upper parts of MS are also used. The slope of the lower part, which contains the components of V818 Tau, is obtained as

$$S_{MS1} = \frac{\Delta M_V}{\Delta(B-V)} = 4.6, \qquad (2.2)$$

from application of a least-square method to the data given by de Bruijne et al. (2001). For the upper part, which contains the components of 70 Tau,

$$S_{MS2} = \frac{\Delta M_V}{\Delta(B-V)} = 6.6. \qquad (2.3)$$

Very similar slopes are found also from the MS of Praesepe, which is very similar to Hyades in many respects. This means that similar difference between models and the stars also exist for this cluster.

3. The mixing-length parameter from α Centauri

We have relatively very accurate both seismic and non-seismic data for α Cen A and B. The compatibility of these two data sets is discussed in detail by Miglio and Moltaban (2005) and Yıldız (2007) While the seismic constraints imply that age of the system is about 5.6–5.9 Gy, the non-seismic constraints give 8.9 Gy. The values of the mixing-length parameters for both components are rather different. While α_A is larger than α_B for the non-seismic constraints, for the seismic constraints, α_B is larger than α_A. The former result is compatible with Eq. 2.1, the latter result, however, is possible if mixing-length parameter changes with evolution (time). The mixing-length parameter can be written as functions of stellar parameters, which changes with time. Yıldız et al. (2006) give two different expressions for α:

$$f_1 = 2.5 - \rho_{bcz}\left(\frac{2.7}{T_{6,bcz}}\right)^4 - \rho_{ph}\left(\frac{1.4}{T_{5,ph}}\right)^4, \qquad (3.1)$$

$$f_2 = \frac{3.25}{\rho_{ph}^{0.8}}\left(\frac{m_{bcz}}{r_{bcz}^2}\right)^{0.5}\left(\frac{T_{6,bcz}}{3.8}\right)^4. \qquad (3.2)$$

The subscripts bcz and ph represent the base of convective zone and the photosphere, respectively. The radius of the base of the convective zone (r_{bcz}) and the mass inside the sphere with this radius (M_{bcz}) are in solar units. Both of these expressions are in good agreement with Eq. 2.1 near the ZAMS. In the later phase of MS evolution of α Cen A, however, f_1 is an increasing function of time but f_1 is an decreasing function.

4. The mixing-length parameter from the eclipsing binary FL Lyr

In order to test variability of the mixing-length parameter in some other systems, double lined eclipsing binaries with late type components are among the most suitable systems. One of these sytems is FL Lyr. Masses and radii of its components are accurately determined (Andersen 1991) from light curve and radial velocity. The Fundamental properties of the component stars are listed in Table 1. Age of the system is from Pols et al. (1997) and Z is computed from calibration of photometric properties by using bolometric correction tables of Lejeune et al. (1998).

We construct models for these stars with variable mixing-length parameter and try to fit the models to the fundamental properties derived from light curve and radial velocity curve. In the calibration process, for a given hydrogen abundance X, we obtain

Table 1. The properties of FL Lyr A and B (Andersen 1991). The age is from Pols et al. (1997). The metallicity is computed from the calibration of photometric properties by using bolometric correction tables of Lejeune et al. (1998).

Star	$B-V$	$\log(M/M_\odot)$	$\log(R/R_\odot)$	$\log(g)$	$\log(T_{\rm eff})$	$\log(L/L_\odot)$	M_V	age(y)	Z
A	0.52	1.218 ± 0.016	1.282 ± 0.028	4.308 ± 0.020	3.789 ± 0.007	0.32 ± 0.03	3.98 ± 0.09	2.37e9	0.0177
B	0.78	0.958 ± 0.012	0.962 ± 0.028	4.453 ± 0.026	3.724 ± 0.008	-0.18 ± 0.04	5.40 ± 0.10	2.37e9	0.0187

Table 2. For given initial hydrogen abundance (X=0.70) and α values ($\alpha_A = \alpha_B = 1.82$), time t is found for FL Lyr A and B, at which luminosity of model is equal to the observed value. The logarithmic time difference $(t_A - t_B)/t_A$ for different values of Z is given in the fourth column.

Z	t_A (My)	t_B (My)	$\Delta t/t_A$
0.014	130	60	0.54
0.021	2750	5080	-0.85
0.015	480	484	-0.01

reference models for both components in order to determine sensitivity of luminosity of each components to heavy element abundance and find time at which $L(t) = L_{\rm obs}$. The difference between the times required for the agreement of models with observed luminosities is a function of Z. The time differences for some Z values are listed in Table 2. The time difference between the ages of the component stars is negligibly small when Z = 0.015. The age of the system is then 480 My. From the calibration of radii around this age, we find that $\alpha_A = 1.30$, $\alpha_A = 1.05$ and t = 500 My. For X = 0.718, models with Z = 0.0127 give the same and very similar mixing-length parameters: $\alpha_A = 1.27$, $\alpha_A = 0.97$. As a result, for different combinations of X and Z we find that for also this system mixing-length parameter is an increasing function of stellar mass

In Fig. 1, evolutionary tracks of FL Lyr A and B are plotted in HR diagram. While FL Lyr A is slightly evolved, FL Lyr B is very close to the ZAMS line.

5. The mixing-length parameter from the eclipsing binary V442 Cyg

The components of this system are F-type stars. Applying the same method as for FL Lyr, we find that Z = 0.0142 and age of the system is about 1.73 Gy, for X = 0.705. From the calibration of radii, the mixing-length parameters of V442 Cyg A and B are found as $\alpha_A = 3.46$ and $\alpha_A = 2.29$. Very similar values are found for different combinations of X and Z. So, for also this system, we obtain that the mixing-length parameter is increasing function of stellar mass. Furthermore, the values of mixing-length parameter for V442 Cyg B is compatible with the value derived from expression for Hyades.

6. Conclusion

In order to fit model radii of late type stars to their observed radii, one of possible ways is that the mixing-length parameter is an increasing function of stellar mass around ZAMS. For the later phase of MS evolution, while the same result is found from analysis of the eclipsing binary V442 Cyg, the seismic and and non-seismic constraints of α Cen give contrary results. For the later phases, old eclipsing binaries in old open clusters can be a useful lab (for example, V12 binary system in NGC 188).

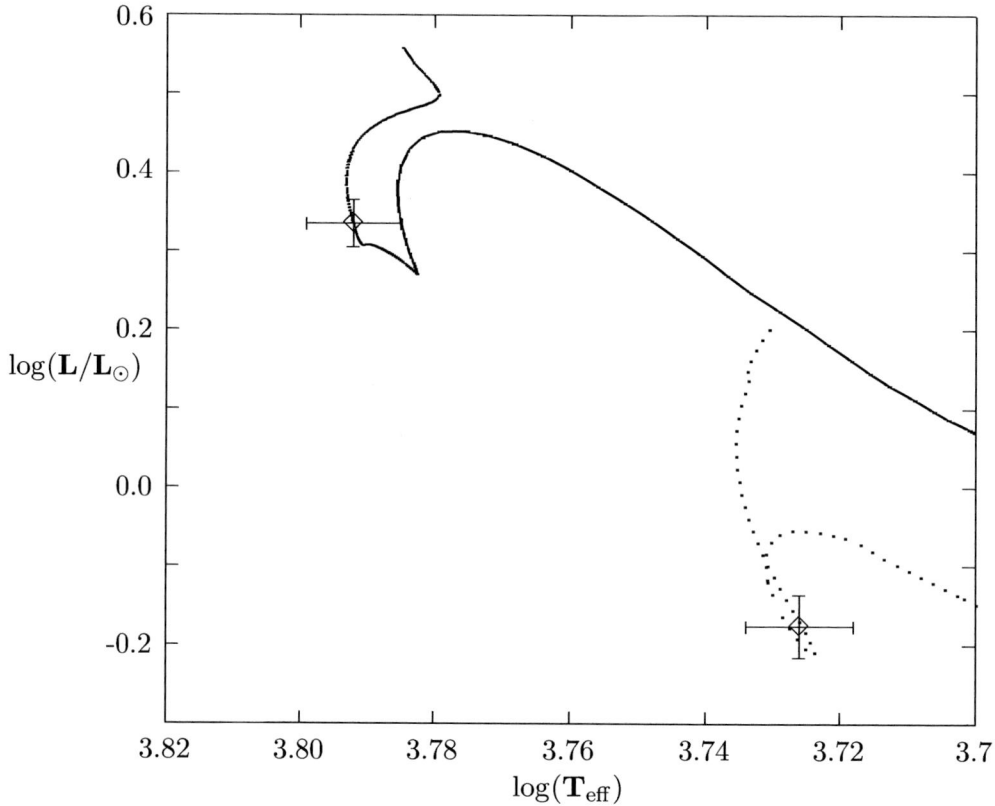

Figure 1. HR diagram for FL Lyr A and B. The models are computed with Z = 0.015 and X = 0.7. The agreement is obtained at t = 500 My.

References

Andersen, J. 1991, *A&AR* 3,91
Bohm-Vitense, E. 1958, *Zs. Ap.* 46, 108
de Bruijne, J. H. J., Hoogerwerf, R., & de Zeeuw, P. T. 2001, A&A, 367, 111
Lebreton, Y., Fernandes, J., & Lejeune, T. 2001, *A&A* 374, 540
Lejeune, T., Cuisinier, F., & Buser, R. 1998, A&AS, 130, 65
Miglio, A., Montalban, J., 2005, A&A, 441, 615
Pols, O. R., Tout, C. A., Schroder, K-P., Eggleton, P. P., & Manners, J. 1997, *MNRAS* 289, 869
Popper, D. 1980, *ARA&A* 18, 115
Popper, D. 1997, *AJ* 114, 1195
Yıldız, M., 2006, MNRAS, 368, 1941
Yıldız, M., 2007, MNRAS, 374, 1264

Discussion

ZAHN: When you analyze your binary stars, do you assume the same chemical composition for both components?

YILDIZ: Of course. One of the main assumptions about binary systems is that their component stars have the same initial chemical composition.

LUDWIG: Which presicion do you obtain for the mixing length parameters considering the observational uncertainties (in colors, chemical composition, radii, etc.)

Remark: Intrinsically, the mixing length parameter is a function of the stellar atmospheric parameters ($T_{\rm eff}$, gravity, chemical composition), and perhaps radius if sphericity effects become important. Hence, one should rather use these as independent variables instead of age, mass and luminosity to describe its functional dependence.

Yildiz: The problem here is that the models of late type stars with standard theory are inadequate to refer observed radii. The considered stars are among the well known stars whose fundamental properties are relatively very accurate. However, for the stars in Hyades, for example, the slopes of main sequence in two intervals used as main constraints on models are very precise. Of course, there are some other possible mechanisms in order to restore the disagreement between the model and observed radii, but mixing-length parameter is perhaps the first among them to test.

Convection is one of the most complicated phenomena in the nature. Regarding this point in mind, the main point here is to show the variability of the mixing-length parameter without any prejudice and use more complicated expression if required.

The speaker, M. Yıdız, is presenting his work.

Deathzones and exponents: A different approach to incorporating mass loss in stellar evolution calculations

Lee Anne Willson[1]

[1]Department of Physics and Astronomy, Iowa State University, Ames, IA 50010, USA
email:lwillson@iastate.edu

Abstract. Observations tend to select mass loss rates near the critical rate, $\dot{M}_{crit} = M\dot{L}/L$. There are two reasons for this. In some situations, such as near the tip of the AGB, the mass loss rate is very sensitive to stellar parameters. In this case, stars with $\dot{M} \ll \dot{M}_{crit}$ have dust-free, hard-to-measure mass loss rates while stars with $\dot{M} \gg \dot{M}_{crit}$ do not survive very long and thus make up a small fraction of any sample. Selection effects dominate the fitting of empirical formulae; observations of mass loss rates tell us more about which stars are losing mass than about how a star loses mass. In other situations, such as for some of the stars along the RGB, a steady state situation occurs where the loss of mass leads to a decrease in mass loss rate while the evolutionary changes lead to an increase; the result is a steady state with $\dot{M} = \dot{M}_{crit}$. To determine the envelope mass and composition at the end of a phase of intensive mass loss requires stellar evolution models capable of responding on a time scale $\sim t_{KH}$ and thus, a new generation of stellar modeling codes.

Keywords. Stars: mass loss, stars: evolution

Stars lose mass, but only some of the time. We can distinguish mass loss, where the rate is high enough to affect the evolution of a star, from a stellar wind that carries only very small amounts of mass away. In most situations, mass loss occurs when $\dot{M}/M$ is not too much smaller than $\dot{L}/L$ or $\dot{R}/R$ or $\dot{T}_{eff}/T_{eff}$, and over a limited range of parameters for a given star (Willson 2000, Willson 2006, Willson 2007).

Known epochs of mass loss for stars with masses from about 0.8 to about 8 M_{Sun} include the upper asymptotic giant branch (AGB) and, at least for some stars, the first ascent red giant branch (RGB). The stars near the tip of the AGB that are losing mass we identify as Mira variables; because the atmospheric structure is different for stars with strong winds, these stars have large visual magnitude variations as they pulsate. The stars with mass loss rates $> 10^{-5} M_{Sun}/yr$ are also known as OHIR stars, IR bright stars with OH masers, as they have opaque circumstellar outflows.

Stellar evolution modeling has sought to include mass-loss by way of formulae found by fitting observations of mass loss rate expressed as functions of luminosity L, radius R, effective temperature T_{eff}, pulsation period P, or combinations of these. Other formulae have been proposed from theoretical studies of a particular mass loss mechanism. Mechanisms that have been modeled in particular detail include pulsation/dust mass loss for red giants (Bowen 1988, Bowen & Willson 1991, Höfner 2007a, Höfner 2007b), and line driven winds for hot stars (Castor et al. 1975, Abbott 1982, Kudritzki et al. 1989).

Formulae derived in the usual manner by fitting mass loss rates vs. stellar parameters are not particularly useful for stellar evolution models, for two reasons: (a) they are dominated by severe selection effects; and (b) they are calibrated for a limited volume of parameter space where often at least one parameter (usually M and/or R) cannot be well constrained by observations. For discussion of these points, see Willson (2000, 2006, 2007).

Here, I will discuss some general characteristics of mass loss episodes, then illustrate my main points with an analysis of mass loss on the AGB and the RGB.

1. Why observations favor $\dot{M} \approx M(\dot{X}/X)$

Observations tend to yield $\dot{M} \approx M\dot{X}/X$ (where X is L or R or T_{eff} depending on the direction the star is evolving in the HR diagram) in two situations: (a) when the mass loss rate depends steeply on stellar parameters, often with positive feedback, or (b) when mass loss changes the situation so as to reduce the mass loss rate but evolutionary changes increase $\dot{M}$ - negative feedback or steady state mass loss.

In the first case, when mass loss rates depend steeply on stellar parameters, there is a narrow range of stellar parameters for which $log\dot{M} = log(M\dot{X}/X) \pm 1$. Below this range, mass loss rates have little effect on the evolution and may be hard to detect. Above this range, mass loss quickly removes the envelope of the star. Thus we will preferentially select and tabulate rates near $\dot{M}_{crit} = M\dot{X}/X$. This situation tends to apply where mass loss increases R and mass loss rate increases with R - such as near the tip of the AGB until the envelope mass is very small and the star leaves the AGB.

In this first case, it is useful to define the *deathline* as where $\dot{M} = M\dot{X}/X = \dot{M}_{crit}$ and the *deathzone* as where $log\dot{M} = log\dot{M}_{crit} \pm 1$. Nearly all the mass loss of evolutionary significance will take place in the deathzone when the mass loss rate depends steeply on stellar parameters.

In the second case, when the loss of a little mass changes the situation so as to reduce the mass loss rate, but evolutionary changes increase $\dot{M}$, then the mass loss rate will tend to the steady state value $\dot{M} = M\dot{X}/X$ ($X = L$ or R or T_{eff}) depending on the direction of evolution in the HR Diagram). In this case, the rate at which the star evolves due to internal processes determines the mass loss rate, and to predict the mass loss rate we need only know the rate of evolution, $\dot{X}/X$, for each mass M from a stellar evolution code - we do not need to know the mechanism to predict the mass loss rate, as the steady $\dot{M}/M = \dot{X}/X$.

The mass loss terminates, in the first case, when the star has lost enough mass that the mass loss rate no longer increases with decreasing mass. For example, when an AGB star's envelope mass goes below a small value, the radius decreases with decreasing M. However, it will take the star a time on the order of the Kelvin-Helmholtz time (t_{KH}) to adjust to a change in its mass, and thus the mass loss process will "overshoot" the simplest estimate by some amount. An order-of-magnitude estimate for the overshooting is $\dot{M}_{max} \times t_{KH}$; however, for such important questions as the evolution of post-AGB stars. the surface layers on white dwarfs, and the structure of pre-SNIIs, it will be necessary to use evolutionary models capable of responding on times shorter than t_{KH} to derive $dlogR/dlogM$ for the process. Such codes are becoming available - for example, there is the Djehuty code being developed at LLNL (Bazán *et al.* 2003).

Finally, I note that a similar logic has long been applied to the evolution of binary systems with Roche lobe overflow, where the mass exchange is modeled using information about the evolutionary changes in each star and the effect of mass exchange, and sometimes also mass loss from the system, on the orbits of the two stars involved. (See Hjellming & Webbnik 1987, Webbink 1976.) What I propose is we consider a similar approach to mass loss more generally, particularly since reliable observational determination of $\dot{M}(L, R, M, Z)$ is not available, and theoretical computations are not yet reliably predictive of such a formulae either. Both for complete understanding of the binary evolution and the single star evolution with mass loss, however, we need to be able to study the response of a star to changes of mass on time scales comparable to or less than the Kelvin-Helmholtz time. One significant difference is, though, that mass-loss on a dynamical time-scale, expected in some binary star systems, is not likely for the single star case. The difference is that in the binary case the orbit evolves on a dynamical time scale, so when mass loss leads to a decrease in the size of the orbit relative to the stellar radius this initiates very fast exchange. For the single star case we expect the changes to take place on time scales between the thermal or Kelvin-Helmhotlz timescale t_{KH}, and the nuclear time scale.

2. Examples of the two cases

2.1. Mass loss at the tip of the AGB

This case is one I have reviewed a number of times; see Willson (2000), Willson (2006), Willson (2007), Willson (2008). Because this is the main mass loss episode for most stars ($M = 0.8$ to 8 M_{Sun}), there have been quite a few attempts to produce reliable mass loss laws from observations or from theory. Examples are reviewed in Willson (2007) and Willson (2008) taken from Vassiliadis &Wood (1993), van Loon et al. (2005), Bryan, Volk & Kwok (1990), Blöcker (1995), Schröder and Cuntz (2005), Baud & Habing (1983), and Wachter et al. (2002). We have argued (Bowen & Willson 1991, Willson 2000, Willson 2006, Willson 2007) that the episode is short-lived because the mass loss rate is very sensitive to stellar parameters. As the star evolves in L and R, $\dot{M}$ increases from too small to notice to large enough to kill the star quickly in the course of just 2 to $4x10^5$ years. The mass loss prescription based on our 1995 grid of models = BW in what follows. Note that Blöcker (1995) used an earlier grid of models where the regulation of the driving amplitude in the models had not yet been set to an energy condition; for the 1995 grid and subsequent modeling we have implemented a maximum power condition that results in a steeper dependence on L, R, and M (Willson 2000). A recent exploration with an independent code also provides support for this choice (LAWMa08).

If we reduce all the mass loss laws to a common basis, $\dot{M}(L, M, \alpha, Z)$, where $\alpha = \ell/H_p$, using evolutionary tracks with period-mass-radius relations and the definition of T_{eff} as needed, then we can see (Figure 1 of Willson (2007), Figures 1 & 2 of Willson (2008)) that the relations all cross within a modest range of $\log \dot{M}$ and $\log L$, but that the slopes $d \log Mdot/d \log L$ (taken along the evolutionary track in this figure) are quite varied. This is what we expect if (a) the mass loss rate is highly sensitive to stellar parameters and (b) observations mainly pertain to stars within an order of magnitude or so of the critical rate, as noted above. Thus I have concluded that *observations of mass loss rates tell us which stars are losing mass but (mostly) do not tell us much about how a given star will lose mass*. The one possible exception to this rule is the Vassiliadis &Wood (1993) = VW fit for Miras, $log \dot{M} = -11.4 + 0.123 P$. Because they used a variable that is easily and accurately observed, pulsation period P, and because they restricted the fit to a homogeneous class of objects, Miras, they obtained a fit with a steep dependence of mass loss rate on stellar parameters near the deathline that is probably closer to having the right slope near the deathline than any of the other observational calibrations. Even in their case, however, not having observational determinations of the stellar masses leads to underestimating the sensitivity of the mass loss rate to the combined effects of increasing R and decreasing M.

Theoretical models suffer also from selection effects. These include the choice of L, R or T_{eff} for which to run models; the choice of modeling parameters (such as mixing length or R for a given L, M, for a giant); "piston" amplitude for most pulsating atmosphere modeling; dust condensation formulation; handling of cooling and heating by interaction with the radiation field; and treatment of the radiative transfer problem for a dynamical atmosphere). Some choices will produce more shallow slopes and others, a steeper slope. For example, the piston amplitude may be taken to have the same velocity amplitude $\triangle v$, the same $\triangle v^2$, or the same peak power (taking into account work done on the atmosphere by the piston at some phases and by the atmosphere on the piston at other phases).

Typically, the power going into driving the wind, roughly $\frac{1}{2}\dot{M}(v_{esc}^2 + v_\infty^2)$, is many orders of magnitude below the stellar luminosity, and so, it is a small residual of much more energetic processes. Without violating energy conservation the energy associated with mass loss can easily vary around a typical $10^{-5}L_*$ (corresponding to the Reimers' relation) by several orders of magnitude. The choice of which models to run and how to model the processes giving rise to mass loss can influence the derived formula, just as the selection of which stars to study can affect the empirical relations.

The mass loss process along the AGB is an example of positive feedback - as M decreases, R increases, and both changes increase the mass loss rate. The duration of the mass loss phase is thus determined by how steeply the mass loss rate depends on M, L, R and/or T_{eff}. Reducing the variables for a given M to a single one, L, taken along the evolutionary track, and finding the slope at the deathline, gives values that may be compared directly with each other at a given L near the deathline. The duration of the mass loss phase is $< (L/\dot{L})/(d\log \dot{M}/d\log L) \approx 10^6 years/(d\log \dot{M}/d\log L)$. The duration is $> 10^6$ years for most of the formulae we have examined, but $\sim 4x10^5$ years for the VW and BW mass loss formulae; observations support a value closer to 2 to $4x10^5$ years (Jura and Kleinmann 1992).

When the envelope mass reaches a critical value that depends on the core mass, further reduction in the envelope mass will decrease the radius of the star (e.g. Frankowski 2003, Iben & Renzini 1983). Now, both evolution and mass loss have the same effect - reduction of the envelope mass. Evolution also increases L, but relatively slowly, so the mass loss rate should quickly go to zero as the star shrinks at nearly constant L. Again, we expect the shrinkage to take at least t_{KH}, so we expect the mass loss to end with an envelope mass smaller than the critical value obtained from quasi-equilibrium stellar models. Note that t_{KH} for these extreme giants is only a few decades, while the dynamical time is around one year; Ostlie & Cox (1986) (Figure 4) showed that the growth rates of pulsation modes were quite sensitive to the ratio t_{dyn}/t_{KH} and relatively large for small ratios, so some erratic pulsation may be expected at/near the end of this mass loss episode.

2.2. Mass loss along the RGB

From the appearance of cluster HR diagrams with well-populated horizontal branches it is apparent that cluster stars of a given initial mass do not all lose the same amount of mass along the RGB. Recent observations by Origlia et al.(2007) have confirmed that the mass loss rates of red giants in 47 Tuc are not all the same at a given L, with 10 to 30% of red giants at a given L showing an IR excess. These also show that mass loss occurs within a single cluster at a range of values of L. The rates satisfy $\dot{M} \approx \dot{M}_{crit}$ also for the RGB stars. Because this mass loss occurs over a range of L for a given initial M, this is most likely to be a case of a steady state mass-loss process, i.e. one where the loss of a little mass leads to changes that reduce the mass loss rate. Since, on the RGB, decreasing M increases R, and since mass loss rates generally increase with R, this is unlikely to be the result of a mechanism operating for a single star. We are investigating a mechanism for mass loss that involves a low mass companion with a star with relatively rapidly increasing radius (subgiant or giant); in this case, the companion can't enforce synchronous rotation, and therefore the standard Roche lobe analysis does not apply. For details, see Wang *et al.* (this volume) and Struck *et al.* (in prep).

3. Recipe for handling mass loss in stellar evolution calculations

Above I have argued that (a) we do not have a reliable mass loss law for important episodes of mass loss in the life of many stars; and (b) typically, interesting mass loss happens near $\dot{M} = \dot{M}_{crit}$ where $\dot{M}_{crit}$ may be determined directly from evolutionary models. This suggests a very different approach for incorporating mass loss into stellar evolution models:

(*a*) Include mass loss only where observations say it occurs - do not extrapolate from limited ranges of stellar parameter space.

(*b*) Where mass loss matters, use duration, amplitude, and distribution data to estimate the exponents (slopes $d \log \dot{M}/d \log X$ where $X = L, R, T_{eff}$ or a combination taken along the dominant direction of evolution.

(*c*) Calculate, from models, $M\dot{L}/L = \dot{M}_{crit}$. See how the observed mass loss rates are distributed with respect to this critical rate.

  [i] If the distribution is broad, and the stars are located near or at an extreme value of stellar parameters, suspect a "deathline/deathzone" situation and proceed accordingly.

  [ii] If the distribution is narrow, and the stars distributed over a range of stellar parameters, suspect a steady-state mass loss process, and evaluate possible mechanisms accordingly.

(*d*) Compute a grid of models without massloss that cover all the parameter space occupied by stars in the appropriate population.

(*e*) Where mass loss is likely, use an appropriate code to determine the (local) exponents $dlogR/dlogM$, $dlogL/dlogM$, and $dlogT_{eff}/dlogM$. Based on the argument, above, that the mass loss time scale will be close to the nuclear time scale at steady state or at the deathline, and normally $t_{dyn} \ll t_{KH} \ll t_{nuclear}$, an ordinary evolutionary code may be used to find these exponents for mass loss rates up to the critical value, while a code capable of modeling processes faster than the nuclear scale will be needed for the high-mass-loss end of the deathzone.

(*f*) Supplement the no-mass-loss grid with the results of the mass loss analysis. This may truncate the evolution, or shift subsequent evolution to a new track, or introduce a dispersion in the results to be expected from an initially homogeneous population (with heterogeneous rotation, duplicity or planet families).

4. Conclusions and the future

To a great extent we can separate two hard problems: The incorporation of mass loss into stellar evolution, and the determination of mass loss rates and mass loss mechanisms for various classes of star. This parallels the logic that has been used to model the evolution of binary star systems, but instead of a condition of Roche lobe overflow to determine when mass loss occurs, we will use observations of which stars are losing mass at interesting rates, and constraints from the observed duration, amplitude, and distribution of mass loss rates in each episode. This approach will suffice to derive initial-final mass relations, and thus also to compute the colors or composite spectra of populations of stars. It will not suffice to tell us the precise envelope mass at which the mass loss process stops, however, nor do traditional stellar evolution models suffice. For such nuances we will need to study the response of a star to mass loss with codes capable of following changes on time scales shorter than the Kelvin-Helmholtz time.

References

Abbott, D. ~C. 1982, *ApJ*, 259, 282
Baud, B. & Habing, H. J. 1983 *aap*, 127, 78–83
Bazán, G., et al. 2003, 3D Stellar Evolution, 293, 1
Blöcker, T. 1995, *aap*, 297, 727
Bowen, G. H. & Willson, L. A. 1991, *ApJ*, 375, L53
Bowen, G. H. 1988, *ApJ*, 329, 299
Bryan, G., Volk, K., & Kwok, S. 1990, *ApJ*, 365, 301–211
Castor, J. I., Abbott, D. C., & Klein, R. I. 1975, *ApJ*, 195, 157
Frankowski, A. 2003, *aap*, 406, 265
Höfner, S., Gautschy-Loidl, R., Aringer, B., & Jørgensen, U. G. 2003, *aap*, 399, 589
Höfner, S. & Anderson, A. 2007, *A&A*, 465, L39-L42
Höfner, S. 2007, Why Galaxies Care About AGB Stars: Their Importance as Actors and Probes, 378, 145
Hjellming, M. S., & Webbnik, R. F. 1987, *ApJ*, 318, 794
Iben, I., Jr. 1984, *ApJ*, 277, 333
Iben, I., Jr. & Renzini, A. 1983, *ARAA*, 21, 271–342
Jura, M. & Kleinmann, S. G. 1992, *ApJS*, 79, 105–121
Kudritzki, R. P., Pauldrach, A., Puls, J., & Abbott, D. C. 1989, *aap*, 219, 205
Mattson, L., Wahlin, R., H'ofner, S., & Eriksson, K. in press, *A&A*, ArXiv 0804.2482
Origlia, L., Rood, R. T., Fabbri, S., Ferraro, F. R., Fusi Pecci, F., & Rich, R. M. 2007, *ApJL*, 667, L85
Ostlie, D. A. & Cox, A. N. 1986, *ApJ*, 311, 864
Reimers, D. 1975, *Mem. Soc. Roy. Sci. Liège*, 8, 369
Schröder, K.-P. & Cuntz, M. 2005, *ApJ*, 630, L73
van Loon, J. T., Cioni, M.-R. L., Zijlstra, A. A., & Loup, C. 2005, *aap*, 438, 273
Vassiliadis, E. & Wood, P. R. 1993, *ApJ*, 413, 641
Volk, K. and Kwok, S. 1988, *ApJ*, 331, 435-462
Wachter, A., Schröder, K.-P., Winters, J. M., Arndt, T. U., & Sedlmayr, E. 2002, *aap*, 384, 452
Webbink, R. F. 1976, *ApJ*, 209, 829
Willson, L. A. 2000, *ARAA*, 38, 573
Willson, L. A. 2006, ESO Astrophysics Symposia, Springer, Planetary Nebulae Beyond the Milky Way, 99
Willson, L. A. 2007, ASP Conference Series, Vol. 378, Why Galaxies Care About AGB Stars: Their Importance as Actors and Probes, p211
Willson, L. A. 2008, ASP Conference Series, The BIggest Baddest Coolest Stars, in press.
Willson, L. A. & Kim, A. 2004, ASP Conf. Ser. 313: Asymmetrical Planetary Nebulae III: Winds, Structure and the Thunderbird, 313, 394

Discussion

PODSIADLOWSKI: I completely agree with you that AGB stars must have a deathline, and there is a simple theoretical reason for it, as pointed out many years ago by Han, Podsiadlowski and Eggleton (1994), going back to a suggestion by Ziolkowski and Paczynski, and that is that at this point the AGB star envelopes start to have positive binding energies. This explains both the initial to final mass relation and the white-dwarf mass distribution. Indeed, I think that this must be the ultimate cause for the Mira variability.

WILLSON: Several things happen near the AGB tip – the question is which happens first. The initial-final mass relation is not the best constraint – final L is better since L-M_c relations can be shown to be wrong.

WOITKE: What can we learn from the structure of planetary nebulae, where the final mass loss history is imprinted? Due to thermal pulses, the stars may cross the "death zone" several times during the final evolution, producing multiple shells (e.g., Wachter *et al.*; Schröder *et al.*).

WILLSON: (1) I think there is useful information in these observations, but we are not yet ready to extract the meaning. (2) We need more sophisticated models to extract the information of $\dot{M}$-history from PN. There are complications due to binaries.

The speaker, L.A. Willson (right), is arguing with P. Woitke (face not visible) and Q. Wang in front of a poster paper.

Stellar Evolution from AGB to Planetary Nebulae

Sun Kwok[1]

[1]Department of Physics, University of Hong Kong, Hong Kong, China
email: sunkwok@hku.hk

Abstract. Planetary nebulae are formed by an interacting winds process where the remnant of the AGB wind is compressed and accelerated by a later-developed fast wind from the central star. One-dimensional dynamical models have successfully explained the multi-shell (bubble, shell, crown, haloes) structures and the kinematics of planetary nebulae. However, the origin of the diverse asymmetric morphology of planetary nebulae is still not understood. Recent observations in the visible, infrared, and the submillimeter have suggested that the AGB mass loss becomes aspherical in the very late stages, forming an expanding torus around the star. A fast, highly collimated wind then emerges in the polar directions and carves out a cavity in the AGB envelope to form a bipolar nebula. Newly discovered structures such as concentric arcs, 2-D rings, multiple lobes, and point-symmetric structures suggest that both the slow and fast winds may have temporal and directional variations, and precession can play a role in the shaping of planetary nebulae. In this paper, we review the latest observations of planetary nebulae and proto-planetary nebulae and discuss the various physical mechanisms (rotation, binary, magnetic field, etc) that could lead to the observed morphologies.

Keywords. Stars: AGB and post-AGB, Planetary nebulae: general, Stars: evolution

1. Introduction

Planetary nebulae (PNe) represent a short ($\sim 10^4$ yr) phase of stellar evolution between the asymptotic giant branch (AGB) and white dwarfs. The central stars of PNe are remnants of the electron-degenerate C-O cores of their AGB progenitors, having lost most of their H envelopes due to mass loss on the AGB. They maintain their energy output through H-shell burning and evolve with constant luminosity from low to high temperature across the H-R diagram. When their entire H envelope is consumed by a combination of nuclear burning and mass loss, their luminosities begin to decrease and the central star gradually cool to become white dwarfs. This basic scenario was outlined by Paczyński (1971), and was confirmed by detailed evolutionary tracks calculations by Schönberner (1979) and Wood & Faulkner (1986).

2. What is a planetary nebula?

Traditionally, PNe are observationally defined as optical nebulae with a central star with some degree of symmetry and a strong emission-line spectrum with no or very weak continuum. This definition does not take into account modern observational properties such as dust emission in the infrared, X-ray continuum from hot gas, emission lines from molecules, etc. However, even an expanded observational definition is not enough because symbiotic stars share many of these same properties and there is significant confusion between PNe and symbiotic stars in the literature (Kwok 2003). It is clear that we need to incorporate a theoretical element into our definition. For example, we can define PNe as ionized circumstellar shells showing some degree of symmetry surrounding a hot, compact star evolving from the AGB to the white dwarf phase (Kwok 2000). This definition will distinguish PNe from symbiotic stars or novae, which are binary systems undergoing mass exchange. While the hydrogen envelope mass of a PN central star is being depleted by nuclear burning and mass loss, and therefore constantly evolving, a symbiotic star maintains its energy source through accretion, and is therefore stationary in evolution. It should be noted, however, a post-outburst nova or symbiotic nova also evolves to the blue similar to a PN. Unless the outburst (H ignition) is observed, it may also be difficult to distinguish a symbiotic nova from a PN.

3. The existence of planetary nebulae

Since PNe represent material ejected during the AGB phase and later ionized by the increasingly hot central star, the existence of PNe requires that the dynamical (expansion) age of the nebula be comparable to the evolutionary age of the central star. A star with higher core mass leaves the AGB with a smaller H envelope but burns at a higher rate. As a result, the transition time between the AGB and PN stage is highly dependent on the core mass. While a high-mass star will evolve so fast that it only illuminates the nebula for a very short time, a star with low core mass will evolve too slowly to ionize the circumstellar nebula before it dissipates into the ISM. Consequently, not all stars that evolve past the AGB will become PNe.

Due to strong mass loss on the AGB, stars with main-sequence masses as high as 8 $M_\odot$ can lose their entire H envelope before the core mass reaches the Chandrasekhar limit of 1.4 $M_\odot$. From the initial mass function, one can estimate that 95% of all stars that have evolved from the main sequence during the lifetime of the Galaxy will end their lives as white dwarfs. A majority (30-90%, Drilling and Schönberner 1985) of these stars will have passed through the PN stage. Figure 1 shows a simulated distribution of post-AGB stars on the H-R diagram. We can see that all the high core-mass stars are on the cooling track, and the horizontal parts of the evolutionary tracks are mostly populated by PNe with core masses around 0.6 $M_\odot$.

4. Planetary nebulae as an evolving dynamical system

While an impulsive ejection appears intuitively obvious for the formation of PNe, the discovery of large-scale mass loss on the AGB led to the realization of the importance of AGB mass loss on the origin of PNe and the formulation of the interacting stellar winds (ISW) model (Kwok *et al.* 1978, Kwok 1982). In addition to the well-observed shell, the ISW model predicts the following components: a fast wind from the central star, a low-density halo representing the unshocked AGB wind, and a high-temperature bubble representing the shocked fast wind. The *IUE* satellite has observed P Cygni profiles corresponding to wind velocities of several thousands km s^{-1} in many central stars of PNe, confirming that fast winds are indeed common in PNe. Evidence for the existence of haloes outside the PN shells were found by CCD imaging and by the detections of dust and molecular envelopes by infrared and millimeter-wave observations. Extended diffuse X-ray emission from the hot bubble was detected by *ROSAT* and by *CHANDRA* observations.

A PN is a dynamical system whose evolution is tightly coupled to the evolution of the central star through a changing rate of stellar Lyman continuum output and photoionization, and by a changing mass loss rate and wind velocity from the central star and wind interaction. The appearance and the structure of PN therefore reflect the coupled dynamical and ionizational evolution of the nebula. The time-dependent nature of PN

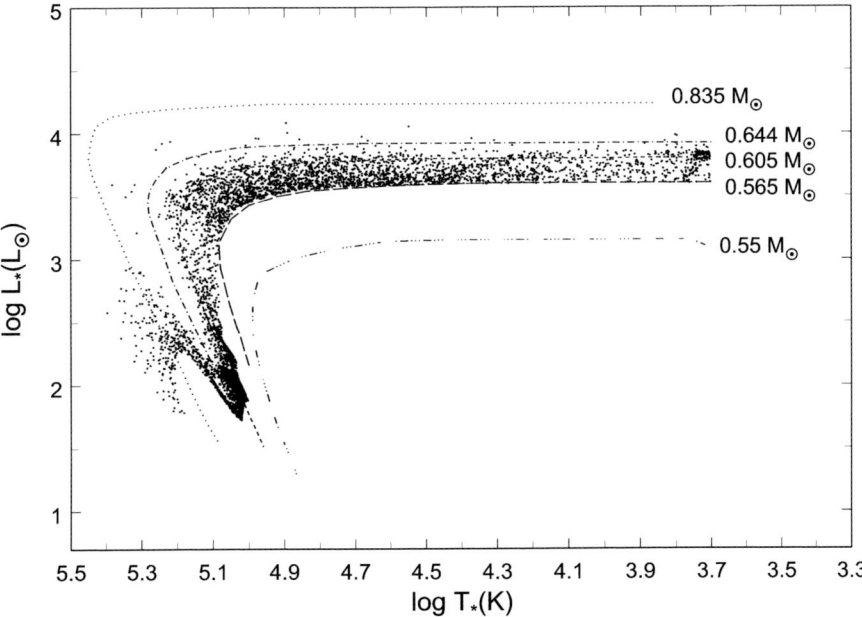

Figure 1. A simulated distribution of post-AGB stars in the H-R diagram. A random sample of 9951 stars with initial masses between 0.95 and 8 M$_\odot$ are generated using the Salpeter initial mass function and the initial mass-final mass relationship of Weidemann (2000). They are distributed in equal time intervals between $t = 0$ (set at $T_e = 3.7$) and 40,000 yr along H-shell burning evolutionary tracks. Tracks with masses between the five core masses calculated by Schönberner (1981, 1983) are derived from interpolation. Stars on the left of $\log T_e = 4.5$ are PNe and those on the right are PPNe.

evolution was incorporated into many of the 1-D spherically symmetric treatments of the ISW model (e.g., Marten & Schönberner 1991; Frank 1994; Mellema 1994). High dynamical range imaging observations of PNe with CCD detectors have revealed a much more complicated structure of PN than the classical picture of a shell plus central star. In addition to a bright shell, a low-surface brightness outer structure ("crown") and an extensive faint halo can be seen (Guerrero et al. 2000, Frank et al. 1990). While such multiple shell structures are difficult to understand in the classical model, they can be reproduced when the evolution of the central star is incorporated in the ISW model (Mellema 1994, Steffen et al. 1998, Corradi et al. 2000). In a 1-D model, the "shell" can be identified as the high-density shell compressed by the hot bubble, the "crown" as the extent of the ionization front, and the halo as the remnant of the AGB wind (Schönberner et al. 2005). It is important to note that these structures can change with time as the star evolves.

5. The intrinsic 3-D structure of PNe

Beginning with the work of Curtis (1918), there have been many attempts in classifying the morphologies of PNe (e.g. , Stanghellini et al. 1993, Manchado et al. 1996). However, all classification schemes suffer from the problem of sensitivity dependence where a deeper exposure can reveal fainter structures which change the classification of the PNe. For example, the waist of a bipolar nebula could be classified as elliptical if the bipolar lobes are too faint to be detected. NGC 650-1, Sh 1-89, and SaWe 3 are some of the cases where their bipolar nature were discovered as the result of deep CCD imaging.

The apparent morphology is also dependent on the lines of different ions observed due to ionization structures and stratification effects. Furthermore, projection effects due to viewing orientation may affect the morphological classification. It is clear that the examination of the apparent morphology alone is not sufficient to obtain the true intrinsic structure of PN. Kinematic data are necessary to separate various components projected on the same positions in the sky.

When the above factors are taken into consideration, it becomes clear that the fraction of bipolar nebulae is much higher than the fraction given in classification schemes based on apparent morphology. As a first order approximation, we may consider that the 2-D structure of a PN consists of the following components:

- A low-density spherical halo representing a remnant of the AGB wind.
- An ionization-bounded, dust-obscured torus representing an equatorial outflow during the late stages of AGB evolution.
- Two density-bounded polar lobes representing cavities created by fast outflows and the subsequent photoionization of the circumstellar material.

Under this model, a PN viewed near pole on will appear elliptical, with the torus seen as a shell and the lobes seen as envelopes surrounding the shell. Detailed kinematic studies have shown that several of the well-observed ring-like PN are in fact bipolar (NGC 6720, Bryce et al. 1994, Steffen et al. 2008; NGC 7027, Latter et al. 2000; NGC 3132, Monteiro et al. 2000). When viewed near edge on, the PN will appear to have a bipolar or butterfly shape. Although the fraction of bipolar PNe based on apparent morphology is relatively small ($\sim$15%), the true fraction could be much higher.

Beyond this torus-bipolar lobes basic structure, recent high-resolution images obtained with the *HST*, have discovered a number of additional microstructures:

- FLIERS: FLIERS (fast low-ionization emission regions) are pairs of small, bright knots of low excitation gas found along the major axes of PNe (Balick et al. 1998). Examples of well-defined FLIERS can be found in NGC 3242, NGC 6826, and NGC 7009.
- Collimated outflows: some PNe and PPNe (e.g. Hen 3-401, Sahai et al. 1999a) have extreme bipolar (cylindrical) shapes, suggesting that their morphology is shaped by a collimated outflow. The direct imaging of bipolar lobes emerging from a circumstellar disk in the PPN IRAS 17106−3046 (Kwok et al. 2000) suggests that disks could play a role in the collimation of the bipolar flows.
- Multipolar and point-symmetric structures: point-symmetric pairs of knots in an S-shape structure, or sometimes referred to as bipolar, rotating, episodic jets (BRET), have been seen in a number of PNe (e.g. KjPn8, López et al. 1995; NGC 6884, Miranda et al. 1999). Some PNe have been found to have more than one polar axis, suggesting that the outflow direction has changed with time (e.g. NGC 2440, López et al. 1998; M1-37 and He2-47, Sahai 2000). High-resolution mid-infrared imaging has found that the dust torus in the PPN IRAS 17441−2411 is in fact mis-aligned by 23° with the optical bipolar axis, suggesting that the torus may have precessed over time (Volk et al. 2007).
- Concentric arcs: concentric circular arcs have been observed in both PNe (e.g. Hb 5, NGC 6543, NGC 7027) and PPNe (e.g. AFGL 2688, IRAS 17150−3224). These arcs are of almost perfectly circular in shape, and have relatively uniform separations of $\sim 10^2$ yr (Kwok et al. 2001). Such time intervals are too short for thermal pulses ($\sim 10^4$ yr) and too long for pulsations ($\sim 10^0$ yr). Similar arcs have been detected in the carbon star IRC+10216, suggesting that these features originate in the AGB phase. The coexistence of these perfectly circular features with bipolar lobes suggests that the arcs are projections of undisturbed spherical shells on the sky. Possible mechanisms for the creation of such arcs include dynamical instability in the gas-dust coupling in the AGB outflow (Deguchi

1997), perturbation by a binary companion (Mastrodemos & Morris 1999), and magnetic cycle (Soker 2000, Garia-Segura *et al.* 2001).

- Two-dimensional co-axial rings: multiple rings perpendicular to the bipolar axis have been found in several PNe, including MyCn18 (Sahai *et al.* 1999b) and NGC 6881 (Kwok & Su 2005) and Hb 12 (Kwok & Hsia 2007). Such rings may be created by a periodic or episodic fast wind interacting with the circumstellar medium.

6. Shaping of PNe and the origin of the asymmetry

The fact that AGB circumstellar envelopes are generally spherically symmetric and PNe are highly asymmetric suggests that a drastic morphological transformation must have occurred in the post-AGB evolution. The discovery of PPNe (Kwok 1993) offers the opportunity to pinpoint the earliest stage of this transformation. Early models of PN shaping relied on the asymmetry of the AGB and the ISW process to amplify such asymmetry to create bipolar morphology observed in PNe (Balick 1987, Mellema & Frank 1995). According to this scenario, PNe morphology should be age-dependent with the most evolved objects being most extreme in bipolar form and the youngest members being much less so. However, imaging of PPNe have found many PPNe to possess bipolar morphology, implying that the shaping process occurs during the PPN phase, well before the photoionization of the nebulae (Kwok *et al.* 1996, 1998). Some PPNe and young PNe have highly focused lobes, suggesting that the fast wind is being collimated by a disk. The misalignment between the infrared dust torus and the optical bipolar lobes seen in the PPN IRAS 17441−2411 suggests that some of such collimating disks may be precessing. Hydrodynamic models have shown that FLIERs and BRETs could be naturally produced by the ISW process if the mass loss rate and velocity of the fast wind are functions of both time and direction (Steffen *et al.* 2001). The change in outflow direction could be the result of rotation and magnetic fields (Garcia-Segura *et al.* 1999).

7. Conclusions

From the above discussions, it is clear that the study of circumstellar matter such as PNe can reveal a great deal about the physical processes within stars. The detection of concentric arcs suggests that the AGB wind may have discrete shell structures. The fact that many PPNe possess highly bipolar structures suggest that the morphological transformation is by a highly collimated fast wind. The observations of multiple co-axial 2-D rings suggest that this fast wind may be episodic. The existence of point-symmetric structure suggests that the fast wind may be precessing. These observations raised a number of questions on their physical origins, e.g., what roles do binary evolution, magnetic fields, or rotation play in the launching, collimation, and precession of the fast outflow? Are these circumstellar asymmetry in any way related to the asymmetry generated by convection, turbulence, and magnetic fields in the interiors of AGB stars? All these questions are relevant to the development of realistic models of the structure and evolution of stars.

References

Balick, B. 1987, *AJ*, 94, 671
Balick, B., Alexander, J., Hajian, A. R., Terzian, Y., Perinotto, M., & Patriarch, P. 1998, *AJ*, 116, 360
Bryce, M., Balick, B., & Meaburn, J. 1994, *MNRAS*, 266, 721

Corradi, R. L. M., Schönberner, D., Steffen, M., & Perinotto, M. 2000, *A&A*, 354, 1071
Curtis, H. D. 1918, *Publ. Lick Obs.*, Vol. XIII, Part III, p. 57
Deguchi, S. 1997, in *IAU Symp. 180: Planetary Nebulae*, H. J. Habing & H. J. G. L. M. Lamers (eds.), Kluwers, p. 151
Drilling, J. S. & Schönberner, D. 1985, *A&A*, 146, L23
Frank, A. 1994, *AJ*, 107, 261
Frank, A., Balick, B., & Riley, J. 1990, *AJ*, 100, 1903
Garcia-Segura, G., Langer, N., Rózyczka, M., & Franco, J. 1999, *ApJ*, 517, 767
Garcia-Segura, G., López, J. A., & Franco, J. 2001, *ApJ*, 560, 928
Guerrero, M. A., Miranda, L. F., Manchado, A., & Vázquez, R. 2000, *MNRAS*, 313, 1
Kwok, S. 1982, *ApJ*, 258, 280
Kwok, S. 1993, *Ann. Rev. Astr. Ap.*31, 63-92
Kwok, S. 2000, *The Origin and Evolution of Planetary Nebulae*, Astrophysics Series No. 33, Cambridge University Press
Kwok, S. 2003, in *Symbiotic Stars Probing Stellar Evolution*, R. L. M. Corradi, R. Mikolajewska and T. J. Mahoney (eds.), ASP Conf. Ser. 303, p. 428
Kwok, S., Hrivnak, B. J., & Su, K. Y. L. 2000, *ApJ*, 544, L149
Kwok, S., Hrivnak, B. J., Zhang, C. Y., & Langill, P. L. 1996, *ApJ*, 472, 287
Kwok, S. & Hsia, C. H. 2007, *ApJ*, 660, 341
Kwok, S., Purton, C. R., & FitzGerald, M. P. 1978, *ApJ*, 219, L125
Kwok, S. & Su, K. Y. L. 2005, *ApJ*, 635, L49
Kwok, S., Su, K. Y. L., & Hrivnak, B. J. 1998, *ApJ*, 501, L117
Kwok, S., Su, K. Y. L., & Stoesz, J. A. 2001, in *Post-AGB Objects as a Phase of Stellar Evolution*, R. Szczerba & S. K. Górny (eds), Kluwer, p. 115
Latter, W. B., Dayal, A., Bieging, J. H., Meakin, C., Hora, J. L., Kelly, D. M., & Tielens, A. G. G. M. 2000, *ApJ*, 539, 783
López, J. A., Meaburn, J., Bryce, M., & Holloway, A. J. 1998, *ApJ*, 493, 803
López, J. A., Vázquez, R., & Rodriguez, L. F. 1995, *ApJ*, 455, L63
Manchado, A., Guerrero, M. A., Stanghellini, L., & Serra-Ricart, M. 1996, *The IAC Morphological Catalog of Northern Galactic Planetary Nebulae*, (IAC:Tenerife)
Marten, H. & Schönberner, D. 1991, *A&A*, 248, 590
Mastrodemos, N. & Morris, M. 1999, *ApJ*, 523, 357
Mellema, G. 1994, *A&A*, 290, 915
Mellema, G. & Frank, A. 1995, *MNRAS*, 273, 401
Miranda, L. F., Guerrero, M. A., & Torrelles, J. M. 1999, *ApJ*, 117, 1421
Monteiro, H., Morisset, C., Gruenwald, R., & Viegas, S. M. 2000, *ApJ*, 537, 853
Paczyński, B. 1971, *Acta Astr.*, 21, 417
Sahai, R. 2000, *ApJ*, 537, L43
Sahai, R., Bujarrabal, V., & Zijlstra, Z. 1999a, *ApJ*, 518, L115
Sahai, R. et al. 1999b, *AJ*, 118, 468
Schönberner, D. 1979, *A&A*, 79, 108
Schönberner, D. 1981, *A&A*, 103, 119
Schönberner, D. 1983, *ApJ*, 272, 708
Schönberner, D., Jacob, R., Steffen, M., Perinotto, M., Corradi, R. L. M., & Acker, A. 2005, *A&A*, 431, 963
Soker, N. 2000, *ApJ*, 540, 436
Stanghellini, L., Corradi, R. L. M., & Schwarz, H. E. 1993, *A&AS*, 279, 521
Steffen, M., Szczerba, R., & Schönberner, D. 1998, *A&A*, 337, 149
Steffen, W., López, J. A., & Lim, A. 2001, *ApJ*, 556, 823
Steffen, W., López, J. A., Koning, N., Kwok, S., Riesgo, H., Richer, M. G., & Morisset, C. 2008, in *Asymmetrical Planetary Nebulae IV*, Corradi, R. L. M., Manchado, A., Soker, N. (eds.), in press
Volk, S., Kwok, S., & Hrivnak, B. J. 2007, *ApJ*, 670, 1137
Weidemann, V. 2000, *A&A*, 363, 647
Wood, P. R. & Faulkner, D. J. 1986, *ApJ*, 307, 659

Discussion

G. HENSLER: What are spectra telling us about kinematics and excitation mechanisms of the PNe substructures?

S. KWOK: While it is now possible to do spectroscopy on the shell, crown and even the haloes (and therefore able to obtain kinematic information on these structures), the arcs and the rings are too faint to do spectroscopy.

P. WOITKE: Could you comment on the possibility that the rings around PNe are actually not concentric, but narrowly wiggled spirals? A companion star that wobbles the mass losing star could then provide an explanation. With the exception of IRC+10216, the rings of all PNe seem to be exactly concentric.

S. KWOK: The only AGB star that has been found to have arcs is IRC+10216. There are some suggestions that the arcs in this object is part of a spiral pattern. However, the PPN, and PNs, the arcs are highly concentric. Binary perturbations have indeed been suggested as a cause of the existence of these uniformly spaced concentric arcs.

J. VINK: I have two questions: (1) Could you put some number on the incidence of bipolarity among the class of PNe, and (2) you mentioned the asphericity stars during the p-AGB phase. Would you be able to say at which T_{eff} this happens?

S. KWOK: In early morphological classification schemes, the fraction of bipolar nebulae is estimated to be $\sim 15\%$. However, when sensitivity and orientation factors are taken into account, its fraction is considerably higher.

From the existence of highly bipolar pre-planetary nebulae, the initiation of a fast, collimated wind must be very early, may be right at the end of the AGB. Pre-planetary nebulae are typically of special type F or G, so the fast wind could begin as early as $T_e \simeq 5000$ K.

From left: S. Kwok (the speaker), G. Houdek, and F. Kupka

AGB star models. Results from 1D stellar evolution and multi-dimensional hydrodynamics simulations

Falk Herwig†

[1]Keele Astrophysics Group, School of Physical and Geographical Sciences, Keele University, Staffordshire ST5 5BG, UK
email: fherwig@astro.keele.ac.uk

Abstract. In this review I am discussing the current state of simulating the internal evolution of AGB stars. Recent work on AGB stars include the effect of rotation, magnetic fields and internal gravity waves, as well as thermohaline mixing induced by the ^{3}He $+^3$He pp-chain reaction. Hydrodynamic simulations of the interior convection of AGB stars are now becoming available, giving insights to convective boundary mixing, for example for He-shell flash convection. At very low metallicity convective-reactive events are encountered in AGB stars (as well as in massive stars), and the necessity of hydrodynamic simulations to address this difficult phase of stellar evolution is emphasized.

Keywords. Stars: AGB and post-AGB, stars: rotation, stars: magnetic fields

1. Introduction

Asymptotic Giant Branch (AGB) stars are the premier nuclear production phase of low- and intermediate mass stars. Their evolution has been extensively studied through one-dimensional, spherically symmetric stellar evolution simulations and numerous reviews are available that summarize the results, including, for example, Iben & Renzini (1983), Lattanzio (1989), Blöcker et al. (2000) and more recently Herwig (2005).

An important goal of AGB stellar evolution simulations is to provide the thermodynamic evolution that allows the detailed determination of nucleosynthesis and resulting yields, which is for example reviewed by Lattanzio & Boothroyd (1997). A detailed account of the s process in AGB stars has been provided by Busso et al. (1999). Yields based on stellar evolution tracks have been published, for example, by Forestini & Charbonnel (1997) and more recently by the Monash group, culminating in a complete yield set for a wide range of initial masses and metallicities including the light elements up to sulfur (Karakas & Lattanzio 2007). Their results and comparison to previous work reaffirms the two dominant factors in providing accurate yields: mixing and mass loss.

Nucleosynthesis predictions of AGB stellar evolution are confronted with a wide variety of observables. Among these are isotopic ratios of pre-solar grain, now increasingly from multiple measurements of single grains (e.g. Barzyk et al. 2006; Marhas et al. 2007). AGB nucleosynthesis is observable intrinsically in AGB stars (McSaveney et al. 2007, for a recent example), as well in the post-AGB stars, for example in planetary nebulae (Sterling & Dinerstein 2008) and the H-deficient post-AGB stars (Werner & Herwig 2006). AGB nucleosynthesis yields are included in galaxy chemical evolution models (e.g. Travaglio et al. 2001; Recchi 2007) and are now helping to address questions such as the "missing satellites" problem of cosmological simulations (Fenner et al. 2006).

† Present address: Dept. of Physics and Astronomy, University of Victoria, Victoria, BC, Canada

In this paper I will cover work that appeared in the last couple of years, as previous work is reviewed in some detail in the above mentioned and similar references. In Sect. 2 I will discuss the situation of mixing and mass loss in the context of one-dimensional stellar evolution models. Sect. 3 is devoted to simulations in the very low metallicity regime. In both sections I will highlight which problems can now be addressed using 3D hydrodynamic simulations of parts of the star, and this topic will be addressed in Sect. 4, to be followed by some concluding remarks.

2. Progress in 1D AGB evolution models

2.1. Convection

AGB star evolution simulations and the resulting observable predictions are extremely sensitive to the mixing physics, and in this regard AGB stars are no different than for example massive stars. And as in other types of stars several physical processes contribute to the overall mixing.

Foremost of course is convection with fluid velocities, for example in the He-shell flash convection, of several km/s. Nuclear physics measurements are now providing the data to analyse s-process branchings that are sensitive to the details of the convective flow in the He-shell flash convection zone (Reifarth et al. 2004; Mohr et al. 2007; Heil et al. 2008), and a more realistic treatment of convective flows than possible through the mixing-length theory is needed to take full advantage of these new measurements.

The role of mixing at convective boundaries has been extensively discussed over the past years, e.g. Mowlavi (1999), or for a more recent summary Herwig (2005). The bottom line is that the important third dredge-up can only be obtained in stellar evolution models with some kind of mixing at the bottom of the convective boundary, and various physical models and numerical algorithms are employed. Also, of course, the ^{13}C pocket as the source of neutrons in the s-process depends on some mixing triggered by convection at the time of the third dredge-up. While it is clear that the source of this convective boundary mixing must be a hydrodynamic instability the exact nature of this mixing is still unclear. But in any case, the effect of this instability must be taken into account during the dredge-up phase of AGB stars of all masses and metallicities (see Sect. 3).

In Fig. 1 I show core-mass evolutions for several AGB tracks with certain choices for the overshoot parameter at the bottom of the pulse-driven convection zone (= He-shell flash convection zone) $f_{\rm PDCZ}$, and the bottom of the envelope conception during the third dredge-up, $f_{\rm CE}$. The former parameter is indicated by hydrodynamic He-shell flash convection simulations (Herwig et al. 2007), while the latter reproduces the observed s-process strength of intrinsic AGB stars (Lugaro et al. 2003). Stellar evolution models show that the choices for convective boundary mixing sensitively affect yields, for example through the thermal pulse strength and the dredge-up efficiency.

The properties of envelope convection determines the yields and quantitative evolution as well. This has been pointed out by Boothroyd & Sackmann (1988) in the context of the mixing-length theory, and by Mazzitelli et al. (1999) when applying the full-spectrum turbulence theory instead to AGB stars. The synthetic AGB models by Marigo et al. (1996) considered the mixing-length parameter uncertainty as well. Nevertheless all other contemporary AGB stellar evolution models are calculated with a mixing-length parameter calibrated to reproduce the parameters of the sun although there is no reason that that mixing-length parameter should be universally applicable to all convection zones in all phases of stellar evolution. Results by Porter & Woodward (2000) and McSaveney et al. (2007) hint at a possibly larger mixing length parameter compared to the one that best

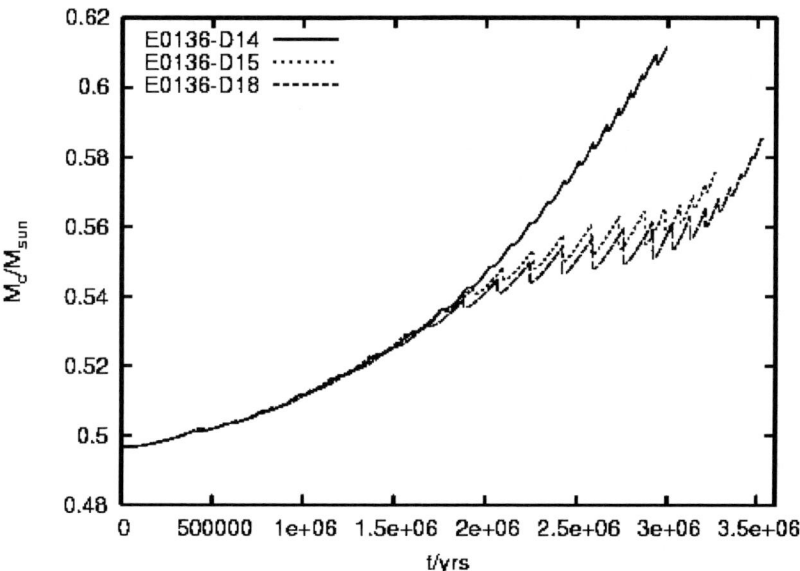

Figure 1. Core mass evolution for initial mass $2\,\mathrm{M}_\odot$ and $Z = 0.01$ for three choices of convective parameters. D14: $f_{\mathrm{PDCZ}} = 0.008$, $f_{\mathrm{CE}} = 0.128$ and $\alpha_{\mathrm{MLT}} = 1.7$; D15: same as D14 but $\alpha_{\mathrm{MLT}} = 2.5$; D18: same as D15 but revised nuclear reaction rate for $^{14}N(p,\gamma)$ ($0.64 \times$ NACRE, Herwig & Austin 2004).

reproduces the solar parameters. Figure 1 shows the core mass evolution for cases with sun-calibrated mixing-length parameter α_{MLT} (case D14) and increased envelope convection efficiency. The effect is significant as deeper third dredge-up (as in the D15 and D18 sequence) implies a larger enrichment in processed material. The calibration of α_{MLT} for AGB stars certainly deserves more attention, for example through well observed giants in the Magelanic Clouds. Additionally, α_{MLT} needs to be determined from convective envelope simulations like those by Porter & Woodward (2000), maybe with updated input physics, or more recently by Freytag & Höfner (2008).

I will discuss the particular difficulties of convective boundaries at very low metalicity in Sect. 3.

2.2. Other mixing sources

As reviewed in Herwig (2005) there is numerous observational evidence that AGB stars experience some mixing through the radiative layer between the convective envelope and the H-burning shell (see Uttenthaler et al. 2007, for a recent example). Recently, two physical processes have been investigated in this context.

Eggleton et al. (2006) reported a long overlooked, but absolutely necessary source of mixing, resulting from the reduction of the mean molecular weight through the nuclear reaction $^3\mathrm{He}(^3\mathrm{He},2p)^4\mathrm{He}$ which creates locally some boyuancy leading to mixing. The authors make quite correctly the point that the mixing from this μ-effect is robust and has to happen. It is not optional and not based on rather uncertain phsyics assumption. As shown in this conference this μ mixing can have observable effects in AGB stars as well (this vol., Cantiello & Langer 2008), and it remains to be seen if there is still a need for additional physical processes as a source of mixing between the H-shell and the bottom of the convection zone, such as magnetic field induced mixing proposed by Busso et al. (2007), in order to accomodate observations. It is interesting to note that the μ-mixing may be very efficient at extremely low metallicity as ^{3}He is a primordial BBN isotope and relatively overabundant at extremely low metallicity.

Rotation has been long suspected to play an important role in AGB stars, but this seems not to be the case, at least according to current simulations of AGB stars including the effect of rotation. Accordingly, rotation is not responsible for the formation of the ^{13}C pocket†, and indeed seems to rather prohibit the s process during the interpulse phase even if a ^{13}C pocket has been assumed to form through some other mixing process (Herwig et al. 2003). Recently, Suijs et al. (2008) reported that the additional inclusion of magnetic fields leads to an enhanced angular momentum transport and eventually reduced AGB core spin rates, bringing simulations to better agreement to asterioseismologically observed white dwarf rotation rates. Whether magnetic fields can also resolve the difference between rotating model s-process predictions and observations quantitatively remains to be seen.

In addition to convection, thermohaline μ-mixing, rotation and magnetic fields, the contribution of internal gravity waves (Press 1981) to angular momentum redistribution and mixing has been investigated. We certainly observe the excitation of these waves in the vicinity of He-shell flash convection in 2D and 3D compressible hydrodynamic simulations (see Sect. 4, Herwig et al. 2006). Again, this type of mixing may be related to the formation of the ^{13}C pocket (Denissenkov & Tout 2003) or to the global redistribution of angular momentum or chemical species (Talon & Charbonnel 2008).

2.3. Mass loss and opacities

In recent years progress has been also made in the area of mass loss. While it has been pointed out that the dist-driven wind scenario is in trouble for stars with C/O < 1 (Höfner 2007) progress has been made for the C-stars through more realistic treatment of radiation in the wind models (Höfner et al. 2003) and by implementing a more realistic treatment of the pulsations now from full 3D hydrodynamic simulations (Freytag & Höfner 2008).

As an example, recent applications of the new generation of tools to study mass loss in C stars suggests that the energy injection of pulsations is of greater importance than metallicity, and that C-stars at low metallicity should therefore have similar mass loss as their more metal-rich counterparts (Mattsson et al. 2008).

In order to couple these new mass loss rates to stellar evolution calculations chemistry dependent molecular opacities need to be included to ensure the right surface parameters, and examples of such calculations include those by Cristallo et al. (2007) with an emphasis on metal-poor AGB stars. It seems that the new mass loss simulations together with a consistent treatment of low-temperature opacities puts a solution of the mass loss problem, as least for C-stars within reach.

2.4. Nuclear physics

Nuclear reaction rate uncertainties effect both the yields directly as well as indirectly through the structural evolution. For example, the downward revision of the $^{14}N(p,\gamma)^{15}O$ reaction (e.g. LUNA Collaboration et al. 2006) leads to a notable increase of the third dredge-up efficiency (Herwig & Austin 2004; Weiss et al. 2005; Herwig et al. 2006). Other work has focused on the yield effect of nuclear reaction rate uncertainties, for example for p-capture rates of the NeNa and MgAl chains during HBB in intermediate-mass AGB stars (Izzard et al. 2007).

† The ^{13}C pocket reported by Langer et al. (1999) is too small in mass to account for observations, as shown by Herwig et al. (2003).

3. Evolution at very low metallicity

The evolution of AGB stars at very low metallicity has received enourmous attention over the past years, mostly due of course to the continuing discovery of more interesting very metal-poor stars (Beers & Christlieb 2005). As many of the C-enhanced stars of extremely low metallicity (CEMP) are believed to be companions of and polluted by former AGB stars (e.g. Lucatello et al. 2005; Lugaro et al. 2008; Stancliffe et al. 2007, see also Pols, this vol.) we are confronted with the investigation of stellar evolution at very low metal content in stars in general, and in AGB stars especially.

One feature is particularly noteworthy, as it causes serious problems for the traditional one-dimensional, spherically symmetric stellar evolution approach. At very low metal content the entropy barrier between He- and H-shell burning layers is reduced and mixing of H into very hot ^{12}C-rich He-shell burning ash material may trigger violent flash-like convective-reactive phases. Such events can be found in massive stars (e.g. non-rotating: Heger & Woosley 2008), and sometimes induced by rotation (Hirschi et al. 2005; Ekström et al. 2008).

In AGB stars two variants of the convective-reactive theme are known. The H-ingestion flash in low-metallicity AGB stars involves entrainment of H into the He-shell flash convection zone (Fujimoto et al. 2000; Herwig 2003; Iwamoto et al. 2004; Campbell 2007, and many older references there). The hot dredge-up (Herwig 2004; Goriely & Siess 2004) is encountered if even only a small fraction of the convective boundary mixing that was introduced to create a H-^{12}C partial mixing zone for the ^{13}C is also included at the bottom of the convective envelope during the dredge-up in slightly more massive ($M > 2 \ldots 3 \, \mathrm{M_\odot}$) AGB stars of very low metallicity ([Fe/H] < -2). In both cases the commonly used mixing-length theory to treat convection and derive a diffusion coefficient for mixing is not applicable as it does not account for injection of nuclear energy on a hydrodynamic time scale of the convective flow. Woodward et al. (2008) discuss in some detail this problematic situation, which requires 3-dimensional hydrodynamics simulations to be properly addressed.

4. 3D hydrodynamic simulations of AGB interiors

Stellar hydrodynamics has overwhelmingly been concerned with the stellar surface layers, atmospheres and shallow envelopes (e.g. Nordlund 1982; Stein & Nordlund 1998; Porter & Woodward 2000) because the Mach numbers are larger near the surface, making explicit, compressible simulations feasible.

But convective boundaries in surface convection zones behave radically different compared to the deep stellar interior. For example, the hydrodynamic simulations of surface convection in A-type stars and white dwarfs (Freytag et al. 1996) show that coherent convective systems are accelerated in the unstable zone and then transition into the neighbouring formally stable layer only to reverse direction after crossing a significant fraction of a pressure scale height. The situation is very different in the deep interior of AGB stars. Plane-parallel "box-in-a-star" hydrodynamic simulations of He-shell flash convection in two and three dimenstions (Herwig et al. 2006) show rather stiff convective boundaries. While convective systems occupy the entire vertical span of the unstable zone of approx. 4500km in these simulations the convective bounadries are poorly resolved at grid size of 20–50km. Nevertheless, careful analysis of simulatons with a resolution of 13.5km showed that overshooting at the bottom of the convection zone is as expected small (Freytag & Herwig 2008; Herwig et al. 2007), and to first order in agreement with stellar evolution constraints from H-deficient post-AGB stars (Werner & Herwig 2006).

The upper He-shell flash convection boundary is not important in normal thermal pulses, but in those AGB stars of very low metallicity and on the post-AGB where the H-ingestion flashes take place, the upper boundary is of utmost importance as it is here that the H-entrainment will take place. Woodward *et al.* (2008) reported preliminary simulations zooming in with a grid of up to 1024^3 on the 100km above and below the top convective boundary. These simulations show a significant entrainment of H into the C-rich intershell. It is the competition of mixing accross the bounadry and larger-scale convective transport that will determine the inhomogeneities of the fuel-mix, and eventually determine the properties of the H-burning front. For example, larger horizontal inhomogeneities of mixed-down H will lead to a more patchy energy generation and a broader H-burning front, which may in the end be permeable for mixing of the resulting ^{13}C to the bottom of the He-shell flash convection zone. Here the lifetime of ^{13}C against α capture, and thereby neutron release, is only seconds. By definition spherically-symmetric simulations can not account for any of these horizontal inhomogeneities, which makes them unsuitable for predictive simulations of the convective-reactive phases of evolution in stars, including the hot dredge-up discussed in Sect. 3.

5. Conclusions

In this review I have discussed some aspects of AGB stellar evolution that are currently been worked on. Significant progress has been made in better understanding the effect of the various physical processes that may be responsible for mixing. In some areas we are now (or very soon) getting better agreement with observations, for example in models including rotation and additional angular momentum transport. However, predicted white dwarf rotation rates are still typically higher by an order of magnitude compared to observed rotation rates, even including the effect of magnetic fields. Then, on the underlying physics there is in some areas no consesus yet, as demonstrated recently by the debate over the Tayler-Pitts-Spruit dynamo supposed to provide substantial mixing and angular momentum transport (Zahn *et al.* 2007). Furthermore, the effect of some processes discussed maybe virtually indistinguishable in observations. Internal gravity waves and magnetic fields may have similar consquences, making it maybe at this time difficult to validate. However, the recent trajectory of progress clearly indicates that at least in the area of convection we should be able to finally resolve issues surrouding the mechanism and efficiency of overshooting in the next couple of years. This, together with progress in mass loss, molecular opacitiy and the thermohaline μ-mixing will already provide for a significantly improved new standard model of AGB stars which should emerge in the near future.

Acknowledgements

I have to thank Bernd Freytag for his most important contributions to our collaborative work on He-shell flash convection. I am indebted to Paul Woodward, David Porter and their students for an extremely enjoyable, ongoing collaboration. I would like to thanks Marco Pignatari for his suggestions on nucleosynthesis and grains. This research was supported by a Marie Curie International Reintegration Grant withing the 6th European Community Framework Programme, grant MIRG-CT-2006-046520 as well as through the Joint Institute for Nuclear Astrophysics, NSF grant PHY 02-16783.

References

Barzyk, J. G., Savina, M. R., Davis, A. M., Gallino, R., Pellin, M. J., Lewis, R. S., Amari, S., & Clayton, R. N. 2006, New Astronomy Review, 50, 587
Beers, T. C. & Christlieb, N. 2005, ARAA, 43, 531
Blöcker, T., Herwig, F., & Driebe, T. 2000, in The changes in abundances in AGB stars, ed. F. D'Antona & R. Gallino, Mem. Soc. Astron. Ital., in press, astro-ph/0002455
Boothroyd, A. I. & Sackmann, I.-J. 1988, ApJ, 328, 671
Busso, M., Gallino, R., & Wasserburg, G. J. 1999, ARA&A, 37, 239
Busso, M., Wasserburg, G. J., Nollett, K. M., & Calandra, A. 2007, ApJ, 671, 802
Campbell, S. W. 2007, PhD thesis, Monash University, Australia, http://www.asiaa.sinica.edu.tw/simcam/work/phd/thesis-SimonCampbell-A4singleside-6Jul07.pdf
Cantiello, M. & Langer, N. 2008, ArXiv e-prints, 806
Cristallo, S., Straniero, O., Lederer, M. T., & Aringer, B. 2007, ApJ, 667, 489
Denissenkov, P. A. & Tout, C. A. 2003, MNRAS, 340, 722
Eggleton, P. P., Dearborn, D. S. P., & Lattanzio, J. C. 2006, Science, 314, 1580
Ekström, S., Meynet, G., Chiappini, C., Hirschi, R., & Maeder, A. 2008, ArXiv e-prints, 807
Fenner, Y., Gibson, B. K., Gallino, R., & Lugaro, M. 2006, ApJ, 646, 184
Forestini, M. & Charbonnel, C. 1997, A&AS, 123, 241
Freytag, B. & Herwig, F. 2008, ApJ, in prep
Freytag, B. & Höfner, S. 2008, A&A, 483, 571
Freytag, B., Ludwig, H.-G., & Steffen, M. 1996, A&A, 313, 497
Fujimoto, M. Y., Ikeda, Y., & Iben, I., J. 2000, ApJ Lett., 529, L25
Goriely, S. & Siess, L. 2004, A&A, 421, L25
Heger, A. & Woosley, S. E. 2008, ArXiv e-prints, 803
Heil, M., Winckler, N., Dababneh, S., Käppeler, F., Wisshak, K., Bisterzo, S., Gallino, R., Davis, A. M., & Rauscher, T. 2008, ApJ, 673, 434
Herwig, F. 2003, in CNO in the Universe, ASP Conf. Ser. astro-ph/0212366
Herwig, F. 2004, ApJ, 605, 425
—. 2005, ARAA, 43
Herwig, F. & Austin, S. M. 2004, ApJ Lett., 613, L73
Herwig, F., Austin, S. M., & Lattanzio, J. C. 2006, Phys. Rev. C., 73, 025802
Herwig, F., Freytag, B., Fuchs, T., Hansen, J. P., Hueckstaedt, R. M., Porter, D. H., Timmes, F. X., & Woodward, P. R. 2007, in Astronomical Society of the Pacific Conference Series, Vol. 378, Why Galaxies Care About AGB Stars: Their Importance as Actors and Probes, ed. F. Kerschbaum, C. Charbonnel, & R. F. Wing, 43
Herwig, F., Freytag, B., Hueckstaedt, R. M., & Timmes, F. X. 2006, ApJ, 642, 1057
Herwig, F., Langer, N., & Lugaro, M. 2003, ApJ, 593, 1056
Hirschi, R., Meynet, G., & Maeder, A. 2005, A&A, 443, 581
Höfner, S. 2007, in Astronomical Society of the Pacific Conference Series, Vol. 378, Why Galaxies Care About AGB Stars: Their Importance as Actors and Probes, ed. F. Kerschbaum, C. Charbonnel, & R. F. Wing, 145–+
Höfner, S., Gautschy-Loidl, R., Aringer, B., & Jørgensen, U. G. 2003, A&A, 399, 589
Iben, Jr., I. & Renzini, A. 1983, ARA&A, 21, 271
Iwamoto, N., Kajino, T., Mathews, G. J., Fujimoto, M. Y., & Aoki, W. 2004, ApJ, 602, 378
Izzard, R. G., Lugaro, M., Karakas, A. I., Iliadis, C., & van Raai, M. 2007, A&A, 466, 641
Karakas, A. & Lattanzio, J. C. 2007, Publications of the Astronomical Society of Australia, 24, 103
Langer, N., Heger, A., Wellstein, S., & Herwig, F. 1999, A&A, 346, L37
Lattanzio, J. 1989, in Evolution of Peculiar Red Giants, ed. H. R. Johnson & B. Zuckermann, IAU Symp. 106 (Cambridge University Press), 161
Lattanzio, J. C. & Boothroyd, A. I. 1997, in Astrophysical Implications of the Laboratory Study of Presolar Materials, ed. T. Bernatowitz & E. Zinner (AIP Conf. Ser.), 85
Lucatello, S., Tsangarides, S., Beers, T. C., Carretta, E., Gratton, R. G., & Ryan, S. G. 2005, ApJ, 625, 825

Lugaro, M., de Mink, S. E., Izzard, R. G., Campbell, S. W., Karakas, A. I., Cristallo, S., Pols, O. R., Lattanzio, J. C., Straniero, O., Gallino, R., & Beers, T. C. 2008, A&A, 484, L27

Lugaro, M., Herwig, F., Lattanzio, J. C., Gallino, R., & Straniero, O. 2003, ApJ, 586, 1305

LUNA Collaboration, Lemut, A., Bemmerer, D., Confortola, F., Bonetti, R., Broggini, C., Corvisiero, P., Costantini, H., Cruz, J., Formicola, A., Fülöp, Z., Gervino, G., Guglielmetti, A., Gustavino, C., Gyürky, G., Imbriani, G., Jesus, A. P., Junker, M., Limata, B., Menegazzo, R., Prati, P., Roca, V., Rogalla, D., Rolfs, C., Romano, M., Rossi Alvarez, C., Schümann, F., Somorjai, E., Straniero, O., Strieder, F., Terrasi, F., & Trautvetter, H. P. 2006, Physics Letters B, 634, 483

Marhas, K., Hoppe, P., & Ott, U. 2007, "Meteoritics & Planetary Science", 42, 1043

Marigo, P., Bressan, A., & Chiosi, C. 1996, A&A, 313, 545

Mattsson, L., Wahlin, R., Höfner, S., & Eriksson, K. 2008, A&A, 484, L5

Mazzitelli, I., D'Antona, F., & Ventura, P. 1999, A&A, 348, 846

McSaveney, J. A., Wood, P. R., Scholz, M., Lattanzio, J. C., & Hinkle, K. H. 2007, MNRAS, 378, 1089

Mohr, P., Käppeler, F., & Gallino, R. 2007, Phys. Rev. C., 75, 012802

Mowlavi, N. 1999, A&A, 344, 617

Nordlund, A. 1982, A&A, 107, 1

Porter, D. H. & Woodward, P. R. 2000, ApJS, 127, 159

Press, W. H. 1981, ApJ, 245, 286

Recchi, S. 2007, in Astronomical Society of the Pacific Conference Series, Vol. 378, Why Galaxies Care About AGB Stars: Their Importance as Actors and Probes, ed. F. Kerschbaum, C. Charbonnel, & R. F. Wing, 364–+

Reifarth, R., Käppeler, F., Voss, F., Wisshak, K., Gallino, R., Pignatari, M., & Straniero, O. 2004, ApJ, 614, 363

Stancliffe, R. J., Glebbeek, E., Izzard, R. G., & Pols, O. R. 2007, A&A, 464, L57

Stein, R. F. & Nordlund, A. 1998, ApJ, 499, 914

Sterling, N. C. & Dinerstein, H. L. 2008, ApJS, 174, 158

Suijs, M. P. L., Langer, N., Poelarends, A.-J., Yoon, S.-C., Heger, A., & Herwig, F. 2008, A&A, 481, L87

Talon, S. & Charbonnel, C. 2008, A&A, 482, 597

Travaglio, C., Gallino, R., Busso, M., & Gratton, R. 2001, ApJ, 549, 346

Uttenthaler, S., Lebzelter, T., Palmerini, S., Busso, M., Aringer, B., & Lederer, M. T. 2007, A&A, 471, L41

Weiss, A., Serenelli, A., Kitsikis, A., Schlattl, H., & Christensen-Dalsgaard, J. 2005, ArXiv Astrophysics e-prints

Werner, K. & Herwig, F. 2006, PASP, 118, 183

Woodward, P., Herwig, F., Porter, D., Fuchs, T., Nowatzki, A., & Pignatari, M. 2008, in American Institute of Physics Conference Series, Vol. 990, First Stars III, 300–308

Zahn, J.-P., Brun, A. S., & Mathis, S. 2007, A&A, 474, 145

Discussion

F. KUPKA:

(1) Considering that the physical conditions for which the f-parameter model was derived for and the He-shell flash scenario, with all these new and different processes going on in that case – why do you still use the f-parameter model, since its physical basis for that scenario is certainly even less well founded than MLT? It cannot be more than just a mathematical fit formula.

(2) As you have shown with one of your movies, if one does global simulations for this problem, on has to account for large and complex inhomogeneities in the physical conditions at a given radius. It would help to have a table to compare the scales of the box-in-a-star simulations to the global ones. To tie the two types of simulations together to eventually derive predictive models is quite some work – which you may have already studied?

F. HERWIG:

(1) Freytag et al. (1996) showed, based on their simulations, that (i) the overshooting fluid flows lead to diffusive spreading of hot particle ensembles and (ii) that the corresponding diffusion coefficients decrease exponentially with the distance from the convective boundary. They determined the exponential delay parameter "f" for A type and white dwarf convection from their simulations as f = 0.25 and 1.0, respectively. Already from their paper you can see that although both conventional zones show a similar qualitative exponential overshoot behavior, which are quantitatively represented by different f. It is therefore not at all surprising that once we put this mathematical expression for the overshoot behavior into a stellar evolution code we have get again different f-values at different convective boundaries. In some cases we can not even be certain that the physics of convective boundary mixing is the same as in Freytag's simulation. In the deep interior we might rather encounter turbulent entrainment (see Casey Meakin's talk). In that case the f-parameter model still gives you a convenient tool to estimate the effect of convective boundary mixing in 1D-stellar evolution. As a time and depth dependent overshooting implementation it allows us for example to study convective boundary mixing for the formation of the ^{13}C-pocket on the hot dredge-up situation in massive, low-Z AGB stars. So, I think it is quite a useful concept. So why do I still use the simple f-parameter model for He-shell flash convection? The first step to either determine f or improve the mixing model was to start doing the He-shell flash convection simulations. I think we still have to do more work on those. But Bernd Freytag has done a preliminary determination of f-values at the top and bottom of the convective shell, and I showed the results. Interestingly, the f-value we found for the bottom of the He-shell flash convection zone does not obviously violate observables. But we have to check this more carefully.

(2) Yes, I should have explained the geometric setup of the box-in-a-star simulations. The information is fully given in Herwig et al. (2006).

Norber Langer (left) and Falk Herwig (the speaker) debating in the poster room.

Hydrodynamic simulations of the core helium flash

Miroslav Mocák, Ewald Müller
Achim Weiss and Konstantinos Kifonidis

Max-Planck-Institut für Astrophysik, Postfach 1312, 85741 Garching, Germany
email: mmocak@mpa-garching.mpg.de

Abstract. We desribe and discuss hydrodynamic simulations of the core helium flash using an initial model of a 1.25 $M_\odot$ star with a metallicity of 0.02 near at its peak. Past research concerned with the dynamics of the core helium flash is inconclusive. Its results range from a confirmation of the standard picture, where the star remains in hydrostatic equilibrium during the flash (Deupree 1996), to a disruption or a significant mass loss of the star (Edwards 1969; Cole & Deupree 1980). However, the most recent multidimensional hydrodynamic study (Dearborn et al. 2006) suggests a quiescent behavior of the core helium flash and seems to rule out an explosive scenario. Here we present partial results of a new comprehensive study of the core helium flash, which seem to confirm this qualitative behavior and give a better insight into operation of the convection zone powered by helium burning during the flash. The hydrodynamic evolution is followed on a computational grid in spherical coordinates using our new version of the multi-dimensional hydrodynamic code HERAKLES, which is based on a direct Eulerian implementation of the piecewise parabolic method.

Keywords. Stars: evolution – hydrodynamics – convection

1. Introduction

First results on the core helium flash were gained from one-dimensional hydrostatic numerical simulations of a 1.3 $M_\odot$ star (Z = 0.001) (e.g., Schwarzschild & Härm 1961). During the flash, the star underwent a thermal runaway due to the ignition of helium under degenerate conditions in its center. It reached a peak at maximum core temperature of $\sim 3.5\,10^8$ K and total energy generation rate of $\sim 10^{12} L_\odot$. The calculations were redone later with better numerical techniques and improved treatment of major physical processes (Sweigert & Gross 1978) and although the ignition of helium occured off-center due to neutrino processes, they did not change the general picture mentioned earlier. It turns out, that the typical e-folding times for the energy release from helium burning become as low as hours at the peak of the flash, and therefore are comparable to convective turnover times. Thus, the usual assumptions used in simple descriptions of convection in one-dimensional hydrostatic calculations (e.g. instantaneous mixing) do not have to be valid any longer. Previous attempts to relax these assumptions by allowing for hydrodynamic flow remained inconclusive (Edwards 1969; Deupree 1996; Dearborn et al. 2006). Using a modified version of the HERAKLES code (Kifonidis et al. 2003) which is capable of solving the hydrodynamic equations coupled to nuclear burning and thermal transport in up to three spatial dimensions, we want to deepen our understanding of the convection during the core helium flash at its peak investigating it by means of two-dimensional and three-dimensional hydrodynamic simulations.

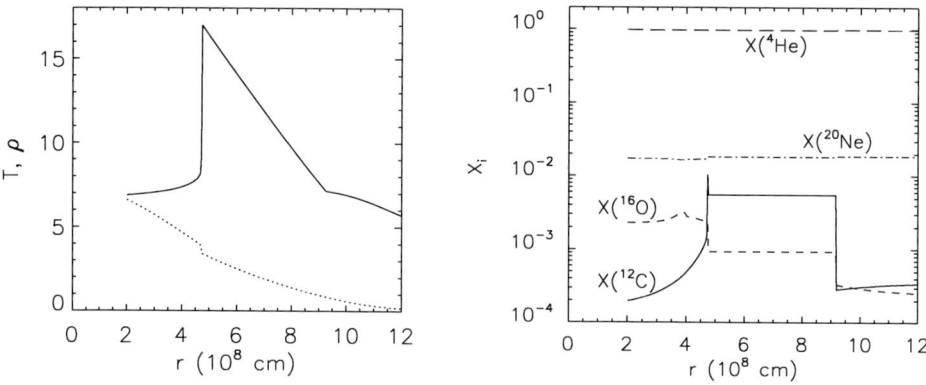

Figure 1. *Left panel:* Temperature (in 10^7 K, solid) and density (in 10^5 g cm^{-3}, dotted) distribution of the initial model M. *Right panel:* Chemical composition of the initial model M, showing dominant fraction of helium.

Table 1. Some properties of the initial model: total mass M, stellar population, metal content Z, mass M_{He} and radius R_{He} of the helium core ($X(^4He) > 0.98$), nuclear energy production in the helium core L_{He}, maximum temperature of the star T_{max}, and radius r_{max} and density ρ_{max} at the temperature maximum.

Model	M [$M_\odot$]	Pop.	Z	M_{He} [$M_\odot$]	R_{He} [10^9 cm]	L_{He} [$10^9 L_\odot$]	T_{max} [10^8 K]	r_{max} [10^8 cm]	ρ_{max} [10^5 g cm^{-3}]
M	1.25	I	0.02	0.38	1.91	1.03	1.70	4.71	3.44

2. Initial setup

The initial model was obtained with the stellar evolution code GARSTEC (Weiss & Schlattl 2007). Some of its properties are listed in Table 1. The temperature, density and composition distribution of the model is depicted in Figure 1. The model encompasses a white dwarf-like degenerate structure with an off-center temperature maximum resulting from plasma- and photo-neutrino cooling and a central density of about 7 10^5 g cm^{-3}. The isothermal region in the center of the helium core is followed by almost discontinuous jump in temperature up to $T_{max} \sim 1.7\ 10^8$ K and convection zone driven by the superadiabatic temperature gradient. The model is composed mostly of helium ^{4}He with an abundance X(^{4}He)> 0.98. The remaining composition of the stellar model is ^{1}H, ^{3}He, ^{12}C, ^{13}C, ^{14}N, ^{15}N and ^{16}O. For our hydrodynamic simulations we adopt the abundances of ^{4}He, ^{12}C and ^{16}O from the initial model, since the triple-α reaction dominates the nuclear energy production rate during the flash. The remaining composition is assumed to be adequately represented by a gas with a mean molecular weight equal to that of ^{20}Ne.

3. Hydrodynamic simulations

Table 2 summarizes some characteristic parameters of our two-dimensional (2D) and three-dimensional (3D) simulations that are based on model M. They were performed on an equidistant spherical grid encompassing 95% of the helium core's mass except for a central region with a radius of r = 2 10^8 cm, which was excised in order to allow for larger timesteps.

All our 2D and 3D models undergo initially (t < 1200 s) a common evolution where convection sets in after roughly 1000 s. During this phase, hot bubbles appear in the

Table 2. Some properties of the two and three-dimensional simulations: number of grid points in radial (N_r) and angular (N_θ, N_ϕ) dimension, radial (Δr in 10^8 cm) and angular ($\Delta\theta, \Delta\phi$) resolution, characteristic length scale l_c (in 10^8 cm) and velocity v_c (in 10^6 cm s^{-1}) of the flow, respectively, expansion velocity at the position of temperature maximum v_{exp} (in cm s^{-1}), entrainment rate v_{ent} of the outer convective boundary (in m s^{-1}), typical convective turnover time t_o and maximum evolution time t_{max} (in s), respectively.

run	$N_r \times N_\theta \times N_\phi$	Δr	$\Delta\theta$	$\Delta\phi$	l_c	v_c	v_{exp}	v_{ent}	t_o	t_{max}
DV2	180×90	5.55	2.°	-	4.7	1.03	−6.	7.	910	30000
DV4	360×240	2.77	0.75°	-	4.7	1.52	+92.	14.	620	60000
TR	$180 \times 60 \times 60$	5.55	1.5°	1.5°	4.7	0.7	+6.	7.	1340	5300

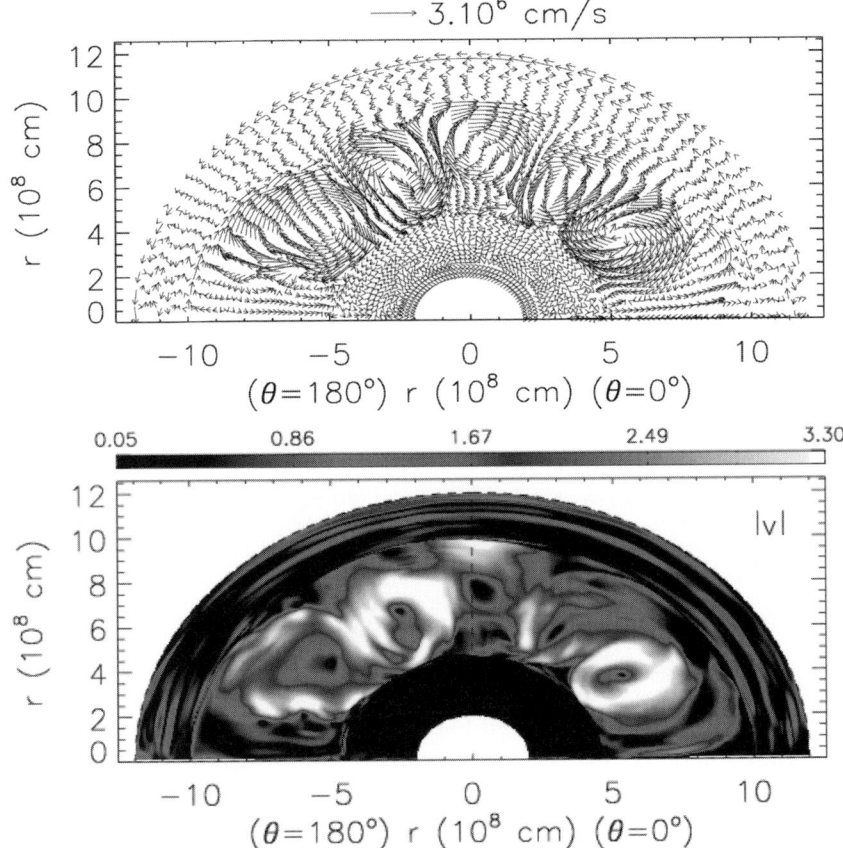

Figure 2. Snapshots of the evolved convection at 60000 s in model DV4, showing the velocity field (upper panel), and the velocity amplitude $|\mathbf{v}|$ in 10^6 cm s^{-1} (bottom panel), respectively.

region where helium burns in a thin shell (r ∼ 5 10^8 K). After ∼ 200 s, they cover complete height of the convective region and reach a steady state with several upstreams (or plumes) of hot gas carrying the released nuclear energy away from the burning region, thereby inhibiting a thermonuclear runaway.

Fully evolved convection (t > 1500 s) in 3D is significantly different than in 2D, since the shape of turbulent streams which transport energy is totally distinct. However, the amount of energy which needs to be transported by the convection in order to prevent

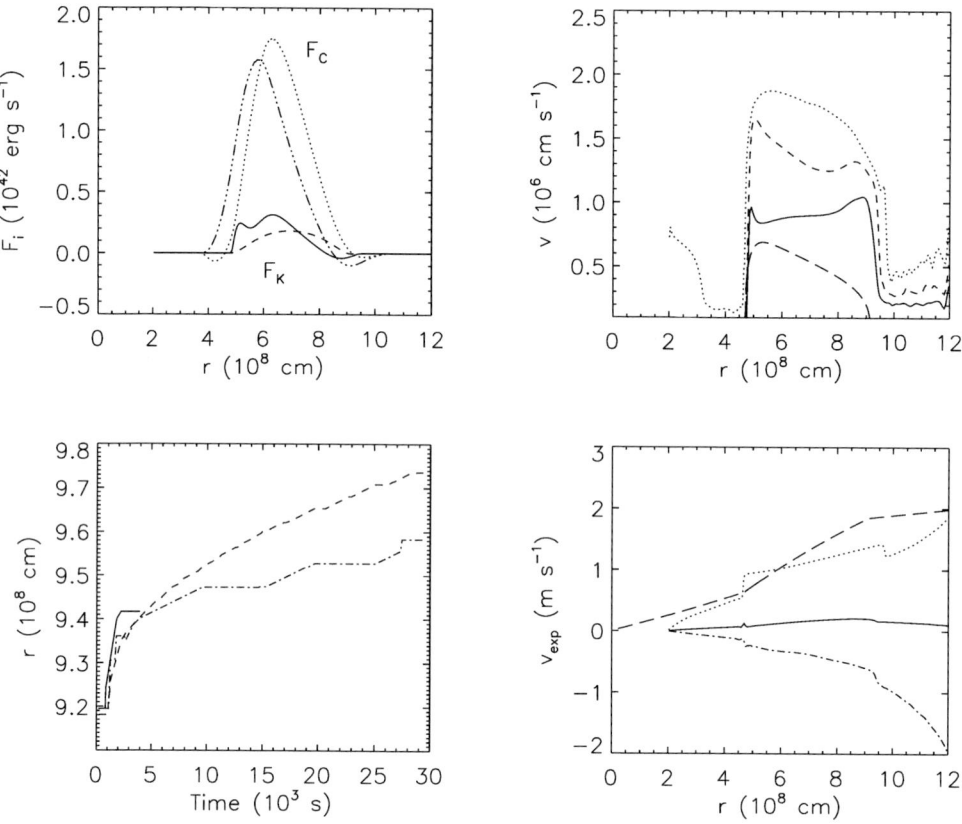

Figure 3. *Left upper panel:* Snapshot of convective F_C (dash-dotted) and kinetic F_K (solid) flux (Hurlburt *et al.* 1986) in model DV2 (black) and TR (red) averaged over a few convective turnover times. *Right upper panel:* The r.m.s convection velocity in model DV2 (dash-dotted), DV4 (dashed) and TR (solid) overplotted with the convective velocities predicted by mixing-length theory (long-dashed blue). *Bottom left panel:* Temporal evolution of the outer convective boundary in model DV2 (dash-dotted), DV4 (dashed) and TR (solid), respectively. *Bottom right panel:* Expansion velocity v_{exp} in model DV2 (dash-dotted), DV4 (dashed) and TR (solid) together with the expansion velocities of the initial stellar model (long-dashed blue).

a thermonuclear runaway during the flash is in both cases similar. The resulting typical convective velocities are therefore much higher in 2D than in 3D (Fig. 3).

The structural differences between 2D and 3D flows are clearly visible in the distribution of the kinetic flux across the convection zone (Fig. 3). The typical evolved 2D flows contain well defined vortices (Fig. 2) with their central regions never interacting with the region of the dominant nuclear burning above the T_{max}. This results in a reduced kinetic flux between $5\,10^8$ cm $<$ r $<$ $6\,10^8$ cm, since the gas in that region, on average, is located at bottom of convective vortices, and thus does not experience any strong radial flow. On the other hand, the distribution of the kinetic flux in 3D is rather smooth, and the flow structures tend to be also smaller than in 2D. This is apparent when comparing Figure 4 with Figure 2. The 2D structures (vortices) have an angular size of around $40°$. The structures in the 3D are column shaped, with a smaller angular size. The convective and kinetic flux is lower in 3D than in the 2D, but the total energy production is about 20 % higher in 3D (because no symmetry restrictions are imposed, and due to the strong dependence of the triple-α reaction rate on the temperature). The convective and kinetic flux carry together more than 90 % of the energy produced by the burning. The 3D

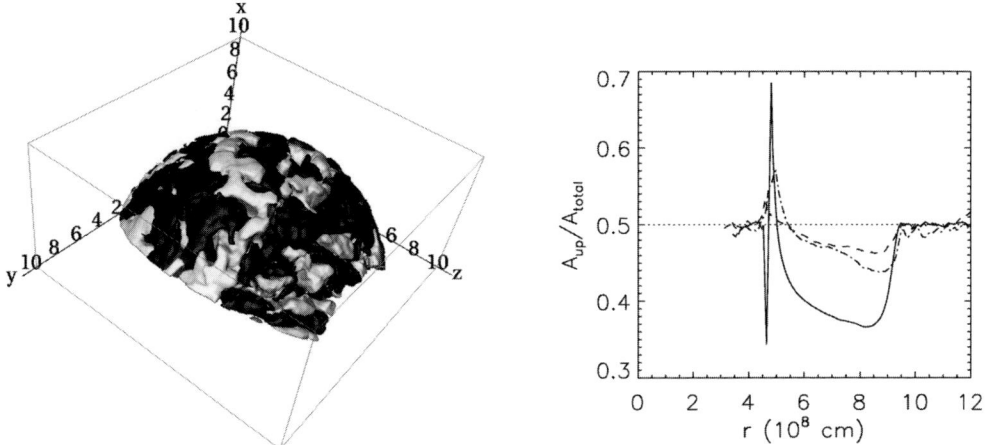

Figure 4. *Left panel:* Isosurface of a radial velocity field of model TR at t = 4150 s. The yellow color marks a positive velocity of +6 10^5 cm s^{-1} (the upflow streams) and the blue color a negative velocity of −6 10^5 cm s^{-1} (downflow streams). Axis tickmarks are in units 10^8 cm. *Right panel:* Fractional volume occupied by upflow and downflow streams in model DV2 (dashed-dotted), DV4 (dashed) and TR (solid), respectively.

velocities qualitatively match the velocities predicted by the mixing-length theory better than in 2D, where the velocities are clearly overestimated. Figure 3 shows that they also depend on resolution, being higher in the simulation with the highest resolution.

The extent of the convection zone increases with time. Due to turbulent entrainment (Meakin & Arnett 2007), convective boundaries defined by the Schwarzschild criterium are pushed towards the center of the star, and towards the stellar surface, respectively (Fig. 3). This is in contradiction with the predictions made by (1D) hydrostatic stellar modeling. For ilustration, the temporal evolution of the location of the outer convective boundary is depicted in Figure 3. It is defined as the radius, where the mean carbon abundance $X(^{12}C) \sim 0.002$. The rapid initial jump of the boundary position to $r \sim 9.4\ 10^8$ cm at about ~ 1200 s is due to the first touch of the convective flow on the boundary. Later entrainment is rather steady. The velocity of the outer boundary (entrainment rate) in our models are listed in Table 2. The entrainment involves a few radial zones only over the longest simulation we have performed. Although the ^{12}C abundance distribution stayed discontinuous at boundaries (no evident effect of numerical diffusion is detected), the entrainment rates presented here have to be considered as an order of magnitude estimate only. The entrainment at the inner convective boundary occurs with a rate much smaller than at the outer convective boundary. Therefore it is not discussed further here, since longer simulations are needed for definite statements about its evolution.

A similar feature which 2D and 3D seem to share is the upflow-downflow asymmetry. The downflows cover a much bigger volume in the convection zone than the upflows. The downflows dominate more in 3D. Interestingly, the kinetic flux is always positive in both cases, although the downflows fill almost the whole convection zone. This implies that the downflows are much slower then the upflows.

The expansion velocities $v_{exp} = \dot{M}_r/4\pi r^2 \rho$ are in good agreement with those of initial stellar model only in the 2D model with highest resolution DV4 (Fig. 3). The expansion in the low resolution models does not match at all. Due to the different dynamic properties of the flow in the less resolved models, the spherical mass flow is weaker.

Mixing at the convective boundaries and across the convection zone in 2D and 3D is quite different as well. In 2D, due to the symmetry restriction, every turbulent feature

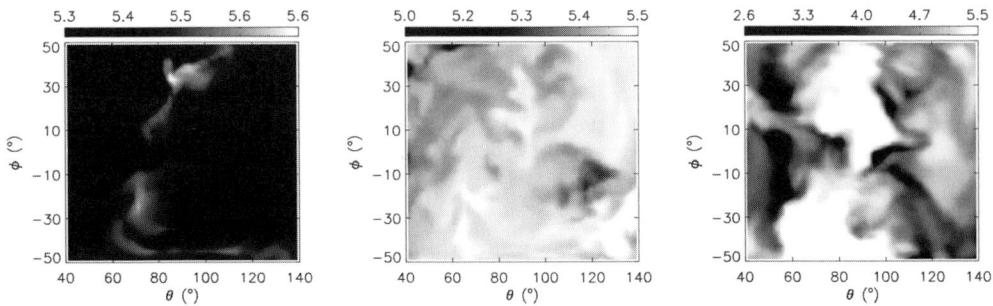

Figure 5. Maps of ^{12}C abundance (in units of 10^{-3}) in a horizontal plane in model TR at t = 5231 s, at different radii: $r_1 = 4.8\ 10^8$ cm (left), $r_2 = 6.5\ 10^8$ cm (middle), $r_3 = 9.3\ 10^8$ cm (right).

is in fact an annulus. Hence, turbulence and mixing can be properly modelled only by means of 3D simulations. The most apparent turbulent structures during the flash, in 3D, look at the bottom of the convection zone like thin hot fibers enriched by carbon and oxygen (ashes from the helium burning). The flow then gets more uniform inside the convective region, but looks more turbulent again at the outer convective boundary (Fig. 5).

4. Conclusions

We find that the core helium flash neither rips the star apart, nor significantly alters its structure. The evolved convection in 3D looks different from that in 2D. Typical convective velocities are higher in 2D than in 3D where they also tend to fit the predictions made by mixing length theory better. Hydrodynamic simulations show the presence of turbulent entrainment, which results in a growth of the convection zone on dynamic time scales.

Acknowledgements

The calculations were performed at the Leibniz-Rechenzentrum of the Bavarian Academy of Sciences and Humanities on the SGI Altix 4700 system. A considerable grant of computer time is thankfully acknowledge.

References

Cole, P. W. & Deupree, R. G., 1980, ApJ, 239, 284
Deupree, R. G., 1996, ApJ, 471, 377
Dearborn, D. S. P., Lattanzio, J. C., & Eggleton, P., 2006, ApJ, 639, 405
Edwards, A. C., 1969, MNRAS 146, 445
Hurlburt, N. E., Toomre, J., & Massaguer, J. M., 1986, ApJ, 311,563
Kifonidis, K., Plewa, T., Janka, H-Th., & Müller, E., 2003, A&A, 408, 621
Meakin, C. A. & Arnett, D., 2007, ApJ, 667, 448
Sweigart, A. V. & Gross, P. G., 1978, ApJS, 36, 405
Schwarzschild, M. & Härm, R.,1962, ApJ, 136, 158
Weiss, A. & Schlattl, H., 2007, Ap&SS, 341

Discussion

CHAN: What happens when $t < 0$? Your calculation shows that all these dynamical processes start and grow real quick after $t = 0$. How to ensure that things are so quiescent as assumed in your initial setup?

MOCAK: Although my simulations show differences in compare to the predictions made by the classical 1D stellar evolution calculations, the differences are not so big. However, what really looks to be treated in wrong way is that turbulent entrainment or "overshooting" at the boundaries of the convection zone. One possibility of how to solve it could be to treat the regions between convectively stable and unstable layers in the classical 1D simulations as a dynamic convective boundary is suggested by Meakin & Arnett, 2007, 667, 448.

The chairman M. Asplund (left) is reminding the speaker N. Langer about the time limit.

A snapshot of the meeting: from left, C. Charbonnel, L. Siess & M. Mocak.

Non-LTE Spectral Analysis of Extremely Hot Post-AGB Stars: Constraints for Evolutionary Theory

Thomas Rauch[1], Klaus Werner[1], Marc Ziegler[1], Lars Koesterke[2] and Jeffrey W. Kruk[3]

[1]Institute for Astronomy and Astrophysics, Kepler Center for Astro and Particle Physics, Eberhard Karls University, Sand 1, 72076 Tübingen, Germany
e-mail: rauch@astro.uni-tuebingen.de

[2]Texas Advanced Computer Center, University of Texas, Austin, TX 78712, USA

[3]Department of Physics and Astronomy, Johns Hopkins University, Baltimore, MD 21218, USA

Abstract. Spectral analysis by means of Non-LTE model-atmosphere techniques has arrived at a high level of sophistication: fully line-blanketed model atmospheres which consider opacities of all elements from H to Ni allow the reliable determination of photospheric parameters of hot, compact stars. Such models provide a crucial test of stellar evolutionary theory: recent abundance determinations of trace elements like, e.g., F, Ne, Mg, P, S, Ar, Fe, and Ni are suited to investigate on AGB nucleosynthesis. E.g., the strong Fe depletion found in hydrogen-deficient post-AGB stars is a clear indication of an efficient s-process on the AGB where Fe is transformed into Ni or even heavier trans iron-group elements. We present results of recent spectral analyses based on high-resolution UV observations of hot stars.

Keywords. astronomical data bases: miscellaneous, atomic data, line: identification, stars: abundances, stars: AGB and post-AGB, stars: atmospheres, stars: early-type, stars: evolution, (stars:) white dwarfs

---

1. Introduction

In the last decades, our picture of post-AGB stellar evolution has been greatly improved. The "standard" evolution, i.e. the hydrogen-rich sequence, has been understood in the early eighties of the last century by comparison of spectral analysis and evolutionary models. The spectral analysis of the hottest post-AGB stars with effective temperatures ($T_{\rm eff} > 100,000\,\rm K$) was hampered by the lack of appropriate model atmospheres which considered deviations from the local thermodynamic equilibrium (LTE) playing an important role in the photospheres of these stars. The development of such Non-LTE model atmospheres is briefly summarized in Sect. 2.

Spectral analyses of hot post-AGB stars have shown then that about a quarter of these are hydrogen-deficient (Werner & Herwig 2006). The "born-again post-AGB star" scenario by Iben *et al.* (1983) is – in general – able to explain the evolution of these stars. A final thermal pulse (TP, re-ignition of the helium shell) brings the star back to the AGB and it has a second, helium-burning post-AGB phase. However, the mechanism to dispose of the entire hydrogen-rich envelope was unclear. Present evolutionary calculations which consider the mixing and burning processes during the helium-shell flash in detail, are able to explain the hydrogen deficiency.

The amount of remaining hydrogen is depending on the particular time when the TP occurs. Still on the AGB (AGB Final Thermal Pulse, AFTP), the masses of the hydrogen-rich envelope and the helium-rich intershell layer (Fig. 1) are about equal, being roughly

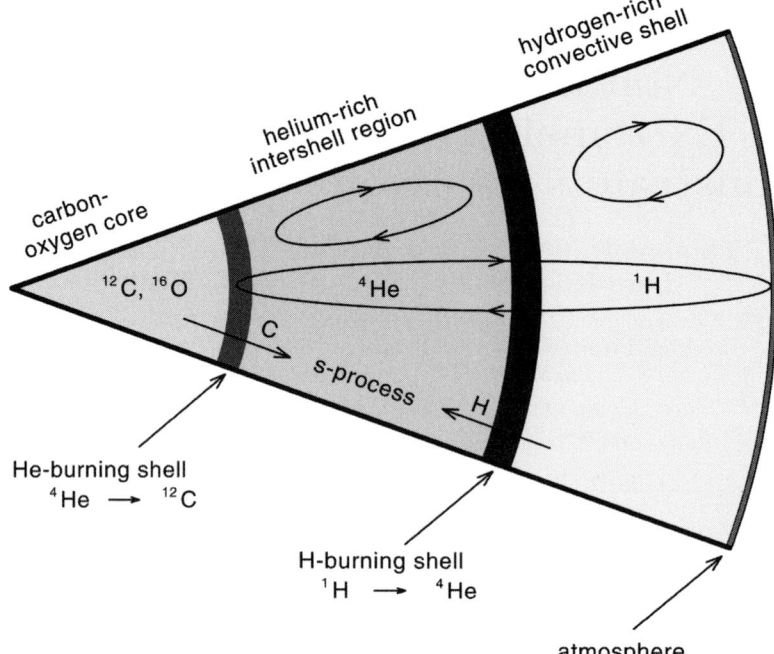

Figure 1. Inner structure of an AGB star. Note that the convection zone from the surface to the bottom of the helium-burning shell is established during a TP only.

$10^{-2}\,M_\odot$. The TP mixing results in $\approx 40\,\%$ hydrogen content (by mass) at the stellar surface. This is detectable in the observations. After the departure from the AGB, the mass of the hydrogen-rich envelope is much lower ($M_H \approx 10^{-4}\,M_\odot$). If the TP occurs at still constant (high) luminosity (Late Thermal Pulse, LTP), i.e. nuclear burning is still "on", the mixing during the flash reduces the surface hydrogen to $\approx 1\,\%$ which is not detectable at the high surface gravity ($\log g \approx 5 - 8$ in cm/sec^2). A TP at already declining luminosity, i.e. there is no entropy barrier due to the hydrogen-burning shell, will mix down the hydrogen-rich envelope to the bottom of the now re-ignited helium shell where hydrogen is burned and the star becomes hydrogen free.

The mixing process during the TP brings intershell matter up to the stellar surface and, thus, allows a direct view on it. Spectral analysis allows then to conclude on details of both, mixing as well as nuclear processes like, e.g. the s-process, in AGB stars. This is a crucial test for stellar evolutionary models.

2. Model atmospheres and atomic data

The spectral analysis of hot stars requires suitable Non-LTE model atmospheres†. Such models are available since Werner (1986, 1989) presented the first calculation based on the accelerated lambda iteration (ALI) techniques. For more details see, e.g., Hubeny (2003); Lanz & Hubeny (2003); Rauch & Deetjen (2003). Presently, our Tübingen NLTE Model Atmosphere Package‡ (*TMAP*, Werner *et al.* 2003; Rauch & Deetjen 2003) is

† For cool stars with spectral type B ar later LTE modeling may be adequate but there are always Non-LTE effects in any star, which are important at least if high-resolution and/or high/energy observations are analyzed.
‡ http://astro.uni-tuebingen.de/~rauch/TMAP/TMAP.html

capable to calculate plane-parallel and spherical, chemically homogeneous, Non-LTE model atmospheres in radiative and hydrostatic equilibrium and considers opacities of all species from hydrogen to nickel (Rauch 1997, 2003).

In the last two decades, TMAP has been successfully employed for the analysis of hot post-AGB stars (e.g. Jahn et al. 2007; Rauch et al. 2007). Space-based observatories like, e.g., FUSE† and HST‡, have provided high-resolution, high-S/N UV spectra which have shown that we arrived at a severe limitation due to the lack of reliable atomic and line-broadening data for highly ionized species. This is a challenge for atomic physics.

3. Selected results of spectral analyses

In this section, we will highlight some recent important results of our analyses. For a more detailed review on the spectroscopy of hydrogen-deficient post-AGB stars see Werner & Herwig (2006).

Fluorine (F VI λ 1139.50 Å) has been identified for the first time in FUSE observations of post-AGB stars (Werner et al. 2005). F is produced during helium burning on the AGB via $^{14}N(\alpha,\gamma)^{18}F(\beta^+)^{18}O(p,\alpha)^{15}N(\alpha,\gamma)^{19}F$. Our spectral analysis has shown that the F abundance in hydrogen-deficient PG 1159-type stars is about 200× solar (cf. Asplund et al. 2005). This result confirms the F intershell abundances predicted by evolutionary models (Lugaro et al. 2004).

Neon (Ne VII λ 3644.6 Å) was first identified by Werner & Rauch (1994) in optical observations of PG 1159 stars. It is worthwhile to note that this line was detected in a high-temperature discharge plasma as a weak blend on a strong Ne II line already many years ago (?) und is frequently used as a calibration line in the laboratory (König et al. 1993) – unfortunately this was not known in astronomy before and demonstrates the need for improvement and extension of atomic-data databases. Ne is synthesized in the helium-burning shell by the $^{14}N(\alpha,\gamma)^{18}F(\beta^+)^{18}O(\alpha,\gamma)^{22}Ne$ chain. The determined photospheric abundance is 2% by mass (11× solar, Werner & Rauch 1994), well in agreement with early evolutionary models of Iben & Tutukov (1985).

In FUSE observations of PG 1159 stars, we could firstly identify Ne VII λ 973.3 Å (Werner et al. 2004) – a stellar wind can form a strong Ne VII P Cygni profile (Herald et al. 2005) – the closely located C III λ 977.3 Å line is much too weak at the relevant temperature regime (Fig. 2).

Recently, Kramida & Buchet-Poulizac (2006) presented reliable atomic data of Ne VIII. This enables us to calculate the Ne VIII spectrum of hot post-AGB stars and we could identify Ne VIII absorption lines in FUSE observations (Werner et al. 2007b).

The identification of Ne VII and Ne VIII lines provides a new sensitive tool for very hot stars to determine T_{eff} by an evaluation of the Ne VII/VIII ionization equilibrium. It is worthwhile to note, that T_{eff} of the hottest known helium-rich DO white dwarf, KPD 0005+5106, had to be revised to $T_{\text{eff}} \approx 200\,000$ K (previously 120 000 K) by the identification and modelling of Ne VIII emission lines in its spectrum. These lines were previously thought to be O VIII lines which however could not be of photospheric origin (Werner et al. 2007b).

Argon also provides an ionization equilibrium (Ar VI/VII) for the determination of T_{eff} in very hot stars. Werner et al. (2007a) identified Ar VII λ 1063.55 Å in FUSE observations

† Far Ultraviolet Spectroscopic Explorer
‡ Hubble Space Telescope

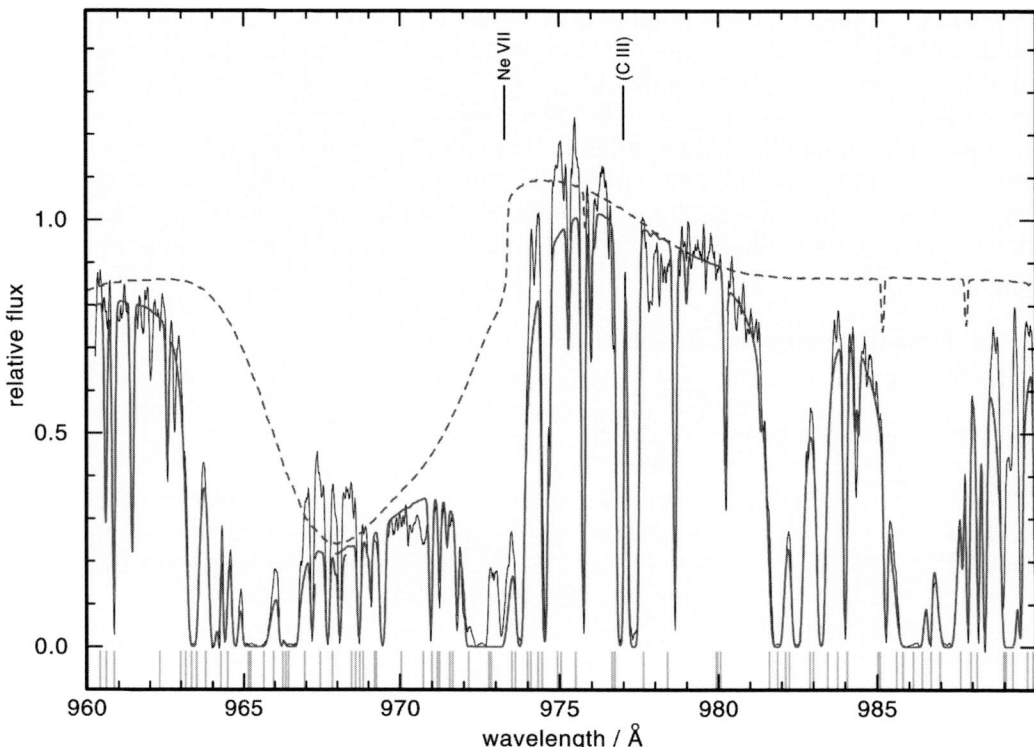

Figure 2. Comparison of the synthetic stellar spectrum (dashed, calculated with the *HotBlast* wind code) and the combined synthetic stellar and interstellar line spectrum around Ne VII $\lambda\,973.3$ Å with the FUSE observation of the CSPN NGC 7094. Note that C III $\lambda\,977.3$ Å is much too weak at $T_{\text{eff}} = 110\,000$ K to explain the observed (Ne VII) profile. The vertical bars at the bottom mark the positions of H$_2$ lines in the interstellar spectrum.

of hot central stars of planetary nebulae (CSPNe) and in white dwarfs. Rauch *et al.* (2007) identified Ar VI lines in a HST/STIS¶ observations of the CSPN LSV 46°21.

Iron is not affected by nuclear burning but its abundance may be reduced due to n-captures in the s-process. Evolutionary calculations predict only a very weak extent of its depletion. In contrast, spectral analyses of FUSE observations with sufficiently high S/N of four hydrogen-deficient post-AGB stars have shown that no iron lines are detectable. Thus, a very strong Fe depletion (at least $1-2$ dex) takes place in the intershell (Miksa *et al.* 2002).

4. One example of spectral analysis: The exciting star of the planetary nebula NGC 7094

The CSPN NGC 7094 is a so-called hybrid PG 1159 star, i.e. it exhibits hydrogen lines in its spectrum. In a Non-LTE spectral analysis Dreizler *et al.* (1995) determined $T_{\text{eff}} = 110\,000$ K, $\log g = 5.7$, and H:He:C:N:O = 0.42:0.51:0.05:(<0.01):(<0.01) in mass fractions. The CSPN NGC 7094 may thus be an AFTP star. Miksa *et al.* (2002) discovered a strong iron deficiency of about two dex. In an on-going analysis of FUSE and HST/STIS observations (Ziegler *et al.* in prep.), we aim to identify nickel lines in order to determine

¶ Space Telescope Imaging Spectrograph

the Fe/Ni abundance ratio which is a indicator for the efficiency of the s-process. We did not succeed in this attempt. It is possible that the s-process has transferred iron into nickel and then into more heavy species. Unfortunately, we cannot search for lines of trans iron-group elements because no atomic data are available.

We used the newly developed *HotBlast* Non-LTE code for spherically expanding atmospheres in order to calculate the P Cygni profile of Ne VII $\lambda\,973.3\,\text{Å}$. *HotBlast* uses as an input the atmospheric structure of our static *TMAP* model to simulate the atmosphere below the "wind region". In the course of our analysis, it turned out that the FUSE observation is strongly contaminated by interstellar line absorption (Fig. 2). Therefore, we employed the programme *OWENS* (cf. Lemoine *et al.* 2002; Hébrard *et al.* 2002) which can simulate interstellar clouds with individual parameters like, e.g., radial velocity, column density in the line of sight, temperature of the gas, and microturbulence velocity. Although we show in Fig. 2 only a qualitative fit of the ISM line absorption, it is obvious that simultaneous modeling of both, the stellar and interstellar line spectrum, is necessary in order to search and identify weak metal lines.

5. Spectral analysis in the 21st century

Spectral analysis by means of Non-LTE model-atmosphere techniques has for a long time been regarded as a domain of specialists. Within the *German Astrophysical Virtual Observatory* (*GAVO*†) project, we have created the VO service *TheoSSA*‡. A VO user may use pre-calculated grids of spectral energy distributions (SEDs, in a pilot phase calculated by *TMAP* for hot, compact stars only) which are ready to use and it may be interpolated between them to match the user-required parameters. This is the easiest way to use synthetic SEDs calculated from Non-LTE model atmospheres. They represent stars much better than still oftenly used blackbody flux distributions for PNe analyses.

If individual parameters are requested which do not fit to an already existing SED in the database, the VO user is guided to *TMAW*¶. With this WWW interface, the VO user may calculate an individual model atmosphere, requesting T_eff, $\log g$, and mass fractions $\{X_i\}$, $i \in$ [H, He, C, N, O] (more species will be included in the future). For this calculation, standard model atoms are used which are provided within the Tübingen Model-Atom Database (*TMAD*‖). Since the VO user can do this without detailed knowledge of the programme code working in the background, the access to individually calculated SEDs is as simple as the use of pre-calculated SEDs – however, the calculation needs some time (depending on the number of species considered, the wall-clock time is ranging from hours to a few days). Standard SEDs of all calculated model atmospheres are automatically ingested into the *GAVO* data base and, thus, it is growing in time.

In case that a detailed spectral analysis is performed, an experienced VO user may create an own atomic data file tailored for a specific purpose considering all necessary species ($i \in$ [H – Ni]) and calculate own model atmospheres and SEDs.

In close collaboration of *GAVO* with the *German Astronomy Community Grid* (*AstroGrid-D*††), the calculations of *TMAW* will be performed on GRID computers in the future. This will allow to calculate small grids of model atmospheres and SEDs on a reasonable timescale.

† http://www.g-vo.org
‡ Theoretical Simple Spectra Access, http://vo.ari.uni-heidelberg.de/ssatr-0.01/TrSpectra.jsp
¶ http://astro.uni-tuebingen.de/~rauch/TMAW/TMAW.shtml
‖ http://astro.uni-tuebingen.de/~rauch/TMAD/TMAD.html
†† http://www.gac-grid.de

Acknowledgements

T. R. is supported by the German Astrophysical Virtual Observatory (GAVO) project of the German Federal Ministry of Education and Research (BMBF) under the grant 05 AC6VTB. This work has been done using the profile fitting procedure *OWENS* developed by M. Lemoine and the FUSE French Team.

References

Asplund, M., Grevesse, N., & Sauval, A. J. 2005, in: *Cosmic Abundances Records of Stellar Evolution and Nucleosynthesis*, eds. T. G. III. Barnes, & F. N. Bash, The ASP Conference Series Vol. 336 (San Francisco ASP), p. 25
Dreizler, S., Werner, K., & Heber, U. 1995, in: *White Dwarfs*, eds. D. Koester, & K. Werner, LNP, 443, 160
Hébrard, G., et al. 2002, ApJS, 140, 103
Herald, J. E., Bianchi, L., & Hillier, D. J. 2005, ApJ, 627, 424
Hubeny, I. 2003, in: *Stellar Atmosphere Modeling*, eds. I. Hubeny, D. Mihalas, & K. Werner, The ASP Conference Series Vol. 288 (San Francisco ASP), p. 51
Iben, I., Jr. & Tutukov, A. V. 1985, ApJS, 58, 661
Iben, I., Jr., Kaler, J. B., Truran, J. W., & Renzini, A. 1983, ApJ, 264, 605
Jahn, D., Rauch, T., Reiff, E., Werner, K., Kruk, J. W., Dreizler, S., & Herwig, F. 2007, A&A, 462, 281
Johnston, W. D. & Kunze, H.-J. 1969, ApJ, 157, 1469
König, R., Kolk, K.-H., & Kunze, H.-J. 1993, Physica Scripta 48, 9
Kramida, A. E. & Buchet-Poulizac, M.-C. 2006, Eur. Phys. J. D, 39, 173
Lanz, T. & Hubeny, I. 2003, in: *Stellar Atmosphere Modeling*, eds. I. Hubeny, D. Mihalas, & K. Werner, The ASP Conference Series Vol. 288 (San Francisco ASP), p. 117
Lemoine, M., et al. 2002, ApJS, 140, 67
Lugaro, M., Ugalde, C., & Karakas, A. I., et al. 2004, ApJ, 615, L934
Miksa, S., Deetjen, J. L., Dreizler, S., Kruk, J .W., Rauch, T., & Werner, K. 2002, A&A 389, 953
Rauch, T. 1997, A&A, 320, 237
Rauch, T. 2003, A&A, 403, 709
Rauch, T., & Deetjen, J. L. 2003, in: *Stellar Atmosphere Modeling*, eds. I. Hubeny, D. Mihalas, & K. Werner, The ASP Conference Series Vol. 288 (San Francisco ASP), p. 103
Rauch, T., Ziegler, M., Werner, K., et al. 2007, A&A, 470, 317
Werner, K. 1986, A&A, 161, 177
Werner, K. 1989, A&A, 226, 265
Werner, K. & Herwig, F. 2006, PASP, 118, 183
Werner, K., Deetjen, J. L., Dreizler, et al. 2003, in: *Stellar Atmosphere Modeling*, eds. I. Hubeny, D. Mihalas, & K. Werner, The ASP Conference Series Vol. 288 (San Francisco ASP), p. 31
Werner, K. & Rauch, T. 1994, A&A, 284, L5
Werner, K., Rauch, T., Reiff, E., Kruk, J. W., & Napiwotzki, R. 2004, A&A, 427, 685
Werner, K., Rauch, T., & Kruk, J. W. 2005, A&A, 433, 641
Werner, K., Rauch, T., & Kruk, J. W. 2007a, A&A, 466, 317
Werner, K., Rauch, T., & Kruk, J. W. 2007b, A&A, 481, 807

Dust-driven Winds Beyond Spherical Symmetry

Peter Woitke[1,2]

[1]UK Astronomy Technology Centre, Blackford Hill, EH9 3HJ Edinburgh, Scotland, UK

[2]School of Physics & Astronomy, University of St Andrews, North Haugh, KY16 9SS
St. Andrews, Scotland, UK, email: ptw@roe.ac.uk

Abstract. New 2D dynamical models for the winds of AGB stars are presented which include hydrodynamics with radiation pressure on dust, equilibrium chemistry, time-dependent dust formation theory, and coupled frequency-dependent Monte Carlo radiative transfer. The simulations reveal a much more complicated picture of the dust formation and wind acceleration as compared to 1D spherical wind models. Triggered by non-spherical pulsations or large-scale convective motions, dust forms event-like in the cooler regions above the stellar surface which are temporarily less illuminated, followed by the radial ejection of dust arcs and clumps. These simulations can possibly explain recent high angular resolution interferometric IR observations of red giants, which show an often non-symmetric and highly time-variable innermost dust formation and wind acceleration zone. The dependence of the mass-loss rates on stellar parameters is less threshold-like as used from 1D models, and therefore, it seems quite possible that the phenomenon of dust-driven winds may occur also in less evolved red giants.

Keywords. hydrodynamics, radiative transfer, stars: late-type, stars: winds, stars: mass loss

---

1. State of Art: 1D models

The massive winds of asymptotic giant branch (AGB) stars and red supergiants are of paramount importance for late stages of stellar evolution (see e.g. Siess and Willson, this volume) and undoubtably play an important role in the cycle of matter in galaxies.

But despite 40 years of active research, our theoretical understanding of these winds is still comparably poor. Shock waves created by the stellar pulsation have been identified to trigger the process of dust formation close to the star, and radiation pressure on newly formed dust grains can overcome local gravity. However, the details of the wind driving mechanism are still puzzling. The most advanced computational wind models nowadays include hydrodynamics, chemistry, time-dependent dust formation and radiative transfer with varying degree of sophistication (Winters *et al.* 2000, Höfner *et al.* 2003, Sandin & Höfner 2003, Schirrmacher *et al.* 2003, Jeong *et al.* 2003). Other 1D models rely on the assumption of stationarity (e. g. Ferrarotti & Gail 2006). These simulations can explain the general features of dust-driven winds for the coolest and most luminous AGB stars, in particular the carbon stars, but for slightly warmer or less luminous stars, the 1D models fail to predict the observed mass-loss rates. For example, Mattsson *et al.* (2007) concluded that the Höfner *et al.* models can only be applied to carbon stars with $T_{\rm eff} > 3200\,{\rm K}$ and $C/O > 1.2$. Otherwise, the dust forms too distant from the star or the resulting dust opacity is simply insufficient to drive the wind, respectively.

In contrast, observations tell us that massive dusty winds also exist for less extreme carbon stars (e.g. Hony & Bouwman 2004) as well as for red supergiants (Verhoelst *et al.* 2006), M-type giants and even S-stars (Ramstedt *et al.* 2007). The mass-loss rate rather seems to vanish smoothly as function of stellar parameter toward RGB stars and K giants without a clear cut-off, which would be expected if all these winds were actually dust-driven. The common winds of oxygen-rich AGB stars are particularly hard to understand, if frequency-dependent radiative transfer effects are taken into account, because the peak of the dust opacity around $10\,\mu$m differs strongly from the peak of the stellar radiation around $1\,\mu$m (Woitke 2006b, Höfner & Andersen 2007), in particular with regard to small grains which are supposed to drive the wind.

Regarding the current and future possibilities to monitor directly the dust formation and wind acceleration zones of red giants (AMBER on the VLTI, CHARA, ALMA, or MIRI on the JWST), spherical models actually appear as a non-starter. I have therefore developed the first 2D-models for dust driven-winds, both carbon-rich (Woitke 2006a) and oxygen-rich (Woitke 2006b) with frequency-dependent radiative transfer, which allow for a profound investigation of stability, symmetry breaking and pattern formation. The models can predict what types of spatial dust distributions and wind asymmetries form in red giant winds (shells, arcs, clumps, etc.), how they evolve in time, and how these stars would look alike (lightcurves, SEDs, visibilities, images).

Recently, Freytag & Höfner (2008) have published a combination of two radiation hydrodynamics models: an inner 3D model for the convection and pulsation of the AGB star, and an outer 1D model for the dust formation and wind acceleration. The inner grey 3D model shows large convection cells up to the scale of the star itself along with a strongly non-isotropic irradiation of the circumstellar environment where the dust forms. These simulations provide further motivations to drop the assumption of spherical symmetry and to study the complicated interplay between dynamics, dust formation and radiative transfer in more than one spatial dimension in the 21st century.

2. New models

The 2D models are developed in the frame of the FLASH 2.4 hydrocode (Fryxell et al. 2000) which is an explicit, finite volume, high-order Godunov-type hydro-solver which uses Adaptive Mesh Refinement (AMR) and is fully parallel (MPI). The implementation of chemistry and dust formation into this hydro-solver is quite straightforward (operator splitting, see Woitke 2006a for details). In the C-rich case, the models use the standard theory for the formation of amorphous carbon grains (Gail & Sedlmayr 1988). Concerning the O-rich models, this kinetic description has been extended to model the nucleation, growth and evaporation of dirty dust particles, which consist of numerous small islands of different solid materials like Mg_2SiO_4, SiO_2, Al_2O_3, Fe and TiO_2, see Helling & Woitke (2006) and Helling, Woitke & Thi (2008) for details. A fast equation of state which includes ionization potential of H^+, dissociation potential of H_2, and vibrational and rotational excitation energies of H_2 is implemented (see Woitke 2008).

The coupling to an efficient radiative transfer tool has turned out to be rather difficult. The optical depths range from $\sim 10^4$ at the bottom of the atmosphere to truly zero at large distances from the star. The opaque, dusty regions can be confined into geometrically thin shells or clouds which move in an otherwise quite transparent medium, making long ray tracing unavoidable. The dusty regions scatter, re-emit and cast shadows, thereby illuminating the optically thin regions in between in complicated ways.

Moreover, the efficient absorption and thermal re-emission of radiation by the dust grains leads to a very fast relaxation of the internal dust temperature to the ambient radiation field with cooling timescales of the order of milliseconds to 10^{-1} seconds for small grains (Woitke et al. 1999). This fast relaxation introduces a stiff coupling between hydrodynamics and radiative transfer. All explicit schemes that are based on formal solutions of the radiative transfer problem (e.g. short and long characteristics methods) have the disadvantage to slow down the computational timestep (typically a few 10^4 sec) to this radiative cooling timescale, which is not acceptable in this application.

Therefore, I have used a Monte Carlo code (Niccolini, Woitke & Lopez 2003) to calculate the radiation field *under the auxiliary condition that the dust is in radiative equilibrium*, which de-stiffens the physical problem at hand by the elimination of the shortest characteristic timescale in the system. In contrast to the dust temperature which is hence a direct result of the radiative transfer, the gas temperature is calculated time-dependently via the energy equation with radiative heating and cooling. To my knowledge, this coupling to Monte Carlo radiative transfer is an innovative approach in computational hydrodynamics.

The basis for the radiative transfer treatment are monochromatic molecular gas opacities from the MARCS stellar atmosphere code (Jørgensen *et al.* 1992), extracted by Helling *et al.* (2000), and dust opacities calculated in the Rayleigh limit of Mie theory according to the Jena Optical Data Base (for amorphous carbon, I use the 1000 K sample from pyrolysis experiments (Jäger *et al.* 1998). The frequency space is subdivided into five spectral bands with two opacity distribution points in each band, resulting in altogether 5×2 effective wavelengths sampling points. High and low opacity values are pre-tabulated for each spectral band in such a way that the band-mean Planck and Rosseland opacities are properly represented. The details will be explained in Woitke (2008).

3. Results

As an illustrative example, a carbon star with stellar mass $M_\star = 1 M_\odot$, stellar luminosity $L_\star = 5000 L_\odot$, effective temperature $T_\star = 2500$K, solar metallicities ($\epsilon_{He} = 10.99$, $\epsilon_O = 8.87$) except carbon (C/O=1.4), pulsational period $P = 1$yr and piston amplitude $\Delta u = 3$km/s is simulated. In order to focus on the effects of multi-dimensionality, a sub-sonic *p-mode-like* non-radial stellar pulsation is considered as inner boundary condition by means of the following modified piston approach

$$u(r_{in}, \theta, t) = \Delta u \sin\left(\frac{2\pi t}{P}\right) \cos(2\theta), \qquad (3.1)$$

where u is the velocity at the inner boundary r_{in} and θ is the latitude angle. Although the calculated mean wind properties are quite similar to a 1D spherical model with equal parameters (mass loss rate $\langle \dot{M} \rangle = 7 \cdot 10^{-7} M_\odot$/yr, final outflow velocity $\langle v_\infty \rangle = 20$ km/s, dust-to-gas mass ratio $\langle \rho_d / \rho_g \rangle = 3 \cdot 10^{-4}$, the model yields quite a different picture of the dust and wind formation close to the star (see Fig. 2).

A non-spherical stellar atmosphere (let it be caused by non-radial pulsations or convective motions) illuminates its environment in a likewise non-uniform way. For example, an oblate-shaped star releases its luminosity preferentially along the poles, because the additional gas above the stellar equator blocks the straight escape path of the photons (see Fig. 1, lower plot). Therefore, the gas just above the stellar atmosphere changes its temperature as the star changes its shape from prolate to spherical to oblate to spherical, and so on, in this model. Consequently, there are always thermodynamical conditions present in spatially limited regions above the stellar surface that are just favorable for new dust formation (high densities & low temperatures). This leads to a non-uniform dust production, sometimes astonishingly close to the star, followed by radial ejections of dust arcs or smaller cap-like clumps due to radiation pressure which expand tangentially like mushroom clouds (see Fig. 1, upper plot).

Figure 2 shows a comparison between a high-resolution (20 milli-arc-sec) speckle interferometry image of IRC+10216 in the K-band from the KECK telescope (Monnier 2002) and a Monte-Carlo simulated monochromatic image for $\lambda = 2.2 \mu$m based on a calculated

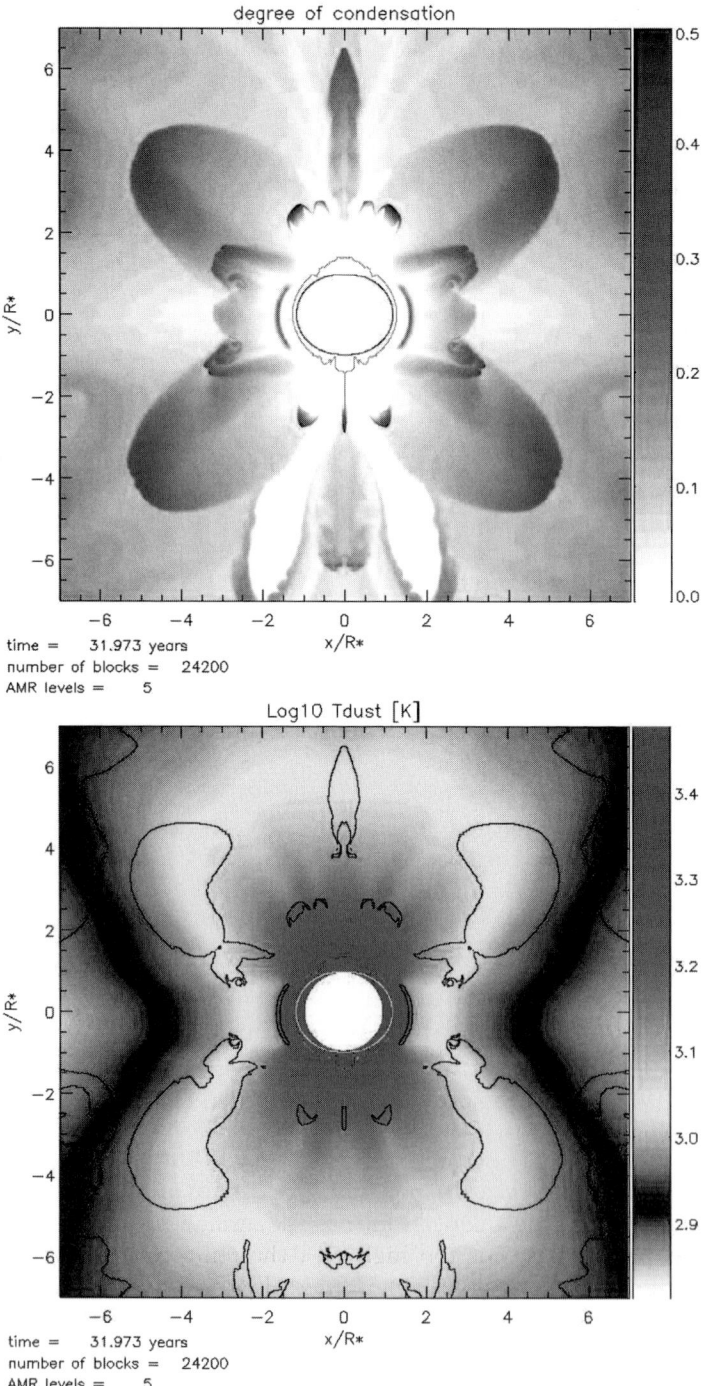

Figure 1. Snapshot of the innermost 13×13 stellar radii after 32 years of simulation. The upper plot shows the degree of condensation $f_{\rm cond}$ (0 = dust-free, 1 = complete condensation) and the lower plot the calculated dust temperature $T_{\rm d}$. The blank circle in the center is not modelled. The inner red/white contour line shows a density of $\rho = 10^{-10}\,{\rm g/cm^3}$ inside the star, the outer blue contour line marks $\tau_{\rm std} = 1$. The black contour lines in the lower plot encircle the dusty regions with $f_{\rm cond} = 0.15$. Note the asymmetric illumination by the deformed star, the shadows of the dust clouds and the new dust formation event just above the equator.

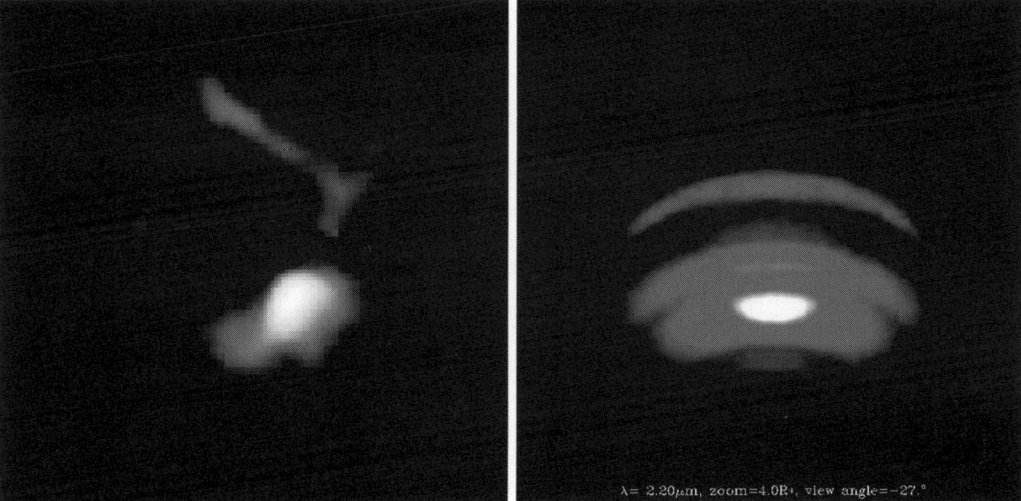

Figure 2. Comparison of observation and theory. IRC+10216 in the K-band from the KECK telescope with speckle interferometry (angular resolution 20 milli-arc-sec, J. Monnier, 2002, priv. comm.) and a simulated image from a 2D wind model at $\lambda = 2.2\,\mu$m. The bright spot in the simulated image originates from a hole in the dust shell. Although the star is situated in the center of this image, it is mostly obscured by dust at $2.2\,\mu$m. The stellar disk would be clearly visible at $10\,\mu$m in the model.

model structure. The images of the 2D model show a strong dependency on time and inclination angle. Sometimes, the calculated dust distribution around the star possesses a hole which, if viewed from a particular direction, produces a bright spot in the image, because the hot material close to the star becomes visible.

4. Summary and Conclusions

The wind driving mechanism of red giants and supergiants is still a matter of debate. For radiation pressure on dust, two conditions must be fulfilled:

(1) The gas close to the star must provide, at least temporarily, the necessary thermodynamical conditions for the gas-to-dust phase transition to occur: sufficiently low dust *and* gas temperatures (about 1000 K to 1600 K, depending on the dust grain material) and sufficiently high densities ($\gtrsim 10^9\,\mathrm{cm}^{-3}$).

(2) The opacity of the formed dust integrated over the incident radiation flux must be sufficiently high to cause a radiative acceleration that locally overcomes gravity.

In 1D spherically symmetric models, condition 1 translates into rather robust threshold values for stellar parameters. For too high T_eff, the temperature window shifts to too large distances from the star where the density is already too low (in absence of another driving mechanism). Stellar pulsations help to fulfil condition 1. Condition 2 puts similar constraints on the $L_\star/M_\star$ ratio, and on $\epsilon_\mathrm{C} - \epsilon_\mathrm{O}$ in case of C-stars or on the metallicity in case of O-rich AGB stars. Condition 2 makes it hard to understand why M-type, S-type and C-type AGB stars actually appear so similar in observed mass-loss rates and outflow velocities, although the amount and type of dust should be quite different.

Summarizing these constraints from the 1D models, the dust-driving hypothesis seems to work only for rather low-mass, cool and luminous AGB stars, preferentially carbon stars which can produce the necessary dust opacity in the near IR to efficiently absorb the stellar radiation. Taking into account frequency-dependent radiative transfer effects, the hypothesis already seems to fail for O-rich AGB stars which simply cannot provide the necessary dust opacity in the near IR (Woitke 2006b).

However, deviations from spherical symmetry may help! Stars which undergo violent convection or non-radial pulsation motions on their surfaces provide a rich variety of thermodynamical conditions in the gas just above the photosphere. The radiation leaves the star non-isotropically (see Fig. 1 in Freytag & Höfner (2008), right column) preferentially via the hot surface areas, leaving "cool spots" in between that provide the perfect conditions for dust formation. Once formed, the dust just has to wait until the region of interest is again strongly illuminated, which then produces the necessary combination of high opacity and high luminosity.

In this way, it seems quite possible that even warmer AGB stars may possess dust-driven winds, because extended parts of their surfaces are actually cooler. Thus, the dependency of the mass-loss rate on stellar parameter should be less threshold-like and the range of applicability of dust-driven winds more extended toward less evolved stars as compared to the results of 1D models. In that respect, I disagree with the conclusions of Freytag & Höfner (2008) that the overall spherical expansion of shock waves justifies the application of 1D models, since the inclusion of multi-D radiative transfer effects is crucial for the above line of argumentation.

Acknowledgments: The author acknowledges the support by the UK STFC, rolling grant PP/E001181/1. This work has furthermore been supported by the Dutch NWO COMPUTATIONAL PHYSICS PROGRAMME, grant 614.031.017. The computations have been done on the SARA massive parallel computers in Amsterdam, grant SG-184.

References

Ferrarotti, A. S. & Gail, H.-P., 2006, A&A 447, 553
Fryxell, B., Olson, K., Ricker, P., Timmes, F. X., & Zingale, M., et al., 2000, ApJ 131, 273
Freytag, B. & Höfner S., 2008, A&A 483, 571
Helling, Ch., Winters, J. M., & Sedlmayr, E, 2000, A&A 358, 651
Helling, Ch. & Woitke, P., 2006, A&A 455, 325
Helling, Ch., Woitke, P., & Thi, W.-F., 2008, A&A accepted, astro-ph 0803.4315v1
Höfner, S., Gautschy-Loidl, R., Aringer, B., & J/orgensen, U. G., 2003, A&A 399, 589
Höfner, S. & Andersen, A. C., 2007, A&A 465, L39
Hony, S. & Bouwman, J., 2004, A&A 413, 981
Jäger, C., Mutschke, H., Dorschner, J., & Henning, Th., 1998, A&A 332, 291
Jørgensen, U. G., Johnson, H. R., & Nordlund, Å, 1992, A&A 261, 263
Jeong, K. S., Winters, J. M., Le Bertre, T., & Sedlmayr, E., 2003, A&A 407, 191
Mattsson, L., Höfner, S., Wahlin, R., & Herwig, F., 2007 astro-ph 0705.2315v2
Monnier, J. D., Millan-Gabet, R., & Tuthill, P. G., et al., 2004, ApJ 605, 436
Niccolini G., Woitke P., & Lopez B., 2003, A&A 399, 703
Ramstedt, S., Schoeier, F. L., & Olofsson, H., 2007, astro-ph 0706.2559v1
Sandin, C., & Höfner, S., 2003, A&A 404, 789
Schirrmacher, V., Woitke, P., & Sedlmayr, E., 2003, A&A 404, 267
Gail, H.-P., & Sedlmayr, E., 1988, A&A 206, 153
Verhoelst, T., Decin, L., van Malderen, R., Hony, S., Cami, J., Eriksson, K., Perrin, G., Deroo, P., Vandenbussche, B., & Waters, L. B. F. M., 2006, A&A 447, 311
Winters, J. M., Le Bertre, T., Jeong, K. S., Helling,. Ch., & Sedlmayr, E., 2000, A&A 361, 641
Woitke, P., 1999, in Astronomy with Radioactivities, ed. R. Diehl & D. Hartmann (Schloß Ringberg, Germany: MPE Report 274), 163–174
Woitke, P., 2006a, A&A 452, 537
Woitke, P., 2006b, A&A 460, L9
Woitke, P., 2008, "Monte Carlo Radiative Transfer from $\tau = 0$ to $\tau = \infty$", MNRAS, in preparation

Low-Mass Extremely Metal-Poor Stellar Models: Yields, Uncertainties and the Galactic Halo Stars

Simon W. Campbell[1,2] and J. C. Lattanzio[1]

[1] Academia Sinica Institute of Astronomy and Astrophysics,
P.O. Box 23-141, Taipei, Taiwan 10617
email: simcam@asiaa.sinica.edu.tw

[2] Centre for Stellar and Planetary Astrophysics, School of Mathematical Sciences,
Monash University, Melbourne, Australia 3800
email: john.lattanzio@sci.monash.edu.au

Abstract. We have calculated a set of low-mass ($0.85\,M_\odot \leqslant M \leqslant 3.0\,M_\odot$) zero metallicity and extremely metal-poor ($-6.5 \leqslant [\text{Fe/H}] \leqslant -3.0$) stellar models, including nucleosynthetic yields for 74 species. As far as we are aware these are the first detailed yields in the mass and metallicity range considered. Due to the difficulty in modelling such stars the yields naturally contain numerous uncertainties, and thus present interesting challenges for future stellar modelling. We briefly present some results in the context of the Galactic Halo star observations, and also discuss qualitatively some of the uncertainties in the modelling. We conclude by suggesting that much work is still necessary in this research area. For example, multidimensional fluid dynamics models are needed to simulate the violent proton ingestion events that occur during the core He flash and early TPAGB, observations and theory of mass loss at low metallicities are needed, the effects of reaction rate uncertainties need to be quantified, and low temperature opacities variable in carbon (and nitrogen) need to be included in the models.

Keywords. stars: AGB and post-AGB, stars: evolution

1. Introduction

The discovery of extremely metal-poor stars (EMPs) in the Galactic Halo has renewed interest in the theoretical modelling of Population III and low-metallicity stars (eg. Fujimoto *et al.* 1990, Cassisi *et al.* 1996, Siess *et al.* 2002, Schlattl *et al.* 2002, Picardi *et al.* 2004, Suda *et al.* 2004). Most of these stars show abundances that conform to a simple Galactic chemical evolution line (see Figure 1). However a subset of the EMP stars have been observed to contain large amounts of carbon. These C-rich EMPs (CEMPs) make up a large proportion of the EMP population ($\sim 10 \rightarrow 20\%$; see eg. Beers & Christlieb 2005). This population is also highlighted in Figure 1. Apart from carbon the EMP stars also display variation in a range of other elements (see Beers & Christlieb 2005 for a review of the observations). A number of theories have been proposed to explain the various patterns, ranging from pre-formation pollution via Pop III SNe (eg. Shigeyama & Tsujimoto 1998, Limongi *et al.* 2003) to self-pollution through peculiar evolutionary events (eg. Fujimoto *et al.* 2000, Weiss *et al.* 2004) to binary mass transfer (eg. Suda *et al.* 2004).

In the current study we have undertaken a broad exploration of EMP stellar evolution and nucleosynthesis in the low and intermediate mass regime ($0.85\,M_\odot \leqslant M \leqslant 3.0\,M_\odot$). Our study expands on the previous work in the field. In particular it includes full evolutionary calculations from ZAMS to the end of the thermally-pulsing AGB phase

(TPAGB), as well as chemical yields for 74 nuclear species. With this homogeneous set of models we hope to shed some light on whether or not 1D stellar models can explain some of the EMP halo star observations.

2. Method

Our simulations were performed utilising two numerical codes – a stellar structure code and post-process nucleosynthesis code.

The stellar structure code used was the Monash version of the Monash/Mount Stromlo stellar evolution code (MONSTAR; see eg. Wood & Zarro 1981, Frost & Lattanzio 1996). The code is largely a standard 1D code that utilises the Henyey-matrix method (a modified Newton-Raphson method) for solving the stellar structure equations. Opacities have been updated to those from Iglesias & Rogers (1996) (for mid-range temperatures) and Ferguson et al. (2005) (for low temperatures). For the present study the instantaneous convective mixing routine was replaced by a time-dependent (diffusive) mixing routine (similar to that described by Meynet et al. 2004). This change was necessary due to the violent evolutionary events that occur in models of $Z = 0$ and EMP stars. Convective boundaries were always defined by the Schwarzschild criterion – ie. the search for a neutral convective boundary was not performed and no overshoot was applied. In terms of the resultant chemical yields, the extent of convective zones in this study could thus be considered as conservative, since any extension of the Schwarzschild boundary would lead to even more pollution.

The nucleosynthesis calculations were made with the Monash Stellar Nucleosynthesis code (MONSN), a post-process code. As input it takes the key structural properties of each hydrostatic model from the MONSTAR code (eg. density, temperature, convective velocities). It solves a network of 506 nuclear reactions involving 74 nuclear species (see eg. Cannon 1993, Lattanzio et al. 1996, Lugaro et al. 2004 for more details on this code). Our grid of models includes the following masses: $M = 0.85, 1.0, 2.0, 3.0\,M_\odot$ and metallicities: $[\text{Fe/H}] = -6.5, -5.45, -4.0, -3.0$, plus $Z = 0$.

3. Brief Results: Carbon

3.1. Comparisons with Observations

In Figure 1 we compare the carbon yields from our entire grid of models with the observed [C/Fe] abundances in EMP halo stars. It can be seen that the yields from our models are *all* C-rich. The reason for this is that the low-mass ($M = 0.85$ & $1.0\,M_\odot$) models that would not normally have C-rich yields, due to a lack of third dredge-up (TDU, represented by green dots in 1), undergo proton ingestion episodes (PIEs) and subsequent dredge-up of material that has experienced He burning. In these models this event occurs during the core He flash at the tip of the RGB. In the intermediate mass models ($M = 2.0, 3.0\,M_\odot$) a similar event occurs at the beginning of the TPAGB phase (this time involving proton ingestion into the AGB He shell). In both cases a double spike in luminosity occurs (from the normal He flashes plus the PIEs). For this reason we refer to these events as 'Dual Core Flashes' (DCFs, at the RGB tip, red crosses in Figure 1) and 'Dual Shell Flashes' (DSFs, early TPAGB, blue triangles in Figure 1). It can be seen in Figure 1 that the carbon yields at $[\text{Fe/H}] = -3$, for which there are the most observations, cover the upper envelope of the observations quite well. At $[\text{Fe/H}] = -4$ there is a fair agreement with the observations, although there are less observations to compare with. At $[\text{Fe/H}] = -5.45$ the yields show somewhat more C than is observed, however there are only two stars observed at this metallicity. Another interesting feature seen in this figure

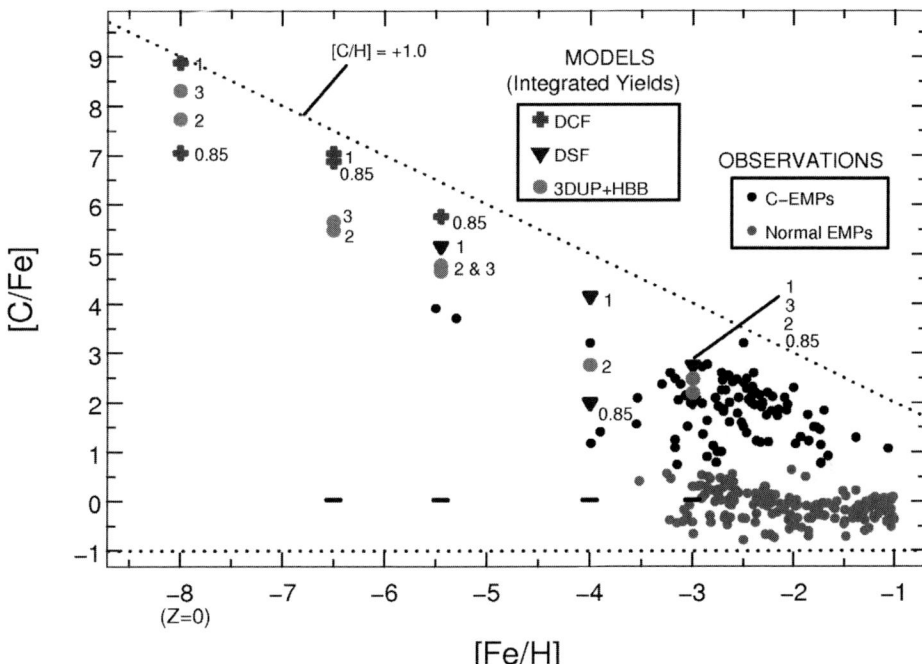

Figure 1. Comparing the carbon yields from all our models with observations of EMP stars. Here we have colour-coded the observations into [C/Fe]-rich (the CEMPs, black dots) and [C/Fe]-normal (normal EMPs, grey/magenta dots). We have defined [C/Fe]-rich by [C/Fe] > +0.7. It can be seen that the normal EMPs scatter around the Galactic chemical evolution line (at [C/Fe] ∼ 0.0). The two most Fe-poor stars known, HE 0107-5240 ([Fe/H] = −5.3, Christlieb et al. 2004) and HE 1327-2326 ([Fe/H] = −5.4, Aoki et al. 2006) are labelled. We note that the most C-rich stars have similar amounts of carbon as the Sun, giving them a Z-defined metallicity that is roughly solar. Observational data sets are from Frebel et al. (2006), Spite et al. (2006), Aoki et al. (2007), Beers et al. (2007), and Cohen et al. (2006). The short horizontal lines indicate the starting composition of the models (in this case they are all at [C/Fe] ∼ 0, except for the $Z = 0$ models). We have plotted the $Z = 0$ model yields at [Fe/H] = −8 for comparison. An upper envelope to the self-pollution of the models – and the observations – is marked by the dotted line at [C/H] = +1.0. The yields from our models are colour- and shape-coded to highlight the different episodes that produced the bulk of the pollution in each yield (see text for details). Numbers beside each yield marker indicate initial stellar mass, in M$_\odot$.

is that the model yields predict [C/Fe] to *continue* increasing towards lower metallicities. Furthermore, taking into account the evolutionary stage at which the surface pollution is gained in the lower mass models ($M = 0.85$, $1.0\,M_\odot$) – ie. the DCF events rather than the AGB – the models also predict a higher *proportion* of C-rich stars at lower and lower metallicities. This is due to the fact that these stars already have self-polluted surfaces during the HB stage – which has a lifetime roughly 1 order of magnitude longer than the AGB phase. We note that, although the lifetimes of all but the least massive models in our grid ($M = 0.85\,M_\odot$) are too short to be still 'living' today, the more massive stars could still have contributed to the chemical enrichment of the CEMP population through binary mass transfer (see eg. Suda et al. 2004).

Last but not least we stress that although the [C/Fe] abundances in the yields compare reasonably with the observations, a 'multi-element' comparison is a much more stringent test of the nucleosynthesis results. We shall make comparisons with the available observed abundance patterns in our upcoming papers.

4. Some Sources of Uncertainties

4.1. *Mass Loss Rates at $Z = 0$ and Very Low Metallicity*

A key problem with modelling EMP giant stars is the unknown driver(s) of mass loss. Currently the dominant theory is that mass is lost from red giant envelopes through radiation pressure acting on grains. Thus, in the EMP or $Z = 0$ regime, mass loss is thought to be negligible. However, due to the peculiar evolution of models of EMP and $Z = 0$ stars, we argue that mass loss is significant in AGB stars at these metallicities. In fact for this study we decided to retain the standard treatment for mass loss (the MONSTAR code uses empirical mass loss formula from Reimers 1975 during the RGB and that of Vassiliadis & Wood 1993 during the AGB), for the following reasons.

The first reason is related to the fact that the RGB phases of $Z = 0$ and EMP models are very short (or non-existent - many of our stars stay blue until the AGB). This is because He is ignited much earlier in EMP stars than in stars of comparable mass at higher metallicities. For the low-mass EMP models that do make it to the core He flash, the luminosity at the tip of the RGB can be up to 1 order of magnitude lower than in Solar metallicity models (also see eg. D'Antona 1982). Since mass loss is strongly dependent on luminosity the RGB mass loss in these stars is thus largely negligible, particularly at the lowest metallicities (although this is less true for [Fe/H] $\gtrsim -4.0$ in terms of the yields this material only has a diluting effect). Therefore the choice of mass-loss formalism is not very important on the EMP RGB, and it is the AGB mass loss that we need to follow properly.

The second reason is that, as mentioned in the Section 3, all these models experience some sort of surface polluting episode – be it from dredged-up material from the DCF or DSF events, or through normal TDU events. Moreover, the Dual Flashes cause pollution before or at the very beginning of the AGB phase, so the surface is polluted for the whole of the AGB. Surface pollution via TDU in the higher mass models occurs over a longer timescale, but we note that the majority of the mass loss occurs towards the end of the AGB, when the luminosity is high and the surface is strongly polluted. All of this has the consequence that the surface of the AGB models usually have metallicities approaching that of the LMC or even Solar (as defined by $Z = 1 - X - Y$ rather than Fe – they are still metal poor in terms of Fe). Thus, since the stellar surfaces have (some of) the ingredients needed to form grains, we argue that using the standard mass loss formula given by Vassiliadis & Wood (1993) is warranted. We also note that metallicity is also indirectly taken into account by the mass loss formulae, since they depend on bulk stellar properties (such as radius, luminosity, pulsation period), which vary significantly with metallicity.

Although we have argued that the mass loss in these EMP AGB stars should be comparable to that of higher metallicity stars, we also suggest that the mass loss should be handled better in future models. This requires observations of the mass-loss of metal-poor stars, as well as improved theoretical models. Some work in the observational direction has recently been reported by Origlia *et al.* (2002) and McDonald & van Loon (2007). Both studies derive mass-loss rates from observations of a selection of globular clusters covering a wide metallicity range ($\sim -2.2 <$ [Fe/H] < -0.6). Interestingly both studies find no dependence of mass-loss on metallicity, although more work is probably needed to confirm this. On the theoretical front Mattsson *et al.* (2008) report that mass loss may be more dependent on pulsation than metallicity, and as such low metallicity stars may still have high mass-loss rates.

4.2. The Dual Flash Events & the Need for Multidimensional Fluid Dynamics Simulations

As mentioned in Section 3 models of EMP and $Z=0$ stars undergo violent evolutionary episodes involving proton ingestion into hot He burning regions. The most severe of these events is the Dual Core Flash. It occurs in low-mass stars with metallicity $[Fe/H] \lesssim -2.5$ (Fujimoto et al. 2000). When the proton-rich material is mixed down into the hot He-flash-driven convective zone it burns at a very high rate. This flash reaches luminosities comparable to the core He flash itself but its onset is much more sudden than that of the core He flash. The rapid onset and short timescale of the H-flash requires high spatial and temporal resolution to simulate. For instance, the time steps in our models during these phases are often as low as $\sim 10^{-3}$ years (sometimes shorter). Such short time steps violate the assumption of hydrostatic equilibrium implicit in our code and thus introduce an uncertainty in our modelling. We are however unsure whether this has any significant effect in this case, but suggest that multidimensional fluid dynamics calculations are needed to check this. Other uncertainties during this challenging phase of evolution have been explored by Schlattl et al. (2001) who performed some tests (using a 1D stellar code) and found that the initial He abundance and atomic diffusion can effect the occurrence of the DCFs. Another argument for the necessity of fluid dynamics calculations arises from the likely inadequacy of the Mixing Length Theory for approximating convection in this regime of high nuclear energy release and turbulent convection (indeed, it was not designed for these conditions). Following these events properly is very important for the resulting yields of these stars, since in many cases it is these events (and the associated mixing up of the processed material) that dominate the pollution of the envelope (although at higher masses the TDU tends to dominate). We note that, since the timescale of evolution for a H-flash is of order years, it is still impossible to follow the entire evolution of the event with current multidimensional fluid dynamics codes. However even following a few hours of the evolution may reveal important features – such as if the H and He convection zones remain separated or merge. We are currently attempting these simulations thanks to new collaborations formed at this IAU Symposium.

4.3. Other Uncertainties

Other uncertainties that may effect the yields of our models include:
- *No rotation included:* EMP/Pop III stars are more compact, so they should rotate faster. This may cause extra mixing and effect the yields.
- *Nuclear reaction rates* are known to contain uncertainties (although some are well constrained). Varying reaction rates within their expected uncertainties can give us an idea of how uncertain the yield is for each species.
- *Low temperature opacities* used in our model (and most stellar codes) do not include the opacity effects of enhanced carbon (or nitrogen) on the cool AGB surfaces. This has feedback effects that can significantly alter the evolution and therefore the yields (see eg. Marigo 2002, Lederer & Aringer 2008, Cristallo et al. 2007).

5. Conclusion

Our study gives a 'zeroth order' set of models and yields for low mass, EMP stars. Although the models can produce the large amounts of carbon needed to explain the CEMPs, we need to compare our yield results with the observed abundance *patterns* for a range of species. As discussed the models are naturally subject to many uncertainties, all of which need to be investigated for future modelling. Finally we stress again that fluid dynamics simulations are urgently needed to assess he properties of the proton ingestion flashes and how this affects the further evolution and yields.

Acknowledgements

We utilised the APAC supercomputer for this work (project $g61$). We would like to thank the organisers of this IAU symposium held in Sanya, China.

References

Fujimoto, M. Y., Iben, I. J., & Hollowell, D. 1990 *ApJ* 349, 580
Cassisi, S., Castellani, V., & Tornambe, A. 1996 *ApJ* 459, 298
Siess, L., Livio, M., & Lattanzio, J. 2002 *ApJ* 570, 329
Schlattl, H., Salaris, M., Cassisi, S., & Weiss, A. 2002 *A&A* 395, 77
Picardi, I., Chieffi, A., Limongi, M., Pisanti, O., Miele, G., Mangano, G., & Imbriani, G. 2004 *ApJ* 609, 1035
Suda, T., Aikawa, M., Machida, M. N., Fujimoto, M. Y., & Iben, I. J. 2004 *ApJ* 611, 476
Schlattl, H., Cassisi, S., Salaris, M., & Weiss, A. 2001 *ApJ*, 559, 1082
Beers, T. C. & Christlieb, N. 2005 *ARA&A* 43, 531
Shigeyama, T. & Tsujimoto, T. 1998 *ApJL* 507, L135
Limongi, M., Chieffi, A., & Bonifacio, P. 2003 *ApJL* 594, L123
Fujimoto, M. Y., Ikeda, Y., & Iben, I. J. 2000 *ApJ* 529, L25
Weiss, A., Schlattl, H., Salaris, M., & Cassisi, S. 2004 *A&A* 422, 217
Christlieb, N., Gustafsson, B., Korn, A. J., Barklem, P. S., Beers, T. C., Bessell, M. S., Karlsson, T., & Mizuno-Wiedner, M. 2004 *ApJ* 603, 708
Aoki, W., Frebel, A., Christlieb, N., Norris, J. E., *et al.* 2006 *ApJ* 639, 897
Frebel, A. Christlieb, N., Norris, J. E., Beers, T. C., *et al.* 2006 *ApJ* 652, 1585
Spite, M., Cayrel, R., Hill, V., Spite, F., *et al.* 2006 *A&A* 455, 291
Aoki, W., Beers, T. C., Christlieb, N., Norris, J. E., Ryan, S. G., & Tsangarides, S. 2007 *ApJ* 655, 492
Beers, T. C., Sivarani, T., Marsteller, B., Lee, Y., Rossi, S., & Plez, B. 2007 *AJ* 133, 1193
Cohen, J. G., McWilliam, A., Shectman, S., Thompson, I., Christlieb, N., Melendez, J., Ramirez, S., Swensson, A., & Zickgraf, F.-J. 2006 *AJ* 132, 137
Frost, C. A. & Lattanzio, J. C. 1996 *ApJ* 473, 383
Wood, P. R. & Zarro, D. M. 1981 *ApJ* 247, 247
Iglesias, C. A. & Rogers, F. J. 1996 *ApJ* 464, 943
Ferguson, J. W., Alexander, D. R., Allard, F., Barman, T., Bodnarik, J. G., Hauschildt, P. H., Heffner-Wong, A., & Tamanai, A. 2005 *ApJ* 623, 585
Meynet, G., Maeder, A., & Mowlavi, N. 2004 *A&A* 416, 1023
Cannon, R. C. 1993 *MNRAS* 263, 817
Lattanzio, J., Frost, C., Cannon, R., & Wood, P. R. 1996 *MmSAI* 67, 729
Lugaro, M., Ugalde, C., Karakas, A. I., Görres, J., Wiescher, M., Lattanzio, J. C., & Cannon, R. C. 2004 *ApJ* 615, 934
Reimers, D. 1975 *Memoires of the Societe Royale des Sciences de Liege* 8, 369
Vassiliadis, E. & Wood, P. R. 1993 *ApJ* 413, 641
D'Antona, F. 1982 *A&A* 115, L1
Origlia, L., Rood, R. T., Fabbri, S., Ferraro, F. R., Fusi Pecci, F., & Rich, R. M. 2007 *ApJL* 667, L85
Mattsson, L., Wahlin, R., Höfner, S., & Eriksson, K. 2008 *A&A* 484, L5
McDonald, I. & van Loon, J. T. 2007 *A&A* 476, 1261
Origlia, L., Ferraro, F. R., Fusi Pecci, F., & Rood, R. T. 2002 *ApJ* 571, 458
Marigo, P., Girardi, L., Chiosi, C., & Wood, P. R. 2002 *A&A* 371, 152
Cristallo, S., Straniero, O., Lederer, M. T., & Aringer, B. 2007 *ApJ* 667, 489
Lederer, M. T. & Aringer, B. 2008 in: *Evolution and Nucleosynthesis in AGB Stars*, American Institute of Physics Conference Series, 1001, 11

Discussion

WALDMAN: Is the mixing of hydrogen into the helium burning layer a consequence of using the Schwarzschild criterion rather then Ledoux?

CAMPBELL: The main factor seems to be that the He ignition occurs so far off-center, close to the H-He discontinuity. The weaker entropy barrier in $Z = 0$ and extremely metal-poor stars also plays a role. I don't know if using the Ledoux criterion would be important or not.

S. Campbell is doing presentation.

The Art of Modelling Stars in the 21st Century
Proceedings IAU Symposium No. 252, 2008
L. Deng & K.L. Chan, eds.

Solar-like oscillations in red giant ε Ophiuchi

S. L. Bi and N. Gai

Department of Astronomy Beijing Normal University, Beijing 100875, China
email: bisl@bnu.edu.cn

Abstract. Asteroseismology is a powerful tool to help determining the internal structure of the stars. Solar-like oscillations have been discovered in the G9.5 red giant ε Ophiuchi, and it opened up a new part of the Hertzsprung-Russell diagram to be explored with asteroseismic techniques. We present the detailed study of the properties of ε Oph including convective overshooting and extra-mixing.

Keywords. stars: oscillations, stars: evolution

1. Introduction

Asteroseismic observations, based on the ground and the MOST space experiment, have shown that solar-like oscillations can be excited stochastically by convection in several G and K-class red giant stars (Frandsen *et al.* 2002; Kjeldsen *et al.* 2005). These stars are very different from main-sequence stars, with highly condensed core and low density envelope. Therefore, red giant stars are extremely difficult to model since they can be very different states of evolution and still be at the same place in the HR diagram.

Very recently, the observations of oscillations in the red giant star ε Ophiuchi have reliably identified the presence of radial modes, and a possible large separations (De Ridder *et al.* 2006; Hekker *et al.* 2006; Barban *et al.* 2007; Kallinger *et al.* 2008). Meanwhile, Hekker *et al.* have found evidence for non-radial oscillations. If such confirmed, this would significantly make progress in understanding the structure and evolution of red giants by seismic analysis.

2. Stellar models

ε Ophiuchi (HD 146791) is a G9.5 giant star of low metallicity, with the visual magnitude of $m_v = 3.24 \pm 0.79$ (Blackwell *et al.* 1990) and the Hipparcos parallax of 30.34 ± 0.79 mas (ESA 1997). Fundamental stellar parameters for ε Oph we adopt are the mean value, e.g. the effective temperature $T_{eff} = 4877 \pm 100K$, luminosity $L/L_\odot = 59 \pm 5$ (De Ridder *et al.* 2006), radius $R/R_\odot = 10.4 \pm 0.45$ (Richichi *et al.* 2005). We have used these observations to constrain a grid of stellar models of ε Oph based on recent physics, with the Yale stellar evolution code (YREC; Guenther *et al.* 1992).

The influence of overshooting and a diffusion of helium and heavy elements on the evolutionary tracks is shown in Figure 1A, with initial heavy metal abundance $Z_i = 0.012$, initial helium abundance $Y_i = 0.28$ and mixing-length parameter $\alpha = 1.75$ calibrated on a solar model. It is found that the effects of overshooting and extra-mixing, ignored in these early studies, are significant during the main-sequence phase, leading to the important consequences in successive evolutionary stage of helium burning. In fact, the presence of overshoot and extra-mixing modify the chemical composition inside the star, and change the boundary between the radiative and convective zones, as well as the structure of the core (Bi, *et al.* 2008).

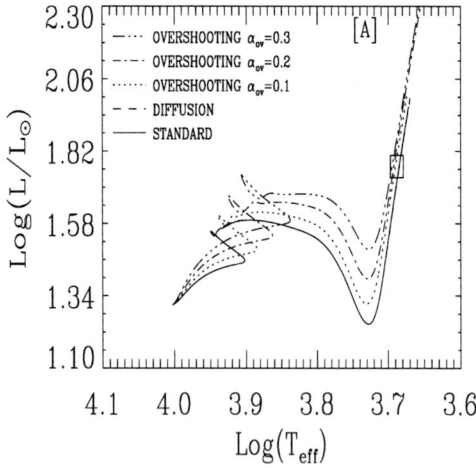 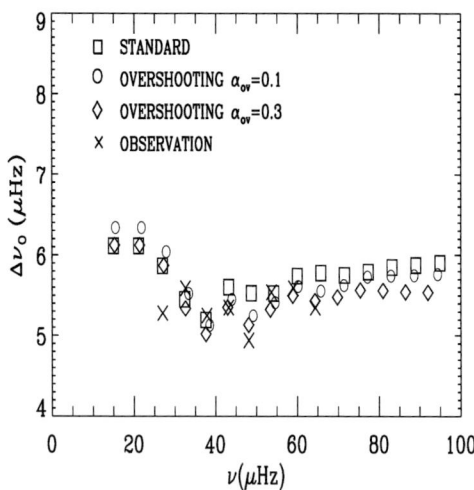

Figure 1. [A]: The effect of the diffusion and core overshooting on the evolution of ε Oph in the HR diagram. The evolutionary tracks are calculated by varying α_{ov} from 0.1 to 0.3, with $M = 2.0 M_\odot$, $Z = 0.012$, to match the observed error box. [B]: Large spacings versus frequency for several models with and without convective overshooting. The symbols correspond to modes of degree $\ell = 0$.

3. Results and discussion

We used the stellar pulsation code of Guenther (1994) to calculate the p-mode eigenfrequencies with harmonic degree $\ell = 0, 1, 2, 3$, corresponding 3 evolutionary stellar models that fall within the observational error box, without and with overshooting respectively. It is interesting that the large spacing of modes with $\ell = 0$ appear to become small with including overshooting as the star evolves towards larger effective temperature, shown in Figure 1B. This indicates that the effects of overshooting on evolution and oscillation properties of the models may be important at red giant stage. In addition, we found, from calculation of the theoretical oscillation spectrum of ε Oph, that the average large spacing for $\ell = 0$ is about $5.5\,\mu Hz$, and near observational value.

Acknowledgements

This work was supported by The Ministry of Science and Technology of the People's republic of China through grant 2007CB815406, and by NSFC grants 10433030, 10773003, and 10778601.

References

Barban, C., Matthews, M. J., De Ridder, J., et al. 2007, A&A, 468, 1033
Bi, S. L., Basu, Sarbani, & Li, L. H. 2008, ApJ 673, 1093
Blackwell, D. E., Petford, A. D., et al. 1990, A&A, 232, 396
De Ridder, J., Barban, C., Carrier, F., et al. 2006, A&A, 448, 689
ESA. 1997, The Hipparcos and Tycho Catalogues (ESA SP-1200) (Noordwijk: ESA)
Frandsen, S., Carrier, F., Aerts, C., et al. 2002, A&A, 394, L5
Guenther, D. B., et al. 1992, ApJ, 387, 372
Guenther, D. B. 1994, ApJ, 422, 400
Hekker, S., Aerts, C., De Ridder, J., Carrier, F. 2006, A&A, 458, 931
Kallinger, T., Guenther, D. B., Matthews, J. M., et al. 2008, A&A, 478, 497
Kjeldsen, H., Bedding, T. R., et al. 2005, ApJ, 635, 1281
Richichi, A., Percheron, I., & Khristoforova, M. 2005, A&A, 431, 773

Thermohaline mixing and fossil magnetic fields in red giant stars

Corinne Charbonnel[1,2] and Jean-Paul Zahn[3]

[1] Geneva Observatory, University of Geneva, ch. des Maillettes 51, 1290 Versoix, Switzerland
email: Corinne.Charbonnel@obs.unige.ch

[2] Laboratoire d'Astrophysique de Toulouse-Tarbes, Université de Toulouse, CNRS UMR 5572,
14 Av. E. Belin, 31400 Toulouse, France

[3] LUTH, Observatoire de Paris, CNRS, Université Paris-Diderot, France
email: Jean-Paul.Zahn@obspm.fr

Abstract. We discuss the occurence and consequences of thermohaline mixing in RGB stars, as well as the possible inhibition of this process by a fossil magnetic field in Ap star descendants.

Keywords. hydrodynamics - instabilities - stars: abundances- stars: evolution - stars: interiors - stars: rotation

1. Thermohaline mixing in RGB stars

Thermohaline mixing has been recently identified as the dominating process that governs the photospheric composition of low-mass bright red giant stars (Charbonnel & Zahn 2007a). As shown in Fig. 1, thermohaline convection indeed simultaneously accounts for the observed behaviour of the carbon isotopic ratio and of the abundances of Li, C and N in the upper part of the red giant branch. These results have been obtained using the prescription recommended by Ulrich (1982) for the turbulent diffusivity produced by the thermohaline instability in stellar radiation zones, with the shape factor sustained by laboratory experiments (Krishnamurthi 2003). Thermohaline mixing also significantly reduces the ^{3}He production with respect to canonical evolution models as required by measurements of ^{3}He/H in galactic HII regions. This solves the so-called "^{3}He problem", which requires that less than $\sim 10\%$ of low-mass stars are producing ^{3}He as predicted by the classical stellar theory (Galli *et al.* 1997; Charbonnel 2002).

2. The peculiar case of "thermohaline deviant stars": Ap star descendants?

However a couple of planetary nebulae, namely NGC 3242 and J320, have been found to behave "classically": slightly more massive than the sun, they are returning fresh ^{3}He to the internal medium, in the amount predicted by classical stellar models (Rood *et al.* 1992; Balser *et al.* 1999, 2006).

To reconcile the ^{3}He/H measurements in Galactic HII regions with the high values of ^{3}He in NGC 3242 and J320, we propose that thermohaline mixing is inhibited by a fossil magnetic field in RGB stars that are descendants of Ap stars (Charbonnel & Zahn 2007b). We obtain a threshold for the magnetic field of 10^4–10^5 Gauss, above which it inhibits thermohaline mixing in red giant stars located at or above the L-bump. Fields of that order are expected in the descendants of Ap stars, taking into account the contraction of their core.

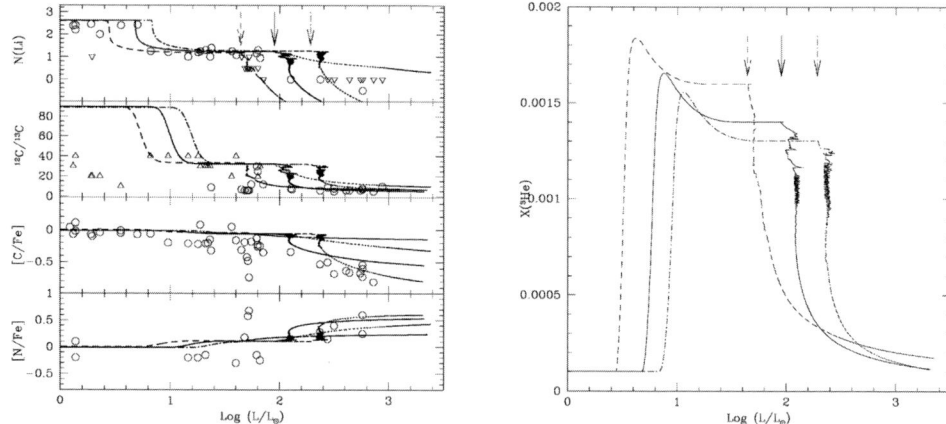

Figure 1. (Left) Evolution of the lithium abundance, of the carbon isotopic ratio, of [C/Fe] and [N/Fe] as a function of the luminosity logarithm for the $0.9 M_\odot$ models with [Fe/H] $= -1.8$, -1.3 and -0.5 (respectively dashed-dotted blue, full line black, and dashed red) computed with thermohaline mixing. The arrows in the upper panel indicate the location of the bump for the three metallicities. Observational data are from Gratton et al. (2000) for field stars in the metallicity range [Fe/H] $\in [-2; -1]$. Blue symbols are for stars with measured [Fe/H] values lower than -1.4, black for stars with $-1.4 \leqslant$ [Fe/H] $\leqslant -1.2$, and red for [Fe/H] higher than -1.2. Circles are actual measurements, open upward triangles are *lower* limits and open downward triangles are *upper* limits. The first modification in the surface abundances is due to the classical first dredge-up. When the star reaches the L-bump, thermohaline mixing strongly modifies the surface abundances, in agreement with the observations. (Right) Evolution of the surface abundance of ^{3}He (in mass fraction). The line symbols are as in the left figure. Figures from Charbonnel & Zahn (2007a).

We conclude that in a large fraction of the descendants of Ap stars thermohaline mixing does not occur. As a consequence these objects must produce ^{3}He as predicted by the standard stellar theory and as observed in the planetary nebulae NGC 3242 and J320. The relative number of such stars with respect to non-magnetic objects that undergo thermohaline mixing is consistent with the statistical constraint coming from observations of the carbon isotopic ratio in red giant stars (Charbonnel & Do Nascimento 1998). It satisfies also the Galactic requirements for the evolution of the ^{3}He abundance.

Acknowledgements

C.C. is supported by the Swiss National Science Foundation (FNS). We acknowledge support from the French Programme National de Physique Stellaire (PNPS/CNRS).

References

Balser, D. A., et al. 1999, *ApJ* 522, L73
Balser, D. A., et al. 2006, *ApJ* 640, 360
Charbonnel, C. 2002, *Nature* 415, 27
Charbonnel, C. & Do Nascimento, J. D. 1998, *A&A* 336, 915
Charbonnel, C. & Zahn, J.-P. 2007a, *A&A* 467, L29
Charbonnel, C. & Zahn, J.-P. 2007b, *A&A* 476, L15
Galli, D., et al. 1997, *ApJ* 477, 218
Krishnamurthi, R. 2003, *J.Fluid Mech.* 483, 287
Rood, R. T., Bania, T. W., & Wilson, T. L. 1992, *Nature* 355, 618
Ulrich, R. K. 1982, *ApJ* 172, 165

The Structure and Kinematics of Envelope around Red Supergiant AH Sco Traced by SiO Masers

X. Chen[1] and Z.-Q. Shen[1]

[1]Shanghai Astronomical Observatory, 80 Nandan Road, Shanghai 200030, PR China
email: chenxi, zshen@shao.ac.cn

Abstract. Observations of 43 GHz $v=1$, $J=1-0$ SiO masers in the circumstellar envelope of the M-type semi-regular variable star AH Sco were performed with the Very Long Baseline Array (VLBA) at 2 epochs in March 2004. These high-resolution VLBA images reveal that the distribution of SiO masers is roughly on a persistent elliptical ring with the lengths of the major and minor axes of about 18.5 and 15.8 mas, respectively, along a position angle of 150°. The 3-dimensional kinematics model-fitting for proper motions and spatial distributions of maser features clearly indicates that the SiO maser shell around AH Sco was undergoing an overall contraction to the star at a velocity of 15 km s^{-1} at a distance of 2.26 kpc to AH Sco due to the gravitation of the central star.

Keywords. (Stars:)circumstellar matter – masers – stars: individual (AH Sco)

1. VLBA observations

Late type stars often exhibit circumstellar maser emission in molecules e.g. OH, H_2O, and SiO. The interferometric observations of these masers would be useful in determining the structure and kinematics of the circumstellar envelop (CSE) and understanding the physical circumstance and mass loss procedure for late type stars. In this paper, we present the first VLBI maps of 43 GHz $v=1$, $J=1-0$ SiO maser emission toward red supergiant AH Scorpii (AH Sco). The observations were performed at two epochs on March 8, 2004 (epoch A) and March 20, 2004 (epoch B) with the VLBA. The data were recorded in left circular polarization in an 8 MHz band and correlated with the FX correlator in Socorro, New Mexico. The correlator output data had 256 spectral channels, corresponding to a velocity resolution of 0.22 km s^{-1}.

2. The spatial structure of the SiO masers

The high resolution VLBI images reveal a persistent elliptical structure of SiO masers around AH Sco during an interval of 12 days (Fig. 1). We characterized this morphology by performing a least-squares fit of an ellipse to the distribution of masers weighted by the flux density of each feature for each of two epochs. The lengths of the major and minor axes were found to be 18.6 and 15.7 mas for epoch A, and 18.4 and 15.9 mas for epoch B, respectively, with the major axis of the ellipse oriented similarly at 150° at both epochs. We also notice that the red-shifted SiO masers lie slightly closer to the center than the blue-shifted masers.

3. The kinematics of the SiO masers

By identifying the matched common maser features that appeared in both epochs, we were able to estimate their proper motions (Fig. 2). We can see that the maser shell shows an overall contraction toward the central star. We have made a 3-dimensional kinematics model-fitting for spatial distribution and proper motion of SiO maser features (see details

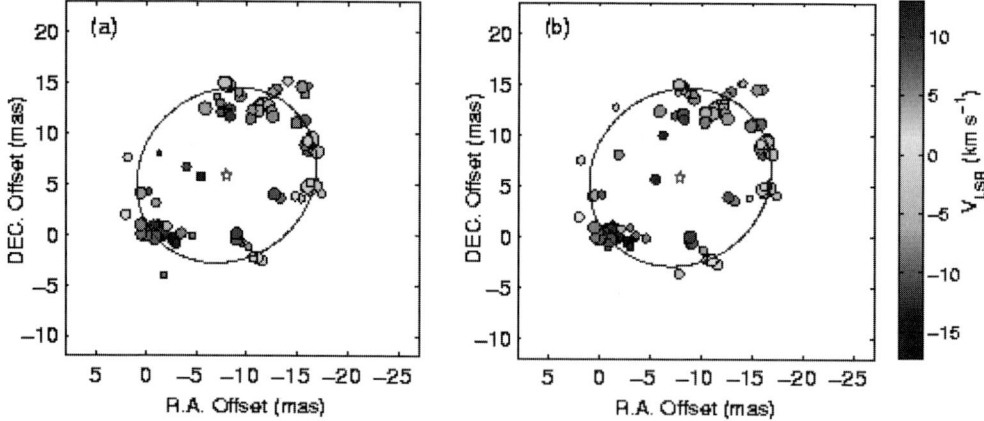

Figure 1. VLBI images of 43 GHz $v = 1$, $J = 1-0$ SiO maser emission toward AH Sco at two epochs. The ellipse indicates the least-squares fit to the maser distribution for each epoch. The fitted center of ellipse model is marked by the red star.

in Chen & Shen 2008). The 3-dimensional kinematics model suggested that the SiO maser shell was undergoing an overall contraction to the star at a velocity of 15 km s^{-1} at a distance of 2.26 kpc to AH Sco due to the gravitation of the central star.

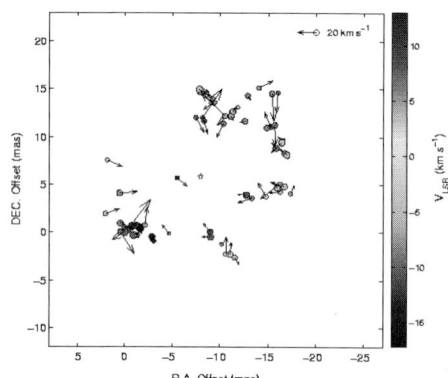

Figure 2. Distribution of proper motion velocity vectors of the matched maser features at a distance of 2.26 kpc.

Interestingly, the optical phase at which the SiO maser shell around the red supergiants contracts is nearly the same as that seen in Mira variables: red supergiant AH Sco ($\phi \approx 0.55$; Chen & Shen 2008); red supergiant VX Sgr ($\phi = 0.75 - 0.80$; Chen et al. 2006), Mira variable R Aqr ($\phi = 0.78 - 0.04$; Boboltz, Diamond & Kemball 1997), Mira variable TX Cam ($\phi = 0.50 - 0.65$; Diamond & Kemball 2003). This infers that the contraction of the SiO maser shell would occur during an optical stellar phase of 0.5-1, which agrees with the theoretical kinematical model results of Humphreys et al. (2002).

Acknowledgement. This work was supported in part by the National Natural Science Foundation of China (grants 10573029, 10625314, and 10633010) and the Knowledge Innovation Program of the Chinese Academy of Sciences (Grant No. KJCX2-YW-T03), and sponsored by the Program of Shanghai Subject Chief Scientist (06XD14024) and the National Key Basic Research Development Program of China (No. 2007CB815405). X. Chen also thanks the support by the Knowledge Innovation Program of the Chinese Academy of Sciences.

References

Boboltz, D. A., Diamond, P. J., & Kemball, A. J. 1997, *ApJ*, 487, L147
Chen, X. & Shen, Z.-Q. 2008, *ApJ*, (in press); astro-ph/08031690
Chen, X., Shen, Z.-Q., Imai, H., et al. 2006, *ApJ*, 640, 982
Diamond, P. J. & Kemball, A. J. 2003, *ApJ*, 599, 1372
Humphreys, E. M. L., Gray, M. D., Yates, J. A., et al. 2002, *A&A*, 386, 256

The Art of Modelling Stars in the 21st Century
Proceedings IAU Symposium No. 252, 2008
L. Deng & K.L. Chan, eds.

© 2008 International Astronomical Union
doi:10.1017/S1743921308022904

A New, Efficient Stellar Evolution Code for Calculating Complete Evolutionary Tracks

A. Kovetz[1,2], O. Yaron[1] and D. Prialnik[1]

[1] Department of Geophysics and Planetary Sciences, Sackler Faculty of Exact Sciences
[2] School of Physics & Astronomy, Sackler Faculty of Exact Sciences, Tel Aviv University, Israel

Abstract. We report on the development of a new stellar evolution code, and provide a taste of results, showing its capability to calculate full evolutionary tracks for a wide range of masses and metalicities. The code is fast and efficient, and is capable of following through all evolutionary phases ,including core/shell flashes and thermal pulses, without any interruption or intervention. It is meant to be used also in the context of modeling the evolution of dense stellar systems, for performing live calculations for both 'normal' ZAMS/PRE-MS models, but mainly for 'non-canonical' stellar configurations (i.e. merger-products). We show a few examples of evolutionary calculations for stellar populations I and II, and for masses in the range 0.25–64 $M_\odot$.

Keywords. stars: evolution, (stars:) Hertzsprung-Russell diagram

1. The Evolution Code

[*Basic Scheme*] The set of differential equations governing the stellar structure and evolution are approximated by finite difference equations and solved by the iterative implicit Newton-Raphson method. Following Eggleton (1971), the equations of structure and composition are all solved simultaneously, together with a mass distribution function - implementing an adaptive mesh. [*Input Physics*] Equations of state include Coulomb and quantum corrections; H and He ionization equilibria, H_2 creation, pair creation and pressure ionization (Pols *et al.* (1995)) are taken into account. Using OPAL opacities (Iglesias & Rogers (1996)) for high temperatures and Ferguson & Alexander *et al.* (2005) for low temperatures, and applying electron scattering (for the highest temperatures) and conductive opacities (when applicable), as obtained from Cassisi & Potekhin *et al.* (2007), we create, for a required metalicity, our own set of opacity tables, covering both the Hydrogen mass fractions and Carbon/Oxygen excesses mass fractions from 0 to $[1 - Z]$. Nuclear reaction rates are from Caughlan & Fowler (1988), and follow H, α and CO - burning. Neutrino losses are according to Itoh *et al.* (1996). Several mass-loss prescriptions have been incorporated; Reimers (1975) and some of it's variations for mostly the advanced (post-RGB) stages. [*Computational details*] Our automatically varying timesteps, determined mainly by limits imposed on the changes allowed during timestep, span a wide dynamic range – from $\lesssim 1\ sec$ (e.g. core He flash) to $\gtrsim 10^8\ years$ (MS). The grid mass shells span a range of $\sim 10^{-15} M_\odot$ (WD atmosphere) to $\gtrsim 10^{-1}\ M_\odot$ (inert stellar core). The typical number of grid points is kept in between 150 and 200; a typical number of timesteps for a complete evolution track is 1–2×10^3; execution time being typically in the range 5–15 minutes.

2. Results

In the four panels of Fig. 1 we show a few representative results of full evolutionary calculations: A calibration solar model, following through all evolutionary stages and

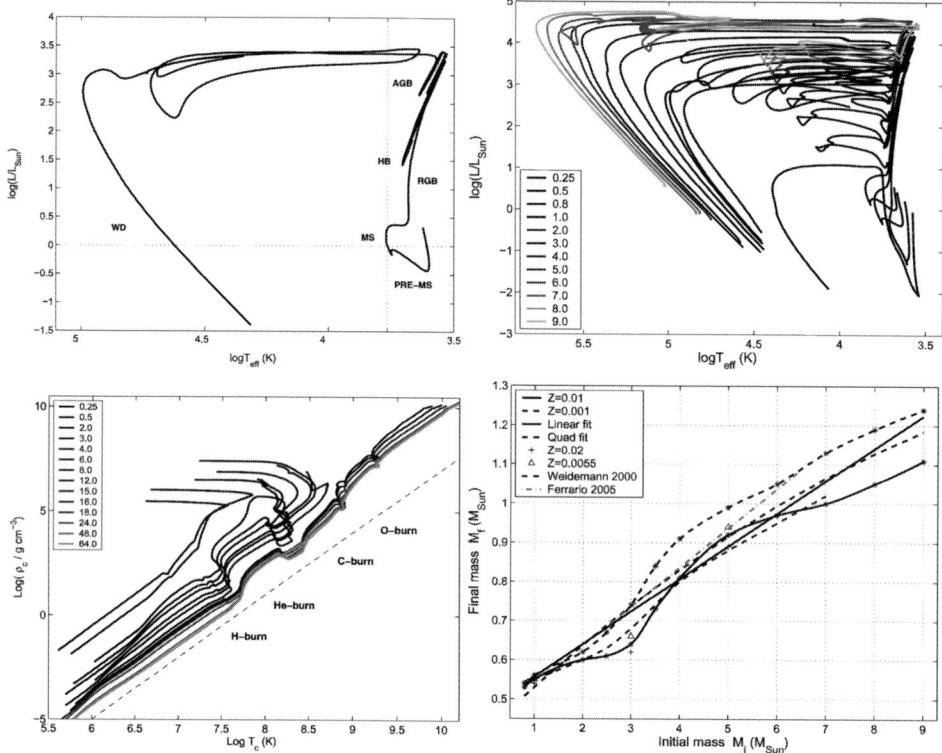

Figure 1. Top left: Solar model - $1M_\odot$, $Z = 0.018$, $Y = 0.29$. Top right: Complete tracks for Pop. II models ($Z = 0.001$), 0.25–9.0 $M_\odot$. Bottom left: Pop. I ($Z = 0.01$) - Evolution of the central stellar density and temperature for masses in the range 0.25–64 $M_\odot$. Dotted line has a slope of 3 (as obtained for the $log\rho_c - logT_c$ relation of hydrostatic equilibrium under ideal gas law). Nuclear burning phases are marked along the tracks. Bottom right: IFMR - Initial vs. final masses for M_i in the range 0.8–9 $M_\odot$, both populations. We show our linear and quadratic fits to all values, and fits of Weidemann 2000 and Ferrario 2005 for comparison.

ending as a cooling 0.55 $M_\odot$ CO WD. Features at 4.5 Gyr match the present sun to an accuracy of 1% or better. We show tracks in $log\rho_c - logT_c$ plane, exhibiting the branching off between intermediate-mass stars that end their lives as WDs, and massive stars that go through advanced nuclear burning stages, ending their lives in dynamic core collapse which is followed by the code until very high central pressures are attained and the adiabatic exponent approaches 4/3 throughout the core. Based on our sequences of results for both Pops. I and II, we also present Initial Final Mass Relationship for initial masses in the range 0.8–9 $M_\odot$, and propose our own fits with comparison to others'.

References

Cassisi, S., Potekhin, A. Y., Pietrinferni, A., Catelan, M., & Salaris, M. 2007 *ApJ* 661, 1094
Caughlan, G. R. & Fowler, W. A. 1988 *Atomic Data and Nuclear Data Tables* 40, 283
Eggleton, P. P. 1971, *MNRAS* 151, 351
Ferguson, J. W., Alexander, D. R., et al. 2005 *ApJ* 623, 585
Iglesias, C. A. & Rogers F. J. 1996, *ApJ* 464, 943
Itoh, N., Hayashi, H., Nishikawa, A., & Kohyama, Y. 1996 *ApJS* 102, 411
Pols, O. R., Tout, C. A., Eggleton, P. P., & Han, Z. 1995, *MNRAS* 274, 964
Reimers, D. 1975 *MSRSL* 8, 369

Lithium isotopes in halo dwarfs

Laurent Piau

CEA-Saclay DSM/IRFU/SAp, L'Orme des Merisiers 91191 Gif-sur-Yvette France

Abstract. We present calculations of ^{7}Li evolution in halo dwarfs during pre-MS and MS. The combination of tachocline mixing, nuclear destruction and microscopic diffusion is investigated. We briefly touch on the question of ^{6}Li.

Keywords. Stars: abundances, stars: evolution

1. General inputs

We exploit the CESAM stellar evolution code (Morel 1997) with the following inputs:
- OPAL equation of state and opacities for a Population II mixture: $[\alpha/\text{Fe}] = 0.3$.
- Nuclear reactions from NACRE (Angulo et al. 1999) and microscopic diffusion of Michaud & Proffitt (1993).

The initial composition is $Y = 0.2479$, $[\text{Fe/H}] = -2$ & $[^7\text{Li}] = 2.6$ (Coc et al. 2004), $[^6\text{Li}] = 1.1$. We assume that the rotation history is similar to Population I solar analogs. The angular momentum (J) loss through magnetized wind and the rotation (Ω) rely on the Kawaler (1988) prescription: $\frac{dJ}{dt} = -K\Omega 3 (\frac{R}{R_\odot})^{1/2} (\frac{M}{M_\odot})^{-1/2}$

K is a constant calibrated to reach the solar rotation in solar models. Rotation induces mixing in the upper radiation zone, just below the base of the convection zone : the tachocline mixing (Spiegel & Zahn 1992). The analytical developments provide the rotationally induced turbulence coefficient: $D_T(r) \sim \nu_H \left(\frac{d}{r_{bcz}}\right) 2 (\tilde\Omega/\Omega) 2 \exp(-2\zeta) \cos 2(\zeta)$ with ν_H the horizontal viscosity, $\tilde\Omega$ the differential rotation. $\zeta = 4.933(r_{bzc} - r)/d$, d = $r_{bzc}(2\Omega/N)^{1/2}(4K/\nu_H)^{1/4}$ the width of the tachocline. r_{bcz} is the radius of the convection zone, N2 = $g[1/\Gamma_1 d\ln p/dr - d\ln \rho/dr]$ is the squared buoyancy frequency, K = $\chi/\rho c_p$ is the radiative diffusivity.

The stars are evolved to 200 Myr (end of pre-MS) and then 13 Gyr (present age). calculations as they need $\sim$1 Gyr to have a significant impact.

2. Lithium evolution

Above $T_{\text{eff}} = 5200\text{K}$ no ^{7}Li pre-MS depletion is computed. Below this temperature the slope of the ^{7}Li-T_{eff} relation does not match the observations. Thus ^{7}Li surface abundances are presumably determined during the MS.

Provided both microscopic and tachocline diffusion are accounted for, ^{7}Li abundances at 13 Gyr fit the observations below $T_{\text{eff}} = 5500\,\text{K}$. A plateau is also reproduced above that temperature but it lies $\sim$0.2 dex above the current observations (figure 1). Despite their fast rotation (5–10 km.s^{-1}) and thus reinforced deep mixing, the warm ^{7}Li poor stars can't be explained in the framework of our calculations. Three out of four of these objects are confirmed spectroscopic binaries (Ryan et al. 2002) the companion being most likely a compact object. Given the shallow convection zones of the ^{7}Li poor stars, a mass transfer of 1 to $3.10^{-2} M_\odot$ of ^{7}Li free matter as the companion was on the giant stage could explain the observed abundances: $[^7\text{Li}] < 1.7$.

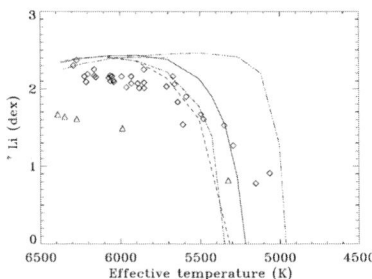

Figure 1. $T_{eff} - {}^{7}Li$ MS relation at 13 Gyr in tachocline models for buoyancy frequency 10 μHz (solid line) and 2 μHz (dashed line) in the tachocline region. The buoyancy frequency varies rapidly with depth near the upper limit of the radiation zone, thus it is a free parameter of the tachocline mixing. The depletion pattern is also provided for five time faster rotation history and buoyancy frequency of 10 μHz (dot-dashed line) and pure microscopic diffusion models (dash-three dotted line). Observations (diamonds) and upper limits (triangles) are from Ryan & Deliyannis (1998), Ryan et al. (1999) and Ryan et al. (2001).

${}^{6}Li$ is destroyed by proton capture at lower temperature than ${}^{7}Li$. Any object now having $T_{eff} < 6200K$ should have experienced a strong pre-MS depletion. For the models with tachocline mixing exhibiting $T_{eff} = 6000K$ at 13 Gyr we predict a $\sim$1 dex depletion.

3. Conclusion

$\Longrightarrow$ The ${}^{7}Li$ abundances mostly result from the MS evolution. Below $T_{eff} = 5500$ K our predictions on ${}^{7}Li$ match the observations. A flat plateau is calculated above this temperature but still lies $\sim$0.2 dex above the current observations if we start the models with the standard BBN abundance $[{}^{7}Li] = 2.6$. This assumption could be reconsidered however (e.g. Piau et al. 2006). A mass transfer of a few percent of solar masses of ${}^{7}Li$ free matter could explain the warm ($T_{eff} > 6000$ K) lithium poor halo dwarfs.

$\Longrightarrow$ We predict a very strong pre-MS and MS ${}^{6}Li$ depletion below $T_{eff} = 6200$ K. This hardly is compatible with the recent $[{}^{6}Li] \sim 1$ measurements in the halo unless the isotope was massively produced before/during the Galactic halo formation. Observationally the ${}^{6}Li$ measurements need confirmation (Cayrel et al. 2007).

References

Angulo, C., et al., Nucl. Phys. A656 (1999)3-187 2005, Proceedings of IAU symposium 228, Ed. V. Hill, P. François, F. Primas. 2006, ApJ, 644, 229
Cayrel, R., Steffen, M., et al. , 2007, A&A, 473, L37
Coc, A., Vangioni-Flam, E., Descouvemont, P., Adahchour, A., Angulo, C., 2004, ApJ, 600, 544
Kawaler, S. D., 1988, ApJ, 333, 236
Michaud, G. & Proffitt, C. R., 1993, Inside the stars , IAU colloquium, 137, ASP Conference series, eds. A. Baglin, & W. W. Weiss vol 40, 426
Morel, P., 1997, A&AS, 124, 597
Piau, L., Beers, T. C. et al. , 2006, ApJ, 653, 300 ApJ 458, 543
Ryan, S. G. & Deliyannis, C. P., 1998, ApJ, 500, 398
Ryan, S. G., Norris, J. E. & Beers, T. C., 1999, ApJ, 523, 654
Ryan, S. G. & Kajino, T., et al., 2001, ApJ 549, 55
Ryan, S. G., Gregory, S. G., Kolb, U., Beers, T. C., & Kajino, T., 2002, ApJ, 571, 501
Spiegel, E. A., Zahn, J. P., 1992, A&A, 265, 106

Surface convection in Population II stars

Laurent Piau[1] and Robert F. Stein[2]

[1] CEA-Saclay DSM/IRFU/SAp, L'Orme des Merisiers 91191 Gif-sur-Yvette France
[2] Michigan State University, Department of Physics & Astronomy East Lansing, MI 48824-2320, USA

Abstract. The initial surface abundances of Population II stars have been altered by the interplay between convection, rotational mixing and diffusion. In particular the shallower the outer convection zone the stronger the diffusion impact. We present preliminary results on constraining the extension of the convection zones of Population II stars thanks to 3D hydrodynamical simulations.

Keywords. Convection, stars: populations II

1. STAGGER

We use the STAGGER code (Stein & Nordlund 1998) to investigate the convective and the radiative energy transfer from ∼0.5 Mm above the photosphere downto ∼3.5 below it. The computational domain extends over 6 by 6 Mm horizontaly. The current grid has 63 points in each direction. We solve the fully compressible equations of hydrodynamics. The specific internal energy and density of the material entering the computational domain from below are adjusted in order to obtain the desired effective temperature in a given gravity field. Because of the cool effective temperatures a significant part of the energy is carried to the surface in form of ionization energy which makes the choice of a realistic equation of state (EoS) important. We use the OPAL2005 EoS for a pure hydrogen/helium mix with Y= 0.2479.

The equation of radiative transfer is :

$$cos\theta \frac{dI_\nu}{(\kappa_\nu + \sigma_\nu)\rho dz} = I_\nu - B_\nu \quad (1.1)$$

With θ the angle from the vertical, κ_ν and σ_ν respectively the absorption and diffusion coefficients, I_ν the specific intensity and B_ν the Planck function.

The equation of radiative heating writes :

$$Q_{rad} = 4\pi\rho \int_0^\infty \kappa_\nu (J_\nu - B_\nu) d\nu \quad (1.2)$$

With J_ν the mean intensity and ρ the density.

For each cell these equations are solved numerically along one vertical ray and four slanted rays. We use the opacity binning method (Nordlund 1982) which is intended to compute the thermal structure when the medium is neither optically thin nor thick. For some time it was successfully employed in the solar case and recently for other stars (Ludwig *et al.* 2006).

2. Method and first comparison

We determine the required log g and T_{eff} by building stellar models with the stellar evolution code CESAM code (Morel 1997). CESAM utilizes the same composition, EoS and opacities as STAGGER. Our models have [Fe/H] = −3 and Y = 0.2479.

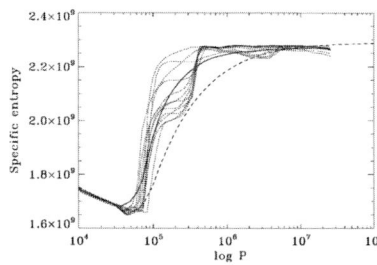

Figure 1. Specific entropy as function of depth (represented by pressure). Black solid line: hydrodynamic model average value. Blue dotted lines spatial fluctuations on the computational box. Green line best fit for a stellar envelope model in terms of α_{mlt} and for the corresponding $T_{eff} = 6700$ K and log g = 4.2.

STAGGER was run for 4 hours of surface convection time for surface conditions near the turn-off and and the lower main sequence. ¿From these computations we have inferred an 'hydrodynamic' entropy jump Δs_{hydro}. Figure 1 shows the specific entropy profiles from both hydrodynamical calculations and mixing length theory calculations.

- Near the turn off at $T_{eff} = 6700$K and log g = 4.2 we find Δs_{hydro} = 5.780 108erg.g^{-1}.K^{-1}.
- Lower on the MS at $T_{eff} = 5700$K and log g = 4.6 we find $\Delta s_{hydro} = 1.702$ 108erg.g^{-1}.K^{-1}.

We then buildt grids of envelope model for various α_{mlt}, T_{eff} and log g and found the envelope models providing the closest Δs_{mlt} to Δs_{hydro}:

- For $T_{eff} = 6772$K and log g=4.24 we have Δs_{mlt}=5.782 108erg.g^{-1}.K^{-1} when and $\alpha_{mlt} = 1.52$.
- For $T_{eff} = 5776$K and log g=4.64 we have Δs_{mlt}=1.705108erg.g^{-1}.K^{-1} when $\alpha_{mlt} = 1.77$.

3. Conclusion

- Hydrodynamical 3D simulations can constrain the phenomenological theories used to model convection in stellar evolution. In the context of active research devoted to the oldest stars, we address this issue for extremely metal poor dwarfs.
- As shown by Ludwig *et al.* (2002) the mixing length theory does not properly describes all the convection properties. Yet obtaining α_{mlt} thanks to the associated specific entropy jump is sufficient to perform stellar evolution.
- Refined EoS, opacity tables and atmosphere structures are required. The mixing length α_{mlt} has been constrained for two typical surface conditions of Population II dwarfs in terms of T_{eff} and log g. Once calibrated the mixing length parameters will be used to perform stellar evolution.

References

Ludwig, H.-G., Allard, F., & Hauschildt, P. H., 2006, A&A, 459, 599
Ludwig, H.-G., Allard, F., & Hauschildt, P. H., 2002, A&A, 395, 99
Morel, P., 1997, A&AS, 124, 597
Nordlund, A, 1982, A&A, 107, 1.
Stein, R. F. & Nordlund, A, 1998, ApJ, 499, 914 Ferguson, J. W., 2006, ApJ, 653, 300 Chan, K. L., Guenther, D. B., 2003, MNRAS, 340, 923 Franois, P., Bonifacio, P., Barbuy, B., Beers, T., and 4 coauthors, 2005, A&A, 430, 655

On MHD rotational transport, instabilities and dynamo action in stellar radiation zones

S. Mathis[1,2], J.-P. Zahn[2] and A.-S. Brun[1,2]

[1]Laboratoire AIM, CEA/DSM-CNRS-Université Paris Diderot, IRFU/SAp,
F-91191 Gif-sur-Yvette Cedex, France
email: stephane.mathis@cea.fr, allan-sacha.brun@cea.fr

[2]LUTH, Observatoire de Paris-CNRS-Université Paris Diderot,
5 Place Jules Janssen, F-92195 Meudon Cedex, France
email: jean-paul.zahn@obspm.fr

Abstract. Magnetic field is an essential dynamical process in stellar radiation zones. Moreover, it has been suggested that a dynamo action, sustained by a MHD instability which affects the toroidal axisymmetric magnetic field, could lead to a strong transport of angular momentum and of chemicals in such regions. Here, we recall the different magnetic transport and mixing processes in radiative regions. Next, we show that the dynamo cannot operate as described by Spruit (2002) and recall the condition required to close the dynamo loop. We perform high-resolution 3D simulations with the ASH code, where we observe indeed the MHD instability, but where we do not detect any dynamo action, contrary to J. Braithwaite (2006). We conclude on the picture we get for magnetic transport mechanisms in radiation zones and the associated consequences for stellar evolution.

Keywords. MHD, Sun: magnetic fields, Sun: interior, stars: magnetic fields, stars: interiors

1. MHD instabilities and possible dynamo in stellar radiation zones

Purely axisymmetric poloidal and toroidal fields are unstable (see Pitts & Tayler 1985 and references therein). Moreover, Spruit (2002) suggests that the instability of such toroidal field could sustain a dynamo in stellar radiation zones. This idea is quite interesting, but we argue that this dynamo cannot operate as he describes it. According to him, the non-axisymmetric instability-generated small-scale field, which has zero average, is wound up by the differential rotation "into a new contribution to the azimuthal field. This again is unstable, thus closing the dynamo loop." But this shear induced azimuthal field has the same azimuthal wavenumber as the instability-generated field, i.e. $m \neq 0$ and predominantly $m = 1$: it has no mean azimuthal component, and thus it cannot regenerate the mean toroidal field that is required to sustain the instability. For the same reason, the instability-generated field cannot regenerate the mean poloidal field, as was suggested by Braithwaite (2006). Therefore, the Pitts & Tayler instability cannot be the cause of a dynamo, as it was described by Spruit and Braithwaite. In fact, the dynamo loop can only be achieved through the azimuthal average of the fluctuation-fluctuation term of the induction equation $<\vec{\nabla}\times(\vec{v}'\times\vec{B}')>_\varphi$ (cf. Zahn, Brun & Mathis 2007).

2. Numerical simulations

We perform 3D-numerical simulations of the problem using the global ASH code (Clune et al. 1999, Brun et al. 2004) to solve the relevant anelastic MHD equations in a spherical shell representing the upper part of the solar radiation zone ($0.35 \leqslant r/R_\odot \leqslant 0.70$) using a resolution of $N_r \times N_\theta \times N_\varphi = 193 \times 128 \times 256$. A detailed discussion of the set-up is

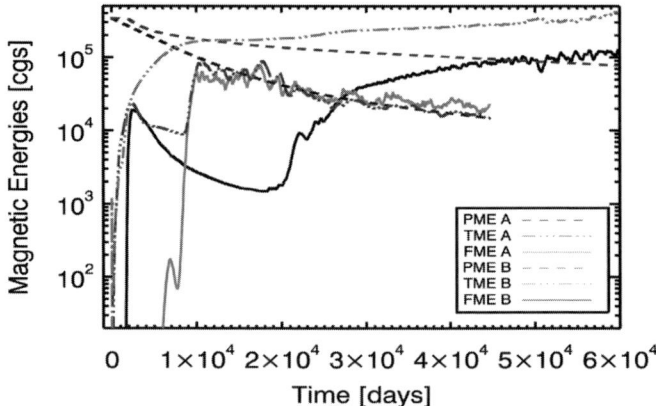

Figure 1. Time evolution of the energies of the mean poloidal (PME), mean toroidal (TME) and non-axisymmetric (FME) components of the magnetic field. Cases A and B refer respectively to higher and lower magnetic diffusivity. Note the steady decline of the poloidal field, which is not affected by the irruption of the $m=1$ Pitts & Tayler instability (at t $\approx$ 8,000 days in case A and $\approx$ 20,000 days in case B). (Zahn, Brun & Mathis 2007, courtesy A&A)

given in Brun & Zahn (2006) and in Zahn, Brun & Mathis (2007). We study the case A discussed in Brun & Zahn (2006) and we performed an additional series of simulations with a lower Ohmic diffusivity (by a factor of 10, case B), in order to reach a higher magnetic Reynolds number in favor of a dynamo. In our simulations (cf. Fig. 1), the α-effect plays a negligible role since no regeneration of the mean poloidal field is found, at least up to the magnetic Reynolds number $Rm = R^2 \Delta\Omega/\eta \sim 10^5$ for Prandtl number $P_m = \nu/\eta = 1$. On the other hand, the β-effect, i.e. the turbulence-enhanced diffusivity, is absent here. Hence, one should not expect much mixing of the stellar material and the magnetic transport of angular momentum is mainly due to the Lorentz torque that leads to Ferraro's law ($\vec{B}\cdot\vec{\nabla}\Omega=0$) (cf. Brun & Zahn 2006). In fact, the smallest resolved scales do not act on the mean poloidal field as a turbulent diffusivity: they seem to behave rather as gravito-Alfvén waves. Finally, there is no sign either of a small-scale fluctuation dynamo. To check this point, we suppressed the mean poloidal field at the latest stage of our simulation. Then, the mean toroidal field decreases rapidly, because it is no longer produced by the Ω-effect, and the instability-generated field accompanies its decline. Thus the fluctuating field does not maintain itself. Therefore, we conclude that in our simulations the Pitts & Tayler instability is unable to sustain a large-scale mean field dynamo, in the parameter domain that we have explored.

References

Braithwaite, J. 2006, *A&A* 449, 451
Brun, A.-S. & Zahn, J.-P. 2006, *A&A* 457, 665
Brun, A.-S., Miesch, M. S., & Toomre, J. 2004, *ApJ* 614, 1073
Clune, T. L. *et al.* 1999, *Parallel Comput.* 25, 361
Mathis, S. & Zahn, J.-P. 2005, *A&A* 440, 653
Pitts, E. & Tayler, R. J. 1985, *MNRAS* 216, 139
Spruit, H. C. 2002, *A&A* 381, 923
Zahn, J.-P., Brun, A.-S., & Mathis, S. 2007, *A&A* 474, 145

The Art of Modelling Stars in the 21st Century
Proceedings IAU Symposium No. 252, 2008
L. Deng& K.L. Chan, eds.

© 2008 International Astronomical Union
doi:10.1017/S1743921308022941

The Rotation of the Solar Radiative zone

S. Turck-Chieze

IRFU/CEA CE Saclay, 91101 Gif sur Yvette cedex, France
email: cturck@cea.fr

Abstract. Dynamical processes are progressively introduced in stellar evolution. In this framework, the Sun is a very specific case where both models and observations have been developed in parallel during the last decade in order to progress on our present insight of solar like stars. In this poster I show the recent progress done on both sides for the rotation of the radiative zone. The present knowledge of the solar rotation profile comes from the detection of acoustic and gravity modes with the instruments GOLF and MDI aboard SoHO. In parallel we study the sensitivity of the theoretical rotation profiles obtained with the CESAM code using different rotation history in the premainsequence.

Keywords. Sun: rotation, sun: general

1. Introduction

Dynamical processes are present in stars all across the Hertzsprung-Russell diagram. They are clearly visible in young stars, these stars are generally rapid rotators which then decelerate after their dissociation from their initial disk. Rotation plays also a major role in massive stars and more specifically in the final stages of evolution. They are generally strong rotators with strong stellar winds.

In fact, we need to introduce the internal dynamical processes (rotation, magnetic field and gravity waves) in the stellar structure equations to describe properly the different stages of evolution. This is a hard work because it supposes to replace the four structural equations by 16 complex equations, but it is necessary because the 3D simulations cannot yet be used to follow the evolution of stars and consequently to confront results to asteroseismic observations. In this poster we limit our ambition to the first step, which consists to compare the present observed rotation profile to the one obtained in introducing the transport of momentum due to the solar rotation.

2. The internal observed rotation profile

After more than 10 years of SoHO observation, we have got very important constraints on the rotation profile of the radiative zone thanks to the GOLF and MDI instruments. The determination of the splittings of a large number of acoustic modes, in particular those which are not influenced by the superficial magnetic field, has definitively established that the part of the radiative zone which is not influenced by the nuclear core, that means the region between 0.25 to 0.68 $R_\odot$ is really flat with invisible latitudinal differential variation, in great contrast with the latitudinal differential profile of the convective zone. Below 0.25 $R_\odot$, only gravity modes may inform on the rotation profile. These modes begin to be observed by the GOLF instrument (Turck-Chieze *et al.*, 2004, Garcia *et al.*, 2007). Two solutions have been extracted from the pattern attributed to an $\ell = 2$, n = 2 mode depending on the number of the observed components : a slightly reduced rotation rate or an increase by about a factor 3 in the central region (Mathur *et al.* 2008). Indeed the solution of a rapid rotating core is compatible with the two kinds

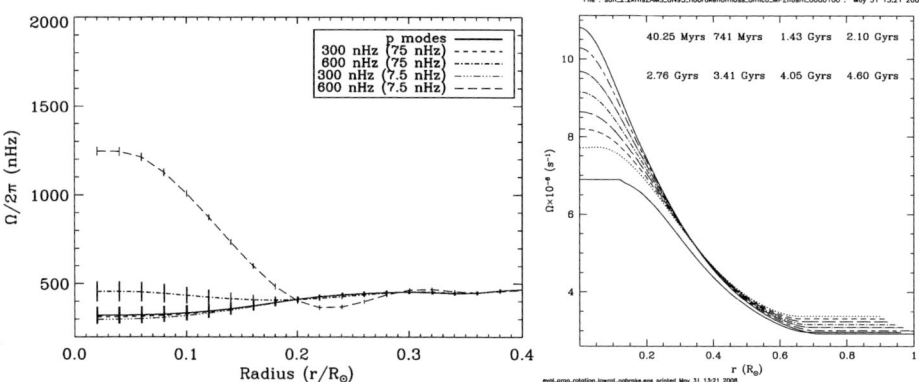

Figure 1. a) Solar observed radiative rotation profile extracted from gravity and acoustic modes using a gravity mode candidate attributed to an $\ell = 2$, n = 3 for two values of splittings (300 and 600 nHz). The values 75 nHz and 7.5 nHz correspond respectively to a possible asymmetry of the components attributed to central magnetic field and to a typical duration of observation. From Mathur (2008). b) Evolution of the theoretical rotation profile supposing initially a flat rotation profile. From Palacios, Nghiem and Turck-Chieze (2008).

of detection (search of individual mode at high frequency, asymptotic behaviour at low frequency). It supposes also that the solar core rotation axis is different from the one of the rest of the radiative zone. This exciting possibility needs to be confirmed with an improved instrument like the GOLF-NG concept. A prototype of this instrument presently observes in Tenerife (Turck-Chieze *et al.* 2006 Adv Spa. Res, 2008 Astron. Notes).

3. The internal theoretical rotation profile

We have computed new solar models from the premainsequence including the transport of angular momentum and chemicals by rotating induced processes (shear turbulence and meridional circulation). We deduce from them the sound speed, density and rotation profiles for a direct comparison with the helioseismic observables. We demonstrate the role of the advection term in following three solar evolutions which differ by their initial conditions: low and high (20km/s and 50 km/s) rotation profiles. They all lead to a differential radial rotation profile in the radiative zone but with significantly different contrast. The observed internal rotation cannot be reproduced in taking into account only these hydrodynamical processes (Fig. 1) as previously mentioned, but we demonstrate the important impact of the initial rotation for a further estimate of the role of the deep magnetic field or of the gravity waves on the transport of momentum or chemicals. These models which are physically more representative of the real Sun than the solar standard or the seismic model slightly degradate the agreement with all the observables (neutrinos and seismology) if one supposes an initial high rotation. These results demonstrate the complexity of the modelling of the solar radiative zone. They call for more theoretical efforts on the dynamical processes in radiative zones.

References

Garcia, R. A., Turck-Chieze, S., Jimenez-Reyes, S. J. *et al.* 2007, *Science*, 316, 1537
Mathur, S., Eff-Darwich, A., Garcia, R. & Turck-Chieze, S. 2008, *A&A*,484, 517
Palacios, Nghiem & Turck-Chieze 2008, *ApJ* submitted
Turck-Chieze, S., Garca, R. A., Couvidat, S. *et al.* 2004, *ApJ* 604, 455; erratum ApJ, 604, 455

The Art of Modelling Stars in the 21st Century
Proceedings IAU Symposium No. IAU 252, 2008
L. Deng & K.L. Chan, eds.

© 2008 International Astronomical Union
doi:10.1017/S1743921308022953

Mass loss from pulsating cool stars

Qian Wang[1], Lee Anne Willson[1] and Steven Kawaler[1]

[1]Department of Physics and Astronomy, Iowa State University, Ames, IA 50010, USA
email: wqinisu@iastate.edu, lwillson@iastate.edu, sdk@iastate.edu

Abstract. It has long been clear that most, if not all, of the mass loss experienced by stars from 0.8 to 8 solar masses occurs near the tip of the AGB and/or the RGB. Evolutionary studies have incorporated empirical mass loss laws but theoretical models suggest quite different dependence of mass loss rate on stellar parameters. We are combining evolutionary model calculations with ISUEVO with mass loss modeling using the Bowen code in a systematic study of final stages of stellar evolution. We mapped the RGB (without steady mass loss) to the "Death Zone" as a function of mixing length, mass, and metallicity. We compared these results with observation data from Origlia. We are investigating a possible mass loss mechanism through companions as a complement to mass loss through pulsation. By the end of the project we expect to provide a reliable prescription for AGB mass loss.

Keywords. Stars: mass loss, stars: AGB and post-AGB

1. Introduction

Stars of mass lower than 2.8 Solar mass experience an episode of mass loss that terminates the nuclear burning inside the star. Although a lot of observations support mass loss in the AGB and RGB stages, there is yet not consensus on a prescription for the mass loss rate as a function of stellar parameters. Here we report some results comparing the mass loss "death zone" derived from the Bowen models (Willson 2007a, Willson 2007b) with the location of the tip of the RGB.

From detailed models, Bowen and Willson(1991), Willson(2000), Willson(2007a), Willson(2007b) have developed a description of the mass loss process in terms of a "cliff", or deathline, and a death zone that includes most of the time that the star spends losing appreciable mass. The mass loss rate is a function of luminosity, radius, mass, and composition (and perhaps also of angular momentum and/or proximity to another star or a planet). Main sequence mass loss rate is $\leqslant 10^{-10}$ solar mass per year for single stars with $M < 2.8 M_{Sun}$, leading to very small integrated mass loss before the star evolves up the red giant branch(RGB). The death line is where $\dot{M}_{crit} = M \dot{L} / L$ and the death zone extends from 0.1 to 10 times this critical mass loss rate.

We have been working on the models for RGB stars with different masses, mixing lengths and metallicities, using the stellar evolution code ISUEVO (Kawaler 2005). By comparing our models with other's work and observational data, we are trying to gain better understanding of the late stages of stellar evolution. For current results regarding AGB mass loss, see Willson(2007b).

2. Mass loss mechanics

After mapping the mass loss for stars on RGB tip, we find only the lowest mass, smaller mixing length (larger R) stars penetrate the death zone. For RGB mass loss, therefore, either the stars are large (mixing length parameter $\leqslant 1$) or RGB mass loss by the pulsation-plus-dust mechanism yields very small net mass loss on the RGB. This

259

leaves two options for RGB mass loss: The mass may be lost during core He flash, dynamically or as a result of increased L and R (penetrating the death zone) or it may be lost along the RGB by some other mechanism perhaps involving some other physics not in the Bowen code. For example, the dust opacity was assumed to be grey while in reality it's not. Recently Woitke(2006) and Höfner(2007) have pointed out that there is a problem driving mass loss with non-gray silicate grains. Höfner and Anderson (Höfner 2007) suggest the solution may be in non-LTE chemistry. We will be updating the Bowen code and running a new grid of models taking this issue into account.

3. Comparison with Tuc 47

The Origlia *et al.* (2002) results show critical mass loss rates, $\dot{M}/M \sim \dot{L}/L$, over a wide range of luminosities on the RGB. This is not consistent with our understanding of pulsation-enhanced red giant winds, with or without dust. Therefore, there must be another mechanism producing this mass loss. We are investigating a mechanism involving low mass companions. To maintain $\dot{M}/M \sim \dot{L}/L$, the orbit must be expanding, i.e. mass must be leaving and not taking a lot of angular momentum with it. Because we are considering m/M to be small, the giant will not be rotating and the Roche Lobe picture therefore does not apply.

4. Conclusion

Applying the death zone analysis from the AGB to the RGB, we find that at most modest amounts of mass can be lost from the normal mechanisms of pulsation and/or dust, and this only from the lowest mass, highest Z stars, before the core Helium flash. Thus RGB mass loss must be associated with the core He flash itself or must involve some mechanism not included in the pulsation-plus-dust Bowen models. We are investigating the effect of low mass companions as one possible mechanism for mass loss along the RGB.

Acknowledgements

We appreciate the opportunity to participate in this simulating and enjoyable conference. We also acknowledge long collaboration with G. Bowen and the use of his code for ongoing studies. NSF grant AST04-56047 to S. Ragland has supported this work; NSF grant AST0708143 is supporting further work on this project.

References

Bowen, G. H. 1988 , *ApJ*, 329, 299
Bowen, G. H. & Willson, L. A. 1991, *ApJ*, 375, L53
Höfner, S. & Anderson, A. 2007, *Astronomy and Astrophysics*, 465, L39–L42
Kawaler, S. 2005, *ApJ*, 621, Issue 1, pp. 432–444
Origlia, L. *et al.* 2002, *ApJ*, 571: 458–468
Struck, C. *et al.* 2004, *MNRAS*, 353, Issue 2, pp. 559–570
Sweigart, A., Greggio, L., & Renzini, A. 1990, *ApJ*, 364: 527–539
Willson, L. A. 2000, *ARAA*, 38, pp. 573–611
Willson, L. A. 2007, *Vienna proceedings, eds. Franz Kerschbaum, Corinne Charbonnel and Robert Wing*, page (to appear)
Willson, L. A. 2007, *this volume*
Woitke, P. 2006, *Astronomy and Astrophysics*, 460, L9–L12

"Hot Helium Flashers" – The Road to Extreme Horizontal Branch Stars

O. Yaron[1], A. Kovetz[1,2] and D. Prialnik[1]

[1]Department of Geophysics and Planetary Sciences, Sackler Faculty of Exact Sciences
[2]School of Physics & Astronomy, Sackler Faculty of Exact Sciences, Tel Aviv University, Israel

Abstract. Observational and theoretical investigations, performed especially over the last two decades, have strongly attributed the far-UV upturn phenomenon to low-mass, small-envelope, He-burning stars in Extreme Horizontal Branch (EHB) and subsequent evolutionary phases.

Using our new stellar evolution code – a code that follows through complete evolutionary tracks, Pre-MS to cooling WD – without any interruption or intervention, we are able to produce a wide array of EHB stars, lying at bluer ($T_{eff} \geqslant 20,000$ K) and less luminous positions on HRD, and also closely examine their post-HB evolution until the final cooling as White Dwarfs.

HB morphology is a complex multiple parameter problem. Two leading players, which seem to possess the ability to affect considerably positions of HB, are those of: 1.Helium abundance, and 2.mass-loss efficiency on the first giant branch. We focus here on the latter; thus, EHB stars are produced in our calculations by increasing the mass-loss rate on the RGB, to a state where prior to reaching core He flash conditions, only a very small H-rich envelope remains. The core flash takes place at hotter positions on the HRD, sometimes while already descending on the WD cooling curve. We show preliminary results for a range of initial masses ($M_{ZAMS} = 0.8 - 1.1\ M_\odot$) and for metallicities covering both populations I and II ($Z = 0.01 - 0.001$). The [M,Z] combinations have been chosen such that the masses would be above and close to typical MS turnoff masses (e.g. the estimation of $M_{TO} \simeq 0.85$ for NGC 2808), and also so that the ages at HB are of order of 10 ± 5 Gyr.

Keywords. stars: evolution, stars: horizontal-branch, stars: mass loss

1. Numerical Computations

Our new evolution code (Kovetz et al. 2008, in prep.) is capable of following through complete evolutionary calculations for a wide range of masses and metallicities. It incorporates up-to-date input physics (EOS, opacities, mass-loss recipes etc...), and, following Eggleton (1971), simultaneously solves the equations of structure and composition with a mass-distribution function, implementing an adaptive mesh.

Mass-loss on the RGB is according to the original Reimers (1975), whereas for post core He-burning stage we have chosen to use one of Bloecker (1995) expressions.

We executed a sequence of runs for each [M,Z] combination; adopting within each sequence increasing values of the mass-loss efficiency parameter – η_{Reim}. In all runs a core He-flash took place, whether normal – at tip of RGB, or delayed – at hotter positions; the end-state of all calculations being a cooling WD.

2. Results

In the left panels of Fig. 1 we show a sequence of evolutionary tracks for a Pop. II $M = 0.9\ M_\odot$ model with increasing mass-loss rates. For the lower mass-loss rates, a normal core He flash occurs at the tip of the RGB, whereas for the higher rates – thus smaller M_{env} by the onset of the core flash – we shift to obtaining "Post-Tip-Flashers",

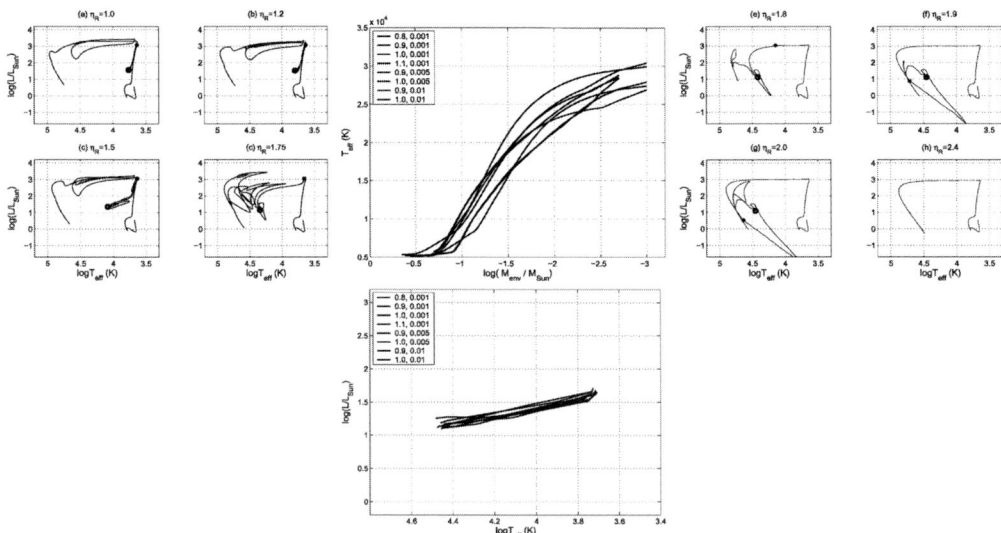

Figure 1. Left: Complete tracks on HRD – $M = 0.90\ M_\odot$, $Z = 0.001$, $Y = 0.24$ – for increasing mass-loss rates ($\eta_{Reim} = 1.0 - 2.4$). Red dots mark the onset of core He flash; blue circles mark HB positions. Age at HB is 10.1 Gyr. Bluest HB, obtained in panel (g), is at $T_{eff} = 2.85 \times 10^4$ K. Panels (d),(e) are "Post-Tip-Flashers", (f),(g) are clear "WD-Flashers"; panel (h) – no He-ignition, resulting in a He-WD. Top right: T_{eff} at HB vs. envelope mass M_{env} (at onset of core-He-flash), for the examined [M,Z] combinations (see legend). Bottom right: HB positions on HRD for the [M,Z] combinations, for increasing mass-loss rates, right to left.

followed by "WD-Flashers", till reaching a situation for which He-ignition does not take place. As apparent, the bluest EHB stars do not evolve to normal Asymptotic Giant Branch, but rather immediately to hotter (UV-bright) positions (e.g. "AGB-manque"). In some cases (and dependent on the invoked post-HB mass-loss recipe), an extended set of multiple last shell flashes may take place, before settling on the WD cooling curve.

With close agreement to former studies – e.g. Cruz *et al.* (1996), Brown *et al.* (2001), and based on our yet preliminary set of results, we can conclude the following:

– For both Populations and the examined masses, it is found possible to produce EHB stars by means of increasing the amount of mass that is lost during RGB.

– A transition from normal to delayed core flash occurs for M_{env} around $1 - 2 \times 10^{-2}\ M_\odot$.

– As apparent from the top right panel of the figure, envelope masses (at onset of core-flash) of below 0.01 $M_\odot$, all yield T_{eff} in excess of 20,000 K while on HB. (A max. T_{eff} of $\sim 30,000$ K was obtained for Pop. II with M_{env} of 0.001 $M_\odot$.)

– For a given initial mass; the lower the metallicity – Z, the higher the mass-loss rate required for obtaining significantly delayed core He flash.

– HB positions of all [M,Z] sequences form a tightly packed thick band, covering the extent of HB from T_{eff} of $\sim 5,000$ to over $\sim 30,000$ K.

References

Bloecker, T. 1995 *A&A* 297, 727
Brown, T. M., Sweigart, A. V., Lanz, T., Landsman, W. B., & Hubeny, I. 2001 *Apj* 562, 368
D'Cruz, N. L., Dorman, B., Rood, R. T., & O'Connell, R. W. 1996 *Apj* 466, 359
Eggleton, P. P. 1971 *MNRAS* 151, 351
Reimers, D. 1975 *MSRSL* 8, 369

Dust Size Effect On IR Colors Of AGB Stars

Huan Wang,[1] B. W. Jiang[1] and R. Szczerba[2]

[1] Department of Astronomy, Beijing Normal University, Beijing 100875, P.R. China
email: whbnu@mail.bnu.edu.cn, bjiang@bnu.edu.cn

[2] Nicolaus Copernicus Astronomical Center, PL-87-100 Torun, Poland
email: szczerba@ncac.torun.pl

Abstract. With the Mie theory and the radiative transfer model, we studied the effect of dust size on the infrared color indexes concerning special filters used in the space infrared missions and typical filters in the near-infrared, of AGB stars with typical oxygen-rich and carbon-rich dust shells. It is found the most affected bands are the near-infrared bands JHK and the Spitzer IRAC bands, meanwhile the wavebands with reference wavelength longer than $10\,\mu$m is little affected. The effect increases fast with the mass loss rate. We also discussed the potential to distinguish the O-rich and C-rich dusts, and the difference in IR colors between the AGB stars and other IR sources like YSOs and galaxies.

Keywords. stars: AGB and post-AGB, stars: mass-loss, (stars:) circumstellar matter

1. Introduction

The recent launch of space missions such as SPITZER and ASTRO-F boosts our knowledge in the infrared (IR) wave bands of celestial bodies significantly. In the mid-IR wave bands, these missions have such a high sensitivity and spatial resolution that they can detect individual AGB stars in Local Group galaxies. To understand the properties of the AGB stars in local galaxies, the correct explanation of photometric results is necessary. With the optical constants and radiative transfer models, the color indexes related to the special filters in the space missions were calculated by Gronewegen (2006). But with only one dust size of $0.1\,\mu$m, his results needs to be refined since it is established that the circumstellar dust has a wide distribution of size and the size of dust would influence the optical properties and therefore the dust temperature that decides the IR colors.

2. Calculation and Results

To study the effect of dust size on the IR colors of AGB stars, the dust opacity is calculated from the Mie theory and the radiative transfer is solved with the dust shells taken into account (Steffen *et al.*, 1997) for a series of dust radius, i.e., a = 0.005, 0.01, 0.035, and from 0.05 to 0.9 μm with a step of 0.05 μm. The 0.035 μm is a reference size, by which the emergent SED is the same as a commonly adopted power-law size distribution of $a^{-3.5}$ between 0.005 and 0.25 μm. This size is derived from density-weighted radius and confirmed in the calculation. The species of dust includes 100% silicate (Laor & Draine 1993) for an oxygen-rich shell and 100% amorphous carbon (AMC) (Jaeger *et al.* 1998) for a carbon-rich shell. To concentrate on the size effect, the stellar parameters are universal and typical for the AGB stars: $L = 4390 L_\odot$, $T_{\rm eff} = 3000$K, $d = 8.5$kpc, $\dot{M} = 10^{-6} M_\odot/$yr, where the distance refers to the center of the Galaxy and the mass loss rate is typical for an AGB star with moderate mass. The calculated SED is folded with the filters' response for the expected observational magnitude in the corresponding bands. All the filters of the missions which produce abundant data are tested, including 2MASS/JHKs, Spitzer/IRAC/MIPS, Astro-F/IRC/MIR-L/MIR-S/MIR-FIS, IRAS, MSX/ACDE.

It is found that the wave bands affected most by the dust size are the near-IR bands and Spitzer/IRAC bands, while the wavebands with reference wavelength longer than

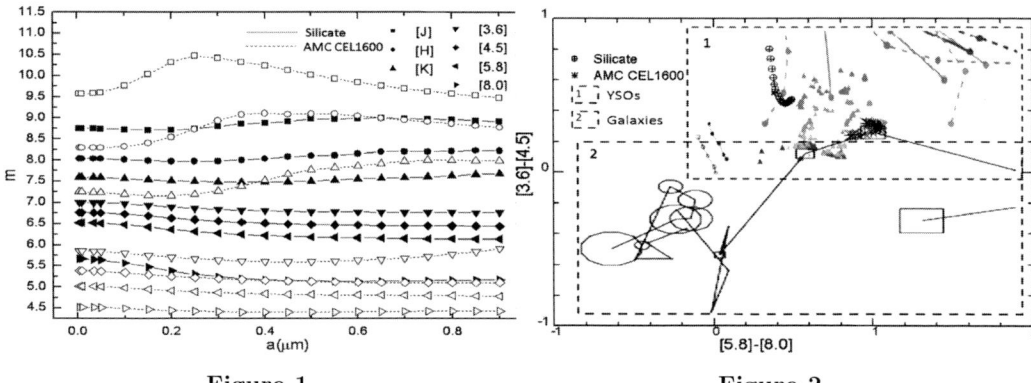

Figure 1.   **Figure 2.**

10 μm are little affected by the dust size in the selected range. This result is consistent with the rule that dust interacts mainly with the radiation at a wavelength comparable to its size, i.e. $\lambda \sim 2\pi a$. In Figure 1, the change of apparent magnitudes is shown for the wavebands which vary significantly with the dust size. The variations in the JHKs and Spitzer IRAC bands are 0.98 mag, 0.82 mag and 0.31 mag at J, H and [3.6]. Consequently, the colors related to these bands may vary by similar orders or even larger. It should be pointed out that the order of variation increases fast with the mass loss rate. For a mass loss rate of $5 \times 10^{-6} M_\odot$/yr, the variation with dust size is exceedingly large.

To distinguish the silicate and carbon dust has long been pursued. The colors which have large difference for these two species of dust are J-K, J-H, [5.8]–[8.0] and [3.6]–[4.5].

Another purpose of such calculation is to separate AGB stars from other IR sources, mainly the young stellar objects (YSOs) and extra-galaxies. In Figure 2 are compared the IRAC colors of AGB stars with them, where the colors of YSOs and galaxies are taken from Allen *et al.* (2004) and Sajina *et al.* (2005). It can be apparently seen that there is a good separation between the AMC and the silicate dust. Meanwhile the AGB stars can also be separated from the galaxies. But they are mixed with YSOs in particular with thick CSEs in this color-color diagram.

3. Summary

(a) The dust size affects the colors related to the wavelength shorter than 10 micron, in particular the near-IR bands such as the JHK and Spitzer/IRAC bands.

(b) The dust size effect increases with the mass loss rate. It is negligible when $\dot{M} < 10^{-7} M_\odot$/yr, but increases by 0.6 mag in J-K at the rate of $5 \times 10^{-7} M_\odot$/yr.

(c) The effect of O-rich and C-rich dust on the IR colors is different, which provides a potential to distinguish the two chemistry type stars from the IR colors, with the Spitzer/IRAC colors suggested.

(d) The ranges of Spitzer/IRAC color indexes are calculated to be different from extra-galaxies.

Acknowledgements: This work is supported by the The Ministry of Science and Technology of the People's republic of China through grant 2007CB815406.

References

Groenewegen, M. A. T. 2006, *AA* 448, 181
Steffen, M., Szczerba, R., Menshchikov, A., & Schoenberner, D. 1997, *AApS* 126, 39
Laor, A., & Draine, B. 1993, *ApJ* 402, 441
Jager, C., Mutschke, H., & Henning, Th. 1998, *AA* 332, 291
Allen, L. E., *et al.* 2004, *ApJS* 154, 363
Sajina, A., Lacy, M., & Scott, D. 2005, *ApJ* 621, 256

The age-metallicity relation in the thin disk

Ji Li, Bo Liang and Weishi Fan

College of Physics and Information Engineering, Hebei Normal University, Shijiazhaung
050016, P. R. China
email: liji@mail.hebtu.edu.cn

Abstract. With two stellar sample A and B, the age-metallicity relation (AMR) in the Galactic thin disk is investigated. The results show two different AMRs: one is a nearly flat AMR from photometric analysis of sample A, the other is an obvious AMR derived from spectroscoipic analysis of sample B.

Keywords. Galaxy: solar neighbourhood, Galaxy: disk, Galaxy: kinematics and dynamics, Galaxy: evolution, Stars: fundamental parameters

1. Introduction

Many studies have found there is a clear relation between the ages and the metallicities of the solar neighbourhood disk stars (e.g. Twarog 1980; Meusinger et al. 1991; Ng & Bertelli 1998; Rocha-Pinto et al. 2000). In contrast to this Edvardsson et al. (1993) found no particular evidence for an AMR in the solar neighbourhood and other two large sample investigations(Feltzing et al. 2001(F01), Nordström et al. 2004(N04)) confirmed this.

Recently, Bensby et al. (2004) investigated a sample of 229 nearby thick disk stars and found there is indeed an AMR in the thick disk. Such a clear AMR is also confirmed by Haywood (2006). The question then arises: could it be so that the lack of a relation between ages and metallicities for stars in the solar neighbourhood is in fact a population effect? Is there a similar AMR in the thin disk as that in the thick disk? In the study presented here we will address the question of a relation between ages and metallicities for stars that are kinematically selected to resemble the thin disk closely.

2. Stellar sample

To obtain an accurate and reliable AMR, a comprehensive and unbiased stellar sample is important. Our stellar sample includes two samples. Sample A consists of 4007 nearby stars selected from the common stars of two large sample works F01 and N04 by requiring the common stars showing consistent ages within the difference of ± 3Gyr. Sample B consists of 641 stars from 13 spectroscopic works (Edvardsson et al. (1993), Nissen & Schuster 1997, Jehin et al.1999, Fulbright 2000, Mishenina & Kyukh 2001, Reddy et al. 2003, Reddy et al. 2006, Bensby et al. 2005, Grraton et al. 2003, Jonsell et al. 2005, Brewer et al. 2006, Fuhrmann et al. 2004, Gehren et al. 2004).

3. Results

The population membership of the sample stars were determined with pure kinematical criteria(Bensby et al. 2004). Figure 1 shows the AMR for 3856 thin disk stars from sample A, where the ages are the average values corresponding to that from F01 and N04, [Fe/H] are photometric metallicities from N04. Figure 2 shows the AMR of 434 thin disk stars

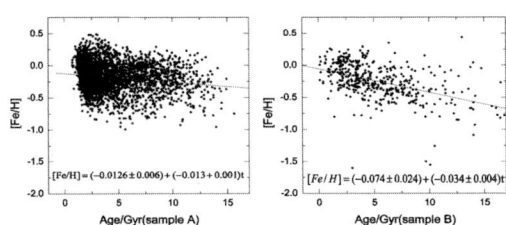

Figure 1. AMR in thin disk stars.

from sample B, where the stellar ages are our re-calculations using Y^2-isochrones (Yi et al. 2003), [Fe/H] are the spectroscopic metallicities adopted from the original references.

4. Discussion and conclusion

Figure 1 shows there is an age-metallicity relation present in the Galactic thin disk either for sample A or sample B, but the AMR derived from sample B is more declining than that derived from sample A. This result means that the AMR of stars in the thin disk suffering from a significant population effect. To fully understand the inward nature of the AMR in thin disk, more complete and accurate observations are needed, and the methods used to compute stellar ages and their uncertainties, as well as the choice and verification of stellar models are far more urgent issues at present.

Acknowledgements

Supported by the National Natural Science Foundation under grant number 10778616.

References

Bensby, T., Feltzing S., & Lundström I. 2004, *A&A* 421, 969
Bensby, T., Feltzing S., & Lundström I., et al. 2005, *A&A* 433, 185
Brewer, Mary-Margaret & Carney Bruce W. 2006, *AJ* 131, 431
Edvardsson, B., Andersen, J., Gustafsson, B., et al. 1993 *A&A* 275, 101
Feltzing, S., Holmberg, J., & Hurley J. R. 1983, *A&A* 377, 911
Fuhrmann, K. 2004, *AN* 325, 3
Fulbright, J. P. 2000, *AJ* 120, 1841
Gehren, T., Liang, Y. C., & Shi, J. R. et al. 2004, *A&A* 413, 1045.
Gratton, R. G, Carretta E., & Claudi R., et al. 2003, *A&A* 404, 187
Haywood, M. 2006, *MNRAS* 371, 1760
Jehin, E., Magain, P., & Neuforge, C., et al. 1999, *A&A* 341, 241
Jonsell, K., Edvardsson, B., & Gustafsson, B., et al. 2005, *A&A* 440, 321
Meusinger, H., Srecklum, B., & Reimann, H-G. 1991, *A&A* 245, 57
Mishenina, T. V. & Kovtyukh, V. V. 2001, *A&A* 370, 951
Ng Y. K. & Bertelli, G. 1998, *A&A* 329, 943
Nordström B., Mayor M., & Andersen, J., et al. 2004, *A&A* 418, 989
Reddy, B. E., Tomkin, J., & Lambert, D. L., et al. 2003, *MNRAS* 340, 304
Reddy, B. E., Lambert, D. L., & Prieto, C. A. 2006, *MNRAS* 367, 1329
Rocha-Pinto, H. J., Maciel, W. J., & Scalo, J., et al. 2000, *A&A* 358, 850
Twarog, B. A. 1980, *ApJS* 44, 1
Yi, S., Demarque, P., & Kim, Y. C., et al. 2003, *ApJS* 144, 256

The Variability Of RSG : HV2576

Ming Yang and B. W. Jiang

Department of Astronomy, Beijing Normal University, Beijing, 100875, China
email: myang@mail.bnu.edu.cn and bjiang@bnu.edu.cn

Abstract. The Harvard Variable HV2576, as a red supergiant in the Large Magallanic Cloud, has a complex light variation, which is not explained well. The existing one-period non-linear pulsation model deviated clearly from its light curve. We tried to fit the light curve by a superposition of several harmonic pulsations. By using the PDM, PERIOD04 and SparSpec codes to analyze the light variation, two periods with 525 and 261 days are well established. Furthermore, the 261-day period is found to change according to the wavelet analysis. In addition, the noise obeying the 1/f law. Based on all these facts, we suggest that the light variation of HV2576 may be due to huge convection cells that interplay with oscillation or the third very long period.

Keywords. stars: oscillations, stars: supergiants, stars: variables: other

1. Introduction

Red supergiants (RSGs) are evolved, moderately massive (10–30 $M_\odot$) He-burning stars. They have long been known for their brightness variation which is usually attributed to the radial pulsation. Moreover, an additional irregular variation occurs as well which may be caused by huge convection cell. The Harvard Variable HV2576 is a red supergiant in the Large Magallanic Cloud. It has a complex light variation. Wood (2006) tried to fit its light curve by a one-period, non-linear pulsation model, but the model deviates clearly from observed light curve. So we try to find out the reliable periods.

2. Data Analysis

The photometric data of HV2576 is taken from the MACHO database, which covers a duration of about 8 years. We used PERIOD04 (Lenz & Breger 2005) to carry out standard iterative sine wave fitting (Fig. 1) together with the error-weighted analysis. From the power spectra, two distinct peaks are identified, corresponding to the periods 525 and 261 days with amplitude of 0.747 mag and 0.751 mag respectively. There would be several more periods from the PERIOD04 analysis if only judged from the S/N of the peaks. To clarify the true periods from spurious ones, the PDM (Phase Dispersion Minimization, Stellingwerf 1978) and SparSpect (Bourguignon, Carfantan, & Bohm 2007) codes are used as well. Both

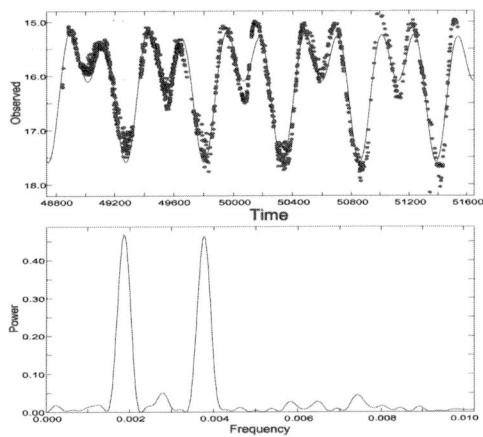

Figure 1: V-band light curve fitting with two periods and the power spectra derived from PERIOD04

confirmed the 525-day period, which is consistent with the 530-day period obtained by Wood (2006). On the other hand, the 261-day period is present in these two methods as well, although it was not suggested in Wood (2006). From the consistency of these three methods, this 261-day period is taken as a real secondary period. Besides, the SparSpec code yields a third period of 174-day if error is weighted, but it is insignificant with unweighted error, neither in PERIOD04, thus abandoned.

With the two periods of 525-day and 261-day, the light curve is fitted as shown in Fig. 1, with the amplitude of 0.747 mag and 0.751 mag and the phase of 0.514 and 0.629 respectively. The fit is much improved than the one-period non-linear model, but still has some departure from the observed light curve. By eyes, the light curve shows a trend of change both in the amplitude and period. To check whether such trend is true, the WWZ (Weighted Wavelet Z-trasform) from AAVSO analysis is performed. The wavelet map shown in Fig. 2 again clearly re-confirm the two distinct periods derived above and indicate that the 261-day period has a trend to change. But the reason why the amplitude and period slowly grow is not clear. One guess is that the convection or a third long period may lead to such situation.

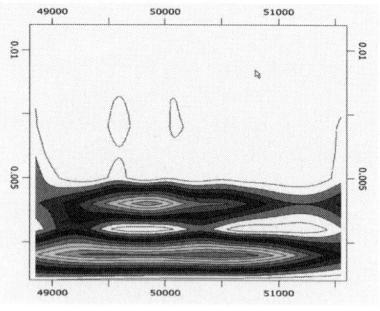

Figure 2: Contour wavelet map

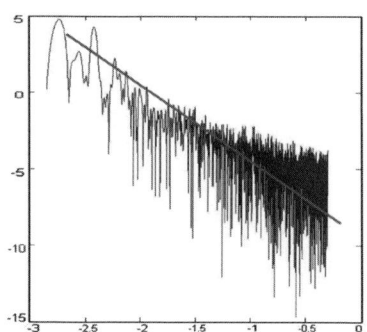

Figure 3: Power density spectrum in log-log axis shows the 1/f noise

We also analyzed whether there is a 1/f noise component in the power spectra which imply the irregular variation owing to large convective cells, similarly to solar granulation background. In Fig. 3, we plot PDS in log-log axis. The trend of 1/f noise can be observed.

3. Conclusion

Analyzing the light curve of HV2576 with the powerful time-series analysis applications, two periods of 525 and 261 days are well established. In addition, the amplitude and period are found to vary. We suggest that the light variation of HV2576 may be due to huge convection cells that interplay with oscillation or the third long period.

Acknowledgements

This work is supported by the NSFC grant No.10778601.

References

Wood, P. R. 2006, *Mem. S.A.It.* 77, 76
Kiss, L. L., Szabo, Gy. M., & Bedding, T. R. 2006, *Mon. Not. R. Astron. Soc.* 372, 1721
Templeton, M. 2004, *JAAVSO* 32, 41
Stellingwerf, R. F. 1978, *ApJ* 224, 953
Lenz, P. & Breger, M. 2005, *CoAst.* 146, 53
Bourguignon, S., Carfantan, H., & Bohm, T. 2007, *AA* 462, 379

The Dynamics Of Galactic Globular Cluster

Chen Ding

National Time Service Center, Chinese Academy of Science. Xi'an 710600, China
email: ding@ntsc.ac.cn

Abstract. We have used the Hubble Space Telescope (HST) to measure proper motion of the globular cluster NGC 6656 (M22) with respect to the background bulge stars and its internal velocity dispersion profile. With the space velocity of $(\Pi, \Theta, W) = (184\pm3, 209\pm14, 132\pm15)\,\mathrm{km\,s^{-1}}$, we also calculate the orbit of the cluster. The central velocity dispersion in both components of the proper motion of cluster stars is $16.99\,\mathrm{km\,s^{-1}}$. We derive the mass-to-ration $(M/L)\sim1.7$ which is relatively higher than the past works.

Keywords. astrometry, globular clusters: general

NGC6656 (M22) is one of the nearest globular clusters from the Sun, being about 3 kpc distant (Peterson & Cudworth 1994; Harris 2003). The metallicity of this cluster is rather low ([Fe/H] = -1.62 ± 0.08, Richter, Hilker, & Richtler 1999). It should have belong to the population of "halo" clusters (Zinn 1985) but its galactic orbit (Douphole et al. 1996; Dinescu, Girard, & van Altena 1999) reveal some ambiguous features of "disk", or "bulge" (Côté 1999) clusters. Proper motions give us two more components to each star's six-dimensional phase-space coordinate. When combined with its radial velocity and position, the phase space coordinate of a star can be fully determined. It can lead to greater understanding of the structure and evolution of globular clusters.

Our proper motions are based on data of three epoch observations taken with Wide Field Planetary Camera 2 (WFPC2) aboard the HST. The first epoch data consist of four frames taken as part of program GO 5404 on 7 April 1994 through the F502N filter. The second epoch data are taken by program GO 7615 from 22 February to 15 June 1999 through F814W and F606W filter. This program also took a few observations in February 2000, which provide an additional epoch data that strengthen the proper-motion measurements.

Thanks to the well overlap of all epoch frames, we have enough stars in each CCD field that it is not necessary to combine different CCDs to obtain the more stars. We just match the same CCD of all epochs. Taking a frame taken in 2000 as the reference, we then construct a matched star list and calculate the relative proper motions of stars in each CCD field.

By means of the Gaussian fit along the RA and Dec axis, respectively, we obtained the relative proper motion of NGC 6656 with respect to the bulge of $\mu_\alpha \cos\delta = 10.19 \pm 0.20\,\mathrm{mas\,yr^{-1}}$ and $\mu_\delta = -3.34 \pm 0.10\,\mathrm{mas\,yr^{-1}}$. Giving the radial velocity of NGC 6656, $V_r = -148.9 \pm 0.4\,\mathrm{km\,s^{-1}}$ and the Sun motion of $(U_\odot, V_\odot, W_\odot) = (10, 15, 8)\,\mathrm{km\,s^{-1}}$,[2,18] the rotational velocity of the Local Standard of Rest $V_{LSR} = 220\,\mathrm{km\,s^{-1}}$ and the solar distance to the galactic center 8.0 kpc,[19] and the cluster distance $d = 3.2 \pm 0.3\,\mathrm{kpc}$ from the Harris'[2] catalog of clusters, we got the three components of space velocity for NGC 6656 with respect to the bulge: $(\Pi, \Theta, W) = (184 \pm 3, 209 \pm 14, -132 \pm 15)\,\mathrm{km\,s^{-1}}$. Errors in velocities include the uncertainties in the proper motions, radial velocities, and the distance of the cluster.

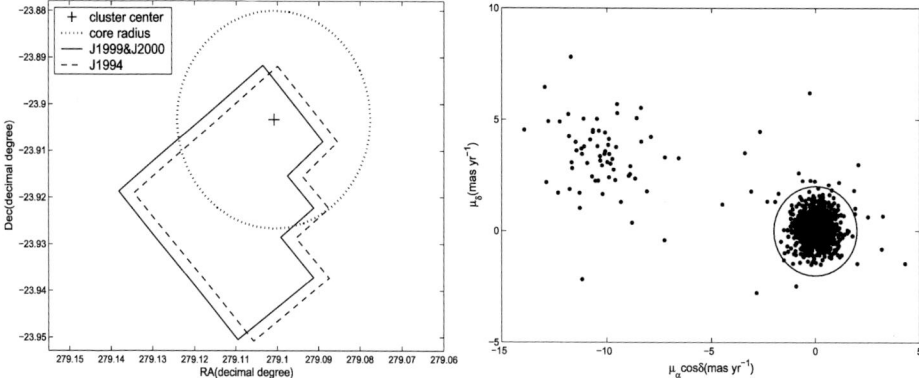

Figure 1. [A]: Location of HST observations of NGC 6656 from different epochs. [B]: Vector-point-diagram of relative proper motions in the J2000 equatorial coordinate system, the stars in the circle of radius $2\,\mathrm{mas\,yr^{-1}}$ centered at $(0, 0)$ are considered to be the cluster stars, while the stars with $\mu_\alpha \cos\delta < -5\,\mathrm{mas\,yr^{-1}}$ are mainly bulge stars.

Using three epoch WFPC2 images, we measured the relative proper motion of NGC 6656 with respect to the bulge and the internal velocity dispersion of the cluster. Combining the result of proper motion with the motion of the Sun, rotation of the Local Standard of Rest and radial velocity of the cluster, we derived the 3-dimensional motion of the cluster. Given the gravitational potential function of our Galaxy, the orbit of the cluster during the past 10 Gyr was calculated. From the results we found, NGC 6656 has a moderate eccentricity and relatively lower obliquity orbit and is near its perigalacticon now. More observations are needed to confirm the results presented here: longer time baseline can further improve the precision of the proper motion; a larger field will contain more cluster and field stars to give a more distinct separation.

Acknowledgements

We would like to thank Philip Warner and Frank Valdes from STScI for the many help on the data reduction, and Jay Anderson, Jon Holtzman, William E. Harris, Peter Stetson and Gordon Drukier for useful discussion and comments.

References

Anderson, J. & King, I. R. 2003 *Publication of the Astronomical Society of the Pacific* 115, 113
Harris, W. E., 2003 http://physun.mcmaster.ca/ harris/mwgc.dat
Trager, S. C., Djorgovski, S., & King, I. R., 1993 *Structure and Dynamics of globular clusters*, ed. S G Djorgovski and G Meylan (San Francisco: ASP) 347,
Richter, P, Hilker, M., & Richtler, T., 1999 *Astron. Astrophys.* 350, 476
Zinn, R. 1985 *Astrophys. J.* 293, 424
Peterson, R. C. & Cudworth, K. M. 1994 *Astrophys. J.* 420, 612
Dinescu, D. I., Girard, T. M., & van Altena, W. F., 1999 *Astron. J.* 117, 1792
Allen, C. & Santillan, A., 1991 *Rev. Mexicana. Astron. Astrof.* 22, 255
Drukier, G. A., Bailyn, C. D., van Altena, W. F., & Girard, T. M. 2003 *Astron. J.* 125, 2559
Peterson, C. J. & King, I. R., 1975 *Astron. J.* 80, 427

# Mass loss and evolution of hot massive stars

Jorick S. Vink[1]

[1] Armagh Observatory, College Hill, Armagh, BT61 9DG, Northern Ireland, United Kingdom
email: jsv@arm.ac.uk

Abstract. We discuss the role of mass loss for the evolution of the most massive stars, highlighting the role of the predicted *bi-stability* jump that might be relevant for the evolution of rotational velocities during or just after the main sequence. This mechanism is also proposed as an explanation for the mass-loss variations seen in the winds from Luminous Blue Variables (LBVs). These might be relevant for the quasi-sinusoidal modulations seen in a number of recent transitional supernovae (SNe), as well as for the double-throughed absorption profile recently discovered in the Hα line of SN 2005gj. Finally, we discuss the role of metallicity via the Z-dependent character of their winds, during both the initial and final (Wolf-Rayet) phases of evolution, with implications for the angular momentum evolution of the progenitor stars of long gamma-ray bursts (GRBs).

Keywords. Stars: winds, stars: mass loss, stars: supergiants, stars: rotation, stars: supernovae

1. Introduction

Mass loss has a major effect on the evolution of stars of all initial masses, however its effect is most prominent for the more massive stars due to their large luminosities – in close proximity to the Eddington limit. Mass loss is relevant both in terms of evolutionary pathways as well as the properties of the pre-supernova (SN) circumstellar environments. In this contribution, we first discuss mass-loss predictions that are relevant for predicting the forward evolution of massive stars. There are actually two aspects that need to be accounted for: (i) the loss of *mass* as winds "peel off" the star's outer layers (Conti 1976), but as massive stars start their evolution as rapid rotators, also (ii) the associated loss of *angular momentum* (e.g. Langer 1998).

Towards the end of the main sequence massive stars encounter the so-called bi-stability jump, for which we discuss the implications of the loss of angular momentum. For the final stages, the evolution of angular momentum is particularly relevant for our understanding of the long gamma-ray burst (GRB) phenomenon, as the popular collapsar model (MacFadyen & Woosley 1999) requires the core of the progenitor star to be rapidly rotating before collapse (but see also Lee & Ramirez-Ruiz 2006). A key parameter in the story is that of metallicity, due to the Z-dependence of radiation-driven winds during both the main sequence and evolved Luminous Blue Variable (LBV) and Wolf-Rayet (WR) phases of massive star evolution.

In the canonical scenario of massive stars, LBVs represent a transitional phase between the main sequence and the WR phase, however we also discuss the possibility for a new evolutionary paradigm in which the variable winds of LBVs might betray themselves as the *direct* progenitors of SNe.

2. Mass loss predictions

The evolution of a massive star, with $M > 30\ M_\odot$ is largely determined by the strength of their winds, which depends on the luminosity (L), mass (M), and metallicity Z.

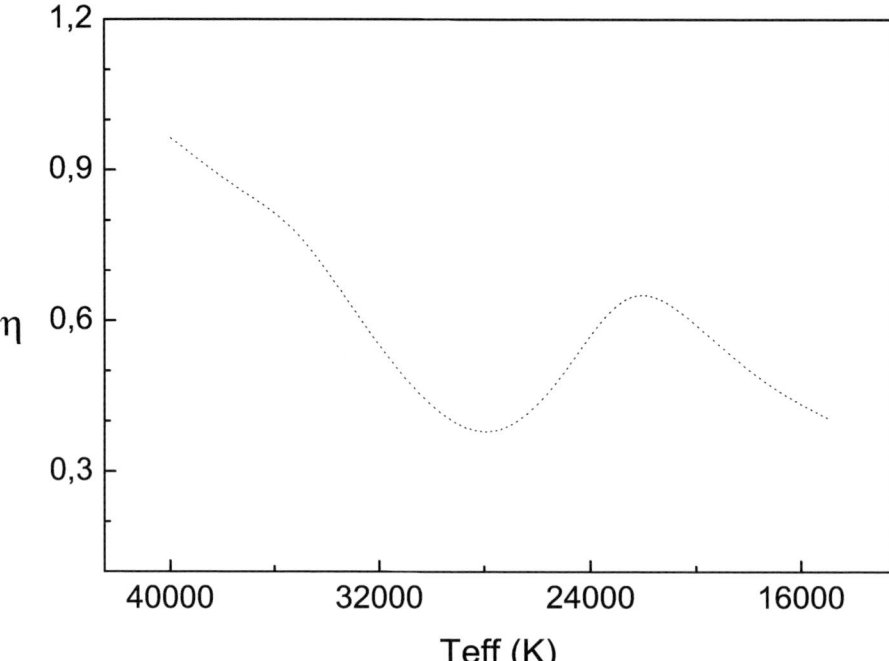

Figure 1. Predicted mass-loss behaviour as a function of effective temperature. Note the position of the local maximum at about 22000 K, coincident with the position of the empirical bi-stability jump at spectral type B1 (Lamers et al. 1995).

The Fe-group elements are particularly efficient scatterers of photons at specific line frequencies responsible for the amount of mass loss, whilst the CNO elements set the wind terminal velocity (Vink et al. 1999). There are currently two basic methods in use for predicting the mass-loss rates from massive stars, which have been reviewed in detail by Vink (2006). In short, the first method concerns the modified-CAK (Castor et al. 1975) method (Kudritzki & Puls 2000), the second one involves the Monte Carlo approach (Abbott & Lucy 1985, Vink et al. 2000). Both methods have their pros and cons. In the first approach, the wind hydrodynamics are more or less self-consistently solved for (albeit using depth-dependent force multiplier parameters), however multi-line scattering is not accounted for. This aspect is included in the second approach, where the line acceleration is calculated for all radii, although most Monte Carlo predictions do not properly account for the wind hydrodynamics (but see Vink et al. 1999).

Vink et al. (1999,2000) predicted the mass-loss rates of OB supergiants as a function of the stellar parameters (L, M, and $T_{\rm eff}$) including multiple scatterings on line and continuum opacity with a Monte Carlo approach and their mass-loss rates were computed as a function of the wind *terminal velocity* – a parameter accurately (within 10%) retrievable from ultraviolet P Cygni line profiles. This is in stark contrast to empirical *mass-loss rates* which are highly uncertain (by factors up to 3-10) due to uncertainties in wind ionization and wind inhomogeneities (see Hamann et al. 2008 for a recent overview on the issue of wind clumping). The Vink et al. mass-loss rates are found to scale as:

$$\dot{M} \propto L^{2.2} \, M^{-1.3} \, T_{\rm eff}^{\ 1} \, (v_\infty/v_{\rm esc})^{-1.3} \quad (2.1)$$

showing that $\dot{M}$ depends rather steeply on the stellar luminosity ($L^{2.2}$). The reason for this is that brighter stars have denser winds and Monte Carlo predictions yield an increasingly larger mass-loss rate than modified-CAK predictions. The Vink et al. mass-loss recipes are widely used in models for massive star evolution (e.g. Meynet & Maeder 2003, Limongi & Chieffi 2006, Eldridge & Vink 2006, Brott et al. in prep.).

Figure 1 depicts how the predicted mass-loss rates vary as a function of $T_{\rm eff}$ and thus how a massive star may find its wind change during its coarse of evolution. The predictions are expressed in terms of the wind momentum efficiency, or wind performance number, $\frac{\dot{M} v_\infty}{L/c}$. The figure shows a declining wind efficiency with $T_{\rm eff}$. At high temperatures ($\sim 40\,000$ K) the wind momentum is large due to the fact that the radiative flux and the opacity have a good "match" with respect to their wavelength distribution. However, when $T_{\rm eff}$ drops, the stellar flux moves away from its maximum towards lower (optical) wavelengths, which results in an ever-growing mismatch between the flux and the ultraviolet line opacity. At $\sim 25\,000$ K, a sudden mass-loss discontinuity is noted. This is due to an increased Fe opacity when Fe IV recombines, and the more abundant Fe III lines provide most of the line force in the inner wind (Vink et al. 1999). This "bi-stability jump" (Pauldrach & Puls 1990) may recently have been confirmed in radio data that appear to confirm the presence of a local maximum (Benaglia et al. 2007, Markova & Puls 2008), however it should also be noted that the predicted values below the temperature of the jump appear to be much larger than those found from empirical modelling by up to a factor of 10 (Vink et al. 2000, Trundle & Lennon 2005, Crowther et al. 2006).

The bi-stability jump where winds change from a low $\dot{M}$, fast wind, to a high $\dot{M}$, slow wind may comprise an important ingredient for stellar evolution calculations when stars evolve off their main-sequence positions towards the lower $T_{\rm eff}$ part of the Hertzsprung-Russell diagram (HRD). This is not only relevant for their *mass* loss, but also for the associated loss of *angular momentum*.

3. Angular momentum loss

Massive stars rotate rapidly at birth (with $v_{\rm rot} \simeq 200$-300 km s^{-1}) and remain relatively rapid rotators throughout their main-sequence lifetimes. Obviously, $v_{\rm rot}$ decreases due to the angular momentum loss via stellar winds, which implies that the effects are largest at the highest initial masses and luminosities, and metallicities. Furthermore, when the objects evolve off the main sequence, they swell up to become (super)giants, and $v_{\rm rot}$ is anticipated to drop due to the increase in stellar radius (e.g. Hunter et al. 2008). However, is this the entire story or is the bi-stability jump also of relevance?

Figure 2 shows a recent figure from Markova & Puls (2008) showing how the rotational velocity of Galactic OB supergiants depends on spectral type. It can be noted that $v \sin i$ drops from ~ 100 km s^{-1} to ~ 50 km s^{-1} close to spectral type B1 – the position of the bi-stability jump (Lamers et al. 1995, Crowther et al. 2006). As we are interested in checking whether the predicted jump in mass loss by a factor of five at the bi-stability jump (Vink et al. 1999, 2000) might potentially explain the steep drop in rotation due to the loss of angular momentum evolutionary tracks were computed with this in mind (Brott et al. in prep.). Figure 3 shows both the Vink et al. (2000) mass-loss rates (dotted line) and the predicted rotational velocity (solid line) of a Galactic 40 $M_\odot$ star which had a initial rotational velocity of 265 km s^{-1} on the zero-age main-sequence (ZAMS). In order for the angular momentum removal to be maximal, a rather large overshooting parameter of 0.335 of a pressure scale-height was employed, as this provides a long time interval on the MS for the mass loss to be most efficient.

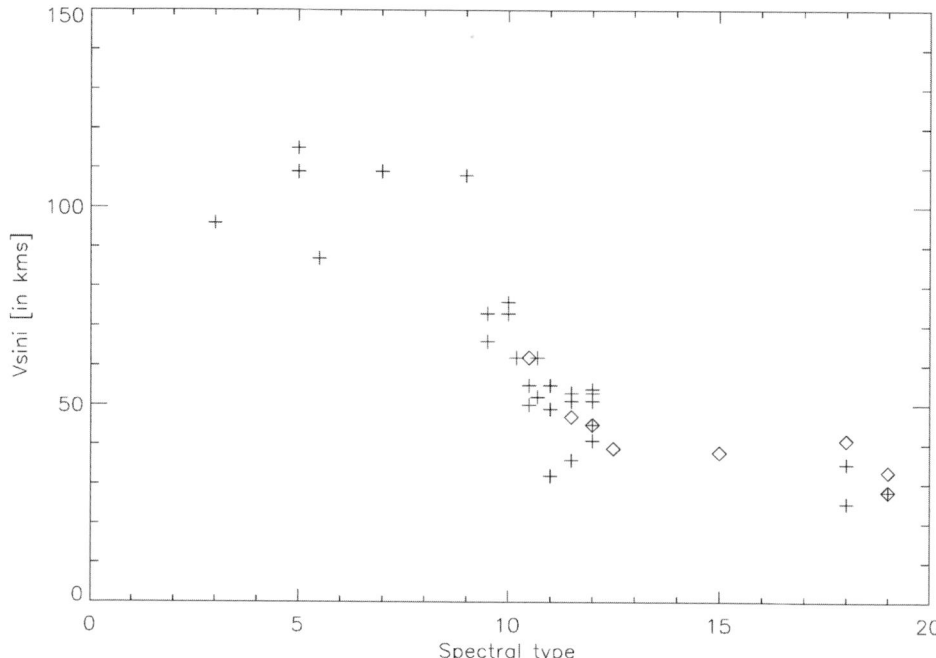

Figure 2. The projected rotational velocity, $v\sin i$ of Galactic B supergiants as a function of spectral type, starting with O3 to O9 and then switching to B0 (at 10) and to A0 (at 20). It can be noted that $v\sin i$ drops from $\sim$100 km s^{-1} to $\sim$50 km s^{-1} at around spectral type B0, which is close to the spectral type of the bi-stability jump B1 (Lamers et al. 1995). The figure has been adapted from Markova & Puls (2008).

As an aside, we note that the drop in rotational velocities at a specific temperature is reminiscent of a situation encountered for Horizontal Branch stars (e.g. Behr et al. 2000). Vink (2007a) recently reviewed whether the steep jump at 10 000 K could be due to the onset of radiation-driven winds as a result of the higher metal content for objects warmer than the jump temperature.

4. Could the changing winds of Luminous Blue Variables change the evolutionary paradigm?

Above we considered the effects of the bi-stability jump for main-sequence stars evolving from hot to cool temperatures. However, even more dramatic effects might occur for objects that have already lost a large fraction of their initial mass, finding themselves in close proximity to the Eddington limit. These Luminous Blue Variables change their effective temperatures on a variety of timescales (Humphreys & Davidson 1994, Vink 2008). The micro variations are not noticably different from small-amplitude variations in other BA supergiants. At the other extreme, we find the super-outburst of objects such as Eta Car, sometimes referred to as SN-impostors, when observed in other galaxies (Van Dyk et al. 2000, Maund et al. 2006). We note that only two of these super-outbursts have been identified in the Milky Way (the eruptions of Eta Car in the 19th and P Cyg in the 17th century). The mass-loss rates involved in these giant eruptions are of order 0.1 $M_\odot$yr^{-1} and are too large to be explained by line acceleration. However, continuum-driven winds may well be able to provide the necessary driving (Smith & Owocki 2006). Most typifying for the class are the S Dor variations, where objects vary on timescales

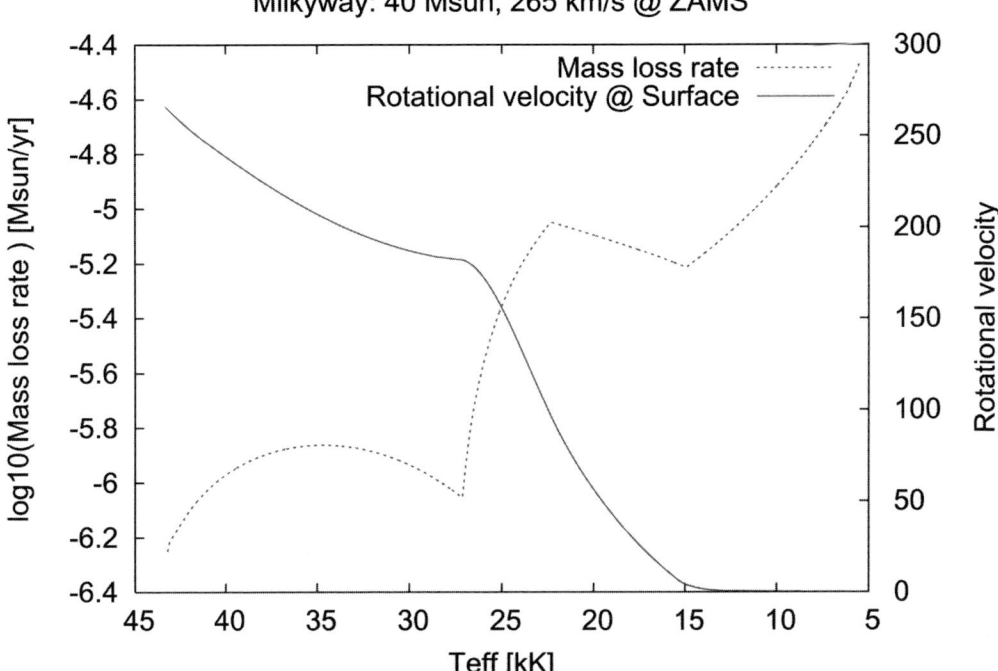

Figure 3. Mass-loss rate (dotted line) and rotational velocity (solid line) of a Galactic 40 $M_\odot$ star which had a initial rotational velocity of 265 km s^{-1} on the ZAMS. A rather large overshootig parameter of 0.335 of a pressure scale-height was employed. The mixing efficiency of fc=0.0228 represents an efficiency factor with which the diffusion coefficients of the different rotational mixing processes are multiplied (see Hunter *et al.* 2008, Brott *et al.*, in prep. for further details.) Note that the second mass-loss increase below 15 000 K is due to a swith from the Vink *et al.* (2000) recipe to a calibration by Nieuwenhuizen & de Jager (1990) – consistent with the position of the second bistability jump predicted by Vink *et al.* and observed by Lamers *et al.* (1995) around spectral type A0.

of years to decades. When LBVs such as AG Car – one of the prototypes – change their radii on their S Dor timescales, they show large mass-loss variations (Stahl *et al.* 2001). Such variable wind behaviour has qualitatively been reproduced by radiation-driven wind models of Vink & de Koter (2002). We anticipate that this type of wind behaviour may result in a circumstellar medium consisting of concentric shells with varying densities, which may have ramifications for the end-points of massive stars. Kotak & Vink (2006) suggested that the quasi-periodic modulations seen in the radio lightcurves of some supernovae (SNe), such as 2001ig (Ryder *et al.* 2004) and 2003bg (Soderberg *et al.* 2006) may indicate that LBVs could be the *direct* progenitors of some SNe.

At first this seems to contradict stellar evolution calculations, which do not predict LBVs to explode, and such a scenario was until recently considered "wildly speculative" (Smith & Owocki 2006). However, the intruiging supernova 2006jc showed a giant eruption just 2 years prior to explosion (Pastorello *et al.* 2007, Foley *et al.* 2007) which may add some confidence to the Kotak & Vink suggestion that LBVs might explode. There have been a number of other studies suggestting that LBVs may explode. Gal Yam *et al.* (2007) reported the detection of a most luminous progenitor of SN 2005gl. Although the properties of the potential progenitor star are consistent with that of an LBV, a hypergiant cannot be classified as an LBV until it has shown S Dor or Eta-Car-type variability (Humphreys & Davidson 1994, Vink 2008). Another interesting hint that LBVs may explode arises from the similarities in LBV nebula morphologies and the circumstellar medium of SN 1987A (Smith 2007). Finally, one the most luminous supernova ever recorded, SN 2006gy (e.g. Ofek *et al.* 2007) may also have been an Eta Carinae type LBV (Smith *et al.* 2007).

As current state-of-the-art stellar evolution calculations do not predict LBVs to explode, this represents a major unresolved problem in the physics of massive stars. In the generally upheld picture for the evolution of the most massive stars, LBVs are considered "transitional" objects in a phase before entering the He-burning WR stage, by the end of which the WR star is expected to explode as a type Ib/c supernova. The reason for the common (Conti 1976) scenario:

$$O \rightarrow \text{LBV} \rightarrow \text{WR} \rightarrow \text{SN},$$

is that LBVs are He (and N) rich compared to O stars, yet H-rich (thus He-poor) compared to WR stars. The situation is even more complex as there is also a group of high-luminosity late-type WR stars which are H-rich and seem closely related to the classical LBVs in quiescence. In particular, we note that the R127 was a late-type WN star (Of/WN9), before it went into outburst. Nonetheless, the picture of a relatively short-lived, some 10^4 yrs, core H-burning LBV phase prior to a more extensive spell of a few times 10^5 yrs core He-burning WR phase seemed well established – until recently.

Whilst the quasi-sinusoidal modulations in the radio lightcurves of transitional SNe may possibly also be explained by alternative scenarios†, it might be relevant that the *same* underlying mechanism, i.e. wind bi-stability, might account for wind-velocity variations seen spectroscopically in the SN SN 2005gj (Trundle *et al.* 2008). Here, the variable winds are inferred from double P Cygni components (see Fig. 4) which appear almost identical to those seen in the Hα profiles of the well-known S Dor variables AG Car and HD 160529. It should also be noted that the timescales and the spectroscopically measured wind velocities of SN 2005gj, with $v_\infty \simeq$ 100-200 km s^{-1}, are consistent with those of LBVs, whilst they are yet again not consistent with those of the much slower RSG winds ($\sim$10 km s^{-1}), or the much faster WR winds ($\simeq$1000-5000 km s^{-1}).

5. Mass loss as a function of metallicity

Metallicity is a key parameter in the physics of stars and star-forming galaxies, largely via metallicity-dependent stellar winds. We compare the predictions of O star mass-loss rates (Vink *et al.* 2001) with recent empirical mass-loss rates from Mokiem *et al.* (2007) using the so-called wind momentum-luminosity relationship (Kudritzki & Puls 2000) for Galactic, Large Magellanic Cloud, and Small Magellanic Cloud O stars in the Z range

† Although there are other explanations for these radio modulations, none of these are entirely satisfactory. Ryder *et al.* (2004) suggested the modulations might be due to a WR pinwheel system where a secondary star perturbs the circumstellar medium of the primary WR star. Although this remains possible (though the inferred radial spacings in Ryder *et al.* (2004) are incorrect by a factor of 10, see Kotak & Vink 2006), the fact that SN 2003bg is so similar to SN 2001ig led Soderberg *et al.* (2006) to suggest the modulations are more likely due to a variable WR wind of a single star, resulting in concentric shells. However, such variability has never been observed in WR stars. This shortcoming was alleviated with the S Dor LBV suggestion of Kotak & Vink. Finally, we mention the possibility that the radio modulations might be due to a variable wind of a pulsating red supergiant (RSG; Heger *et al.* 1997), however the problem with such a scenario is that a RSG is H-rich, whilst SN 2003bg was first classified as a SN Ic. The fact that SNe 2001ig and 2003bg showed exactly the opposite transitional behaviour between type I and II, or H-rich or H-poor, was an extra reason for Kotak & Vink to consider LBVs as possible progenitors, as LBVs are H-rich compared to WR stars, but H-poor compared to RSGs.

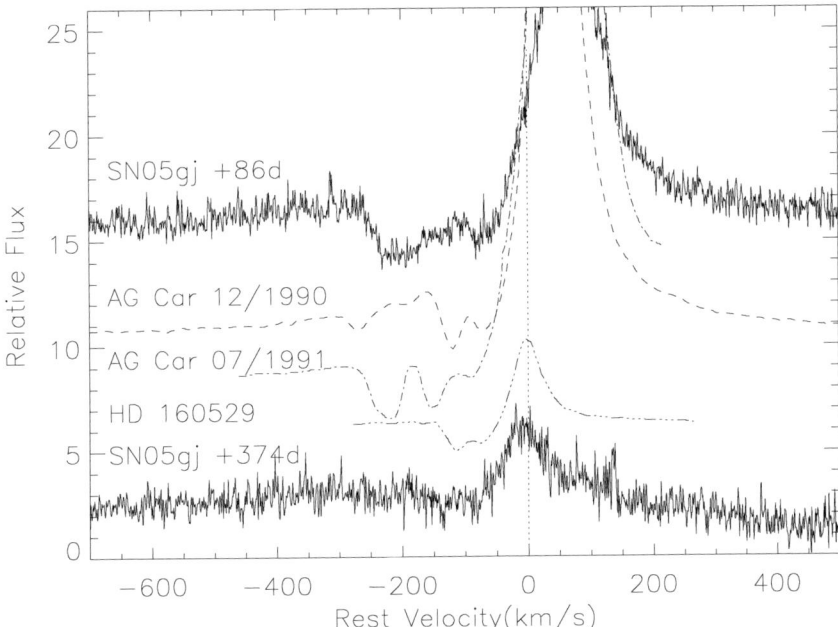

Figure 4. Multiple absorption components seen in the P Cygni Hα profile of SN 2005gj (top) in comparison to the LBVs AG Car and HD 160529. The figure has been taken from Trundle et al. (2008).

from solar to only 20% solar. We note that for all three galaxies, the empirical rates are somewhat larger (by a factor ∼2) than predicted by Vink et al. (see Fig.5). As the empirical rates are most likely affected by wind clumping (e.g. Bouret et al. 2003, Martins et al. 2005, Hamann et al. 2008), the empirical rates are likely maximal. Therefore, when we assume a modest clumping factor corresponding to an empirical $\dot{M}$ reduction of a factor ∼2, the empirical rates show very good agreement with theory. The wind clumping factor however remains an unsolved problem in stellar astrophysics and if the true wind clumping is larger than assumed, with empirical $\dot{M}$ overestimates of ∼10 as some studies suggest (e.g. Fullerton et al. 2006), the current mass-loss predictions might also be too large. This is certainly an important topic for future investigation.

Massive stars lose mass at even higher rates during the more evolved LBV and WR phases. During the latter phase, the outer layers become strongly chemically enriched, which may potentially modify mass loss through winds. Vink & de Koter (2005) investigated the mass loss versus Z dependence for late-type WR stars using a Monte Carlo approach (see Fig. 6). Despite the overwhelming presence of carbon at all Z, $\dot{M}$ does *not* show a Z-independent behaviour (as was generally assumed previously), but WR mass loss depends strongly on the iron (Fe) opacity, just like for O stars (Vink et al. 1999, Gräfener & Hamann 2008).

Furthermore, although the $\dot{M}$ versus Z dependence is consistent with a power-law decline in the observable Universe down to $\log Z/Z_\odot \sim -3$, it flattens off for extremely low Z models. The reason is that carbon, nitrogen, oxygen, hydrogen and helium take over the driving from Fe which dominates the higher Z domain. The strong Z-dependence of WR winds where the WR $\dot{M}$ drops by orders of magnitude, might represent a key result for the high incidence of long-duration GRBs at low metallicity. The favoured progenitors of long GRBs are thought to be rapidly rotating WR stars. However, most Galactic WR

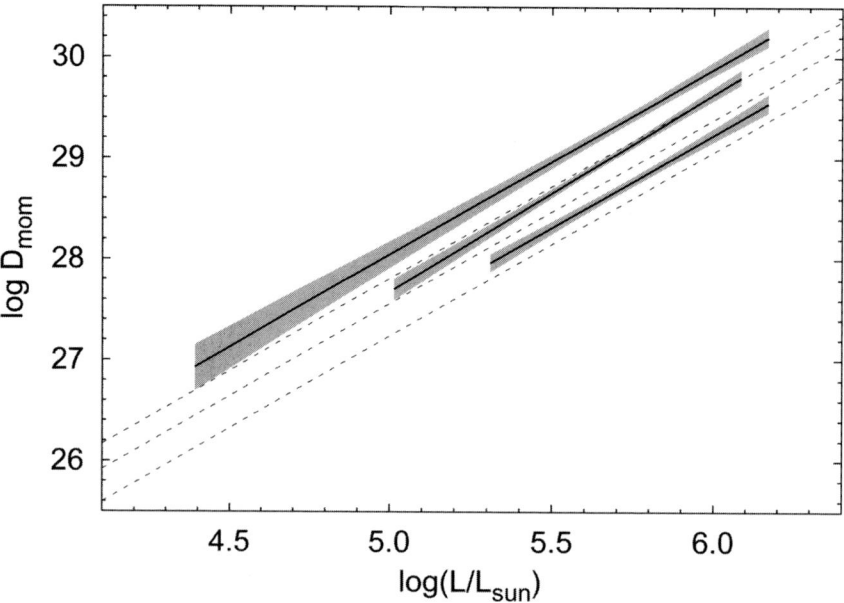

Figure 5. This figure represents a confrontation between O star mass-loss predictions (dashed line) and recent empirical mass-loss rates in the form of the so-called wind momentum-luminosity relationship for Galactic (top), LMC (middle) and SMC (bottom) O stars. The figure is from Mokiem *et al.* (2007).

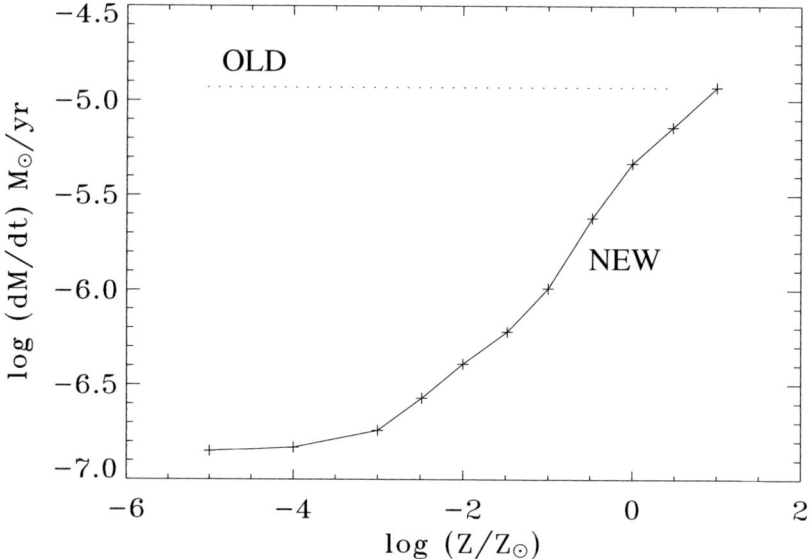

Figure 6. Z-dependent mass-loss predictions for Wolf-Rayet stars (Vink & de Koter 2005). Albeit the overwhelming presence of carbon for all Z, the new WR mass-loss rate does *not* show Z-independent behaviour as assumed previously (dotted line). The new computations show that WR mass loss depends strongly on iron (Z) — a *key* result for predicting a high occurrence of long-duration GRBs at low Z.

stars are slow rotators, as stellar winds probably remove the necessary stellar angular momentum, potentially posing a challenge to the collapsar model for GRBs.

Observational data however indicate that GRBs occur predominately in low metallicity (Z) galaxies (e.g. Le Floc'h *et al.* 2003, Prochaska *et al.* 2004, Vreeswijk *et al.* 2004, Modjaz *et al.* 2008), which may resolve the problem: lower Z leads to less mass loss, which may inhibit angular momentum removal, allowing WR stars to remain rotating rapidly until collapse (Yoon & Langer 2005, Woosley & Heger 2006). As a test of this scenario, Vink (2007b) performed a linear spectropolarimetry survey of WR stars in the low Z environment of the LMC and found an incidence of line polarisation effects in LMC WR stars as low as that of the Galactic sample of Harries *et al.* (1998). This suggests that the threshold metallicity where significant differences in WR rotational properties occur is below that of the LMC (at $Z \sim 0.4\ Z_\odot$), possibly constraining GRB progenitor channels to this upper metallicity.

6. Conclusions

We presented theoretical mass-loss rates and their implications for the peeling off and angular momentum loss of massive stars during the various evolutionary stages from the main sequence, to the LBV, and WR phases. The role of the bi-stability jump at an effective temperature of $\sim$25 000 K was discussed in the context of the observed drop in rotational velocities in this part of the Hertzsprung-Russell diagram as well as for the variable winds of LBVs, suggesting these objects could be in a *direct* pre-SN state of massive star evolution.

Acknowledgements

We would like to thank all our collegues and friends in the field of massive stars and in particular to Alex de Koter and Rubina Kotak. Special thanks also to Ines Brott and Norbert Langer for computing the evolution of the rotation rate with time and temperature.

References

Abbott, D. C. & Lucy, L. B. 1985, ApJ 288, 679
Benaglia, P., Vink, J. S., Marti, J., Maiz Apellaniz, J., Koribalski, B. & Crowther, P. A. 2007, A&A 467, 1265
Behr, B. B., Cohen, J. G., & McCarthy, J. K. 2000, ApJ 531, L37
Bouret, S.-C., Lanz, T. & Hillier, D. J. 2003, ApJ 595, 1182
Castor, J., Abbott, D. C. & Klein, R. I. 1975, ApJ 195, 157
Conti, P. S. 1976, MSRSL 9, 193
Crowther, P. A., Lennon, D. J. & Walborn, N. R. 2006, A&A, 446, 279
Eldridge, J. J. & Vink, J. S. 2006, A&A, 452, 295
Foley, R. J., Smith, N., Ganeshalingam, M., *et al.* 2007, ApJ 657, L105
Fullerton, A. W., Massa, D. L. & Prinja, R. K. 2006, ApJ 637, 1025
Gal-Yam, A., Leonard, D. C., Fox, D. B., *et al.* 2007,, ApJ 656, 372
Gräfener, G. & Hamann, W.-R. 2008, A&A 482, 945
Hamann, W.-R., Oskinova, L. M., & Feldmeier, A. 2008, Clumping in Hot-Star Winds
Harries, T. J., Hillier, D. J., & Howarth, I. D. 1998, MNRAS 296, 1072
Heger, A., Jeannin, L., Langer, N., & Baraffe, I. 1997, A&A 327, 224
Humphreys, R. M. & Davidson, K. 1994, PASP 106, 1025
Hunter, I., Lennon, D. J., Dufton, P. L., Trundle, C., Simon-Diaz, Smartt, S. J., Ryans, R. S. I., & Evans, C. J. 2008, A&A 479, 541
Kotak, R. & Vink, J. S. 2006, A&A 460L, 5

Kudritzki, R. P., Puls, J. 2000, ARA&A 38, 613
Lamers, H. J. G. L. M., Snow, T. P., & Lindholm, D. M. 1995, ApJ 455, L269
Langer N., 1998, A&A 329, 551
Lee, W. H. & Ramirez-Ruiz, E. 2006, ApJ 641, 961
Le Floc'h, E., Duc, P.-A., Mirabel, I. F., et al. 2003, A&A 400, 499
Limongi, M., Chieffi, A. 2006, ApJ 647, 483
MacFadyen, A. I. & Woosley, S. E. 1999, ApJ 524, 262
Martins, F., Schaerer, D., Hillier, D. J., et al. 2005, A&A 441, 735
Markova, N. & Puls, J. 2008, A&A 478, 823
Maund, J. R., Smartt, S. J., Kudritzki, R.-P, et al. 2006, MNRAS 369, 390
Meynet, G. & Maeder, A. 2003, A&A 404, 975
Modjaz, M., Kewley, L., Kirshner, R. P., et al. 2008, AJ 135, 1136
Mokiem, M. R., de Koter, A., Vink, J. S. 2007, et al., A&A 473, 603
Nieuwenhuijzen, H. & de Jager, C. 1990, A&A 231, 134
Ofek, E. O., Cameron, P. B., Kasliwal, M. M., et al. 2007, ApJ 662, 1129
Pastorello, A., Smartt, S. J., Mattila,S., et al. 2007, MNRAS 377, 1531
Pauldrach, A. W. A. & Puls, J. 1990, A&A 237, 409
Prochaska, J. X., Bloom, J. S., Chen, H.-W., et al. 2004, ApJ 611, 200
Ryder, S. D., Sadler, E. M., Subrahmanyan, R., et al. 2004, MNRAS 349, 1093
Smith, N. 2007, AJ 133, 1034
Smith, N. & Owocki, S. P. 2006, ApJ 645, 45
Smith, N., Li, W., Foley, R.J, & et al. 2007b, ApJ, 666, 1116
Soderberg, A. M., Chevalier, R. A., Kulkarni, S. R., & Frail, D. A. 2006, ApJ 651, 1005
Stahl, O., Jankovics, I., Kovacs, J., et al. 2001, A&A 375, 54
Trundle, C. & Lennon, D. J. 2005, A&A 434, 677
Trundle, C., Kotak, R., Vink, J. S., & Meikle, W. P. S. 2008, A&A 483, 47
Van Dyk, S. D., Peng, C. Y., King, J. Y., et al. 2000, PASP 112, 1532
Vink, J. S. 2006, ASPC 353, 113 (astro-ph/0511048)
Vink, J. S. 2007a, in: "Unsolved Problems in Stellar Physics: A Conference in Honor of Douglas Gough, AIPC 948, 389
Vink, J. S. 2007b, A&A 469, 707
Vink, J. S. 2008, in: "Eta Carinae", eds R.M. Humphreys & K. Davidson, Springer
Vink, J. S. & de Koter, A. 2002, A&A 393, 543
Vink, J. S. & de Koter, A. 2005, A&A 442 587
Vink, J. S., de Koter, A., & Lamers, H. J. G. L. M. 1999, A&A 345, 109
Vink, J. S., de Koter, A., & Lamers, H. J. G. L. M. 2000, A&A 362, 295
Vink, J. S., de Koter, A., & Lamers, H. J. G. L. M. 2001, A&A 369, 574
Vreeswijk, P. M., Ellison, S. L., Ledoux, C., et al. 2004, A&A 419, 927
Woosley, S. E. & Heger, A. 2006, ApJ 637, 914
Yoon, S.-C. & Langer, N. 2005, A&A, 443, 643

Discussion

TURCK-CHIEZE: What are the limitations of the mass loss calculation as far as the opacities are concerned: do you use detailed spectra? do you introduce radiative acceleration from theoretical estimates, is the knowledge of these quantities determinant in the calculation?

VINK: As far as the opacities are concerned: we include all elements from H-Zn (1-30) in our line list in Monte Carlo. Before we compute models at zero metallicity, we need a more comprehensive dynamical framework. We have successfully solved the wind dynamics for a few cases, but we want to be able to provide this over the entire HRD.

MEYNET: The argument for a Z dependance of $\dot{M}$ during the WR phase is quite convincing and I agree with you that it may have an impact on the way the WC to WN ratio varies as a function of Z. But when comparing WR populations with models you should not only compare the way the WN/WC ratio varies with Z but also show the WR/O number ratio varies with metallicity.

VINK: I agree with you. The point is that there are many physical mechanisms (rotation, B field, mass loss, binarity) responsible for how an O star becomes WR. To make comprehensive stellar and population synthesis models all these should of course be included. But I think we are still in a phase where we need to identify the relevant physical processes and test these for "simple" cases. To explain the drop in WC to WN, the mass loss versus Fe metallicity dependence may well be the first order effect. That was the point of that study.

BELCZYNSKI: This is a comment on G. Meynet question → that the wind mass loss rates predicted in your simulations may reproduce/give metallicity dependence for W-R stars, but they will significantly underproduce the number of W-R stars. That problem/issue can be avoided if one takes into account (quite easy) → formation of W-R stars in binaries (envelope removal in RLOF): in fact most massive stars are in close binaries and are subject to RLOF.

VINK: Again the point was not how to make WR stars. There are several routes, mass loss, quasi-homogeneous evolution, and, indeed, RLOF in binaries. So to get the complete picture you need to consider all these possibilities for a range of masses, metallicities, binary parameters, $v \sin i$, etc. Do you include mixing? B-field? etc. I do not think theory is ready for that. Does not mean we should not try!

PODSIADLOWSKI: I think that one has to be careful of what one calls an LBV. Some people call everything that shows variable mass loss an LBV. I do not think that this is very helpful. There are other physical mechanisms that causes eruptive events, e.g, binary mergers. These are theoretically predicted to be quite common and transient events in the near future are expected to detect these.

VINK: I do not want to exclude the possibility of binary mergers, they may even be quite common in dense systems. I am interested in your predictions and numbers as to how common they are. I do want to say though find there is evidence for blue supergiants with variable T_{eff} and variable winds and some of these have shown giant η Car eruptions, for instance P Cygni in 1600.

A snapshot of the meeting. Faces recognizable from let to right: S.K. Yi, Y. C. Kim, C. Q. Luo, J. Krticka, K. L. Chan

A snapshot of the meeting. Faces recognizable from let to right: G.H.Zhong, E. Caffau, H. J. Wang, Houdek, X. F. Chen

The influence of inhomogeneities on hot star wind model predictions

J. Krtička[1], L. Muijres[2], J. Puls[3], J. Kubát[4] and A. de Koter[2]

[1] Institute of Theoretical Physics and Astrophysics, Masaryk University, Brno, Czech Republic

[2] Astronomical Institute, "Anton Pannekoek", University of Amsterdam, Amsterdam, The Netherlands

[3] Universitätssternwarte München, München, Germany

[4] Astronomický ústav, Ondřejov, Czech Republic

Abstract. We study the effect of wind inhomogeneities (clumping) on O star wind model predictions. For this purpose we artificially include clumping into our stationary NLTE wind models. As a result of the inclusion of optically thin clumps the radiative line force is increased compared to corresponding unclumped models, with a similar effect on either the mass-loss rate or the terminal velocity. When the clumps are allowed to be optically thick in continuum, on the other hand, the radiative force and consequently the mass-loss rate decreases alternatively.

Keywords. Stars: mass loss, stars: early-type, hydrodynamics

1. Introduction

From the beginning of the quantitative studies of the stellar wind of hot stars it seemed clear that the mass-loss should significantly influence the evolution of these stars. The evolution of massive stars from the main sequence to the final supernova explosion was considered to be driven mainly by the intense winds of these stars. The classical studies of hot star winds culminated by the application of NLTE models both for the determination of hot star wind properties from the observational data (Puls *et al.* 1996) and for the prediction of wind mass-loss rates (Vink *et al.* 2000, Pauldrach *et al.* 2001, Krtička & Kubát 2004). These models generally assumed that the stellar wind can be described in a first approximation by the smooth, basically spherically symmetric outflow. Generally, there was a satisfactory agreement between theoretical predictions and observations.

The picture of smooth outflow was known to be inadequate due to the existence of the inherent instability connected with the radiative driving (Owocki & Rybicki 1984, Feldmeier *et al.* 1997). From the observational point of view, the existence of wind inhomogeneities (or wind "clumping") is not so apparent. The reason is that most diagnostical features depend on the *product* $\sqrt{C_c}\dot M$, where the "clumping factor" $C_c \geqslant 1$ relates the density inside the clump ρ^+ with the mean wind density $\langle\rho\rangle$,

$$\rho^+ = C_c \langle\rho\rangle. \qquad (1.1)$$

Consequently, spectra from winds with a large clumping factor but small mass-loss rate can mimic those from winds with weak clumping but large mass-loss rate. If $C_c > 1$, then the mass-loss rates derived from such diagnostics are overestimated by factor of $\sqrt{C_c}$. Fortunately, some spectral properties may be used to break this degeneracy, and to estimate the value of C_c. For example, Martins *et al.* (2005b) derived clumping a factor of $10 - 100$ for studied Galactic O-stars. If these values of clumping factor are real, this would imply a decrease of the estimated mass-loss rate by factors of 3–10.

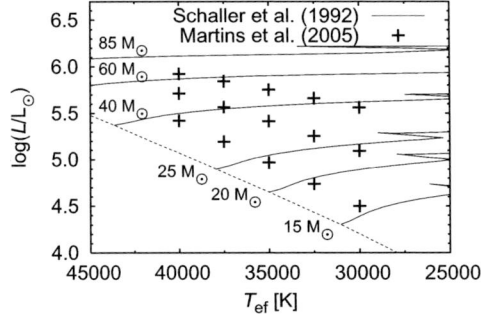

Figure 1. Parameters of studied stars in the HR diagram. Overplotted are the evolutionary tracks from Schaller *et al.* (1992)

Stimulated by these studies, we discuss here the influence of inhomogeneities on the mass-loss rate predictions.

2. Wind models

For our study we used the spherically symmetric, stationary NLTE wind models (Krtička & Kubát 2004). These models solve the equations of statistical equilibrium together with the equations of radiative transfer. The calculated occupation numbers are used to derive the radiative force (in the Sobolev approximation) and the radiative heating/cooling terms. This enables us to obtain the radial stratification of velocity, density and temperature in the wind and finally to predict the wind mass-loss rate $\dot M$ and terminal velocity v_∞.

The wind models are calculated for O-type stellar parameters based on the recent calibrations by Martins *et al.* (2005a, see also Fig. 1).

3. Influence on statistical equilibrium equations: clumping

Wind inhomogeneities may influence the statistical equilibrium equations due to change of the electron density. This effect can be roughly accounted for within the clumping approximation, which is widely used to study the wind spectra, e.g., Martins *et al.* (2005b).

To include the effect of clumping into our models we modified the equations of statistical equilibrium, by using an electron density $\rho_e^+ = C_c \langle \rho_e \rangle$, opacity $\langle \chi \rangle = \chi^+/C_c$, and emissivity $\langle \eta \rangle = \eta^+/C_c$. The superscript + denotes values inside the (homogeneous) clumps and the quantities inside brackets corresponding volume averages.

3.1. Influence of clumping

To investigate the influence of clumping on the stellar wind we have calculated a wind model of an O-type giant at $T_{\text{eff}} = 35\,000\,\text{K}$, assuming the wind to be smooth ($C_c = 1$) close to the star ($r < 2R_*$), and to be clumped ($C_c = 10$) in the outer regions ($r > 2R_*$).

In accord with our assumptions, the presence of clumping leads to an increase of the electron density inside the clumps. Consequently, the recombination rates become higher and the wind becomes less ionized (see Fig. 2). Since lower ions are able to accelerate the stellar wind more efficiently than the higher ones (due to a larger number of driving lines), the radiative force increases, which, in our case, leads to an increase in wind velocity.

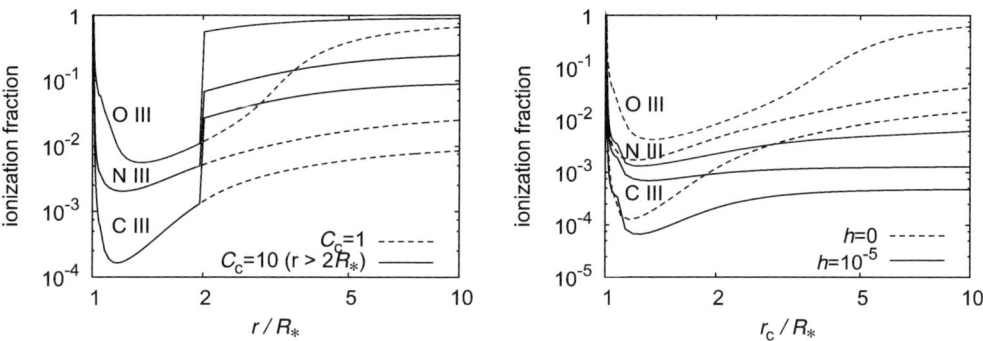

Figure 2. Left: Influence of clumping that starts at $r = 2R_*$ on the ionization fractions of selected ions. Right: Influence of porosity on the ionization fractions.

3.2. Radially constant clumping factor

The influence of clumping on the wind parameters depends on the radial onset of clumping. If clumping starts above the critical point (below which the mass-loss rate is determined), then the terminal velocity increases. On the other hand, if clumping starts below the critical point, then the wind mass-loss rate becomes larger.

In case of radially constant clumping factor, the mass-loss rate increases significantly (on average by a factor of 2 for $C_{\mathrm{c}} = 10$), and the predicted wind-momentum is also higher than for a smooth wind at the same parameters (see Fig. 3).

4. Porosity

Wind inhomogeneities may also influence the wind radiative transfer if the individual clumps are optically thin in continuum. This effect is termed as a porosity (Owocki et al. 2004), and can be introduced into the wind models by decreasing the continuum opacity,

$$\chi_{\mathrm{eff}} = \frac{\langle\chi\rangle}{1 + rh\langle\chi\rangle},$$

where hr is the porosity length. The continuum emissivity has to be modified by the same factor.

The porosity leads to a significant increase of the wind ionization (see Fig. 2 for the results calculated for $C_{\mathrm{c}} = 10$, and $h = 0$ and $h = 10^{-5}$). The increased ionization causes a decrease of the radiative force, and also of the mass-loss rate if the wind is porous below the critical point (see Fig. 3 for the results for radially constant C_{c} and h).

5. Influence on the radiative force

Up to now we discussed *indirect* influence of inhomogeneities on the line radiative force caused by the change of the wind ionization. As the radiative force depends on the velocity via the Doppler effect, the presence of inhomogeneities may influence the radiative force also *directly*. However, numerical simulations (Owocki & Rybicki 1984, Feldmeier et al. 1997) show that on average the instabilities do not directly affect the line radiative force. Consequently, we neglected the direct influence of inhomogeneities on the line radiative force.

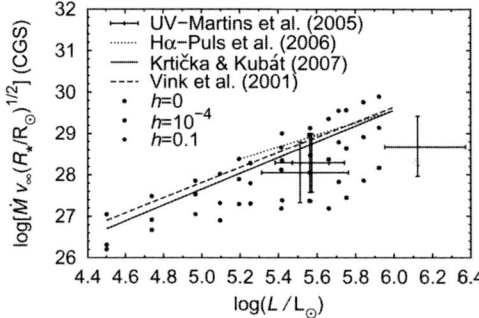

Figure 3. Influence of wind inhomogeneities on the modified wind-momentum (for $C_{\rm c} = 10$).

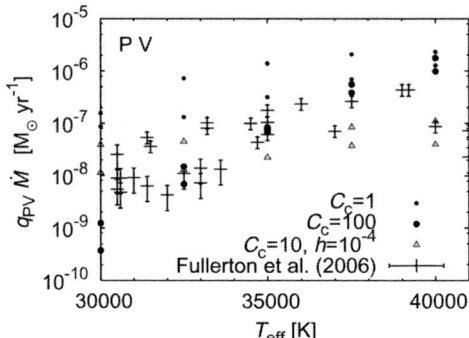

Figure 4. Influence of different types of wind inhomogeneities on the P V ionization.

6. Discussion

We conclude that wind inhomogeneities may influence the predicted wind parameters. An approach which roughly corresponds to the results of time-dependent simulations (i.e., inhomogeneities that "start" above the critical point) does not improve the agreement between theory and observations. On the other hand, if the inhomogeneities start below the critical point, then the effects of clumping and porosity may provide a better agreement between theory and observations, both in terms of modified wind momentum (Fig. 3) and P V ionization fractions (Fig. 4).

Acknowledgements

This research was supported by grants grants GA ČR 205/07/0031, and 205/08/0003.

References

Feldmeier, A., Puls, J., & Pauldrach, A. W. A. 1997, *A&A* 322, 878
Krtička, J. & Kubát, J. 2004, *A&A* 417, 1003
Martins, F., Schaerer, D., & Hillier, D. J. 2005a, *A&A* 436, 1049
Martins F., Schaerer, D., Hillier, D. J., et al. 2005b, *A&A* 441, 735
Owocki, S. P., Gayley, K. G., & Shaviv, N. J. 2004, *ApJ* 616, 525
Owocki, S. P. & Rybicki, G. B. 1984, *ApJ* 284, 337
Pauldrach, A. W. A., Hoffmann, T. L., & Lennon, M. 2001, *A&A* 375, 161
Puls, J., Kudritzki, R.-P., et al. 1996, *A&A* 305, 171
Schaller, G., Schaerer, D., Meynet, G., & Maeder, A. 1992, *A&AS* 96, 269
Vink, J. S., de Koter, A., & Lamers, H. J. G. L. M., 2000, *A&A* 362, 295

Discussion

WOITKE: What confines your "clumps"? In reality, there will be rather turbulence leading to filaments, that could be mistaken for "clumps".

KRTIČKA: The process of radiative acceleration has been shown to cause instabilities, which is different from "normal" turbulence. Need 3D hydro with line transfer to reveal the nature of these structures.

J. Krticka is presenting his work.

Stellar evolution models with mass loss and turbulence

M. Vick[1,2], G. Michaud[1] and O. Richard[2]

[1] Département de Physique, Université de Montréal, Montréal, PQ, H3C 3J7
email: mathieu.vick@umontreal.ca, michaudg@astro.umontreal.ca

[2] GRAAL, Université Montpellier II, CNRS, Place E. Bataillon, 34095 Montpellier Cedex, France
email: Olivier.Richard@graal.univ-montp2.fr

Abstract. Although chemical separation is generally accepted as the main physical process responsible for the anomalous surface abundances of AmFm stars, its exact behavior within the interior of these stars is still uncertain. We will explore two hydrodynamical processes which could compete with atomic diffusion: mass loss and turbulence. We will also discuss the extent to which separation occurs immediately below the surface convection zone as well as the extent to which separation occurs below 200,000 K. To do so, self-consistent stellar models with mass loss and turbulence where calculated using the Montreal stellar evolution code and compared to observations of A and F stars. It is shown that to the precision of observations available for F stars, a mass loss rate of $2\times10^{-14}\ M_\odot \cdot \mathrm{yr}^{-1}$ is compatible with observations and that no turbulence is then required.

Keywords. stars: evolution, stars: abundances, stars: mass loss, turbulence

1. Introduction

On the main sequence, most if not all slowly rotating and non magnetic A and early type F stars are believed to have anomalous surface abundances. These chemically peculiar stars are classified as AmFm stars (Preston 1974) and typically have observed underabundances of CNO, Sc and Ca as well as overabundances of iron peak elements and rare earths (see Cayrel, Burkhart, & Van't Veer 1991). In these stars, the surface convection zone is sufficiently shallow for the effect of a relatively slow process, such as atomic diffusion, to manifest itself at the the stellar surface. Potentially more efficient processes, such as convective, rotational or turbulent mixing as well as large scale processes such as mass loss, can slow the effect of atomic diffusion, thereby reducing surface abundance anomalies. In order to quantitatively reproduce observed surface abundances, models with atomic diffusion must consider one of these aforementioned inhibiting processes in order to reduce predicted anomalies which are larger (by a factor of 2-5) than those observed (Turcotte, Michaud, & Richer 1998).

In Richer, Michaud, & Turcotte (2000), diffusion models which include turbulent mixing are able to reproduce the observed surface abundance anomalies for AmFm stars of many open clusters. These same models have also successfully explained observed abundances in Pop II (Richard, Michaud et & Richer 2005, Korn et al. 2006) and horizontal branch stars (Michaud, Richer, & Richard 2007, Michaud, Richer, & Richard 2008). In the AmFm models as well as the Pop II models, mixing is enforced in such a way that chemical separation occurs deeper than 200,000 K. Around this temperature, an iron convection zone naturally appears for models of at least 1.5 $M_\odot$ (this also depends on the amount of mixing which tends to suppress it). In contrast, another explanation for

AmFm stars suggests that chemical separation is to occur below the surface convection zone (SCZ). Both these scenarios will be discussed.

The effects of mass loss on the surface abundances of these stars have also been previously investigated by Michaud et al. (1983), Michaud & Charland (1986), Alecian (1996) and Leblanc & Alecian (2008). However, a further investigation is warranted since these studies analyzed static stellar models which included a limited number of elements.

In the following we will look at the effects of mass loss and turbulence both at the surface and within the stable radiative zone of these stars. We will start by detailing our models (§2) as well as our treatment of mass loss (§3). The following sections will compare the effects of mass loss and turbulence on the internal profiles (§4) and on surface abundances (§5).

2. Stellar Evolution Models

These 1D stellar evolution models were calculated as described in Turcotte et al. (1998b) (see also Richard, Michaud, & Richer 2001 and references therein). Models were evolved from the homogeneous pre-main sequence up to the bottom of the sub giant branch with the abundance mix listed in Table 1 of Turcotte et al. (1998b). Radiative accelerations are calculated at each time step and at each mesh point for 28 chemical species. The atomic data is taken from the OPAL database (Iglesias & Rogers 1996). The Rosseland mean opacity is also continuously updated which means that the treatment of chemical transport is completely self-consistent. The corrections for redistribution of momentum are from Gonzalez et al. (1995) and LeBlanc, Michaud & Richer (2000). The atomic diffusion coefficients are taken from Paquette et al. (1986). Semi-convection is included as described in Richard et al. (2001), following Kato (1966), Langer, El Eid & Frick (1985) and Maeder (1997). The Krishna-Swamy $T - \tau$ relation (Krishna-Swamy 1966) is used as the boundary condition in the atmosphere. The corresponding value of the mixing length parameter, which is calibrated to fit the radius and luminosity of the Sun is $\alpha = 2.096$ (model H of Turcotte et al. 1998b) with the initial mass fraction of He set to $Y_0 = 0.27769$. Models were calculated from $1.30\,M_\odot$ to $1.55\,M_\odot$, with mass loss rates ranging from 10^{-14} to $10^{-12}\;M_\odot \cdot \mathrm{yr}^{-1}$.

3. Treatment of Mass Loss

The mass loss is assumed spherical and chemically homogeneous. Mass loss is introduced as described in Charbonneau (1993). The resulting transport equation is:

$$\rho \frac{\partial c}{\partial t} = -\nabla \cdot [-\rho D \nabla \ln c + \rho(\mathbf{U} + \mathbf{U}_w)c] + \rho(S_{nuc} + S_w)c, \qquad (3.1)$$

with a Neumann condition imposed at the surface and with $\mathbf{U}_w$ and S_w defined as:

$$\mathbf{U}_w = \begin{cases} v_w \hat{\mathbf{e}}_\mathbf{r} & \text{under the SCZ,} \\ 0 & \text{in the SCZ;} \end{cases} \qquad (3.2)$$

$$S_w = \begin{cases} 0 & \text{under the SCZ,} \\ \frac{\dot{M}}{M_{ZC}} & \text{in the SCZ.} \end{cases} \qquad (3.3)$$

Here, c is the time and depth dependent composition, ρ density, D the total diffusion coefficient, $\mathbf{U}$ the total velocity field, $\mathbf{U}_w$ wind velocity, M_{CZ} the mass of the SCZ, $\dot{M}$ the mass loss rate, S_{nuc} a source/destruction term linked to nuclear reactions and S_w a

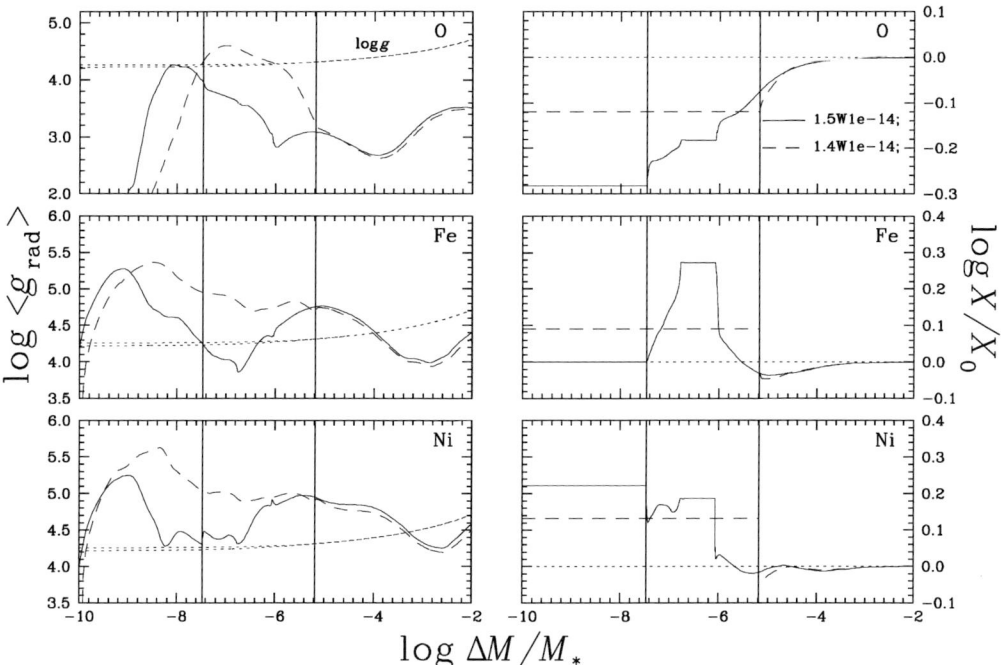

Figure 1. *Left panel* Radiative accelerations (continuous and dashed lines) of O, Fe and Ni and gravity (dotted line) for two models of different masses at 500 Myr. *Right panel* Corresponding internal concentration of the three species. The models are indentified in the right panel. The number to the left W in the model name indicates the mass and the number to the right is the mass loss rate in $M_\odot \cdot yr^{-1}$. The vertical lines show the position of the hydrogen-helium surface convection zone for the respective models.

destruction term linked to mass loss. The mass of the star is also continuously updated so that all quantities which depend on stellar mass are correctly calculated.

4. Internal structure

The anomalous surface abundances of AmFm stars are the result of chemical transport below the surface convection zone which is modulated by the competition between gravity and radiative accelerations. In Figure 1 we see how radiative accelerations and gravity vary with the mass coordinate for two models of different masses ($1.40\,M_\odot$, $1.50\,M_\odot$) with a mass loss rate of $10^{-14}\,M_\odot \cdot yr^{-1}$. When g is always greater than $g_{\rm rad}$, the element tends to sink toward the interior of the star, thus creating an underabundance at the surface, as is the case for oxygen. On the other hand, when an element is supported by the radiative field throughout most of the stellar interior (and more importantly below the surface convection zone), we can expect an overabundance of that element at the surface (as for Ni). In other cases, elements accumulate where $g \sim g_{\rm rad}$. The abundance at the surface will be determined by the position of this accumulation and the bottom of the surface mixing zone (which, in the presence of turbulence, is an extension of the SCZ), since the material at its base is mixed to the surface. For the $1.50\,M_\odot$ model, $g_{\rm rad}$(Fe) dips below gravity just under the surface convection zone which causes an accumulation of iron in that region. This accumulation triples the local metal content, which leads to an increase of opacity and the appearance of an iron convection zone (which can be seen in

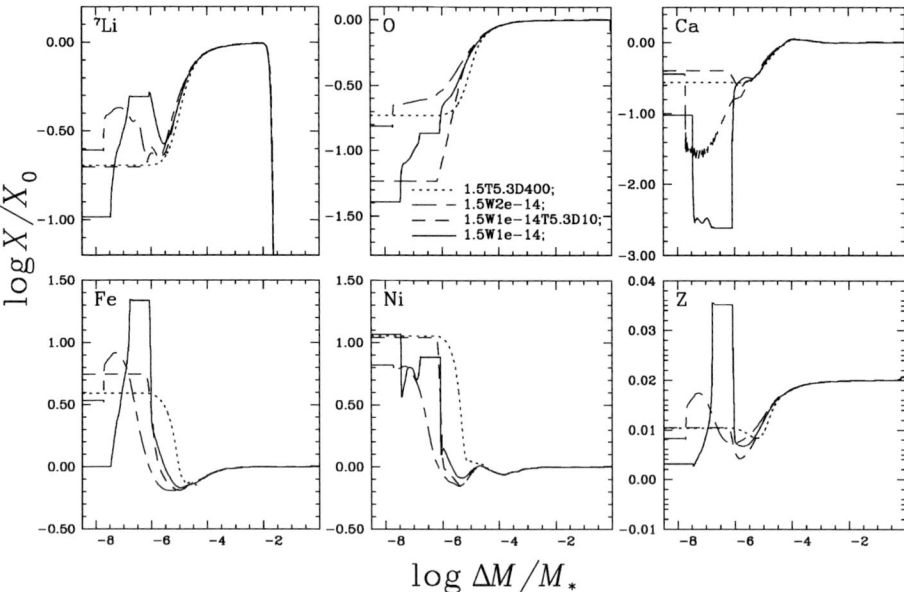

Figure 2. Abundance profiles for 4 models with turbulence and/or mass loss at 500 Myr. In the model name, the quantity to the right of T is the log of the temperature down to which abundances are completely homogenized by turbulence and that to the right of D is the coefficient multiplying the turbulent diffusion coefficient (see Richer et al. 2000). The dip in lithium around $\log \Delta M/M_* = -2$ is due to nuclear burning. Some numerical noise appears in the Ca profile for the 1.5W2e14 model.

the abundance profiles: the plateau between $\log \Delta M/M_* = 6-7$ shows the homogenized abundances due to this convection zone).

In Figure 2, we see the effect of turbulence and mass loss on the abundances profiles. For all models, we see that at 500 Myr, chemical separation implies the outer $10^{-4} M_*$. In the models with mass loss only (1.5W1e-14 and 1.5W2e-14), separation is allowed to occur immediately below the surface convection zone. In models with turbulence (1.5W1e-14T5.3D10 and 1.5T5.3D400), abundances are mixed down to $\log T = 5.3$, which corresponds to $\log \Delta M/M_* \sim -6$. The only model which leads to an iron convection zone is the model with a mass loss rate of $10^{-14} M_\odot \cdot \mathrm{yr}^{-1}$ with no additional turbulence. This model does not generate an advective current which is strong enough to push the accumulation of iron into the surface convection zone. This also leads to a surface iron abundance which is close to its original value. However, if the wind is twice as strong $(2 \times 10^{-14} M_\odot \cdot \mathrm{yr}^{-1}$, the solar value) the iron spike is weakened and pushed toward the surface. There are very significant concentration differences between the various models and this may allow to distiguish them with asterosismology.

5. Surface Abundances

In Figure 3, we compare surface abundances for a $1.50\,M_\odot$ model with observed abundances of τ UMa (Hui-Bon-Hoa 2000) which has an age of approximately 500 Myr (Monier 2005) and $T_{\mathrm{eff}} \sim 7000$ K (van't Veer-Menneret & Mégessier 1996). We tested 2 scenarios: chemical separation below 200,000 K (right panel) and the scenario in which separation occurs immediately below the surface convection zone (left panel). At this age, the surface convection zone is composed of both hydrogen and helium since helium has not yet settled gravitationally. We see that both turbulence and mass loss can effectively reduce

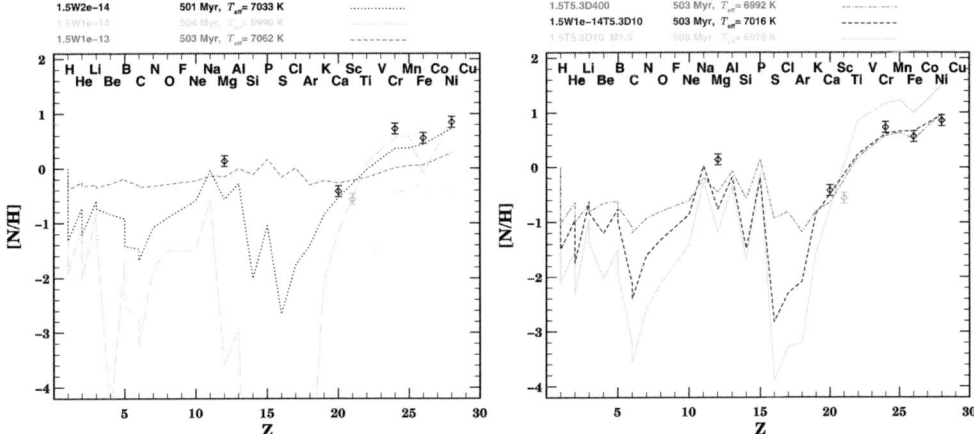

Figure 3. Observed surface abundances of τUma (diamonds) compared to 1.50 $M_\odot$ models. The abundance of Sc is shown, but is not calculated in our models. *Left panel* Models with mass loss only; chemical separation occurs immediately below the surface convection zone. *Right panel* Models with turbulence; chemical separation occurs below 200,000 K.

the surface abundances to the observed levels. The model with the solar mass loss rate (1.5W2e-14) leads to a very good agreement with the data. The model with both mass loss and turbulence (1.5W1e-14D5.3D10) leads to the best fit. The mass loss rate of 10^{-13} $M_\odot \cdot \text{yr}^{-1}$ is too strong to reproduce the observations.

6. Conclusions

With a mass loss rate of the order of the solar mass loss rate we can successfully reproduce the observed abundance anomalies of τUMa. We also find that with sufficiently weak winds and no turbulence, an iron convection zone naturally appears in 1.50 $M_\odot$ models. However, for these models the surface abundance of Fe does not match the observed abundance. It is also shown that mass loss and turbulence affect abundances differently (at the surface and in the interior). It is thus possible that sufficiently precise abundance determinations could constrain the relative importance of each process. This being said, models of greater mass, where abundance determinations are more abundant, would also help in this venture. These are presently being calculated. It is still premature to favor either of the two proposed scenarios as both offer models that are able to adequately reproduce observations within the given errors.

Acknowledgements

This reasearch was partially supported at the Université de Montréal by the Natural Sciences and Engineering Researh Council of Canada. The computational resources were provided by the Réseau Québecois de Calcul de Haute Performance (RQCHP). M.Vick also thanks the Département de physique of l'Université de Montréal as well as the GRAAL of l'Université Montpellier II for financial support.

References

Alecian, G. 1996, *A&A*, 310, 872
Cayrel, R., Burkhart, C., & Van't Veer, C. 1991, in IAU Symposium 145, ed. G.Michaud & A.Tutukov
Charbonneau, P. 1993, *ApJ*, 405, 720

Gonzalez, J.-F., LeBlanc, F., Artru, M.-C., & Michaud, G. 1995, *ApJ*, 297, 223
Hui-Bon-Hoa, A. 2000, *A&AS*, 144, 203
Iglesias, C. A. & Rogers, F. 1996, *ApJ*, 464, 943
Kato, S. 1966, *PASP*, 18, 374
Korn, A. J., Grundahl, F., Richard, O., Barklem, P. S., Mashonkina, L., Collet, R., Piskunov, N., & Gustafsson, B. 2006, *Nature*, 442, 657
Krishna-Swamy, K. S. 1966, *ApJ*, 145, 176
Langer, N., El Eid, M. F., & Fricke, K. J. 1985, *A&A*, 145, 179
LeBlanc, F., Michaud, G., & Richer, J. 2000, *ApJ*, 538, 876
Leblanc, F., & Alecian, G. 2008, *A&A*, 477, 243
Maeder, A. 1997, *A&A*, 321, 134
Michaud, G., Tarasick, D., Charland, Y., & Pelletier, C. 1983, *ApJ*, 269, 239
Michaud, G. & Charland, Y. 1986, *ApJ*, 311, 326
Michaud, G., Richer, J., & Richard, O. 2007, *ApJ*, 670, 1178
Michaud, G., Richer, J., & Richard, O. 2008, *ApJ*, 675, 1223
Monier, R. 2005, *A&A*, 442, 563
Paquette, C., Pelletier, C., Fontaine, G., & Michaud, G. 1986,*ApJS*, 61, 177
Preston, G. W. 1974, *ARAA*, 12, 257
Richard, O., Michaud, G., & Richer, J. 2001, *ApJ*, 558, 377
Richard, O., Michaud, G., & Richer, J. 2005, *ApJ*, 619, 538
Richer, J., Michaud, G., Rogers, F., Turcotte, S., & Iglesias, C. A. 1998, *ApJ*, 492, 833
Richer, J., Michaud, G., & Turcotte, S. 2000, *ApJ*, 529, 338
Turcotte, S., Michaud, G., & Richer, J. 1998a, *ApJ*, 504, 559
Turcotte, S., Richer, J., Michaud, G., Iglesias, C. A., & Rogers, F. 1998b, *ApJ*, 504, 539
van't Veer-Menneret, C. & Mégessier, C. 1996, *A&A*, 309, 879

Discussion

KRTIĆKA: I would suggest to calculate models of Bp stars because there you really expect both winds and chemical peculiarity.

VICK: As you say these would be very interesting to verify. Unfortunately, these hot magnetic stars outline the validity domain of our current models.

WILLSON: Is there an upper limit on $\dot{M}$ from your work?

VICK: Form the models shown, which were for $15 M_\odot$, a mass loss rate of 1×10^{-13} $M_\odot$/yr completely flattens the abundances at the age of comparison (500 Myr). However, for more massive models this limiting mass loss rate increases slightly, just as it decreases for lower masses.

CHRISTENSEN-DALSGAARD: A naive question: what derives the wind in A and F stars?

VICK: This is actually a really good question. Cooler star winds are driven by chromospheric activity or thermal acceleration in hot coronaes (e.g., the solar wind). On the other hand, winds of hotter O and B stars are driven by radiation pressure. The importance of these processes for A and F stars isn't well known. Observational constraints are rather scarce; however, Simon *et al.* (1997) have detected chromospheric activity in stars of up to spectral class A7V. Whether this is the cutoff point between the 2 aforementioned processes, we still don't know.

ZAHN: Is what prevent "normal" A F stars to show abundance anomalies the presence of stars turbulence, linked with fast rotation?

VICK: It would be fair to say that all hot F, and A stars with sufficiently stable radiative zones will eventually show effects of atomic diffusion at their surface. The presence of strong turbulence, whether it is linked to rotation or other processes, will definitely hinder chemical separation. Mixing erases surface effects expected from atomic diffusion.

LANGER: Surface anomalies are also found in slowly rotating massive stars ($\rightarrow\sim 10 M_\odot$). Could those be explained by diffusion?

VICK: If the atmospheres and envelops of these stars are sufficiently stable, we should see effects of atomic diffusion. However, in these stars other processes such as strong mass loss are expected, which would impede slow processes such as microscopic diffusion.

A snapshot of the meeting. Faces recognizable from let to right: S. de Mink, Glebbek, X. F. Chen, O. Yaron.

A meeting snapshot: from left, Licai Deng, Kwing-Lam Chan, Tao Cai, Da Run Xiong.

The most massive AGB stars

Lionel Siess†

Institut d'Astronomie et d'Astrophysique,
Université libre de Bruxelles, 1050 Bruxelles, Belgium
email: siess@astro.ulb.ac.be

Centre for Stellar and Planetary Astrophysics,
School of Mathematical Sciences, Monash University, Victoria 3800, Australia

Abstract. The general properties of stars in the mass range 7-12$M_\odot$, also referred to as *super-AGB* stars, are reviewed and special attention is paid to determine how their mass range depends on the initial metallicity and what fraction of these stars end their life as ONe white dwarfs or explode as electron-capture supernova.

Keywords. Stars: AGB and post-AGB, stars: evolution

1. Introduction

Stars in the mass range 7-12$M_\odot$ experience a peculiar evolution characterized by the off-center ignition of carbon followed by the propagation of a deflagration front toward the center. After the formation of an oxygen-neon (ONe) core, they enter the thermally pulsing SAGB (TP-SAGB) phase during which they experience recurrent thermal instabilities in the helium-burning shell and develop strong winds. The outcome of their evolution is however extremely sensitively to the mass-loss rate and to the efficiency of the third dredge-up during this final evolutionary phase. Depending on these uncertain parameters, the star will either explode as electron-capture supernova (EC-SN) or eject its whole envelope and leave an ONe white dwarf.

From an historical perspective, the first studies of SAGB stars date back to the 80's where the evolution of helium balls, corresponding to the bare core of stars in the mass range 8-12$M_\odot$, were investigated (Miyaji *et al.* 1980, Nomoto 1981; 1984, Habets 1986, Baron *et al.* 1987, Mayle & Wilson 1988). These early simulations demonstrated the key role of electron capture reactions in triggering the collapse of the ONe core. Different aspects of the complex physics of the explosion (e.g. associated with the role of convective mixing, the impact of Coulomb corrections or the presence of residual ^{12}C in the ONe core) was thereafter addressed in several papers (e.g. Canal *et al.* 1992, Gutiérrez *et al.* 1996, 2005) and hydrodynamical simulations were calculated (Dessart *et al.* 2003, Kitaura *et al.* 2006). The first study of the non explosive evolution of SAGB stars from the main sequence up to the thermally pulsing phase was presented in several papers by García-Berro & Iben (1994), Ritossa *et al.* (1996), García-Berro *et al.* (1997), Iben *et al.* (1997) and Ritossa *et al.* (1999). Recently new models become available, revisiting the properties of SAGBs at different metallicities and accounting for mass-loss (Gil-Pons *et al.* 2005,2007; Doherty & Lattanzio 2006; Siess 2006,2007; Poelarends *et al.* 2008).

In this paper, the evolutionary features of super-AGB stars are reviewed. A description of the carbon burning phase and subsequent thermally pulsing super-AGB (SAGB) phase is given and, for stars developing the most massive cores, the process leading to the supernova explosion is briefly discussed. In section 3, we address the question of their

† LS is FNRS research associate

fate, analyzing how the final outcome depends on initial mass, metallicity, mass-loss efficiency and core growth rate.

2. Main features of super-AGB evolution

The evolution of the structure up to the end of central helium burning is very similar to that of intermediate mass stars. H burning takes place in a convective core, the size of which decreases as the increasing He content lowers the central opacity. At H exhaustion, the core contracts and the surface expands leading to the deepening of the convective envelope. During the first dredge-up (1DUP) which only occurs in SAGB stars with $Z > 0.001$, the convective zone penetrates into the previously active H-burning region and brings to the surface the ashes of proton burning, i.e. ^{3,4}He, ^{13}C and ^{14}N, mainly, leading to a drop in the ^{12}C/^{13}C ratio. When the central temperature reaches 10^8 K He ignites at the center. A convective core develops and a large amount of ^{12}C is produced by the 3α reactions. When the central He mass fraction drops below ~ 0.1, the ^{12}C$(\alpha,\gamma)^{16}$O reaction becomes the most rapid reaction and a substantial amount of ^{16}O is produced at the expense of ^{12}C. As studied by Imbriani et al. (2001) and illustrated in Siess (2007) in the context of SAGB stars, the final C/O ratio in the core is very sensitive to the evolution of the core properties near the end of central He burning, and consequently difficult to predict accurately. Typical ^{12}C values range between 0.2 and 0.4.

2.1. Carbon burning phase

When He is depleted at the center, core contraction resumes and as the density increases, neutrinos are further emitted. Energy is removed from the inner regions where the matter becomes degenerate and the temperature inversion moves outward. In low and intermediate mass stars, core contraction releases less energy and stops before the peak temperature reaches the threshold for carbon ignition of $\sim 6 \times 10^8$ K. Then, similarly to the He-flash taking place at the tip of the red giant branch in low mass stars ($M \lesssim 2.25 M_\odot$), carbon ignites off center in conditions of partial degeneracy (the degeneracy parameter $\eta \sim 2-3$). The carbon burning phase proceeds in 2 steps: a flash followed by the propagation of a flame to the center. The flash is more energetic and nuclear luminosities of $10^6 \lesssim L_C/L_\odot \lesssim 2 \times 10^8$ can be reached. Because the plasma is partially degenerate, energy accumulates before the pressure starts rising and a strong temperature gradient develops leading to the formation of a first convective zone (Fig. 1). The energy released during the flash induces a large expansion of the burning shells which quenches the instability. As the carbon luminosity L_C decreases, core contraction resumes and the temperature maximum moves inward and increases again. Eventually carbon reignites but the thermodynamical conditions are now somewhat different: the degeneracy is lower ($\eta \sim 1$) as part of the flash energy was absorbed by the core and the new convective zone that forms develops in a region that was previously occupied by the flash. As a consequence less carbon is made available to power the instability and the resulting luminosity is weaker. The expansion of the core is reduced and convection stays. The star then readjusts toward a steady state configuration (Timmes & Woosley 1992) where all the energy released by carbon burning is carried away by neutrinos ($L_\nu = -L_C$). The propagation of the carbon burning front proceeds as a deflagration where ahead of the convective zone, matter is heated up, nuclearly processed and the ashes subsequently engulfed in the convective zone as the peak temperature moves inward where fuel is more abundant. Typical flame speed are of the order of $10^{-3} - 10^{-2}$ cm s^{-1}. Finally, when the flame reaches the center, convection disappears and carbon burning proceeds radiatively in a shell surrounding the core. From Fig. 1 we also note that the surface is decoupled

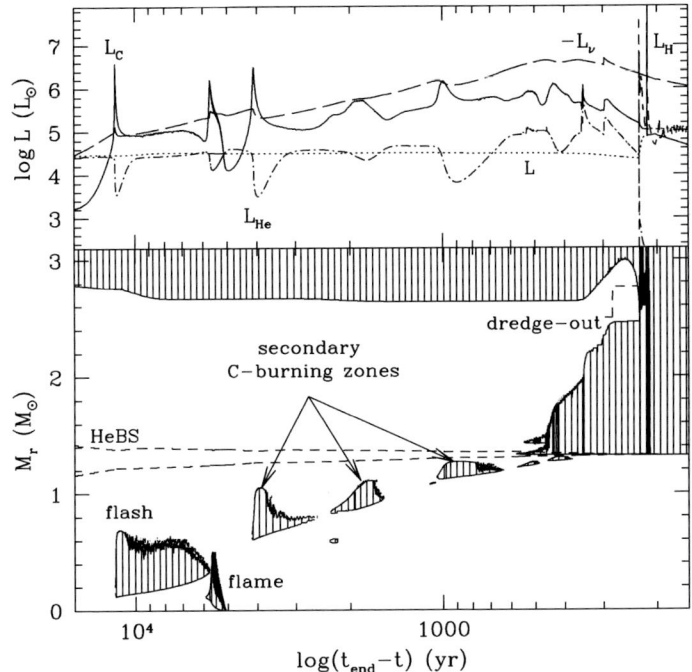

Figure 1. Kippenhahn diagram of a $9.5 M_\odot$ $Z = 0.001$ star during the carbon burning and dredge-out phases. The hatched areas delineate the convective regions and the time is counted backward from the last computed model. The upper panel displays the evolution of the luminosities associated with C (L_C, solid), H (L_H, dashed) and He (L_He, dotted-dashed) burning, with neutrino emission (L_ν, long-dashed) and the surface luminosity L (dotted line) which remains almost constant during this phase.

from the core and is not aware of the structural modifications induced by carbon burning ($L \sim$ constant).

With increasing initial mass, the core temperature is higher and its degeneracy lower. As a consequence the strength of flash and the duration of the flame decrease with increasing mass. Neutrino losses are also weaker in the most massive stars and carbon ignites closer to the center until it reaches the center, in which case the evolution proceeds up to the formation of an iron core in massive stars.

The main nuclear reactions involved in carbon burning are $^{12}\mathrm{C}(^{12}\mathrm{C},p)$ and $^{12}\mathrm{C}(^{12}\mathrm{C},\alpha)$ reactions leading to the production of ^{23}Na and ^{20}Ne. ^{16}O is partially destroyed by (α,γ) reactions but at the same time it is replenished by $^{12}\mathrm{C}(\alpha,\gamma)$ so its abundance remains almost unchanged compared to its value at the end of central He burning. The protons and α particles released by C-burning contribute to the production of ^{24}Mg via $^{23}\mathrm{Na}(p,\gamma)$ and of the ^{25,26}Mg isotopes via (α,n) and (α,γ) reactions on ^{22}Ne, which was synthesized during central He burning. The activation of the neutron source $^{22}\mathrm{Ne}(\alpha,n)$ may also lead to the production of s-elements. At the end of carbon burning, the core is essentially made of ^{16}O ($\sim 50-70\%$), ^{20}Ne ($\sim 15-35\%$) followed by ^{23}Na ($\sim 3-5\%$) plus some residual ^{24,25,26}Mg ($< 3\%$), ^{21,22}Ne and ^{27}Al.

Interestingly, some unburnt ^{12}C may be left in the core after the passage of the flame. Gutiérrez et al. (2005) showed that if the star enters the electron capture (EC) regime traces of carbon as low as 0.015 (in mass fraction) can induce a thermonuclear runaway by igniting explosive oxygen burning. In these circumstances, the supernova explosion leads to complete disruption of the star as in SNIa. In the core, carbon mass fractions as high as ~ 0.04 are found in some of our lowest mass SAGB stars, but they are not expected to follow this fate, provided a binary companion is not influencing their evolution.

2.2. The second dredge-up

At the end of core helium burning, the convective envelope moves inward as gravothermal energy flows from the contracting central regions into the envelope. However, because

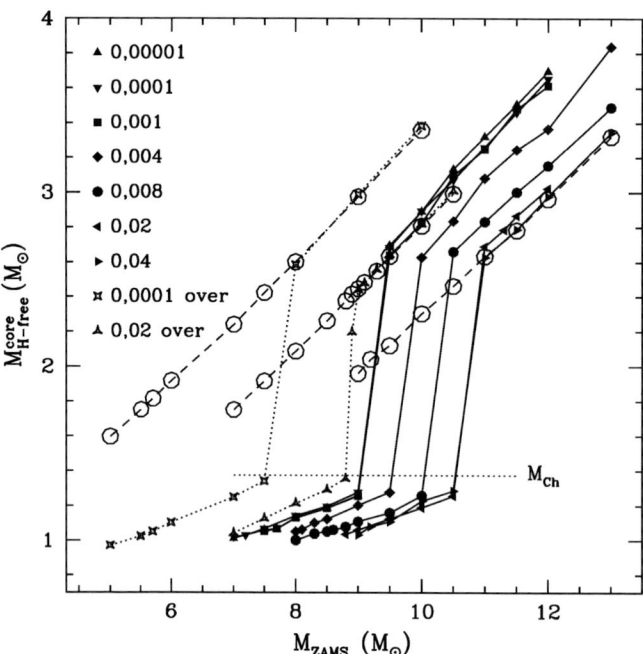

Figure 2. Mass of the H-free core before (dashed line connected by open circles) and after (solid lines) the second dredge-up as a function of the initial mass for various initial compositions. Stars that do not experience the second dredge-up are massive stars that evolve toward core collapse supernova and have H-free cores larger than the Chandrasekhar limit M_{Ch}.

SAGB stars have more massive cores and evolve faster, carbon may ignite before this process is completed. With increasing mass, the second dredge-up (2DUP) occurs later and it is also able to reach deeper layers. In particular, the convective envelope of SAGB stars penetrates below the former location of the H burning shell and brings to the surface large amounts of ^{4}He whose mass fraction increases from $\sim 0.25 - 0.30$ up to ~ 0.45. The 2DUP also contributes to enhance ^{13}C and ^{14}N and decrease the H, ^{12}C and ^{16,18}O surface abundances.

Another characteristic of SAGB evolution is the appearance of the so-called dredge-out phenomenon which takes place at the end of carbon burning when the 2DUP comes to completion (Iben *et al.* 1997, Siess 2007). In the most massive SAGB stars, the energy transferred from the contracting core to the thick He-rich layers that surrounds the ONe core leads to the development of a convective zone in the He burning shell that grows in mass and moves outwards. Eventually this He-driven convective zone merges with the descending envelope in a relatively energetic event as protons are mixed down to where He is burnt. This phenomenon produces a substantial increase in ^{4}He and ^{14}N surface mass fractions.

The 2DUP has another fundamental role: by allowing the envelope to penetrate in the deep stellar interior, it decreases the mass of the H-free core below the Chandrasekhar limit (M_{Ch}). This is illustrated in Fig. 2 where the mass coordinate of the H-exhausted core before (open symbols) and after the completion of the 2DUP is displayed. It is interesting to note that the transition between stars that do and do not experience this deep mixing event occurs abruptly, in a very narrow mass interval of $\lesssim 0.5 M_\odot$. Massive stars avoid the 2DUP and because their core mass exceeds M_{Ch} they are able to activate all nuclear reactions up to the formation of an iron core. SAGB stars could thus be defined as the most massive stars that still experience the 2DUP.

2.3. The thermally pulsing SAGB phase

With the decay of carbon burning and the increasing neutrino emission, the contracting core becomes more degenerate. At the same time, the He burning shell (HeBS) thins out and as it approaches the H rich layers the H-burning shell (HBS) progressively re-ignites.

At this stage of the evolution, the structure of SAGB stars is very similar to that of a lower mass AGB star and the evolution proceeds in an alternate shell burning mode characterized by the development of recurrent convective instabilities in the HeBS. The main difference with "standard" AGB stars resides in the mass and composition of the degenerate core. Because SAGB stars have more massive cores (the minimum mass of an ONe core is predicted to be $\gtrsim 1.05 M_\odot$, Siess 2007), the gravitational pull is stronger and the burning shells are more compressed. They are consequently thinner and their temperature is higher as well. The mass extent of a convective pulse is typically less than a few $10^{-4} M_\odot$ which is more than one order of magnitude smaller that in AGB stars. Owing to the higher temperature in the pulse, when the convective instability develops, the structure reacts more quickly because of the larger contribution of the radiative pressure ($\propto T^4$) to the total pressure. The result is that the thermal pulses are weak (the He generated luminosity $L_{\rm He} \lesssim 10^6 L_\odot$) and their duration relatively short ($\Delta_{\rm pulse} <$ few years). The interpulse duration is also reduced to a few hundred years which implies that the SAGB stars will experience a huge number of thermal pulses, as high as several thousands. But the compressional heating of the internal layers does more than impacting the structure, it also significantly affects the surface abundances, as we shall discuss now.

2.4. Nucleosynthesis

In all our SAGB models, the temperature at the base of the He-driven convective zone exceeds 3.2×10^8 K, allowing for the efficient activation of the ^{22}Ne$(\alpha,$n$)$ neutron source. In principle, this reaction can lead to the synthesis of *s-process* elements but the efficiency of this process may be reduced by the short duration of the neutron irradiation. In addition, because of the large dilution factors involved (the mass of the convective pulse is 4 to 5 orders of magnitude smaller than that of the envelope), the pollution of the surface by third dredge-up (3DUP) episodes is weak and could be hard to detect observationally. It should be noted that He is partially burnt during the thermal pulse and, at the time of the 3DUP, the intershell composition is strongly enriched in ^{22}Ne and ^{25,26}Mg resulting from the α-captures reactions on ^{14}N and ^{22}Ne, respectively.

The main signature of SAGB star nucleosynthesis results from the onset of very efficient proton burning at the base of the convective, also called Hot Bottom Burning (HBB). With envelope temperature ($T_{\rm env}$) ranging between 10^8 and 1.4×10^8 K, all proton burning reactions are activated. The CN cycle works at equilibrium imposing a ^{12}C/^{13}C ratio ~ 4, the ON cycle leaks into the NeNa chain and the two isotopes ^{16,18}O are partially destroyed. The NeNa chain efficiently converts ^{22}Ne into ^{23}Na but above 9×10^7 K, the cycle is broken as (p,γ) reactions onto ^{23}Na become faster that the (p,α) reactions on this element. ^{24}Mg is rapidly depleted to the benefit of ^{25}Mg and ^{26}Al$_g$ is then substantially produced by ^{25}Mg(p,γ). Given the short duration of the TP-SAGB phase ($\tau_{\rm SAGB} \lesssim 10^5$ yr), ^{26}Al$_g$ does not have time to β-decay but above $\gtrsim 1.2 \times 10^8$ K this element suffers heavy proton captures favoring the production of ^{27}Al.

To summarize, because HBB is very strong in SAGB stars, the envelope is efficiently processed by proton burning, increasing the surface abundance of ^{14}N, ^{25}Mg and ^{26,27}Al$_g$ and imposing ^{12}C/^{13}C ~ 4.

2.5. SAGB yields

The yields of SAGB stars are largely unknown and often not accounted for in chemical evolution models despite the relative weight of this stellar population in regards to the IMF. One of the reasons for this lack of data is the huge amount of CPU time needed to follow their full evolution. Assuming, that a 1D stellar evolution code can simulate about 10 thermal pulses a day, given the fact that the star will experience approximately a thousand instabilities, one won't be able to get the uncertain number before at least 3 months! Furthermore, the determination of yields is subject to large uncertainties associated with the efficiency of the third dredge-up, the nuclear reaction rates and, as we shall see, with the thermodynamical conditions at the base of the convective envelope.

The impact of the 3DUP is the result of a subtle competition between on one hand the weak pollution induced by each mixing episode (the envelope mass is up to 10^4–10^5 larger than that of the dredged-up material) and on the other hand on the very large number of thermal pulses. But the problem is more complex and strongly dependent on the initial mass and metallicity. In particular, the imprint of the 3DUP will be stronger at lower metallicity as a consequence of the higher chemical contrast between the intershell and envelope compositions (Siess 2008, in preparation). As a general trend, the 3DUP increases the envelope abundances of ^{12}C, ^{14}N and ^{22}Ne. But this material is subsequently processed by HBB, the efficiency of which is primarily controlled by $T_{\rm env}$. Unfortunately, our poor understanding of convection does not allow us to precisely determine this temperature. Changing the MLT parameter α or using a different convection model has a strong influence on the efficiency of the HBB (Sackmann & Boothroyd 1991, Ventura & d'Antona 2005) through the modification of $T_{\rm env}$. The impact is considerable in SAGB stars and can easily exceed the chemical modifications induced by the presence of the 3DUP (Siess & Arnould 2008). A change in $T_{\rm env}$ mostly affects the nuclides involved in the high temperature proton burning chains. In particular, the abundances of ^{17}O, ^{22}Ne, ^{23}Na, ^{24,26}Mg and ^{26}Al$_g$ can be substantially altered. For example, a 10% uncertainty in $T_{\rm env}$ leads to variations in the ^{26}Al$_g$ mass fraction by a factor ranging between 0.3 and 7 (Siess & Arnould 2008) !

The SAGB yields thus appear to be subject to even larger uncertainties than their lower mass AGB counterparts because of the strong dependence of their surface composition on the efficiency of HBB. Despite these severe limitations, we can reasonably consider SAGB stars as strong contributors of ^{13}C, ^{14}N, ^{17}O, ^{25}Mg and ^{26}Al.

3. The final fate of super-AGB stars

3.1. The mass range of SAGB stars

The masses of SAGB stars range between $M_{\rm up}$, the minimum initial mass for (off-center) carbon ignition and $M_{\rm mas}$ the minimum mass above which the star evolves through all nuclear burning stages up the formation of an iron core. Below $M_{\rm up}$, the evolution leads to the formation of a CO white dwarf (WD) and above $M_{\rm mas}$ the remnant is a neutron star or a black hole depending on the initial mass, composition, mixing and mass-loss rate. The dependence of $M_{\rm up}$ and $M_{\rm mas}$ on the initial metallicity is depicted in Fig. 3 for standard models without core overshooting. The decrease in $M_{\rm up}$ and $M_{\rm mas}$ with Z is due to opacity effects: with less metals, the opacity drops and the temperature and surface luminosity increase. To cope with the structural changes and higher energy demand, the star develops a bigger core which is more prone to ignite carbon (see Siess 2007 for details). It is important to stress that the values of these transition masses are extremely sensitive to the treatment of mixing at the edge of the convective core, the effects becoming critical during central He burning. Taking into account overshooting in the computations increases the size of the core and makes the star behave as if it was more massive. The effect of this extra mixing is significant and shifts the mass range of SAGB stars down by $\sim 2 M_\odot$ (e.g. Siess 2007). This can have significant consequences on chemical evolution models given the strong mass dependence of the IMF. As for massive stars, the evolution of SAGB stars is strongly dependent on the adopted treatment of mixing but also and unavoidably on the numerics (Poelarends et al. 2008).

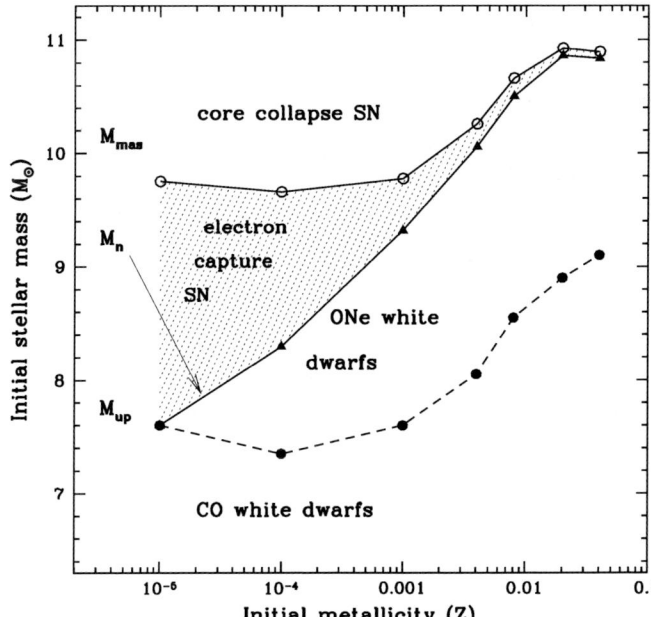

Figure 3. Evolution of the different mass transitions ($M_{\rm up}$, $M_{\rm n}$ and $M_{\rm mas}$) as a function of initial metallicity. Below $M_{\rm up}$, the remnant is a CO white dwarf, between $M_{\rm up}$ and $M_{\rm n}$ the endpoint of the evolution is the formation of an ONe WD. Stars more massive than $M_{\rm mas}$ will proceed through all nuclear burning stages and end their life as core collapse SNe while stars between $M_{\rm n}$ and $M_{\rm mas}$ will experience EC-SN. $M_{\rm n}$ was computed assuming $\dot{M}_{\rm core} = 5 \times 10^{-7} M_\odot\,{\rm yr}^{-1}$ and $\dot{M}_{\rm wind} = 10^{-5} M_\odot\,{\rm yr}^{-1}$.

3.2. The electron capture channel for SAGB stars

After the completion of the 2DUP, the star enters the thermally pulsing AGB phase during which the mass of the degenerate core increases by the accretion of the ashes of H and He shell burning. As the core grows, the central density increases, so does the Fermi energy of the electrons $\varepsilon_{\rm F}$. When $\rho > 10^9$ g cm^{-3} the mass of the core has reached $\sim 1.37 M_\odot$ and $\varepsilon_{\rm F}$ exceeds the threshold energy for electron captures on various elements. These reactions first involve ^{27}Al, followed by ^{25}Mg, ^{23}Na before operating on the abundant ^{20}Ne (Ritossa et al. 1999). At this time, the structure of the core is highly degenerate and pressure is almost entirely supported by the electrons. Thus, when electrons start being substantially captured by ^{20}Ne the pressure drops and core collapse begins. The rapid contraction of the internal layers leads to oxygen ignition and to the conversion of the matter into a nuclear statistical equilibrium (NSE) mixture. The subsequent electron captures on these NSE elements accelerates the collapse that proceeds until the density of nuclear matter is reached (Nomoto 1987). The equation of state stiffens and the inner collapsing shells bounce on the newly formed neutron star. The so-called prompt bounce shock initially progresses outward but it rapidly ($\sim$ 1ms) loses energy by photodissociating and inverting the motion of the infalling shells. It is thanks to the energy deposited by neutrinos that the shock will be revived and the explosion successful (e.g. Dessart et al. 2003, Kitaura et al. 2006, Janka et al. 2008). These explosions are expected to produce sub-luminous type II-P SNe, due to the low H envelope mass ($\lesssim 6-7M_\odot$) and small production of ^{56}Ni (e.g. Bethe & Wilson 1985, Kitaura et al. 2006).

The key point in determining the fate of the SAGB star is whether or not the core mass is able to reach the critical value of $M_{\rm EC} = 1.37 M_\odot$. For instance, if the wind is strong or the core growth somehow reduced (e.g. by deep 3DUPs), the envelope is removed before the central density reaches the threshold for electron capture reactions and the remnant is an ONe WD. On the contrary, if the mass-loss rate is weak, the core can grow up to $M_{\rm EC}$ and a supernova explosion ensues. The fate of the star thus depends on two largely unknown parameters, namely the core growth rate $\dot{M}_{\rm core}$ and the mass-loss rate $\dot{M}_{\rm wind}$. The determination of $\dot{M}_{\rm core}$ is directly related to the efficiency λ of the 3DUP† which depends on our poor understanding of the convective boundaries (see e.g. the contribution of F. Herwig in this proceeding). On the other hand, the knowledge of the mass-loss rate is not better constrained. Since SAGB stars fill the gap between intermediate and massive stars, we have in practice the choice of using mass-loss rate prescriptions that are suited for either stellar class. For typical SAGB stars of $\sim 10^5 L_\odot$, $\sim 1200 R_\odot$ and $T_{\rm eff} \simeq 3000$K, the mass-loss rate ranges between $5.3 \times 10^{-6} M_\odot$ yr^{-1} and 1.5×10^{-4} if one uses the de Jager et al. (1988) or the Vassiliadis & Wood (1993) prescriptions, respectively. The uncertainty in $\dot{M}_{\rm wind}$ is further increased by our poor knowledge of its dependence on the metallicity.

The dividing mass between stars that end their life as ONe WD and those that explode as EC-SN is usually referred to as $M_{\rm n}$ and is determined as follows (see Siess 2007 and Poelarends et al. 2008 for details). For a given set of parameters ($\dot{M}_{\rm core}$, $\dot{M}_{\rm wind}$), we can determine for each star the amount of mass $\Delta M_{\rm core}$ that has been accreted on top of the core during the TP-SAGB phase. From these data, we can then find the initial mass such that, at the end of the evolution when the envelope is removed, the core mass is just equal to the $M_{\rm EC}$, i.e. we search for $M_{\rm n}$ satisfying $M_{\rm core}^{\rm end}(M_{\rm n}) = M_{\rm core}^{\rm eAGB}(M_{\rm n}) + \Delta M_{\rm core}(\dot{M}_{\rm core}, \dot{M}_{\rm wind}) = 1.37 M_\odot$, where $M_{\rm core}^{\rm eAGB}$ is the core mass at the beginning of the TP-SAGB phase. If we set $\dot{M}_{\rm core} = 5 \times 10^{-7} M_\odot$ yr^{-1} (which would correspond to a very inefficient 3DUP characterized by $\lambda \simeq 0$) and impose $\dot{M}_{\rm wind} = 10^{-5} M_\odot$ yr^{-1} and further assume that $\dot{M}_{\rm wind}$ has a metallicity dependence of the form $(Z/Z_\odot)^{1/2}$ (Kudritzki et al. 1987), we obtain for $M_{\rm n}$ the curve shown in Fig. 3. With decreasing metallicity, the mass-loss rate is reduced and a larger fraction of SAGB stars is able to enter the EC-SN regime. We also note that at solar metallicity (with our adopted parameters), the mass window for EC-SN is very small, typically less than $\sim 0.5 M_\odot$ in width. Conversely, at very low Z (below $\sim 10^{-4}$) because of the reduced mass-loss rate, most SAGB stars are expected to follow the EC-SN channel. In this model, the formation of ONe WD is even prevented for $Z \lesssim 10^{-5}$. If we now follow Poelarends et al. (2008) and use a Salpeter IMF to estimate the fraction $f_{\rm EC}$ of EC-SN out of all the SNe we get the following numbers: at solar metallicity, for $M_{\rm mas} = 10.93 M_\odot$ and the previously adopted parameters, we obtain $M_{\rm n} = 10.85 M_\odot$ and $f_{\rm EC} = 1\%$. If instead, we had used a reduced core growth rate of $\dot{M}_{\rm core} = 5 \times 10^{-8} M_\odot$ yr^{-1} (which would have been obtained if $\lambda = 0.9$), $M_{\rm n} = 10.10 M_\odot$ and $f_{\rm EC}$ rises up to $\sim 10\%$. Note that these percentages are weakly affected by the consideration of extra-mixing at the core edge which strongly modifies the mass range of SAGB stars ($M_{\rm up}$ and $M_{\rm mas}$). As illustrated, $f_{\rm EC}$ strongly depends on the 3DUP efficiency but also on the mass-loss rate. With more realistic values of these parameters, Poelarends et al. (2008) find that the current (i.e. at $Z = Z_\odot$) fraction of SNe involving a SAGB progenitor ranges between $\sim 4 - 20\%$.

† By definition and provided the rate of advance of the H burning shell is not affected by the 3DUP, $\dot{M}_{\rm core}(\lambda) = [1 - \lambda] \times \dot{M}_{\rm core}(\lambda = 0)$

4. Conclusion

This overview of SAGB stars has illustrated the wealth of physical processes that pace the evolution of this peculiar stellar population, from the off-center ignition of carbon up to the hydrodynamical phase of the supernova explosion. Among other interesting aspects that were not addressed in this review and which are of fundamental interest is the development of the convective URCA process that starts with the activation of the electron capture reactions (e.g. Ritossa *et al.* 1999). This process is poorly understood and has not yet been fully investigated in our context. It is however of great importance since it sets the conditions for the explosion.

Even if the direct observation of an evolved TP-SAGB star has yet to be confirmed, there are several evidences of their existence. The detection of ONe WD in novae systems (e.g. José & Hernanz 1998) provides certainly the best and most studied observational counterpart of SAGB stars. Actually, ONe WD are very numerous and may be at the center of $\sim 1/3$ of all observed novae (Gil-Pons *et al.* 2003). It is also worth emphasizing that the WD mass distribution (e.g. Liebert *et al.* 2005) shows an extended tail towards higher masses, between $1 \lesssim M_{\rm WD}/M_\odot \lesssim 1.2$ which perfectly fits within the mass range of ONe cores. More recently, the progenitor mass of several sub-luminous type II-P supernovae (e.g. Hendry *et al.* 2005, 2006, Maund *et al.* 2005, Li *et al.* 2007) was estimated. Most interestingly, these studies reveal that these explosions were triggered by stars of $\sim 6-13 M_\odot$, supporting the existence of an EC-SN channel for SAGB stars. Such explosions may also produce neutron stars with low-velocity natal kicks (e.g. Podsiadlowski *et al.* 2004) and be responsible for the formation of high-mass X-ray binaries with long orbital periods and low eccentricities (Pfahl *et al.* 2002). As we have shown previously, because of their relatively high contribution to the IMF, SAGB stars may account for a significant fraction of the currently observed SNII. For the same reason they represent a key ingredient of chemical evolution models and are expected to be strong providers of ^{13}C, ^{14}N, ^{17}O, ^{25}Mg and ^{26}Al as a result of very efficient HBB. In particular, their impact may explain the ^{13}C and ^{14}N enrichments of the interstellar medium observed at low metallicities (Chiappini *et al.* 2005, 2008). From a nucleosynthesis perspective, SAGBs have long been ascribed to be the site of the production of *r-process* elements but this idea is apparently in contradiction with recent consistent hydrodynamical simulations (Hoffman *et al.* 2008) despite previous claims (e.g. Ning *et al.* 2007).

It must be emphasized that all these mass transitions are highly uncertain and strongly dependent on numerous and badly constrained factors such as the mass-loss rate, the properties of the 3DUP, the presence of extra-mixing or the numerics (Eldridge *et al.* 2004, Siess 2007, Poelarends *et al.* 2008). More detailed models of the evolution of SAGB will help clarify these issues and ascertain their role in the chemical evolution of galaxies.

Acknowledgements

The author thanks the organisers for financial support and is most grateful to John Lattanzio for his comments and generous hospitality during the writing of this proceeding. This research was supported under Australian Research Council's Discovery Projects funding scheme (project number DP0877317).

References

Baron, E., Cooperstein, J., & Kahana, S. 1987, ApJ, 320, 300
Bethem, H. A. & Wilson, J. R. 1985, ApJ, 295, 14
Canal, R., Isern, J, & Labay J. 1992, ApJ, 398, 49
Chiappini, C., Ekstrm, S., Meynet, G. *et al.* 2008, A&A, 479, L9

Chiappini, C., Matteucci, F., & Ballero, S. K. 2005, A&A, 437, 429
de Jager, C., Nieuwenhuijzen, H., & van der Hucht K. A. 1988, A&AS, 72, 259
Dessart, L., Burrows, A., Ott, C. D., et al. 2003, ApJ, 644, 1063
Doherty, C. L. & Lattanzio, J. C. 2006, MmSAI 77, 828
Eldridge, J. J. & Tout, C. A. 2004, MNRAS, 353, 87
Gil-Pons, P., García-Berro, E., José, J., Hernanz, M., & Truran, J. W. 2003, A&A, 407, 1021
Gil-Pons, P., Suda, T., Fujimoto, M. Y. & García-Berro, E. 2005, A&A, 433, 1037
Gil-Pons, P., Gutiérrez, J., & García-Berro, E. 2007, A&A, 464, 667
García-Berro, E. & Iben, I. 1994, ApJ, 434, 306
García-Berro, E., Ritossa, C., & Iben, I. 1997, ApJ, 485, 765
Gutiérrez, J., Canal, R., & García-Berro, E. 2005, A&A, 435, 231
Gutiérrez, J., García-Berro, E., Iben, I., et al. 1996, ApJ, 459, 701
Habets, G. M. H. J. 1986, A&A, 167, 61 ApJ, 489, 772
Hendry, M. A., Smartt, S. J., Maund, J. R., et al. 2005, MNRAS, 359, 906
Hendry, M. A., Smartt, S. J., Crockett, R. M., et al. 2006, MNRAS, 369, 1303
Hoffman, R. D., Mller, B., & Janka, H.-T 2008, ApJ, 676, L127
Iben, I., Ritossa, C., & García-Berro, E. 1997,
Imbriani, G., Limongi, M., Gialanella, L., Straniero, O., & Chieffi, A. 2001, ApJ, 558, 903
Janka, H.-T, Mller, B., Kitaura, F. S., & Buras, R. 2008, A&A, in press
José, J. & Hernanz, M. 1998, ApJ, 494, 680
Kitaura, F. S., Janka, H.-Th., & Hillebrandt, W. 2006, A&A, 450, 345
Kudritzki, R. P., Pauldrach, A., & Puls, J. 1987, A&A, 173, 293
Li, W., Wang, X., Van Dyk, S. D., et al. 2007, ApJ, 661, 1013
Liebert, J., Bergeron, P., & Holberg, J. B. 2005, ApJS, 156, L47
Maund, J. R., Smartt, S. J., & Danziger, I. J. 2005, MNRAS, 364, L33
Mayle, R. & Wilson, J. R. 1988, ApJ, 334, 909
Miyaji, S., Nomoto, K., Yokoi, K., & Sugimoto, D. 1980, Publ. Astron. Soc. Japan, 32, 303
Ning, H., Qian, Y.-Z., & Meyer, B. S. 2007, ApJ, 667, L159
Nomoto, K. 1981, IAU Symp., 93, 295
Nomoto, K. 1984, ApJ, 277, 791
Nomoto, K. 1987, ApJ, 322, 206
Pfahl, E., Rappaport, S., Podsiadlowski, P., et al. 2002, ApJ, 574, 364
Podsiadlowski, Ph., Langer, N., Poelarends, A. J. T., et al. 2004, ApJ, 612, 1044
Poelarends, A. J. T., Herwig, F., Langer, N., & Heger, A. 2008, ApJ, 675, 614
Ritossa, C., García-Berro, E., & Iben, I. 1996, ApJ, 460, 489
Ritossa, C., García-Berro, E., & Iben, I. 1999, ApJ, 515, 381
Sackmann, I.-J. & Boothroyd, A. I. 1991, ApJ, 366, 529
Siess, L. 2006, A&A, 448, 717
Siess, L. 2007, A&A, 476, 893
Siess, L. & Arnould, M. 2008, A&A, submitted
Timmes, F. X. & Woosley, S. E. 1992, ApJ, 396, 649
Vassiliadis, E. & Wood, P. R. 1993, ApJ, 413, 641
Ventura, P. & D'Antona, F. 2005, A&A, 431, 279

Discussion

KWOK: Observationally, how do we distinguish SAGB from AGB stars?

SIESS: They are more luminous than AGB stars and probably more Helium rich. Asides from that, their surface composition is very similar to that of massive intermediate mass AGB stars that experience hot bottom burning.

WOITKE: The interpulse timescale of your $M_{initial} = 5 M_\odot$ model matches with $\approx$ 1000 yrs the timescale of concentric rings observed around PPNe and IRC+10216. Could mass loss have reduced the actual mass of IRC+10216 from $5M_\odot \to \leqslant 1.5 M_\odot$ so that it would match?

WILLSON: Might be possible in exceptional cases, but not for the majority of objects with observed concentric rings.

SIESS: It's possible although it would involve a slightly higher mass. Concerning IRC+10216 it has also been proposed by Nearon et al that the rings were formed by the presence of a companion.

C. Meakin is talking to the speaker, L. Siess (right).

Interaction of massive stars with their surroundings

Gerhard Hensler

Institute of Astronomy, University of Vienna, Tuerkenschanzstr. 17, 1180 Vienna, Austria
email: hensler@astro.univie.ac.at

Abstract. Due to their short lifetimes but their enormous energy release in all stages of their lives massive stars are the major engines for the comic matter circuit. They affect not only their close environment but are also responsible to drive mass flows on galactic scales. Recent 2D models of radiation-driven and wind-blown H II regions are summarized which explore the impact of massive stars to the interstellar medium but find surprisingly small energy transfer efficiencies while an observable Carbon self-enrichment in the Wolf-Rayet phase is detected in the warm ionized gas. Finally, the focus is set on state-of-the-art modelling of H II regions and its present weaknesses with respect to uncertainties and simplifications but on a perspective of the requested art of their modelling in the 21st century.

Keywords. Stars: supergiants, (ISM:)H II regions, ISM: kinematics and hydrodynamics, galaxies: evolution

1. Introduction

Massive stars play a crucial role in the evolution of galaxies, as they are the primary source of metals, and they dominate the turbulent energy input into the interstellar medium (ISM) by their massive and fast stellar winds, by the ultraviolet radiation, and by supernova explosions. The radiation field of these stars, at first, photo-dissociates the ambient molecular gas and forms a so-called photo-dissociation region (PDR) of neutral hydrogen. Subsequently, the Lyman continuum photons of the star ionize the H I gas and produce a H II region that expands into the neutral ambient medium.

As these stars have short lifetimes of only a few million years, H II regions indicate the sites of star formation (SF) and are targets to measure the current SF rate in a galaxy. Furthermore, the emission line spectrum produced by the ionized gas allows the accurate determination of the current chemical composition of the gas in a galaxy. Although the physical processes of the line excitation are quite well understood and accurate atomic data are available, so that the spectral analysis of H II regions (see e.g. Stasińska 1979, Evans & Dopita 1987) serves as an essential tool to study the evolution of galaxies, their reliability as diagnostic tool have also to be studied with particular emphasis e.g. to temperature fluctuations (Peimbert 1967, Stasińska 2002) and line excitations.

The simple concept of a uniform medium in ionization equilibrium with the radiation from a massive star (the Strömgren sphere) is successful in describing several global features of H II regions and allows to model the emission line spectrum to a first-order reliability. Since it has long been realized that H II regions are also excerting complex dynamics to the ISM, dynamical modelling of H II regions caused purely by the energy deposit of the stellar radiation field has therefore been started already long ago (see e.g Yorke 1986 and references therein) providing a first insight into the formation of dynamical structures. In addition, also an expanding stellar wind bubble (SWB) with a

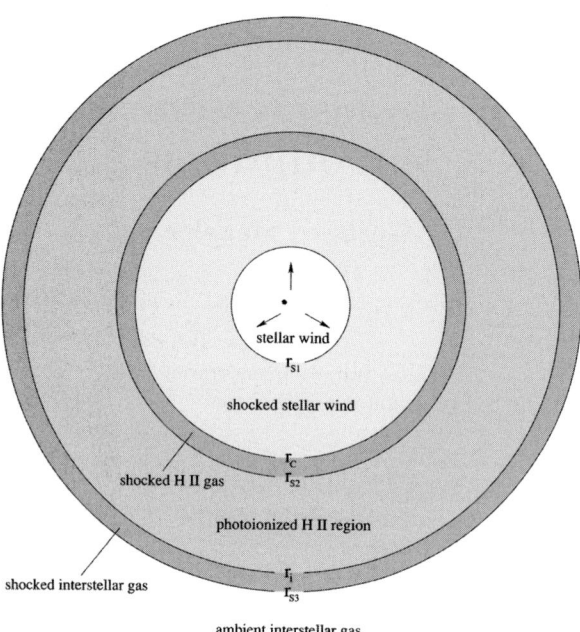

Figure 1. Schematic structure of a SWB as derived by Weaver *et al.* (1977): rs1 marks the position of the reverse shock, rc the contact discontinuity, rs2 the forward shock of the stellar wind bubble, ri the ionization front, and rs3 the forward shock of the HII region expansion.

constant wind power but neglecting its dynamics can be analytically described for the adiabatic phase as the self-similar evolution by Sedov.

In order to approach reality of how a radiation-driven and wind-blown SWB around a massive star interacts with its surroundings one has to take the kinetic energy of the wind and its dissipation into account and has to consider the ionization of the neutral environmental gas where it still exists. Analytically, this can be allowed for in spherical symmetry under the simplifying assumptions of a point source of constant strong wind that interacts with a homogeneous ambient ISM as derived by Weaver *et al.* (1977). This leads to a clear stratification of the surrounding bubble (see fig. 1) from inside out: freely expanding stellar wind, reverse shock, shocked stellar wind, contact discontinuity, SWB forward shock, photo-ionized HII region, ionization front, forward-shocked ambient gas.

Although the analytical and semi-analytical solutions for the evolution of SWBs have been improved over the years as well as the numerical simulations have been done with increasing complexity, like e.g. 2D calculations of SWBs (see e.g. by Różyczka 1985 and a series of papers) and/or combined 1D radiation-hydrodynamical models of HII regions coupled with the dynamical SWB (for references see Freyer *et al.* 2003) a variety of physical effects remains to be included in order to achieve a better agreement of models and observations e.g. with regard to the evolution of the hot phase in bubbles (MacLow 2000, Chu 2000).

2. SWB Models

To improve the insight into the evolution of radiation-driven + wind-blown bubbles around massive stars, we have performed a series of radiation-hydrodynamical simulations with a 2D cylindrical-symmetric nested-grid scheme for stars of masses 15 $M_\odot$ (Kroeger

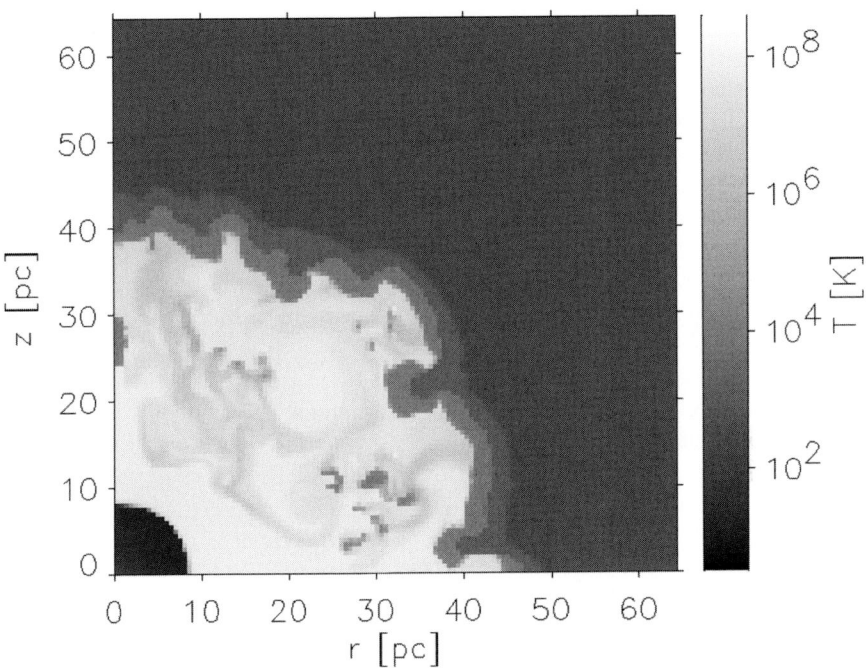

Figure 2. Temperature distriution of stellar wind bubble + HII region surrounding a 60 M$_\odot$ star at an age of 3.3 Myr (Freyer et al. 2003).

et al. in prep.), 35 M$_\odot$ (Freyer et al. 2006), 60 M$_\odot$ (Freyer et al. 2003), and 85 M$_\odot$ (Kroeger et al. 2007). The main issues of these models can be summarized as follows:
1) The HII regions formed around the SWBs have complex structures mainly affected by dynamical processes, like e.g. shell instabilities, vortices, mixing effects, etc. (see fig. 2)
2) The stronger the wind the higher the compresssion of the SWB-surrounding HII region (Hensler et al. 2008), what means that the HII rgion does not increase according to the larger ionizing flux.
3) Finger-like and spiky structures of different densities and temperatures are formed in the photo-ionized region (Freyer et al. 2003).
4) The regions contributing to the HII region emission line spectrum are not solely limited to the photo-ionized shell around the SWB but also form from photo-evaporated gas at the trailing surface of the SWB shock front (Hensler et al. 2008; and see fig. 2).
5) Because dispersion of this cooler photo-evaporated gas into the hot SWB leads to mixing also the stellar material expelled by the wind has to emerge partly in the HII region spectra.
6) As a consequence the metal-enrichment of the wind in the Wolf-Rayet stage which is generally assumed to remain only in the hot SWB for a long time affects the observationally discernible abundance of the HII gas. By these models Kroeger et al. (2006) could prove for the first time that the metal release by Wolf-Rayet stars can be mixed within short timescales from the hot SWB into the warm ionized gas and should become observationally accessible. As the extreme case for the 85 M$_\odot$ star we found a 22% enhancement of Carbon, but neglible amouts for N and O.

Since the occurrence of a WR phase is strongly metal dependent, the enrichment with C should also depend on the average metallicity. This would mean that any radial gradient of C abundance of HII regions in galactic disks is steeper than that of O. And indeed,

Esteban *et al.* (2005) found $d[\log(C/O)]/dr = -0.058 \pm 0.018$ dex kpc^{-1} for the Galactic disk.

7) As expected from the distribution of HII gas the radially projected Hα brightness shows a decrease to the center and a slight brightening to the limb but not as strong as expected according to the increase of the line-of-sight with impact parameter (Freyer *et al.* 2003). This effect depends on the bubble age and evolves from central brightning to a moderate central trough. It also demonstrates both: the neglection of heat conduction and the homogeneous initial density do not allow a sufficient brightning of heat conductive interfaces so that, secondly, only the photo-evaporated backflow can contribute to the Hα luminosity in present models. In reality, condensations which become embedded into the hot SWB are exposed to heat conduction.

8) The sweep-up of the slow red supergiant wind by the fast Wolf-Rayet wind produces remarkable morphological structures and emission signatures which agree well with observed X-ray luminosity and temperature as well as with the limb brightening of the radially projected X-ray intensity profile (for details see Freyer *et al.* 2006).

9) Connected to 1) the higher compression of the pushed HII region leads to a stronger recombination and, by this, a higher energy loss by means of collisionaly excited line emission. This means that the energy transfer efficiencies ϵ's for both radiative as well as kinetic energies remain much lower than analytically derived (more than one order of magnitude) and amount to only a few per mil (Hensler 2007). There is almost no dependence on the stellar mass what contrasts expectations because the energy impact by Lyman continuum photons and by wind luminosity increase with stellar mass. Vice versa, since the gas compression is stronger by a more energetic wind also the energy loss by radiation is more efficient.

3. Caveats and Required Developments for the Art of Modelling in the 21st Century

Nonetheless, words of caution and unfortunately of discouragement have to be expressed here with respect to various aspects:

At first, the stellar evolutionary models are not yet unique but depend on the authors. In order to get a quantifiable comparison of the models by García-Segura *et al.* (1996a, 1996b) with our 2D radiation-hydrodynamical simulations (Freyer *et al.* 2003, 2006) for the 35 and 60 M$_\odot$ studies we used the same stellar parameters. Since no stellar parameters were available from the same group for the 15 and 85 M$_\odot$ models we had to make use of the Geneva models (Schaller *et al.* 1992). A comparison of the age-dependent parameters between the Langer and the Geneva 60 M$_\odot$ models (fig. 3) has revealed enormous differences in the energetics by almost one order of magnitude as well as that the Wolf-Rayet and Luminous Blue Variable stages occur in contrary sequence, respectively (Kroeger 2006).

Secondly, as presented by George Meynet (this conference) the role of stellar rotation was until recently totally underestimated, mainly, because of a lack of model insights. In principle, the radiation-driven stellar wind is smaller at the equator of a rotating star with respect to the poles due to van Zeippel's theorem, while in contrast the centrifugal-driven amount should increase. In addition, the radial and azimuthal redistribution of fresh fuel and already burned material in the stellar interior is uncertain as long as at least 2D stellar evolutionary models are lacking.

Hirschi *et al.* (2004, 2005) showed that taking stellar rotation into account in the Geneva stellar evolution code increases the yields for heavy elements by a factor of

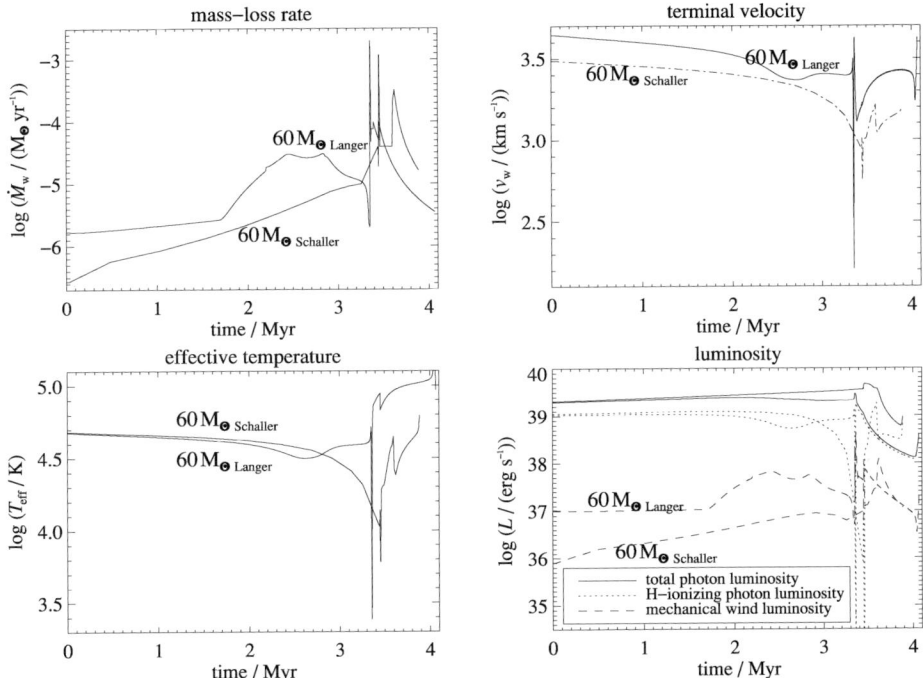

Figure 3. Comparison of the 60 $M_\odot$ star parameters from evolutionary models by Langer García-Segura *et al.* 1996a and Schaller *et al.* (1992) (see line notation). Notice that in the lower-left figure total luminosity by Schaller *et al.* is represented by the upper curve. (from Kroeger 2006)

1.5 – 2.5 for stars between 15 and 30 $M_\odot$. For the more massive stars rotation raises the yields of ^{4}He and other H-burning products like ^{14}N, but the yields of He-burning products like ^{12}C are smaller. Additionally, for stars with $M \gtrsim 60$ $M_\odot$, the evolution differs from that of non-rotating stars by the following manner: Rotating stars enter the WR regime already in the course of their main-sequence. Nevertheless, the WR phase has to be treated in the context of fully self-consistent evolutionary models with meridional circulations.

Already in non-rotating models for metallicities less than solar two effects reduce the heavy element release by WR stars: First, the lower the metallicity the more massive a star has to be to evolve through the WR stages. Therefore, the number of WR stars decreases with decreasing metallicity. Schaller *et al.* (1992) found that with metallicity $Z = 0.001$ the minimal initial H-ZAMS mass for a WR star is > 80 $M_\odot$. At $Z = 0.02$ the minimal initial mass is > 25 $M_\odot$. Second, the lower the metallicity the shorter are the WR lifetimes, and not all WR stages are reached. At solar metallicity WR stars enter all three WR stages (WNL, WNE, WC), whereas at $Z = 0.001$ only the WNL phase is reached (Schaller *et al.* 1992). The WR lifetime of an 85 $M_\odot$ star, e.g., is $t_{WR} = 0.204 \times 10^5$ yr at $Z = 0.001$ and $t_{WR} = 4.008 \times 10^5$ yr at $Z = 0.02$ (Schaller *et al.* 1992).

At third, the quiescent SF as a self-regulated process is a widely accepted concept. The stellar feedback can adapt both signs, positive as a triggering mechanism in a self-propagating manner like in superbubble shells vs. negative as self-regulation. Primarily the correlation between the surface density of disk galaxies' HI gas and the vertically integrated SF rate derived from the Hα flux (Kennicutt 1998) serves as the best proof of a SF self-regulation. Such self-regulation is a plausible process due to the energy release mainly by massive stars as demonstrated by Köppen *et al.* (1995), whereas its level of the SF rate is determined by the deposit of the released energy to the ISM. While the windless i.e. purely photo-ionizing models and analytical results e.g. by Lasker (1967) reach about one percent for the energy transfer efficiency ϵ (see e.g. Freyer *et al.* 2003), the radiation-dirven + wind-blown models by us fall short by more than one order of magnitude even to below 0.1 percent.

As a further aspect, the two-dimensionality of the numerical treatment must be overcome to 3D in order to allow for a proper description of turbulent eddies and small-scale inhomogeneities. In addition this requires also a change in the spatial resolution of the numerical code: the nested-grid strategy has to be changed to a flexible mesh adaptivity (adaptive mesh refinement: AMR). This approach was already developed by Rijkhorst *et al.* (2006) for the publicly available and widely used AMR code FLASH.

Last but not least, massive stars are born in OB associations with separations such low that their wind-driven H II regions should overlap. For observations these colliding SWBs should produce a higher X-ray luminosity than expected from individual massive SWBs. Recent X-ray observations of hot gas even in the Orion complex (Guedel *et al.* (2007)) support this expectation. The modelling of such star-forming regions is highly complex, necessarily 3D, and requires inherently AMR. Only those explorations will enable us to answer the problem of timescales and energetics necessary for the observed gas evacuation of star clusters (Baumgardt & Kroupa 2008).

As one recognizes we are still in the natal phase of understanding and modelling structure and evolution of massive stars and their influence on the ambient ISM. Much before the end of this 21st century revolutionary observational results and numerical models will enlighten our present-day ignorance.

Acknowledgements

The author is gratefully acknowledging contributions and discussions by Tim Freyer, Joachim Köppen, Danica Kroeger, Simone Recci, Wolfgang Vieser, and Harald W. Yorke. Part of the work was funded by the DFG under grants HE 1487/17. G.H. cordially thanks the organizers for the invitation to this conference. The participation was made possible by grants from the conference and from the University of Vienna.

References

Baumgardt, H. & Kroupa, P. 2008, *MNRAS*, 380, 1589
Chu, Y.-H. 2000, *Rev.Mex.Astron.Astrofis.*, 9, 262
Esteban, C., García-Rojas, J., Peimbert, M., *et al.* (2005). *ApJ*, 618, L95
Evans, I. N. & Dopita, M. A. 1987, *ApJ*, 319, 662
Freyer, T., Hensler, G., & Yorke, H. W., 2003, *ApJ*, 594, 888
Freyer, T., Hensler, G., & Yorke, H. W., 2006, *ApJ*, 638, 262
Fukuda, N. & Hanawa, T. 2000, *ApJ*, 533, 911
García-Segura, G. & Mac Low, M.-M. 1995, *ApJ*, 455, 145
García-Segura, G., Mac Low, M.-M., & Langer, N. 1996a, *A&A*, 305, 229
García-Segura, G., Langer, N., & Mac Low, M.-M. 1996b, *A&A*, 316, 133
Guedel, M., Briggs, K. R., Montemerle, T., *et al.* 2007, *Science*, 319, 309
Hensler, G. 2003, *ASP Conf. Ser. Vol.*, 304, eds. C. Charbonnel *et al.*, p. 371
Hensler, G. 2007, *EAS Publ. Ser. Vol.*, 24, eds. E. Ensellem *et al.*, p. 113
Hensler, G., Recchi, S., Kroeger, D., & Freyer, T., 2008, in *"Pathways through an eclectic Universe"*, ASP Conf. Ser., in press, eds. J. Knapen *et al.*, p. 75
Hirschi, R., Meynet, G., & Maeder, A., 2004, *A&A*, 425, 649
Hirschi, R., Meynet, G., & Maeder, A., 2005, *A&A*, 433, 1013

Kennicutt, R. J. 1998, *ApJ*, 498, 541
Köppen, J., Theis, C., & Hensler, G. 1995, *A&A*, 296, 99
Kroeger, D. 2006, PhD thesis, University of Kiel, Germany
Kroeger, D., Hensler, G., & Freyer, T. 2006, *A&A*, 450, L5
Kroeger, D., Freyer, T., Hensler, G., & Yorke, H. W. 2007, *A&A*, submitted
Lasker, B. M. 1967, *ApJ*, 149, 23
MacLow, M.-M. 2000, *Rev.Mex.Astron.Astrofis.*, 9, 273
Peimbert, M. 1967, *ApJ*, 150, 825
Rijkhorst, E.-J., Plewa, T., Dubey, A., & Mellema, G. 2006, *A&A*, 452, 907
Różyczka, M. 1985, *A&A*, 143, 59
Schaller, G., Schaerer, D., Maeder, A., & Meynet, G. 1992, *A&ASuppl.*, 96, 269
Stasińska, G. 1979, *A&ASuppl.*, 32, 429
Stasińska, G. 2002, *Rev.Mex.Astron.Astrofis.*, 12, 62
Weaver, R., McCray, R., Castor, J., *et al.* 1977, *ApJ*, 218, 377; erratum, *ApJ*, 220, 742
Yorke, H. W. 1986, *ARAA*, 24, 49

Discussion

LANGER: To what extent do you need many different stellar evolution input models for your bubble models; it looked as if the energy transfer efficiencies in the 4 very different cases you tried were very similar.

HENSLER: That is correct. From our models the energy transfer efficiency seems not to depend on the stellar mass although the power of radiative and wind releases do. This can be understood, because the stronger stellar wind at larger masses lead to stronger compression of surrounding gas, and, by this, to stronger cooling by collisional excitated emission. Different evolutionary models, however, affect other issues. We explored e.g. that differences in the kinetic energies of the stellar wind (see fig. 3) change the dynamical structure of the 60 $M_\odot$ model in the sense that the fingure-like structures (see Freyer *et al.* 2003) are less pronounced for Schaller *et al.* (2002) parameters. In addition, differences in stellar evolutionary models also change the LBV-WR sequence and the element release and they have stronger effects on the chemical evolution and the self-enrichment of HII regions.

G. Hensler is doing his presentation.

Massive star evolution: from the early to the present day Universe

Georges Meynet[1], Sylvia Ekström[1], Cyril Georgy[1], André Maeder[1], Raphael Hirschi[2]

[1] Observatory of Geneva University, Switzerland
email: georges.meynet@obs.unige.ch

[2] EPSAM, University of Keele, UK
email: r.hirschi@epsam.keele.ac.uk

Abstract. Mass loss and axial rotation are playing key roles in shaping the evolution of massive stars. They affect the tracks in the HR diagram, the lifetimes, the surface abundances, the hardness of the radiation field, the chemical yields, the presupernova status, the nature of the remnant, the mechanical energy released in the interstellar medium, etc... In this paper, after recalling a few characteristics of mass loss and rotation, we review the effects of these two processes at different metallicities. Rotation probably has its most important effects at low metallicities, while mass loss and rotation deeply affect the evolution of massive stars at solar and higher than solar metallicities.

Keywords. stars: abundances, early-type, evolution, mass loss, emission-line, Be, rotation

1. Mass loss due to radiative forces

Radiation triggers mass loss through the line opacities in hot stars. It may also power strong mass loss through the continuum opacity when the star is near the Eddington limit. For cool stars, radiation pressure is exerted also on the dust.

For hot stars, typical values for the terminal wind velocity, v_∞ is of the order of 3 times the escape velocity, *i.e.* about 2000-3000 km/s, mass loss rates are between 10^{-8}-10^{-4} M$_\odot$ per year increasing with the luminosity and therefore the initial mass of the star (Vink *et al.* 2000; 2001). The comparison of mass loss rates for O-type stars obtained by different technics shows sometimes very important differences. For instance, Fullerton *et al.* (2006) using UV line of P^{+4} obtained mass loss rates reduced by a factor ten or more with respect to mass loss determination from radio or Hα determination. Bouret *et al.* (2005) obtained qualitatively similar results to Fullerton *et al.* (2006) but with considerable lower reduction factor (about 3). Such reduction of the mass loss rates during the O-type star phase may have important consequences. Typically a 120 M$_\odot$ loses during its lifetime of 2.5 Myr about 50 M$_\odot$ whith mass loss rates of the order of 2 10^{-5} M$_\odot$ per year. Dividing this mass loss rate by 10, would imply that in the same period, the star would lose only 5 M$_\odot$! Unless stars are strongly mixed (by e.g. fast rotation), or that all WR stars originate in binary systems, it would be difficult to understand how WR stars form with such low mass loss rates.

Stars with initial masses below about 30 M$_\odot$ at solar metallicity evolve to the red supergiant stage where mass loss is enhanced with respect to the mass loss rates in the blue part of the HR diagram (see for instance de Jager *et al.* 1988). In this evolutionary stage, determination of the mass loss is more difficult than in the blue part of the HR diagram due in part to the presence of dust and to various instabilities active in red super giant atmospheres (e.g. convection becomes supersonic and turbulent pressure can no long be ignored). An illustration of the difficulty come from the determination of red supergiant mass loss rates by van Loon et al. (2005). Their study is based on the analysis of optical spectra of a sample of dust-enshrouded red giants in the LMC, complemented with spectroscopic ad infrared photometric data from the literature. Comparison with galactic AGB stars and red supergiants shows excellent agreement for dust-enshrouded objects, but not for optically bright ones. This indicates that their recipe only applies to dust-enshrouded stars. If applied to objects which are not dust enshrouded, their formula gives values which are overestimated by a factor 3-50! In this context the questions of which stars do become dust-enshrouded, at which stage, for how long, become critical to make correct implementations of such mass loss recipes in models.

Stars with initial masses above about 30 $M_\odot$ at solar metallicity may evolve into a short Luminous Blue Variable (LBV) phase. LBV stars show during outbursts mass loss rates as high as 10^{-4}-10^{-1} $M_\odot$ per year. For instance η Carinae ejected near the middle of the nineteenth century between 12 and 20 $M_\odot$ in a period of 20 years, giving an average mass loss rate during this period of 0.5 - 1 $M_\odot$ per year. Such a high mass loss cannot be only radiatively driven according to Owocki et al. (2004). These authors have shown that the maximum mass loss rate that radiation can drive is given by $\dot{M} \sim 1.4 \times 10^{-4} L_6 M_\odot \text{yr}^{-1}$, with L_6 the luminosity expressed in unit of 10^6 $L_\odot$. This means that for $L_6 = 5$ (about the case of ηCar) the maximum mass loss rate would be less than 10^{-3} $M_\odot$ per year, well below the mass loss during the outbursts. These outbursts, which are more shell ejections than steady stellar winds, involve other processes in addition to the effects of the radiation pressure. Among the models proposed let us mention the geyser model by Maeder (1992b), or the reaching of the $\Omega\Gamma$-limit (Maeder & Meynet 2000a).

After the LBV phase, massive stars evolve into the Wolf-Rayet phase, also characterized by strong mass loss rates. Many recent grids of stellar models use the recipe given by Nugis & Lamers (2000) for the WR mass loss rates. These authors deduced the mass loss rates from radio emission power and accounted for the clumping effects.

1.1. Metallicity dependence of the stellar winds

In addition to the intensity of the stellar winds for different evolutionary phases, one needs to know how the winds vary with the metallicity. This is a key effect to understand the different massive star populations observed in regions of different metallicities. This has also an important impact on the nature of the stellar remnant and on the chemical yields expected from stellar models at various metallicities.

Current wisdom considers that very metal-poor stars lose no or very small amounts of mass through radiatively driven stellar winds. This comes from the fact that when the metallicity is low, the number of absorbing lines is small and thus the coupling between the radiative forces and the matter is weak. Wind models impose a scaling relation of the kind $\dot{M}(Z) = \left(\frac{Z}{Z_\odot}\right)^\alpha \dot{M}(Z_\odot)$, where $\dot{M}(Z)$ is the mass loss rate when the metallicitity is equal to Z and $\dot{M}(Z_\odot)$ is the mass loss rate for the solar metallicity, Z being the mass fraction of heavy elements. In the metallicity range from 1/30 to 3.0 times solar, the value of α is between 0.5 and 0.8 according to stellar wind models (Kudritzki et al. 1987; Leitherer et al. 1992; Vink et al. 2001). Such a scaling law implies for instance that a non-rotating 60 $M_\odot$ with $Z = 0.02$ ends its stellar life with a final mass of 14.6 $M_\odot$, the same model with a metallicity of $Z = 10^{-5}$ ends its lifetime with a mass of 59.57 $M_\odot$ (cf. models of Meynet & Maeder 2005 and Meynet et al. 2006 with $\alpha = 0.5$).

During the red supergiant stage, at the moment there is no commonly accepted rule to account for a possible metallicity dependence of the winds. Let us just mention here that according to van Loon et al. (2005), dust-enshrouded objects mass loss appears to be similar for objects in the LMC and the Galaxy. This may have very important consequences for our understanding of metal-poor red supergiant stars. For the LBV's, there is also no real knowledge on how mass loss can depend on metallicity. If the mechanism is mainly triggered by continuum opacity, we can expect that there is only a weak or may be no dependence on the metallicity.

Until very recently, it was considered that the WR mass loss rates did not depend on the initial metallicity i.e. that a WN stars in the SMC, LMC and in the Galaxy would lose mass at the same rate provided they have the same luminosity and the same actual surface abundances. This view has been challenged byVink & de Koter (2005) who find that the winds of WN stars are mainly triggered by iron lines. They suggest a dependence of mass loss on Z (initial value) similar to that of massive OB stars. According to these authors, the winds of WC stars depends also on the iron abundance, but in this case, the metallicity dependence is less steep than for OB stars. Their results apply over a range of metallicities given by $10^{-5} \leqslant (Z/Z_\odot) \leqslant 10$. Very interestingly, they find that once the metal abundance drops below $(Z/Z_\odot) \sim 10^{-3}$, the mass loss of WC stars no longer declines. This is due to an increased importance of radiative driving by intermediate-mass elements, such as carbon. These results have profound consequences for the evolution of stars at low metallicity, affecting the predicted Wolf-Rayet populations (Eldridge & Vink 2006), the evolution of the progenitors of collapsars and long soft Gamma Ray Bursts (Yoon & Langer 2005; Woosley & Heger 2006ab; Meynet & Maeder 2007).

2. Rotation

Rotation induces many processes in stellar interior (see the review by Maeder & Meynet 2000a). In particular, it drives instabilities which transport angular momentum and chemical species. Assuming that the star rapidly settles into a state of shellular rotation (constant angular velocity at the surface of isobars), the transport equations due to meridional currents and shear instabilities can be consistently obtained (Zahn 1992). Since the work by J.-P. Zahn, various improvements have been brought to the formulas giving the velocity of the meridional currents (Maeder & Zahn 1998), those of the various diffusive coefficients describing the effects of shear turbulence (Maeder 1997; Talon & Zahn 1997; Maeder 2003; Mathis et al. 2004), as well as the effects of rotation on the mass loss (Owocki et al. 1996; Maeder 1999; Maeder & Meynet 2000b).

Let us recall a few basic results obtained from rotating stellar models:

1) Angular momentum is mainly transported by the meridional currents. During the Main-Sequence phase, the core contracts and spins up and the envelope expands and spins down. The meridional currents impose some coupling between the two, slowing down the core and accelerating the outer layers. In the outer layers, the velocity of these currents becomes smaller when the density gets higher. As a consequence, the transport of angular momentum from inner to outer regions is less efficient at low metallicity where stars are more compact and thus more dense in the outer layers.

2) The chemical species are mainly transported by shear turbulence (at least in absence of magnetic fields; when magnetic fields are amplified by differential rotation as in the Tayler-Spruit dynamo mechanism, Spruit 2002, the main transport mechanism is meridional circulation, Maeder & Meynet 2005). This process may produce changes of the surface abundances of rotating stars already during the Main-Sequence phase (Heger & Langer 2000; Meynet & Maeder 2000). The shear turbulence is stronger when the gradients of the angular velocity are stronger, i.e. at low metallicity where stars are more compact and where the meridional currents are slower.

In addition to these internal transport processes, rotation also modifies the physical properties of the stellar surface. Indeed the shape of the star is deformed by rotation (a fact which is now put in evidence observationally thanks to the interferometry, see Domiciano de Souza et al. 2003). Rotation implies a non-uniform brightness (also now observed, see e.g. Domiciano de Souza et al. 2005). The polar regions are brighter than the equatorial ones. This is a consequence of the hydrostatic and radiative equilibrium (von Zeipel theorem 1924). The von Zeipel theorem has many very interesting consequences, among them let us mention the followings (see Maeder 1999; Maeder & Meynet 2000b):

• Since the radiative flux and the effective gravity vary as a function of the colatitude, one can define a local Eddington factor as the ratio of the actual flux at that colatitude to the maximum flux allowed at that colatitude (the maximum flux is defined as the flux for which the radiative acceleration compensates for the effective gravity).

• The critical velocity, defined as the value of the rotation velocity at the equator such that the centrifugal acceleration compensates for the net radial attracting force (which results from the gravity in part counterbalanced by an outward radiative acceleration) is different when the stellar luminosity is near or far from the Eddington luminosity.

• The mass loss rates per unit surface varies as a function of the colatitude. Rotation induces wind anisotropies. Polar winds are expected in fast rotating hot stars.

• The line driven mass loss rates are enhanced by rotation.

2.1. Dependence on metallicity of effects induced by rotation

The effects of rotation are not the same at different metallicities. These differences may have different causes:

(a) The distribution of the rotational velocities on the ZAMS may depend on the metallicity. Presently, observations seem to favor high rotation at low Z. For instance, the observed number fraction of Be stars with respect to the total number of B stars is higher at low than at high metallicity (Maeder et al. 1999; Wisniewski & Bjorkman 2006). Be stars are supposed to rotate near the critical velocity and are characterized from an observational point of view by the presence of an equatorial disk expanding outwards where emission lines (responsible for the "e" in Be name) are formed. Another argument supporting the greater number of fast rotators at low metallicity is the fact that the observed surface velocity of stars on the Main-Sequence do appear to be higher at low metallicities. For instance Hunter et al. (2008a) obtain that SMC metallicity stars rotate on average faster than galactic ones (mainly field objects). No difference is found between galactic and LMC stars. Martayan et al. (2007a) find that, for B and Be stars, the lower the metallicity, the higher the rotational velocities (see the review by Meynet et al. 2008 for a more complete review of recent results from large massive star surveys). Let us note that part of this effect may be attributed not to a difference in the initial velocity distribution but to the fact that at low metallicity mass loss rates are weak (see above) and thus remove small amounts of angular momentum from the star.

Let us note that the mechanisms which evacuate the angular momentum from the protostellar cloud during the stellar formation phase may have different efficiencies at low and high metallicity. For instance at high metallicity, the fraction of chemical species easily ionized is higher. This might imply a stronger magnetic coupling between the disk and the star (disk-locking effect) during the pre-MS phases and thus a more efficient spin-down. At low metallicity, it may also occur that the disk is more rapidly photoevaporate by the strong ionizing flux of young massive stars in the vicinity of the nascent star. Indeed, when the metallicity decreases, stars are bluer. This effect may make the disk-locking phase shorter and allow the nascent star to retain more angular momentum†.

(b) The evolution of the surface (equatorial) velocity results of a delicate interplay between the mass loss (which removes angular momentum at the surface) and the meridional circulation (which brings angular momentum from the core to the surface). In the left panel of Fig. 1, we show the internal profile of the radial component of the meridional circulation in four models of 20 $M_\odot$ with $\Omega_{\rm ini}/\Omega_{\rm crit} = 0.50$ at various metallicities‡. Let us focus on the outer cell, which transports the angular momentum outward when $U(r)$ is negative. The amplitude of the meridional circulation is a factor 6 higher in the standard solar metallicity model ($Z = 0.020$) compared to the $Z = 0.002$ model. This factor amounts to 25 when we compare with the $Z = 10^{-5}$ model, and reaches 100 with the $Z = 0$ one. This illustrates the effect of the Gratton-Öpik term ($\propto 1/\rho$) in the expression of the meridional circulation velocity $U(r)$: when the metallicity decreases, stars are more compact, so the density increases, and thus $U(r)$ decreases. In the right panel of Fig. 1, we see the resulting evolution of the equatorial velocity. At standard metallicity, although the amplitude of the meridional circulation is large, the loss of angular momentum through the radiative winds has the strongest effect, and the equatorial velocity slows down. At low or very low Z, the meridional circulation is weak, but the mass loss is so diminished that the models are spinning up. In the case of $Z = 0$ strictly, there is no mass loss to remove mass and angular momentum, but the meridional circulation is so weak that the evolution of $\Omega(r)$ is very close to local angular momentum conservation, $\Omega r^2 =$ constant: because of the natural inflation of the external radius, the surface of the model has to slow down.

(c) A consequence of the above metallicity dependence is that the gradient of the angular velocity in a low metallicity stellar model is steeper than in a corresponding model at high metallicity. Since the shear instability responsible for the mixing of the chemical mixing is stronger when the gradient of Ω is stronger, this implies that a given initial mass star at low metallicity will undergo more mixing than a similar star at high metallicity.

In the following we explore the consequences of rotating stellar models with mass loss in different metallicity environments.

3. Evolution of massive stars with mass loss and rotation at $Z = 0$

Recently Ekström et al. (2008b) computed a set of rotating Pop III massive star models. These authors chose as initial equatorial velocity, a value of 800 km s^{-1} which corresponds to an angular momentum content approximately equivalent to that of massive star models at solar metallicity with rotation velocities of about 300 km s^{-1}. Let us recall that the value of 300 km s^{-1} on the ZAMS produce models with a time averaged velocity on the MS equal to 200 and 250 km s^{-1}, i.e. well in the range of observed values (see e.g. Huang & Gies 2006ab).

Comparing the evolution with and without rotation for Pop III stellar models, the following main differences can be noted:

• At $Z = 0$ models rotate with an internal profile $\Omega(r)$ close to local angular momentum conservation, because of a very weak core-envelope coupling.

---

† In that context it is interesting to note that the rotational velocities of massive stars in clusters is higher than in the field supporting the view that the environment plays also a role in shaping the rotational velocity distribution.

‡ These models are taken from Ekström et al. 2008b

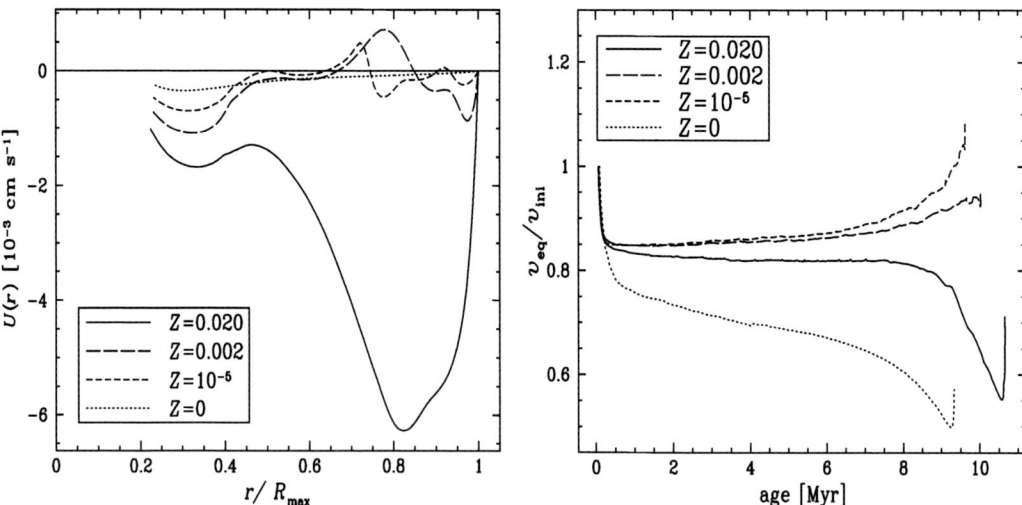

Figure 1. Models of 20 M$_\odot$ at various metallicities, with $\Omega_{\rm ini}/\Omega_{\rm crit} = 0.5$. *Left:* internal profile of $U(r)$, where $u(r,\theta)$ the vertical component of the velocity of the meridional circulation is $u(r,\theta) = U(r)\ P_2(\cos\theta)$. The radius is normalized to the outer one. All the models are at the same evolutionary stage, when the central H mass fraction is about 0.40. *Right:* evolution of the equatorial velocity, normalized to the initial velocity. Figure taken from Ekström *et al.* submitted.

- Rotational mixing drives a H-shell boost due to a sudden onset of CNO cycle in the shell, which leads to high ^{14}N production. This production can be as much as 106 times higher than the production of the non-rotating models. Generally, the rotating models produce much more metals than their non-rotating counterparts.
- The mass loss is very low, even for the models that reach the critical velocity during the main sequence. It may however have an impact on the chemical enrichment of the Universe, because some of the stars are supposed to collapse directly into black holes, contributing to the enrichment only through their winds. While in that case non-rotating stars would not contribute at all, rotating stars may leave an imprint in their surrounding.
- Due to the low mass loss and the weak coupling, the core retains a high angular momentum at the end of the evolution. The high rotation rate at death probably leads to a much stronger explosion than previously expected, changing the fate of the models.

None of our models meet the conditions for becoming WR stars. They end their evolution having kept their envelope, and thus seem to fail in becoming a GRB progenitor as defined in the collapsar model of Woosley (1993). However, we have seen that the rotating 85 $M_\odot$ is likely to undergo pulsational pair instability† and thus lose some mass in this process. If the mass lost is large, it may become a GRB progenitor since the very low core-envelope coupling has maintained a high angular momentum in the core.

4. Evolution of massive stars with mass loss and rotation for $0< Z \leqslant \sim 0.002$

In contrast to the case of massive Pop III stars, which ignite hydrogen through the pp chains, stars beginning their life with a tiny amount of metals (of the order of 10^{-10} in mass fraction) ignite their hydrogen through the CNO cycle. The energy produced by the CNO cycle is sufficient to compensate for the loss of energy by the surface, while that produced by the pp chain is not. In this last case, the star has to extract energy from

† This process has not been followed in the present models.

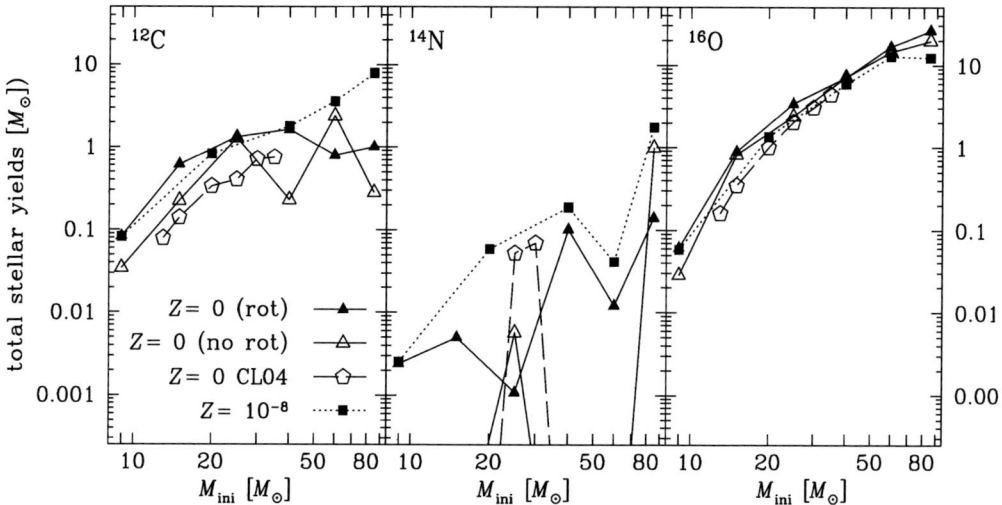

Figure 2. Yields comparison between the non-rotating $Z = 0$ models from Chieffi & Limongi (2004; open pentagons), the rotating $Z = 10^{-8}$ models from Hirschi (2007; filled squares) and those of Ekström et al. (2008b; rotating filled triangles; and non-rotating open triangles) $Z = 0$ models. *Left:* ^{12}C; *centre:* ^{14}N; *right:* ^{16}O. Figure taken from Ekström et al. submitted.

its gravitational reservoir and contracts†. As a consequence the temperature in the core of Pop III stars is in the range of values for He-burning, while the central temperature in non zero metallicity stars is well below. This means also that at the end of the MS phase, the core of the Pop III star will not have to contract a lot in order to activate the reactions of He-burning, while the core of non zero metallicity stars will have to contract significantly. This contraction will create a steep gradient of angular velocity at the border of the core, which will drive a strong mixing, much stronger than in Pop III stars. This is the reason why, although the production of primary nitrogen is already quite significant in the rotating Pop III stars, very metal poor stars are still more efficient producers (for a given initial angular momentum content, see Fig. 2). In Fig. 3 predictions of chemical evolution models using yields of different stellar models are compared with observations of the N/O and ^{12}C/^{13}C ratios at the surface of metal poor halo stars.

Another consequence of the strong mixing in non-zero metallicity stars is the evolution toward the red part of the HR diagram (Maeder & Meynet 2001). In the red part of the HR diagram, a convective zone appears at the surface which dredges-up at the surface great quantities of CNO elements, leading to the loss of a very large amounts of mass. In our 60 $M_\odot$ stellar model with $Z = 10^{-8}$ and $v_{\rm ini} = 800$ km s^{-1}, the CNO content at the surface amounts to one million times the one the star had at its birth (Meynet et al. 2006). Therefore the global metallicity at the surface becomes equivalent to that of a LMC stars while the star began its life with a metallicity which was about 600 000 times lower! If we apply the same rules used at higher metallicity relating the mass loss rate to the global metallicity, we obtain that the star loses about half of its initial mass due to this effect. As shown by Meynet et al. (2006) and Hirschi (2007) the matter released by these winds is enriched in both H- and He-burning products and present striking similarities with the abundance patterns observed at the surface of C-rich Ultra-Metal-Poor-Stars

† The slow contraction stops when the central temperature becomes high enough for triple alpha reactions to be activated. The 3α reactions produce some carbon (of the order of 10^{-10} in mass fraction). From this stage on, hydrogen burns in the core as in more metal rich stars *i.e.* through the CNO cycle.

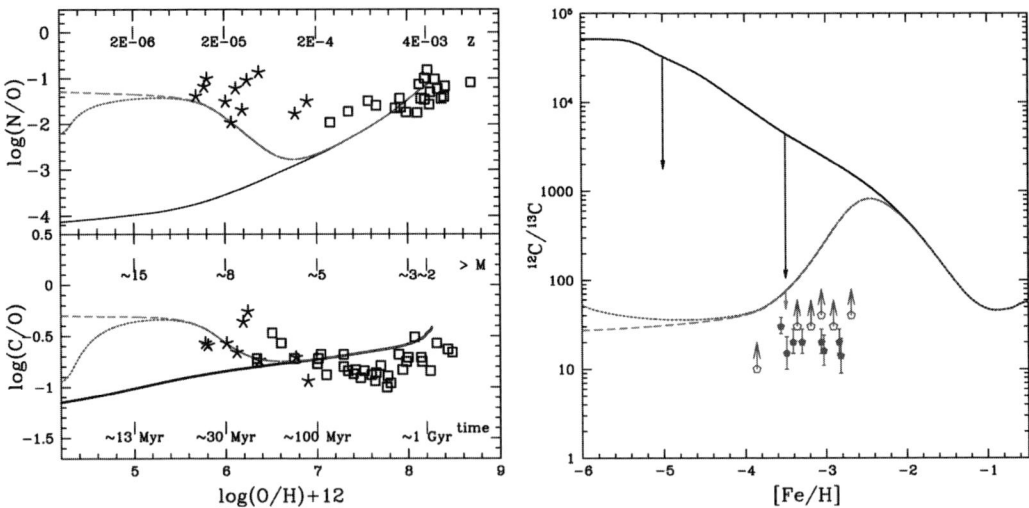

Figure 3. The continuous curve is the chemical evolution model obtained with the stellar yields of slow rotating $Z = 10^{-5}$ models from Meynet & Maeder (2002) and Hirschi et al. (2004). The dashed line includes the yields of fast rotating $Z = 10^{-8}$ models from Hirschi (2007) at very low metallicity. The dotted curve is obtained using the yields of the $Z = 0$ models of Ekström et al. (submitted) up to $Z = 10^{-10}$. *Left:* evolution of the N/O and C/O ratios. Data points are from Israelian et al. (2004; open squares) and Spite et al. (2005; stars). *Right:* evolution of the $^{12}C/^{13}C$ ratio. Data points are *unmixed* stars from Spite et al. (2006): open pentagons are lower limits. The arrows going down from the theoretical curves indicate the final $^{12}C/^{13}C$ observed in giants (after the dredge-up), starting from the initial composition values given by the stellar models (see Chiappini et al. 2008). Figure taken from Ekström et al. submitted.

(CRUMPS). This process does not occur in the present Pop III stellar models. As can be deduced from the explanations above, Pop III stars undergo less efficient mixing during the core He-burning phase. Thus the H-burning shell and the outer radiative envelope are enriched at a lower rate in CNO elements. This tends to delay the apparition of an outer convective zone and to reduce its extension. The increase of the surface metallicity remains very modest and does not lead to strong mass loss.

5. Evolution of massive stars with mass loss and rotation for $Z >\sim 0.002$

Mass loss and rotation play major roles in shaping the populations of massive stars observed in the nearby Universe. For instance predictions of single star models have been obtained for explaining the populations of Be stars (Ekström et al. 2008a), of blue and red supergiants in the SMC (Maeder & Meynet 2001), of Wolf-Rayet stars (Meynet & Maeder 2003, 2005). Fig. 4 (left panel) shows the variation as a function of the metallicity of the number fraction of supernovae having as progenitors WNL, WNE, WC and WO stars. We see that WO star progenitors are predicted to occur only at low metallicities. Smith & Maeder (1991) explained this trend in the following way: when the mass loss rates are low (*i.e.* at low metallicities), more time is needed to remove the H- and He-rich layers and thus when the He-burning core is uncovered, it has reached a more advanced evolutionary stage (*i.e.* more helium converted into carbon and oxygen). Observations show that indeed most of WO stars (6 out of 8) are found in regions with Z inferior to about 0.9 $Z_\odot$. The fact that WO stars are preferentially found at low metallicity has been invoked to associate these stars to the progenitors of long soft GRBs (Hirschi et al 2005).

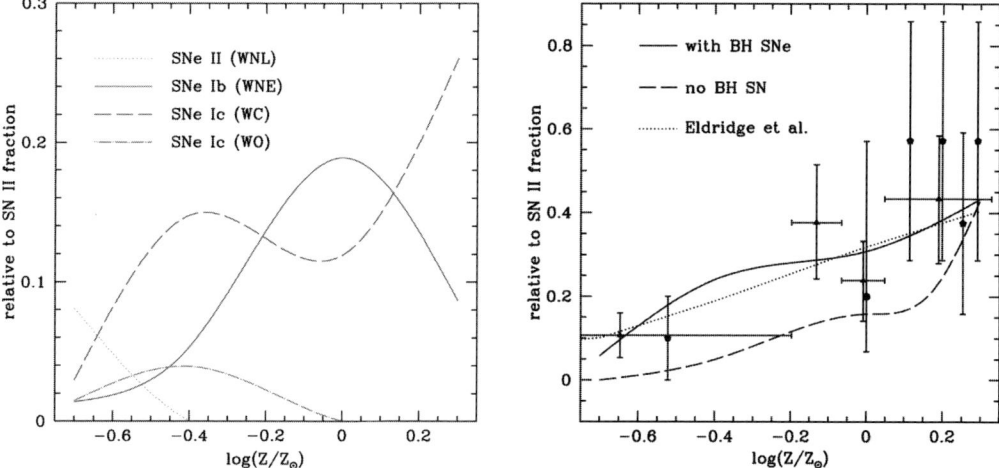

Figure 4. *Left Panel:* Variation as a function of the metallicity of the number fraction of supernovae having as progenitors WNL, WNE, WC and WO stars. The fractions were deduced from the models of Meynet & Maeder (2003; 2005) using a Salpeter IMF. *Right Panel:* Rate of SN Ibc / SN II if all models produce a SN (solid line) or if models producing a black holes do not explode in a SN (dashed line). Pentagons are observational data from Prieto *et al.* (2008), and triangles are data from Prantzos & Boissier (2003). Figure taken from Georgy *et al.* in preparation.

Considering that all models ending their lifetime as a WNE or WC/WO phase will explode as a type Ibc supernova, it is possible to compute the variation with the metallicity of the number ratio of type Ibc to type II supernovae. The result is shown in Fig. 4 (right panel). One sees that this ratio increases with the metallicity. This is due to the fact that at higher metallicity, the minimum initial mass of stars ending their life as WNE, or WC/WO stars is lower than at lower metallicities. Single star model can reasonably well reproduce the observed trend. Models accounting for single and binary channel (but without rotation) are shown as a dotted line (Eldridge *et al.* 2008). They also provide a good fit to the observations. In this last work most of the supernovae originate from the binary channel, leaving little place for the single star scenario.

It might be that the single and binary channels will not predict the same distributions of supernovae among type Ib and Ic's. The observations by Prieto *et al.* (2008) indicate that type Ic's are about twice as frequent as type Ib's at solar metallicity. Georgy *et al* (in preparation) show that rotating single star models can account for this distribution. It would be very interesting to have the predictions of the binary channel for this feature. Hopefully it will allow to better estimate the respective importance of these two channels.

References

Bouret, J.-C., Lanz, T., & Hillier, D. J. 2005, A&A, 438, 301
Chiappini, C., Ekström, S., Meynet, G., Hirschi, R., Maeder, A., & Charbonnel, C. 2008, *A&A*, 479, L9
Chieffi, A. & Limongi, M. 2004, *ApJ*, 608, 405
Domiciano de Souza, A., Driebe, T., & Chesneau, O. 2007, *AJ*, 464, 81
Ekström, S., Meynet, G., Maeder, A., & Barblan, F. 2008, *A&A*, 478, 467
Ekström, S., Meynet, G., Chiappini, C., Hirschi, R., & Maeder, A. 2008b, *A&A*, submitted
Eldridge, J. J. & Vink, J. S. 2006, A&A, 452, 295
Fullerton, A. W., Massa, D. L., & Prinja, R. K. 2006, ApJ, 637, 1025

Heger, A. & Langer, N. 2000, ApJ, 544, 1016
Hirschi, R. 2007, A&A, 461, 571
Hirschi, R., Meynet, G., & Maeder, A. 2004, A&A, 425, 649
Hirschi, R., Meynet, G., & Maeder, A. 2004, A&A, 443, 581
Huang, W. & Gies, D. R. 2006a, ApJ, 648, 580
Huang, W. & Gies, D. R. 2000b, ApJ, 648, 591
Hunter, I., Lennon, D. J., Dufton, P. L. et al. 2008, A&A 479, 541
Israelian, G., Ecuvillon, A., Rebolo, R., Garca-Lpez, R., Bonifacio, P., & Molaro, P. 2004, A&A, 421, 649
de Jager, C., Nieuwenhuijzen, H., & van der Hucht, K. A. 1988, A&AS, 72, 259
Kudritzki, R. P., Pauldrach, A. W. A, & Puls, J. 1987, A&A, 173, 293
Leitherer, C., Robert, C., & Drissen, L. 1992, ApJ, 401, 596
van Loon, J. Th., Cioni, M.-R. L., Zijlstra, A. A., & Loup, C. 2005, A&A, 438, 273
Maeder, A. 1992b, Proc. of the Int. Conf. Amsterdam, Ed. C. de Jager & H. Nieuwenhuijzen, North Holland, p.138.
Maeder, A. 1997, A&A, 321, 134
Maeder, A. 1999, A&A, 347, 185
Maeder, A. 2003, A&A, 399, 263
Maeder, A. & Meynet, G. 2000a, ARAA, 38, 143
Maeder, A. & Meynet, G. 2000b, A&A, 361, 159
Maeder, A. & Meynet, G. 2001, A&A, 373, 555
Maeder, A. & Meynet, G. 2005, A&A, 440, 1041
Maeder, A., Grebel, E. K., & Mermilliod, J.-C. 1999, A&A, 346, 459
Maeder, A., & Zahn, J. P. 1998, A&A, 334, 1000
Martayan, C., Floquet, M., & Hubert, A.-M. et al. 2007, A&A, 472, 577
Mathis, S., Palacios, A., & Zahn, J.-P. 2004, A&A, 425, 243
Meynet, G., Ekstroem, S., & Maeder, A. 2006, A&A, 447, 623
Meynet, G., Ekstrom, S., & Maeder, A. 2008, to be published in the conference proceedings of First Stars III, Santa Fe, arXiv:0709.2275
Meynet, G. & Maeder, A. 2000, A&A, 361, 101
Meynet, G. & Maeder, A., 2002, A&A, 390, 561
Meynet, G. & Maeder, A. 2005, A&A, 429, 581
Meynet, G. & Maeder, A. 2007, A&A, 464, L11
Nugis, T. & Lamers, H. J. G. L. M. 2000, A&A, 360, 227
Owocki, S. P., Cranmer, S. R., & Gayley, K. G. 1996, ApJ, 472, L115
Owocki, S. P., Gayley, K. G., & Shaviv, N. J. 2004, ApJ, 616, 525
Prantzos, N. & Boissier, S. 2003, A&A, 406, 259
Prieto, J. L., Stanek, K. Z., & Beacom, J. F. 2008, ApJ, 673, 999
Smith, L. F. & Maeder, A. 1991, A&A, 241, 77
Spite, M., Cayrel, R., & Plez, B. 2005, A&A, 430, 655
Spite, M., Cayrel, R., & Hill, V. 2006, A&A, 455, 291
Spruit, H. C. 2002, A&A, 381, 923
Talon, S. & Zahn, J. P. 1997, A&A, 317, 749
Trundle, C., Dufton, P. L., Hunter, I. et al. 2007, A&A, 471, 625
Vink, J. S. & de Koter, A. 2005, A&A, 442, 587
Vink, J. S., de Koter, A., & Lamers, H. J. G. L. M. 2000, A&A, 362, 295
Vink, J. S., de Koter, A., & Lamers, H. J. G. L. M. 2001, A&A, 369, 574
Wisniewski, J. P. & Bjorkman, K. S. 2006, ApJ, 652, 458
Woosley, S. E. 1993, ApJ, 405, 273
Woosley, S. E. & Heger, A. 2006a, ApJ, 637, 914
Woosley, S. E. & Heger, A. 2006b, in The Fate of the Most Massive Stars, ASP Conference Series, Vol. 332, Edited by R. Humphreys and K. Stanek. San Francisco: Astronomical Society of the Pacific, p.407
Yoon, S.-C. & Langer, N. 2005, A&A, 443, 643
Zahn, J.-P. 1992, A&A, 265, 115

Discussion

BELCZYNSKI: You said since winds of rotating stars remove less angular momentum, rotating models may help collapsar/GRB scenario. However, you also pointed out that (1) very massive stars (the ones to make black holes, I guess) slow down rapidly; (2) that central parts of fast rotating star are not really rotating fast (even if envelope is spinning close to breakup). So since collapsar model produces GRB, from the collapse of the core, that is required to spin fast, I can not see, how fast initial rotation and anisotropic winds are going to help to produce collapsar?

MEYNET: (1) You first remark (very massive stars slow down rapidly) is correct for solar or higher metallicity stars. This is not true for metal poor stars which have weak stellar winds. Long soft GRB are preferentially observed in low metallicity regions where mass loss does not remove a lot of angular momentum from the surface. (2) The plot showed that indeed only a tiny part of the star rotate at the critical velocity while the interior rotation at a small fraction of the critical velocity. The star is in the MS phase. Now if we isolate, let's say the 3 inner solar mass of the model, we deduce from the model, the angular momentum contained in these 3 $M_\odot$ and compute the rotation of a NS which would contain the same amount of angular momentum, then you would obtain a very high rotation rate, probably much higher than the critical velocity. Said in other words, the fact that at this stage, the core has a rotation rate well below the critical velocity does not mean that its angular momentum content would be unsufficient for forming a collapsar at the presupernova stage.

YI: The helium & N enriched mass loss from extremely metal-poor rotating stars is interesting as a possible solution to the extreme helium population found in GCs. Is there a specific mass range where this happens on does it happen to all massive rotating metal-poor stars?

MEYNET: In the papers by Decressin *et al.* (2006, 2007), you will find the quantity of Helium (newly synthesized) by fast rotating stars of masses between 20 and 200 $M_\odot$ and ejected under the form of a slow wind. This process can occur for a broad range of masses provided the star has at the beginning of its life a sufficient amount of angular momentum and provided it does not undergo too strong mass loss by line driven winds (which would remove angular momentum too rapidly).

Around the Pair Instability Valley – Massive SN Progenitors

Roni Waldman

Racah Institute of Physics, Hebrew University, Jerusalem 91904, Israel
email: waldman@cc.huji.ac.il

Abstract. The discovery of the extremely luminous supernova SN 2006gy, possibly interpreted as a pair instability supernova, renewed the interest in very massive stars. We explore the evolution of these objects, which end their life as pair instability supernovae or as core collapse supernovae with relatively massive iron cores, up to about $3\,M_\odot$.

Keywords. stars: evolution, (stars:) supernovae: general

1. Introduction

The interest in the evolution of very massive stars (VMS), with masses $\gtrsim 100\,M_\odot$, has recently been revived by the discovery of SN 2006gy – the most luminous supernova ever recorded (Smith et al. 2007). This object, having a luminosity of ~ 10 times that of a typical core-collapse SN (CCSN), is probably the first evidence of a pair instability SN (PISN) Woosley et al. (2007). PISN are massive stellar objects, whose evolutionary path brings their center into a region in thermodynamical phase space ($\rho \lesssim 10^6, T \gtrsim 10^9$), where thermal energy is converted into the production of electron-positron pairs, thus resulting in loss of pressure and hydrodynamic instability. This type of supernova was first suggested 40 years ago by Barkat et al. (1967), and since then several works were carried out (e.g. Fraley 1968; Ober et al. 1983; El Eid et al. 1983; Bond et al. 1984; Heger & Woosley 2002; Hirschi et al. 2004; Eldridge & Tout 2004; Nomoto et al. 2005), however the overall interest in this topic has been relatively small, mainly due to lack of observational data.

It was originally believed that stars massive enough to produce PISN could only be found among population III stars with close to zero metallicity ($Z \lesssim 10^{-4}$), and hence only at very high redshift ($z \gtrsim 15$). More recently Scannapieco et al. (2005) discussed the detectability of PISN at redshift of $z \leqslant 6$, arguing that metal enrichment is a local process, therefore metal-free star-forming pockets may be found at such low redshifts. Langer et al. (2007) introduced the effect of rotation into studying this question concluding that PISN could be produced by slow rotators of metallicity $Z \lesssim Z_\odot/3$ at a rate of one in every 1000 SN in the local universe. Furthermore, Smith et al. (2007) point out, that mass loss rates in the local universe might be much lower than previously thought, so that massive stars might be left with enough mass to become PISN. This conclusion is also supported by Yungelson et al. (2008) who extensively discuss the mass loss rates and fates of VMS. It is interesting to note, that SN 2006gy took place in the nearby Universe. Following the discovery of SN 2006gy, Umeda & Nomoto (2008) addressed the question of how much ^{56}Ni can be produced in massive CCSN, while Heger & Woosley (2008) computed the detailed nucleosynthesis in these SNe.

The interest in VMS is further motivated by the discovery of Ultraluminous X-ray Sources (ULX), which can be interpreted as mass-accreting intermediate mass black holes (IMBH) with mass $\sim (10^2 - 10^5)\,M_\odot$. One of the possible scenarios for IMBH formation is by VMS formed by stellar mergers in compact globular clusters (see e.g. Yungelson *et al.* 2008, and references therein). In this context, Nakazato *et al.* (2006, 2007) studied the collapse of massive iron cores with $M \gtrsim 3\,M_\odot$. In their first paper they treat the fate of stars of mass $\geqslant 300\,M_\odot$ which reach the photodisintegration temperature ($\approx 6 \times 10^9 K$) after undergoing pair instability. The entropy per baryon of these models at photodisintegration is $s > 16 k_B$ compared with the classical core-collapse SN with $s \sim 1 k_B$. In the second paper they aim to bridge this entropy gap, corresponding to core masses of $(3-30)\,M_\odot$ but claim that there is a lack of systematic progenitor models for this range, hence they use synthetic initial models for their calculations.

In this work we focus mostly on the mass range $M \lesssim 80\,M_\odot$ (He core mass $M_{He} \lesssim 36\,M_\odot$) immediately below the range which enters the pair instability region, and present a systematic picture of the resulting CCSN progenitors.

2. Method

Since the mass loss rates of stars in this range are highly uncertain, (see e.g. discussion by Yungelson *et al.* 2008), we avoid dealing with this question by following the example of Heger & Woosley (2002), and modeling the evolution of helium cores. Our helium core initial models are homogeneous polytropes composed entirely of helium and metals, with metallicity $Z \approx 0.015$, in the mass range $(8-160)\,M_\odot$. The models were then evolved to the helium zero age main sequence. In the following we will refer to these models as "HeN" where N is the mass of the model. For comparison we evolved also a few models of regular hydrogen stars, beginning from the zero-age main sequence (ZAMS). We will refer to these models as "MN" where N is the mass of the model. All our models have no mass-loss. We argue that as long as the mass loss rate is not so high that it will cut into the He-core, the evolution after the main-sequence phase will be virtually independent of the fate of the hydrogen-rich envelope. We followed the evolution of each model until the star is either completely disrupted (for the PISN case) or Fe begins to photo-disintegrate (for the CCSN case).

We followed the evolution using the Lagrangian one dimensional Tycho evolutionary code version 6.92 (with some modifications), publicly available on the web (the code is described in Young & Arnett 2005). Convection is treated using the well known mixing length theory (MLT) with the Ledoux criterion. In the MLT formulation of Tycho, the value of the mixing length parameter fit to the Sun is $\alpha_{MLT} \approx 2.1$ (Young & Arnett 2005), so we used a value of $\alpha_{MLT} = 2$ in our calculations. The nuclear reaction rates used by TYCHO are taken from the NON-SMOKER database as described in Rauscher & Thielemann (2000).

3. Results and Discussion

Among the He-core models we computed, those in the mass range $M \leqslant 36\,M_\odot$ do not reach pair instability and end their lives as CCSN. Fig. 1 shows the evolution of the central density and temperature (left panel) and the density structure of the pre-SN (right panel), at the moment when the central temperature reaches $7 \times 10^9\,K$. The two extreme models He8 and He36 are shown, as well as M80 which has a He-core mass similar to the He36 model, and M20 – a typical CCSN progenitor. The left panel also shows two He-core models – He80 and He160 that reach pair instability. Note that the two models He36 and M80 are indeed very close to each other. An example of the composition at the pre-SN stage ($T_c = 7 \times 10^9 K$) is shown in Fig. 2. The Fe-core mass is $\approx 3\,M_\odot$, topped by a shell of $\approx 10\,M_\odot$ of Si-group elements. The size of the Fe-core (defined as

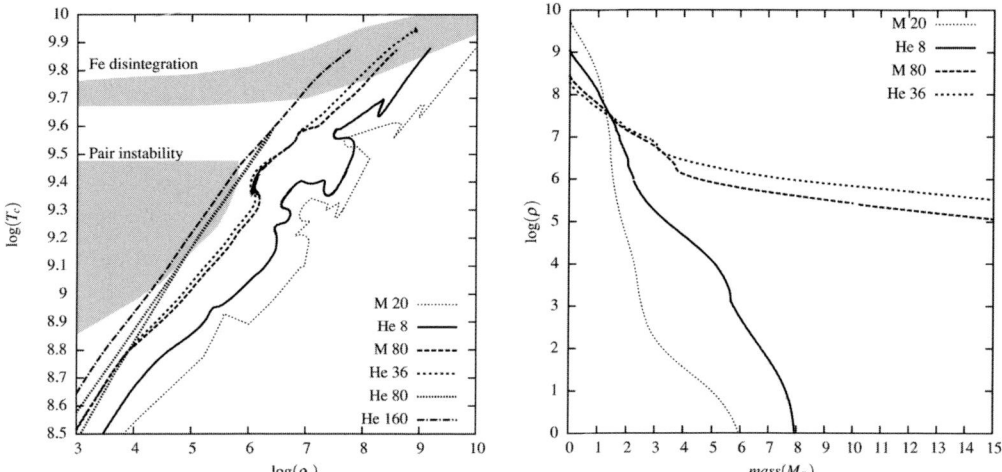

Figure 1. Evolution of the central density and temperature (left panel) and pre-SN density structure (right panel). Each line is labeled "M" for stellar models and "He" for He-core models, followed by the mass of the model.

the mass coordinate where the electron mole fraction $Y_e < 0.49$) and the central entropy per baryon at the pre-SN stage for the whole set of models is shown in Fig. 3. Note that the size of the Fe-core is slightly non-monotonic. The central entropy, is monotonic with mass, but slightly differs between He-core and stellar models.

As can be seen in the above figures, the outstanding features of these massive models compared with the typical CCSN example M20 are:

(a) Lower central density and higher central entropy.

(b) A much shallower density profile.

(c) Relatively large Fe-cores, up to about $3\,M_\odot$, and a large amount (up to about $10\,M_\odot$) of Si-group elements.

These differences might have a considerable impact on the behavior of these models during core collapse and on the outcome of the explosion, a question which we hope to address in the future.

Acknowledgements

I would like to thank Zalman Barkat for many hours of fruitful discussion. I would like to acknowledge David Arnett for providing his TYCHO code for public use.

References

Barkat, Z., Rakavy, G., & Sack, N. 1967, Phys. Rev. Lett., 18, 379
Bond, J. R., Arnett, W. D., & Carr, B. J. 1984, ApJ, 280, 825
El Eid, M. F., Fricke, K. J., & Ober, W. W. 1983, A&A, 119, 54
Eldridge, J. J. & Tout, C. A. 2004, MNRAS, 353, 87
Fraley, G. S. 1968, Ap&SS, 2, 96
Heger, A. & Woosley, S. E. 2002, ApJ, 567, 532
—. 2008, ArXiv e-prints, 803
Hirschi, R., Meynet, G., & Maeder, A. 2004, A&A, 425, 649
Langer, N., Norman, C. A., de Koter, A., Vink, J. S., Cantiello, M., & Yoon, S.-C. 2007, A&A, 475, L19
Nakazato, K., Sumiyoshi, K., & Yamada, S. 2006, ApJ, 645, 519
—. 2007, ApJ, 666, 1140

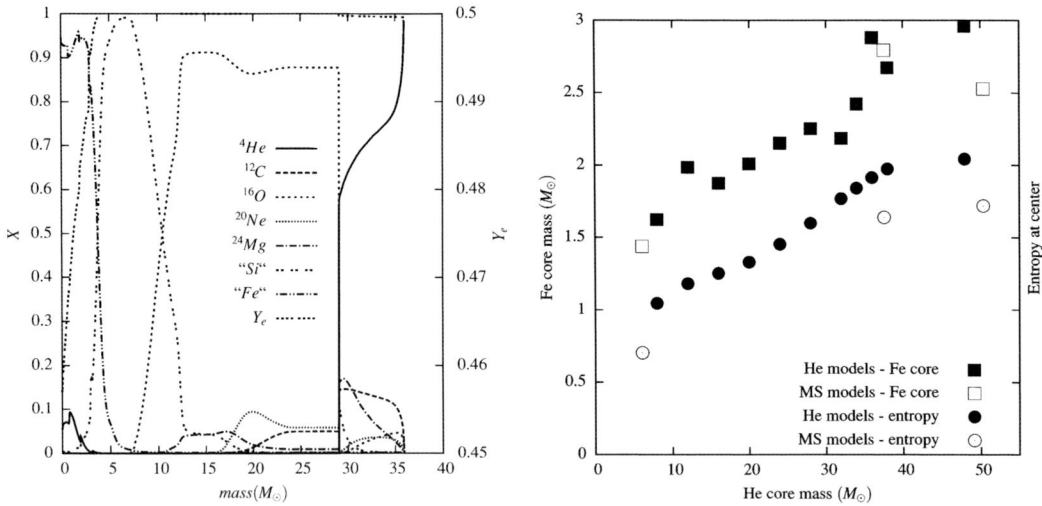

Figure 2. Pre-SN composition of the He36 model. "Si" and "Fe" stand for the total of Si- and Fe-group elements respectively.

Figure 3. Mass of the Fe-core (squares) and central entropy per baryon (circles) for the computed models. Filled shapes designate He–core models, open shapes – stellar models.

Nomoto, K., Tominaga, N., Umeda, H., Maeda, K., Ohkubo, T., Deng, J., & Mazzali, P. A. 2005, in Astronomical Society of the Pacific Conference Series, Vol. 332, The Fate of the Most Massive Stars, ed. R. Humphreys & K. Stanek, 374–+

Ober, W. W., El Eid, M. F., & Fricke, K. J. 1983, A&A, 119, 61

Rauscher, T. & Thielemann, F.-K. 2000, Atomic Data and Nuclear Data Tables, 75, 1

Scannapieco, E., Madau, P., Woosley, S., Heger, A., & Ferrara, A. 2005, ApJ, 633, 1031

Smith, N., Li, W., Foley, R. J., Wheeler, J. C., Pooley, D., Chornock, R., Filippenko, A. V., Silverman, J. M., Quimby, R., Bloom, J. S., & Hansen, C. 2007, ApJ, 666, 1116

Umeda, H. & Nomoto, K. 2008, ApJ, 673, 1014

Woosley, S. E., Blinnikov, S., & Heger, A. 2007, Nature, 450, 390

Young, P. A. & Arnett, D. 2005, ApJ, 618, 908

Yungelson, L. R., van den Heuvel, E. P. J., Vink, J. S., Portegies Zwart, S. F., & de Koter, A. 2008, A&A, 477, 223

Discussion

KUPKA: Are there any observations available for the candidate supernovae you have mentioned in your talk, which would support the element yield rates your progenitor models would predict? Since their interior structure is quite different from the usual core-collapse progenitor models, one would expect also a distinctive pattern in abundances (mass fractions) produced by the subsequent supernovae.

WALDMAN: As far as I know there is no conclusive abundance data that can affirm the identification of these SNe with PISN.

^{60}Fe and Massive Stars

Wei Wang

National Astronomical Observatories, Chinese Academy of Sciences, Beijing 100012, China
email: wangwei@bao.ac.cn

Abstract. Gamma-ray line emission from radioactive decay of ^{60}Fe provides constraints on nucleosynthesis in massive stars and supernovae. We detect the γ-ray lines from ^{60}Fe decay at 1173 and 1333 keV using three years of data from the spectrometer SPI on board *INTEGRAL*. The average flux per line is $(4.4 \pm 0.9) \times 10^{-5}$ ph cm^{-2}s^{-1} rad^{-1} for the inner Galaxy region. Deriving the Galactic ^{26}Al gamma-ray line flux with using the same set of observations and analysis method, we determine the flux ratio of ^{60}Fe/^{26}Al gamma-rays as 0.15 ± 0.05. We discuss the implications of these results for the widely-held hypothesis that ^{60}Fe is synthesized in core-collapse supernovae, and also for the closely-related question of the precise origin of ^{26}Al in massive stars.

Keywords. ISM: abundances, nucleosynthesis

1. Introduction

The radioactive isotope ^{60}Fe is believed to be synthesized through successive neutron captures on Fe isotopes (e.g., ^{56}Fe) in a neutron-rich environment inside He burning shells in AGB stars (^{60}Fe is stored in white dwarfs and cannot be ejected) and massive stars, before or during their final evolution to core collapse supernovae (CCSN). ^{60}Fe can be also synthesized in Type Ia SNe (Woosley 1997). It is also destroyed by the ^{60}Fe (n,γ) process. Since its closest parent, ^{59}Fe is unstable, the ^{59}Fe(n,γ) process must compete with the ^{59}Fe(γ^-) decay to produce an appreciate amount of ^{60}Fe.

The decay chains of ^{60}Fe are shown in Figure 1. ^{60}Fe firstly decays to ^{60}Co, with emitting γ-ray photons at 59 keV, and then decays to ^{60}Ni, with emitting γ-ray photons at 1173 and 1333 keV. The gamma-ray efficiency of the 59 keV transition is only $\sim 2\%$ of those at 1173 and 1333 keV, so the gamma-ray flux at 59 keV is much lower than the fluxes of the high energy lines. The 59 keV gamma-ray line is very difficult to be detected with present missions. Measurements of the two high energy lines have been the main scientific target to study the radioactive ^{60}Fe isotope in the Galaxy.

^{60}Fe has been found to be part of meteorites formed in the early solar system (Shukolyukov *et al.* 1993). The inferred ^{60}Fe/^{56}Fe ratio for these meteorites exceeded the interstellar-medium estimates from nucleosynthesis models, which led to suggestion that the late supernova ejection of ^{60}Fe occurred before formation of the solar system (Tachibana *et al.* 2006). Yet, this is a proof for cosmic ^{60}Fe production, accelerator-mass spectroscopy of seafloor crust material from the southern Pacific ocean has revealed an ^{60}Fe excess in a crust depth corresponding to an age of 2.8 Myr (Knie *et al.* 2004). From this interesting measurement, it is concluded that a supernova explosion event near the solar system occurred about 3 Myr ago, depositing some of its debris directly in the earth's atmosphere. All these measurements based on material samples demonstrate that ^{60}Fe necleosynthesis does occur in nature. It is now interesting to search for current ^{60}Fe production in the Galaxy through detecting radioactive-decay γ-ray lines.

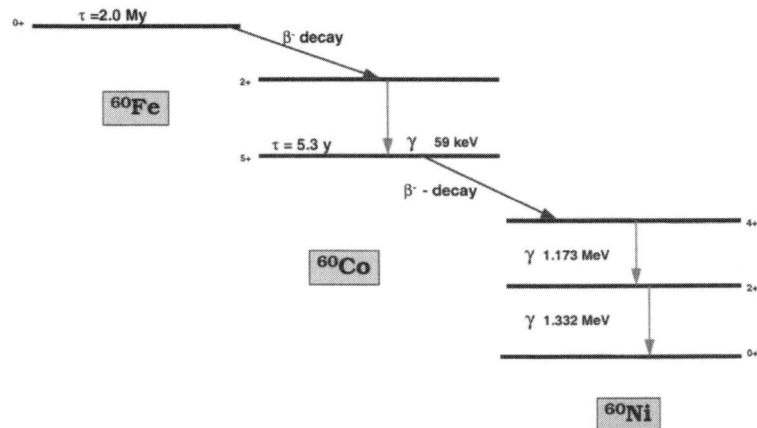

Figure 1. The decay scheme of ^{60}Fe. The mean lifetime is 2×10^6 years. The gamma-ray flux at 59 kev line is $\sim 2\%$ of those at 1173 and 1333 keV.

2. ^{60}Fe emission in the Galaxy

Due to its long decay time ($\tau \simeq 2.2$ My), ^{60}Fe survives to be detected after the supernova ejected it into the interstellar medium, by β-decay via ^{60}Co and γ emission at 1173 keV and 1333 keV – like other radioactive isotopes: ^{44}Ti, ^{56,57}Co, and ^{26}Al. These isotopes provide evidence that nucleosynthesis is ongoing in the Galaxy (The et al. 2006; Diehl et al. 2006). Specially, measurements of ^{60}Fe promise to provide new information about the massive star nucleosynthesis in the late pre-supernova stages.

Gamma-ray signal of ^{60}Fe from the sky is very weak, so there are no confident detections of ^{60}Fe in the Galaxy reported in the previous measurements. Recently, RHESSI reported observations of the gamma-ray lines from ^{60}Fe with an average flux of $(6.3 \pm 5.0) \times 10^{-5}$ph cm^{-2} s^{-1} (Smith 2004).

Now, the spectrometer aboard INTEGRAL (SPI) operates on space. The *INTErnational Gamma-Ray Astrophysics Laboratory* (INTEGRAL) is an European (ESA) Gamma-Ray Observatory Satellite Mission for the study of cosmic gamma-ray sources in the keV to MeV energy range (Winkler et al. 2003). INTEGRAL was successfully launched from Baikonur Cosmodrome (Kazakhstan) on October 17, 2002. The INTEGRAL orbit is eccentric, with an apogee of 153 000 km, a perigee of 9000 km, and a 3 day period. INTEGRAL will continue to work until 2012 approved by ESA. SPI/INTEGRAL consists of 19 high purity germanium detectors which allow for high spectral resolution of ~ 2.5 keV at 1 MeV, suitable for astrophysical studies of individual gamma-ray lines and their shapes, e.g. the 511 keV line, γ-ray lines from radioactivities of ^{44}Ti, ^{26}Al and ^{60}Fe. The basic measurement of SPI consists of event messages per photon triggering the Ge detector camera. We distinguish events which trigger a single Ge detector element only (*single event*, SE), and events which trigger more than two Ge detector elements nearly simultaneously (*multiple event*, ME).

We analyzed the first year of INTEGRAL data to detect the γ-ray lines from ^{60}Fe with an average line flux of $(3.7 \pm 1.1) \times 10^{-5}$ph cm^{-2} s^{-1} (Harris et al. 2005). But the strong background lines near ^{60}Fe lines still contaminate the spectra, which makes this preliminary results questionable. At present, we use three years of INTEGRAL data (from 2003.3 – 2006.3), aiming at a consolidation of the INTEGRAL/SPI measurement of ^{60}Fe gamma-rays.

The newest results on ^{60}Fe gamma-ray lines by INTEGRAL/SPI are presented in Figure 2. All the fluxes given by different databases are consistent with each other. The

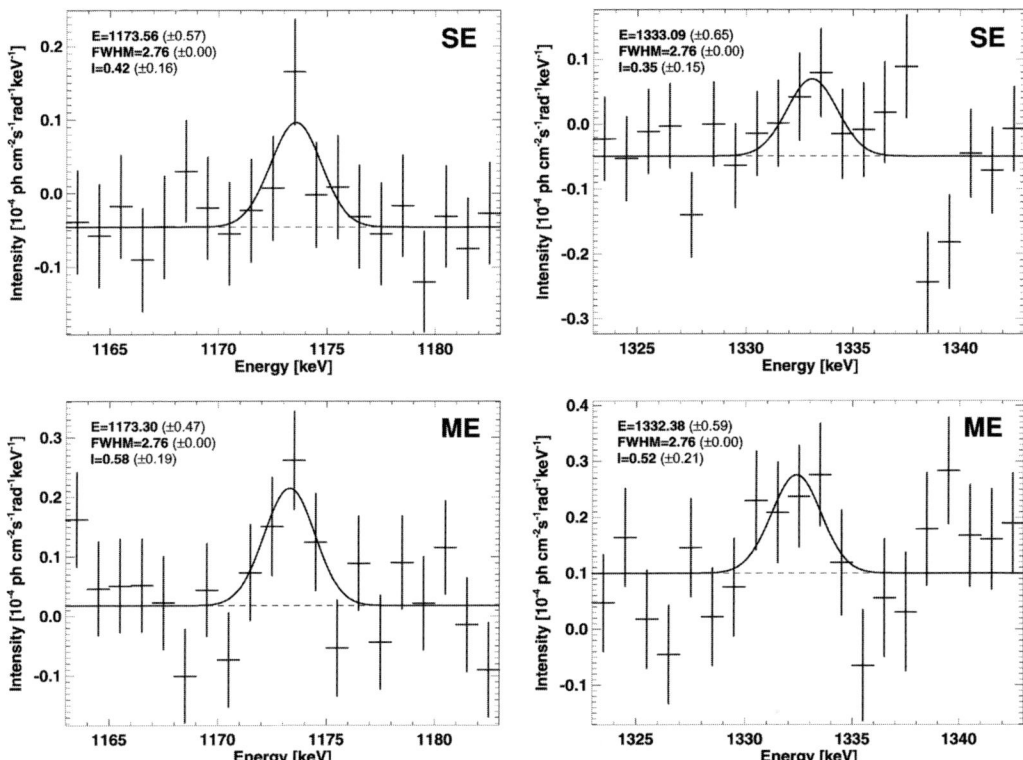

Figure 2. The spectra of two gamma-ray lines of ^{60}Fe from the inner Galaxy: 1173 keV and 1333 keV. We have shown the results both from SE and ME databases. For the SE database, we find a line flux of $(4.2 \pm 1.6) \times 10^{-5}$ ph cm^{-2} s^{-1} rad^{-1} for the 1173 keV line and $(3.5 \pm 1.5) \times 10^{-5}$ ph cm^{-2} s^{-1} rad^{-1} for the 1333 keV line. For the ME database, the line flux is $(5.8 \pm 1.9) \times 10^{-5}$ ph cm^{-2} s^{-1} rad^{-1} for the 1173 keV line and $(5.2 \pm 2.1) \times 10^{-5}$ ph cm^{-2} s^{-1} rad^{-1} for the 1333 keV line.

strong background line at 1337 keV has also been eliminated rather well. Furthermore, a superposition of the four spectra of Figure 2 is shown in Figure 3. The line flux estimated from the combined spectrum is $(4.4 \pm 0.9) \times 10^{-5}$ ph cm^{-2} s^{-1} rad^{-1}. Our significance estimate for the combined spectrum is $\sim 5\sigma$ (Wang et al. 2007).

3. The ratio of ^{60}Fe /^{26}Al

^{26}Al is an unstable isotope with a mean lifetime of 1.04 Myr. ^{26}Al can first decay into an excited state of ^{26}Mg, which de-excites into the Mg ground state by emitting gamma-ray photons with the characteristic energy of 1809 keV . ^{26}Al is produced almost exclusively by proton capture on ^{25}Mg in a sufficiently hot environment. ^{26}Al origin is dominated by massive star and core-collapse supernovae, and small part of ^{26}Al is attributed to AGB stars and novae.

Therefore, ^{26}Al and ^{60}Fe would share at least some of the same production sites, i.e. massive stars and supernovae. In addition both are long-lived radioactive isotopes, so we believe their gamma-ray distributions are similar as well. We derive the ratio of ^{60}Fe /^{26}Al , which can be directly compared with theoretical predictions.

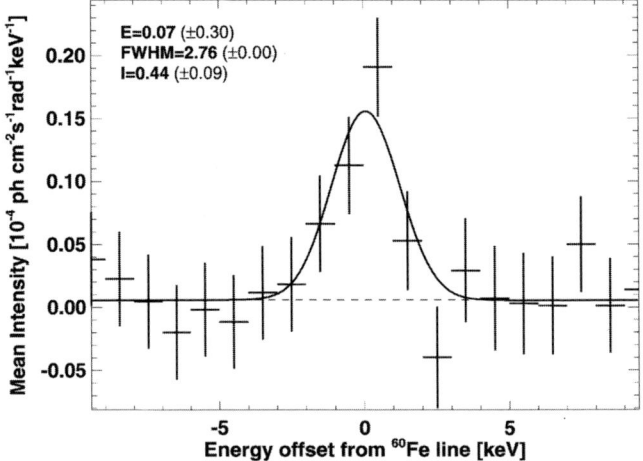

Figure 3. The combined spectrum of the ^{60}Fe signal in the inner Galaxy, superimposing the four spectra of Figure 2. In the laboratory, the line energies are 1173.23 and 1332.49 keV; here superimposed bins are zero at 1173 and 1333 keV. We find a detection significance of 5σ. The average line flux is estimated as $(4.4 \pm 0.9) \times 10^{-5}$ ph cm^{-2} s^{-1} rad^{-1}.

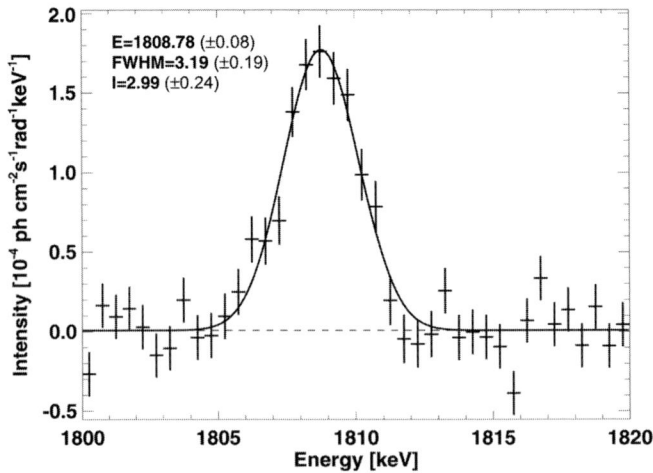

Figure 4. ^{26}Al spectrum derived by INTEGRAL/SPI with 3 years of data. ^{26}Al flux in the Galaxy is $(2.99 \pm 0.24) \times 10^{-4}$ ph cm^{-2} s^{-1} rad^{-1}.

We also obtain the ^{26}Al spectrum in the Galaxy using three years of INTEGRAL data which is shown in Figure 4. ^{26}Al flux is $(2.99\pm0.24)\times10^{-4}$ ph cm^{-2} s^{-1} rad^{-1}. Combining the ^{60}Fe result in Figure 3, we find a flux ratio of ^{60}Fe /^{26}Al of 15 ± 5)%.

Many experiments and efforts were made to measure the ^{60}Fe/^{26}Al flux ratio, and we now provide the most significant detection to date (see Table1 and Figure 5). In the same time, different theoretical models have predicted the ratio of ^{60}Fe/^{26}Al. Timmes *et al.* (1995) published the first detailed theoretical prediction. In their paper, they combined a model for ^{26}Al and ^{60}Fe nucleosynthesis in supernova explosions with a model of chemical evolution, giving a gamma-ray flux ratio $F(^{60}\text{Fe})/F(^{26}\text{Al}) = 0.16 \pm 0.12$. Since 2002, theoreticians have improved various aspects of the stellar-evolution models, including improved stellar wind models and the corresponding mass loss effects on stellar structure and evolution, of mixing effects from rotation, and also updated nuclear cross sections

^{60}Fe and Massive Stars

Table 1. Different measurements of ^{60}Fe/^{26}Al flux ratio

Experiments	$F(^{60}\text{Fe})/F(^{26}\text{Al})$	references
HEAO-3	0.09 ± 0.08	Mahoney et al. 1982
SMM	0.1 ± 0.08	Leising & Share 1994
OSSE	0.21 ± 0.15	Harris et al. 1997
COMPTEL	0.17 ± 0.135	Diehl et al. 1997
GRIS	$< 0.14(2\sigma)$	Naya et al. 1998
RHESSI	0.16 ± 0.13	Smith 2004
SPI	0.15 ± 0.5	this work

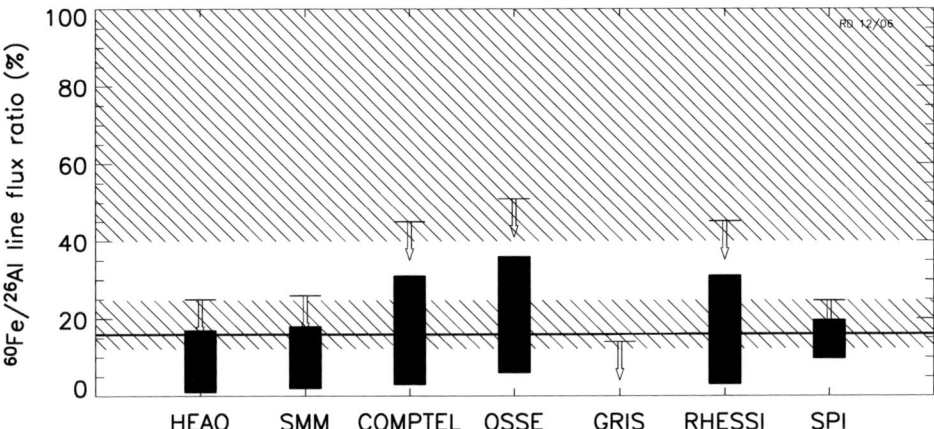

Figure 5. Flux ratio of the gamma-ray lines from the two long-lived radioactive isotopes ^{60}Fe/^{26}Al from several observations, including our SPI result (also see Table 1, from Wang et al. 2007), with upper limits shown at 2σ for all reported values, and comparison with the recent theoretical estimates (the upper hatched region from Prantzos 2004; the straight line taken from Timmes et al. 1995; the lower hatched region, see Limongi & Chieffi 2006). Our present work finds the line flux ratio to be $(15 \pm 5)\%$. See more details in the text.

in the nucleosynthesis parts of the models. As a result, predicted flux ratios ^{60}Fe/^{26}Al rather fell into the range 0.8 ± 0.4 (Prantzos 2004, based on, e.g. Rauscher et al. 2002, Limongi & Chieffi 2003) – such high values would be inconsistent with several observational limits and our SPI result. Limongi & Chieffi (2006) combined their individual yields, using a standard stellar-mass distribution function, to produce an estimate of the ^{60}Fe/^{26}Al gamma-ray flux ratio expected from massive stars. Their calculations yield a lower prediction for the ^{60}Fe/^{26}Al flux ratio of 0.185 ± 0.0625, which is again consistent with the observational constraints.

4. Summary and discussion

Now, we have detected both 1173 keV and 1332 keV lines of ^{60}Fe in the Galaxy (near 5 σ significance) with the 3 years of SPI/INTEGRAL data, which is the best results on detections of ^{60}Fe in the Galaxy, and confirms its existence. The average ^{60}Fe line flux from the inner Galaxy region is $(4.4 \pm 0.9) \times 10^{-5}$ ph cm^{-2} s^{-1} rad^{-1}. From the same observations and analysis procedure applied to ^{26}Al, we find a flux ratio of ^{60}Fe/^{26}Al of $(15 \pm 5.0)\%$.

Though large error bars and uncertainties exist, the original and the latest theoretical prediction of the flux ratio of ^{60}Fe/^{26}Al are consistent with our SPI result. But improvements are needed both in observations and theories. For gamma-ray astronomy, more precise measurements of gamma-ray lines in the Galaxy are required, especially for the ^{60}Fe signals, which may require the more SPI data and the development of next-generation gamma-ray spectrometers/telescopes. Stellar evolution models have potential for improvements in processes related to the production of ^{60}Fe and ^{26}Al, e.g. convective layers in the inner stars, wind models for WR and O stars and the possible effects of stellar rotation (Hirschi *et al.* 2004). The nuclear physics still has serious uncertainties for the productions of ^{26}Al and ^{60}Fe. For example, the cross section of ^{12}C$(\alpha,\gamma)^{16}$O is uncertain, which affects the prediction of both ^{26}Al and ^{60}Fe; the situation of ^{60}Fe is strongly influenced by the cross sections of neutron capture and β-decay which are purely theoretical: no experimental data exist for the ^{59}Fe(n,γ) and ^{60}Fe(n,γ) rates. Therefore, a concerted effort among stellar models, nucleosynthesis theory, and gamma-ray observations is required for a more satisfactory assessment of ^{60}Fe synthesis in the Galaxy.

Acknowledgements

We would like to acknowledge R. Diehl and M. Harris for the discussions.

References

Diehl, R. *et al.* 2006, *Nature* 439, 45
Diehl, R. *et al.* 1997, eds. Dermer, C. D., Strickman, M. S. and Kurfess, J. D., *AIP Conf. Proc.* 410, 1109
Harris, M. J. *et al.* 1997, eds. Dermer, C. D., Strickman, M. S. and Kurfess, J. D., *AIP Conf. Proc.* 410, 1079
Harris, M. J. *et al.* 2005, *A&A* 433, L49
Hirschi, R., Meynet, G., & Maeder, A. 2004, A&A, 425, 649
Knie, K. *et al.* 2004, *Physical Review Letters* 93, 171103
Leising, M. D. & Share, G. H. 1994, *ApJ* 424, 200
Limongi, M. & Chieffi, A. 2003, *ApJ* 592, 404
Limongi, M. & Chieffi, A. 2006, *ApJ* 647, 483
Mahoney, W. A. *et al.* 1982, *ApJ* 262, 742
Naya, J. E. *et al.* 1998, *ApJL* 499, L169
Prantzos, N. 2004, *A&A* 420, 1033
Rauscher, T., Herger, A., Hoffman, R. D. & Woosley, S. E. 2002, *ApJ* 576, 323
Shukolyukov, A. & Lugmair, G. W. 1993, *Science* 259, 1138
Smith, D. M. 2004, *New Astronomy Review* 48, 87
Tachibana, S. *et al.* 2006, *ApJL* 639, L87
The, L. S. *et al.* 2006, *A&A* 450, 1037
Timmes, F. X. *et al.* 1995, *ApJ* 449, 204
Wang, W. *et al.* 2007, *A&A* 469, 1005
Winkler, C. *et al.* 2003, *A&A* 411, L1
Woosley, S. E. 1997, *ApJ* 476, 801

Exploring the nucleosynthesis region of metal-poor Stars

Yuan-Yuan Geng, Dong-Nuan Cui, Jiang Zhang and Bo Zhang

Department of Physics, Hebei Normal University, 113 Yuhua Dong Road, Shijiazhuang 050016, P.R.China
email: gengyuanyuan1982@126.com; zhangbo@hebtu.edu.cn

Abstract. The chemical abundances of the very metal poor double-enhanced stars are excellent information to set new constraints on models of neutron-capture processes at low metallicity. There have been many theoretical studies of s-process nucleosynthesis in low-mass AGB stars. Using the parametric approach based on the radiative s-process nucleosynthesis model, we calculate the following five parameters for a series of metal-poor stars. They are: the mass fraction of ^{13}C pocket q, the overlap factor r, the neutron exposure per interpulse $\Delta\tau$, and the component coefficients that correspond to relative contribution from the s-process and the r-process. We find that the mass fraction of ^{13}C pocket q deduced for the Pb stars is comparable to the overlap factor r, which is about 10 times larger than normal AGB model; $q \sim 0.05$; and the neutron exposure per interpulse $\Delta\tau$ for all Pb stars are about 10 times smaller than the ST case ($\Delta\tau \sim 7.0\text{mb}^{-1}$). Although the two fundamental parameters $\Delta\tau$ and q obtained for the Pb stars are very different from the AGB stellar model, the results of the larger value of q and the smaller value of $\Delta\tau$ can also explain the abundance distribution of the Pb stars. This suggest that the q change to larger than that of normal AGB model. Then, this factor will result in the descent of the density of ^{13}C in the nuclear synthesis region directly. So, the neutron exposure $\Delta\tau$ will also decrease to the same extent. Although the neutron number density in the larger initial mass AGB stars ($m > 3M_\odot$) is high, the neutron irradiation time is shorter, obviously the neutron exposure per interpulse in the AGB stars should be smaller. It is noteworthy that the total amount of ^{13}C in metal poor condition is close to the ST case, which is consistent with the primary nature of the neutron source.

Keywords. Nucleosynthesis, metal-poor Stars

1. Introduction

The elements heavier than the iron peak are made through neutron capture via two principal processes: the r-process and the s-process.

In order to investigate the efficiency and sites of the s- and r-process, the elemental abundances of double-rich stars are particularly useful. There have been many theoretical studies of s-process nucleosynthesis in low-mass AGB stars. Unfortunately, however, the precise mechanism for chemical mixing of protons from the hydrogen-rich envelop into the ^{13}C -rich layer to form ^{13}C-pocket is still unknown. This makes it even harder to understand the particular abundance pattern of the s- and r-process elements found in carbon-rich mental-poor stars.The calculated results and discussion are described in sect.2. The conclusions are given in sect.3.

2. Results and Discussion

There are five parameters in the parametric model on s-process nucleosynthesis: the neutron exposure per pulse, $\Delta\tau$, the mass fraction of ^{13}C pocket in the He intershell q,

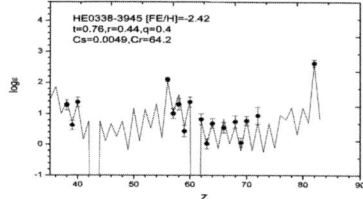 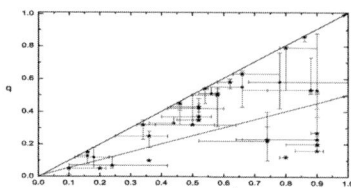

Figure 1. Left: As an example, we show our calculated best-fit results for a CEMP star HE 0338-3945. Right: show our between the mass fraction of ^{13}C pocket in the He intershell q and the overlap factor r is fitted approximately.

overlap factor r, Cs and Cr. We explored the origin of the neutron-capture elements in the double-enhanced stars by comparing the observed abundances with predicted s- and r-process contribution. In the AGB model, the overlap factor r and the neutron exposure per pulse, $\Delta\tau$ are the fundamental parameters. The mass fraction of ^{13}C pocket, q is an important parameter in the radiative s-process nucleosynthesis models.

We find the mass fraction of ^{13}C pocket q deduced for the Pb stars is comparable to the overlap factor r, which is larger than normal AGB model $q \sim 0.05$ about 10 times, and the neutron exposure per interpulse $\Delta\tau$ for all Pb stars are smaller than the ST case ($\Delta\tau \sim 7.0mb^{-1}$) about 10 times. Although the two fundamental parameters $\Delta\tau$ and q obtained for the Pb stars are very different from the AGB stellar model, the results of the larger value of q and the smaller value of $\Delta\tau$ can also explain the abundance distribution of the Pb stars. This suggest that the q change to larger than that of normal AGB model. Then, this factor will result in the descent of the density of ^{13}C in the nuclear synthesis region directly. So, the neutron exposure $\Delta\tau$ will also decrease to the same extent. Although the neutron number density in the larger initial mass AGB stars ($m > 3M_\odot$) is high, the neutron irradiation time is shorter, obviously the neutron exposure per interpulse in the AGB stars should be smaller.

3. Conclusions

It is noteworthy that the total amounts of ^{13}C in metal poor condition are close to the ST case, which is consistent with the primary nature of the neutron source.

Acknowledgements

We would like to acknowledge the useful comments of a referee. This work is supported by the National Natural Science Foundation of China under grant grant no. 10673002.

References

Burbidge, E. M., Burbidge, G. R., Fowler, W. A., & Houle, F. 1957, *Rev. Mod, Phys.* 29, 547
Gallino, R., Arlandini, C., Busso, M, et al. 1998, *ApJ* 497, 388
Zhang, B., Ma, K., & Zhou, G.D 2006, *ApJ* 642, 1075
Cui, W. Y. & Zhang, B. 2006, *MNRAS* 368, 305

Two-fluid models for the winds of OB stars

J. H. Guo

National Astronomical Observatories/Yunnan Observatory, Chinese Academy of Sciences,
P. O. Box 110, Kunming 650011, China.
email: guojh@ynao.ac.cn

Abstract. A two-fluid model to OB stars is investigated, in which the flow is described by a set of two components, One for the dense clumps and the other for the smooth gas. The two components are coupled by friction drag. The velocity structure of clumps is assumed to be beta law, thus, the velocity structure of the second component could be attained by shooting method. The result is compared with the X-ray observations.

Keywords. Stars: supergiants, stars: early-type, hydrodynamics

1. Introduction

Hot stars of spectral O and B have been observed to have X-ray emission. Most of the X-ray emission from early-type stars appears to originate in their winds with temperature of several million degrees (Waldron & Cassinelli 2007). X-ray emission in normal OB stars is generally explained in terms of shock heating of the supersonic, radiation-driven winds (Lucy & White 1980; Cohen *et al.* 2006). Shocks distributed throughout their winds and most likely formed from instabilities of line-driven flow (Lucy & Solomon 1970).

2. The model

For the multiphase hydrodynamical flows, Pistinner & Shaviv (1993) present a formalism at the concept of spatial averages. The advantages of the technique are that it can formulate the equations in a straightforward physical interpretation. We only present the results, for the details reader can refer to Pistinner & Shaviv (1993) and Soo (1989). The smooth gas and clumps are coupled by friction. These clumps are accelerated by stellar radiation and slowed down by gravity and friction force with the smooth gas. The smooth gas is accelerated in terms of drag between clumps and gas. Thus we require the equations of mass and momentum conservation.

Dynamical equations for the two-phase flow can be written as

$$\nabla \cdot (\alpha_g \rho_g v_g) = \nabla \cdot (<\rho_g> v_g) = 0 \quad (2.1)$$

and

$$\nabla \cdot (\alpha_c \rho_c v_c) = \nabla \cdot (<\rho_c> v_c) = 0. \quad (2.2)$$

$$\nabla \cdot [<\rho_g> v_g v_g] = -\nabla P_g + \alpha_c \nabla P_g + <\rho_g> g + \frac{1}{2} C_D \frac{<\rho_C>}{m} \frac{<\rho_g>}{1-\alpha_c} U^2 A \quad (2.3)$$

$$\nabla \cdot [<\rho_c> v_c v_c] = -\alpha_c \nabla P_g + <\rho_c> g - \frac{1}{2} C_D \frac{<\rho_C>}{m} \frac{<\rho_g>}{1-\alpha_c} U^2 A + <\rho_c> g^{rad} \quad (2.4)$$

In this paper we only consider the motion of gas. The velocity structure of clumps is assumed to be β law. Thus we only solve Eqs. (2.1) and (2.3).

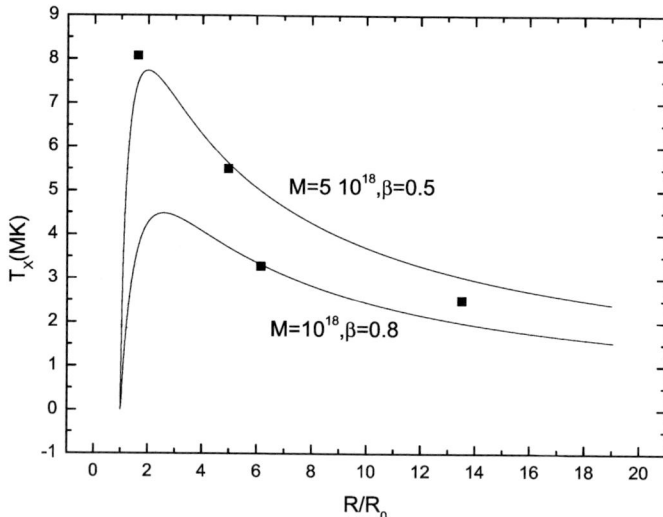

Figure 1. The dependence of T_X on the radius.

The solution of the differential Eqs. (2.1) and (2.3) can be obtained by numerical method. The momentum equation has a singularity at the point where $v_g = a$. The position of singularity is determined when numerator and denominator go to zero simultaneously. To perform the condition, a suitable lower boundary condition is necessary. We apply shooting method to find the lower boundary condition.

The typical O-type star ζ Ori A is ideal candidate to test our X-ray model because it have been study in detail (Feldmeier et al. 1997; Waldron & cassineli 2000). A important test to our model is to compare the X-ray temperature. For the post-shock temperature, we take $T_s = 1.44 \times 10^5 (\frac{U}{100 K M s^{-1}})^2 K$.

3. Results

The spatial X-ray temperature distributions show that the temperature decreases with the increase of radius. Our model predicted the trend. The result also hints that the mass of clump can not greater than 5×10^{18}.

Acknowledgements

This work was supported by the Knowledge innovation Program of Chinese Academy of Sciences.

References

Cohen D. H. et al. 2006, MNRAS, 368, 1905
Feldmeier A., Puls J., & Pauldrach W. A. 1997, A&A, 322, 878
Lucy L. B. & Solomon P. M. 1970, ApJ, 159, 879
Lucy L. B. & White R. L. 1980, ApJ, 241, 300
Pistinner S. & Shaviv G. 1993, 1993, ApJ, 414, 612
Soo S. L. 1989, Particulates and Continuum: Pultiphase Fluid Dynamics (New York:Hemisphere)
Waldron W. L. & Cassinelli J. P. 2000, ApJ, 548, L45
Waldron W. L. & Cassinelli J. P. 2007, ApJ, 668, 456

Tidal disruption of stripped red giants by massive black holes

Y. Lu[1], Y. F. Huang[2], S. N. Zhang[3] and P. Lu[1]

[1] National Astronomical Observatories, Chinese Academy of Sciences, Beijing 100012, China
email: ly@bao.ac.cn

[2] Department of Astronomy, Nanjing University, Nanjing 210093, China

[3] Physics Department and Center for Astrophysics, Tsinghua University, Beijing 100084, China

Abstract. We investigate the tidal disruption of a red giant whose envelope is thought to be stripped off when it passed by a massive black hole. Since the low-density stellar envelope would be lost, the tidal disruption of a red giant by massive black hole is regarded as primarily happening in its core region. The object is called a stripped red giant (SRG). Comparing our results with the three candidate tidal disruption events detected by Chandra in 2001 and 2002, i.e., the X-ray flares of NGC 5905, RX J1242.6-1119A, and RX J1624.9+7554, we argue that the tidal disruption of a stripped red giant is strongly ruled out.

Keywords. Accretion: accretion disks, black hole physics, galaxies: nuclei

1. Introduction

One remarkable sign of black holes residing in the centers of galaxies is the tidal disruption of stars by the black hole (Rees 1998). This provides a uniquely accessible laboratory for studying in detail the connections and interactions between a massive black hole and the stellar system (Alexander 2005, Lu et al. 2003, Lu et al. 2006). The ROSAT All-Sky Survey in 1990–1991 performed an experiment to detect these tidal disruption events with hundreds of thousands of galaxies in the ideal wavelength band. The results show that three galaxies, i.e., RX J1242.6-1119A, RX J1624.9+7554, and NGC 5905, had unusual X-ray flares. Interestingly, later observations of these galaxies by Chandra in 2001 and 2002 showed that their X-ray fluxes continued to decline at a rate consistent with the prediction in the fallback phase of a tidal disruption event (Gezari et al. 2003).

It has been discussed that the tidal disruption of a red giant and a main sequence star could be very different due to their different properties of the stellar structures (Lidskii & Ozernoi(1979)). In this paper, we concentrate on the stripped red giant (SRG) in which only the core of a red giant, which is about 20 percent of the total mass (Luminet & Barbuy(1990)) happens the tidal disruption.

2. Tidal disruption of a stripped red giant

When a star with a given mass of M_* and radius of R_* passes by a massive black hole with a mass of M_{bh}, the star would be tidally disrupted at an average tidal radius of $R_t = 7 \times 10^{12} r_* m_*^{-1/3} M_6^{1/3}$ cm, where $m_* = M_*/M_\odot$, $r_* = R_*/R_\odot$, and $M_6 = M_{bh}/10^6 M_\odot$. $R_\odot$ and $M_\odot$ are the solar radius and mass, respectively. Once a SRG is tidally disrupted by the black hole, the resulting luminosity of the X-ray flares can be approximated as $L_X = \epsilon \dot{M}_{peak} c^2$, where $\epsilon = 5.38 \times 10^{-3} r_*^{-1} m_*^{1/3} M_6^{2/3}$ and $\dot{M}_{peak} \simeq 1.4 r_*^{-3/2} m_*^2 M_6^{-1/2} M_\odot yr^{-1}$. L_x shows that it depends on $\dot{M}$ and ϵ. Given the mass of black holes, L_x is determined by the accretion configuration and the property of disrupted

343

stars. Throughout this discussion we use the term red giants loosely to describe progenitor stars on the first red giant branch, and we adopt the red giant model of (Han *et al.* (2000)) for the evolution of progenitor stars with masses in the range of $1M_\odot < M_G < 3M_\odot$. We define the radius of SRGs by corresponding to the mesh point making the hydrogen-shell burning boundary and its mass by the onset of helium burning, respectively. we derive the mass-radius relation of SRGs as

$$m_S = \begin{cases} 53.11 r_S^2 - 3.29 r_S + 0.37, & \text{if } 0.035 \leqslant r_S \leqslant 0.07756 \\ -16.77 r_S + 0.95, & \text{if } 0.02819 \leqslant r_S \leqslant 0.035 \end{cases} \quad (2.1)$$

where $m_S = M_{SRG}/M_\odot$, $r_S = R_{SRG}/R_\odot$. In the following calculations, we only consider those galaxies hosting massive black holes (i.e. $M_{bh} \geqslant 10^6 M_\odot$) and those with massive black holes not massive enough to swallow a whole star without disruption. We thus ignore the tidal disruption of stars with the mass-radius relation as the lower branch of Eq.(2.1), because the corresponding upper limit of black hole mass is lower than $10^6 M_\odot$, which is the typical lower mass limit for the black hole residing in the Galactic nucleus.

Chandra observations in the 0.1-2.4 keV band (Halpern *et al.*(2004)) show that the peak luminosity of the three flare events in NGC 5905, RX J1242.6-1119A and RX J1624.9+7554 are 4×10^{43} erg/s, 1.6×10^{44} erg/s and 4×10^{43} erg/s. With the prediction of the recent multi-wavelength observations that the black holes residing in these three tidal events are $\sim 1.7 \times 10^8 M_\odot$ (Gezari *et al.*(2003)), $\sim 2 \times 10^8 M_\odot$ (Ferrarese & Merritt(2000)) and $\sim 10^6 M_\odot$ (Grupe *et al.*(1999)), we find that the luminosity calculated from the tidal disruption of a SRG star is too faint to account for the observations.

3. Conclusions

we have investigate the tidal disruption of SRGs by massive black holes. If this is the case, the results show it is ruled out for the three disruption events by comparing with the observations detected by Chandra in the 0.1-2.4 keV band.

Acknowledgements

This research is supported by the National Natural Science Foundation of China (Grants 10273011, 10573021, 10433010, 10625313, 10521001 and 10221001), and by Chinese Academy of Science through project No. KJCX2-YW-T03.

References

Alexander, T. 2005, *Physics Reports* 419, 65
Ferrarese, L. & Merritt, D. 2000, ApJ 539, L9
Frank, J. & Rees, M. 1976, MNRAS 176, 633
Gezari, S., Halpern, J. P., Komossa, S., Grupe, D., & Leighly, K. M. 2003, ApJ 592, 42 (erratum 601, 1159, [2004])
Grupe, D., Thomas, H. C., & Leighly, K. M. 1999, A&A 350, L31
Halpern, J. P., Gezari, S., & Komossa, S. 2004, ApJ 604, 572
Han, Z., Tout, C. A., & Eggleton, P. P. 2000, *MNRAS* 319, 215
Lidskii, V. V. & Ozernoi, L.M. 1979, Sov. Astron. Lett. 5, 16
Loeb, A. & Ulmer, A. 1997, ApJ 489, 573
Lu, Y., Cheng, K. S., & Zhang, S. N. 2003, ApJ 590, 52
Lu, Y., Cheng, K. S., & Huang, Y.F. 2006, ApJ 641, 288
Luminet, J. P. & Barbuy, B. 1990, AJ 61, 219
Rees, M. J. 1988, Nature 333, 523

Study on the spectrum of the injected relativistic protons

Y. P. Wang[1,2], Y. Lu[1] and L. Chen[2]

[1]National Astronomical Observatories, Chinese Academy of Sciences, Beijing 100012, China
email: wangyanping@mail.bnu.edu.cn
[2]Department of Astronomy, Beijing Normal University, Beijing 100875, China

Abstract. About 10 TeV γ-ray emission within 10 pc region from the Galactic Center had been reported by 4 independent groups. Considering that this TeV γ-ray emission is produced via a hadronic model, and the relativistic protons came from the tidal disruption of stars by massive black holes, we investigate the spectral nature of the injected relativistic protons required by the hadronic model. The calculation was carried on the tidal disruption of the different types of stars and the different propagation mechanisms of protons in the interstellar medium. Compared with the observation data from HESS, we find for the best fitting that the power-law index of the spectrum of the injected protons is about -1.9, when a red giant star is tidally disrupted, and the effective confinement of protons diffusion mechanism is adopted.

Keywords. black hole physics- galaxies: jets -Galaxy: center

1. Introduction

The central region of Milky Way is a potential site for the production of effective particle acceleration and copious γ-ray emission. TeV γ-ray emission from the Galactic Center(GC) had been reported by 4 independent groups in recent years: CANGAROO (Tsuchiya *et al.* (2004)), Whipple (Kosack *et al.* (2004)), HESS(Aharonian *et al.* (2004)), and MAGIC (Albert *et al.* (2006)). One possibility for this TeV γ-ray emission source is in the whole diffuse 10 pc region, which is proposed to be related to the massive black hole Sgr A* harbored in our Galactic Center (Aharonian, Neronov (2005)).

There are several radiation mechanisms for the production of TeV γ-ray emission. One of these mechanisms proposed that the TeV γ-rays can be produced indirectly through the processes of π^0-decay when relativistic protons are injected into and interact with the interstellar medium (ISM), called hadronic model (Aharonian, Neronov (2005)). If this is the case, by assuming that the initial injected protons' spectrum follows a power-law plus a high energy cut-off, we can predict the spectrum of the injected relativistic protons by comparing the spectral energy distribution (SED) of the TeV γ-ray emission produced by hadronic model with observation data from HESS.

2. Model for the spectrum of the injected relativistic protons

To calculate the spectrum of the injected relativistic protons, we consider that the TeV γ-ray detected by HESS is produced through the hadronic model, and the injection of the relativistic protons required by the hadronic model came from the jet of the black hole Sgr A* when it captures and tidally disrupts a star(Lu, Cheng & Huang (2006)).

The tidal disruption of stars by massive black hole Sgr A* refers to main sequence (MS) stars and red giants. For these two cases, the total energy carried by the jet is exactly the same as 2.76×10^{51} erg, which is believed to be enough for the production of the injected relativistic protons required by the hadronic model. The definite difference resulting from the two cases are the diffusion timescale of the injected protons: 800 yr for the case of MSs and 2.11×10^4 yr for the red giants, respectively (Lu, Cheng & Huang (2006)).

Three kinds of diffusion mechanisms are involved for the propagation of the relativistic protons when they are injected into and interact with the ISM : (1) the Effective Confinement of Protons (ECP), (2) the Kolmogorov-Type Turbulence (KTT), (3) the Bohm Diffusion (BD). For these three mechanisms, the diffusion coefficient depends on the proton energy, given by the formula of $D(E)=10^{28}(E/1GeV)^\delta \kappa \ cm^2 s^{-1}$. The case of ECP corresponds to $\delta=0.5$ and $\kappa=10^{-4}$, KTT corresponds to $\delta=0.3$ and $\kappa=0.15$, and BD corresponds to $\delta=1.0$ and $\kappa=10^{-2}$.(Aharonian , Neronov (2005))

Giving the initial injected relativistic protons spectrum follows a power-law plus a high energy cut-off, at a given time of t and the source distance of R, the spectral energy distribution of the TeV γ-ray emission produced through the model discussed above can be calculated through $E_\gamma f(E_\gamma) = E_\gamma^2 V \int_0^{R_{size}} \int_{t_{crit}}^{t_{diff}} q_\gamma(E_\gamma, R, t)/S$, where E_γ is the energy of the emitted γ-ray photons, V is the total volume of the ISM, $V = 4\pi R_{size}^3/3$, $R_{size}=10$pc, t_{crit} is the critical time, t_{diff} is is the diffusion time of the injected protons when they propagate in the target(Lu, Cheng & Huang (2006)), q_γ is the emissivity of γ-ray photons, and $S=4\pi(d_{ob})^2$, where d_{ob} is the distance from the observation to the source, and we adopt $d_{ob}=8kpc$ hereafter.

We address the SED of the emitted γ-rays in the following cases: (1) the different power-law index p of injected protons with values of 1.7, 1.8, 1.9, 2.0, 2.1, 2.2; (2)the MSs and red giants are tidally disrupted by massive black hole Sgr A^*; (3)three diffusion mechanisms for the cases of ECP, KTT, and BD. By comparing the theoretical spectrum with the observation data from HESS within the central 10pc of our Galaxy, we find that the theoretical energy distribution can fit best with the observation data when the type of star disrupted by black hole is a red giant and the diffusion mechanism of protons is the Effective Confinement of Protons (ECP). In such a case, we derived that the power-law index of the injected relativistic protons is -1.9.

3. Conclusions and discussion

We have investigated the spectrum of the injected relativistic protons required by the hadronic model to produce the 10 TeV γ-ray emission detected by HESS in the center of our Galaxy. Comparing with the observations, we find that the power-law index of the initial spectrum of the injected protons should be -1.9. This result is based on that the tidal disruption of a red giant by the black hole Sgr A^* is considered and the propagation of protons in the target gas is the ECP scenario.

Acknowledgements

This research is supported by the National Natural Science Foundation of China (Grants 10573021 and 10778716).

References

Aharonian F. A., et al. 2004, *A&A* 425, L13
Aharonian F. A. Neronov A. 2005, *ApJ* 619, 306
Aharonian F. A. Neronov A. 2005, *Ap&SS* 300, 255
Albert J. et al. 2006, *ApJ* 638, L101
Kosack K. et al. 2004, *ApJ* 608, L97
Lu Y., Cheng K. S., & Huang Y. F. 2006, *ApJ* 641, 288
Tsuchiya K. et al. 2004, *ApJ* 606, L115

Advanced test of the model stellar atmospheres: the nature of the light variability of magnetic chemically peculiar stars

J. Krtička[1], Z. Mikulášek[1], J. Zverko[2], J. Žižňovský[2] and P. Zvěřina[1]

[1]Institute of Theoretical Physics and Astrophysics, Masaryk University, Brno, Czech Republic
[2]Astronomical Institute of the Slovak Academy of Sciences, Tatranská Lomnica, Slovak Republic

Abstract. The magnetic chemically peculiar stars exhibit both inhomogeneous horizontal distribution of chemical elements on their surfaces and the light variability. We show that the observed light variability of these stars can be successfully simulated using models of their stellar atmospheres and adopting the observed surface distribution of elements. The most important elements that influence the light variability are silicon, iron, and helium.

Keywords. stars: chemically peculiar, stars: early-type, stars: variables: other, stars: atmospheres, stars: magnetic fields

1. Magnetic chemically peculiar stars

Magnetic chemically peculiar stars belong to one of a few groups of variable stars whose light variations are not very well understood. These stars show inhomogeneous surface distribution of different elements, including helium, silicon or iron. Krtička et al. (2007), Krtička et al. (2008) showed that the redistribution of the emergent flux due to the bound-free (continuum) and bound-bound (line) transitions in the stellar atmosphere is a promising mechanism producing the observed light variations. Consequently, the comparison of observed and simulated light variations can provide an important new test of modern model stellar atmospheres.

2. Light curve calculation

The models of the stellar atmospheres are calculated using the code TLUSTY Hubeny & Lanz (1995), Lanz & Hubeny (2007) assuming a fixed stellar effective temperature and surface gravity. The abundance of chemical elements concerned is set in agreement with the maps of surface elemental distribution derived from spectroscopy.

The emergent fluxes from individual surface elements are calculated using the code SYNSPEC. From these fluxes the magnitudes of the star in individual filters of Strömgren photometric system are derived. Finally, these simulated light variations are compared with the observed ones. We stress that we do not use any free parameter to fit the observed light curves.

We use the ephemeris derived by Adelman (1997b) for HD 37776 and by Krtička et al. (2008) for HR 7224.

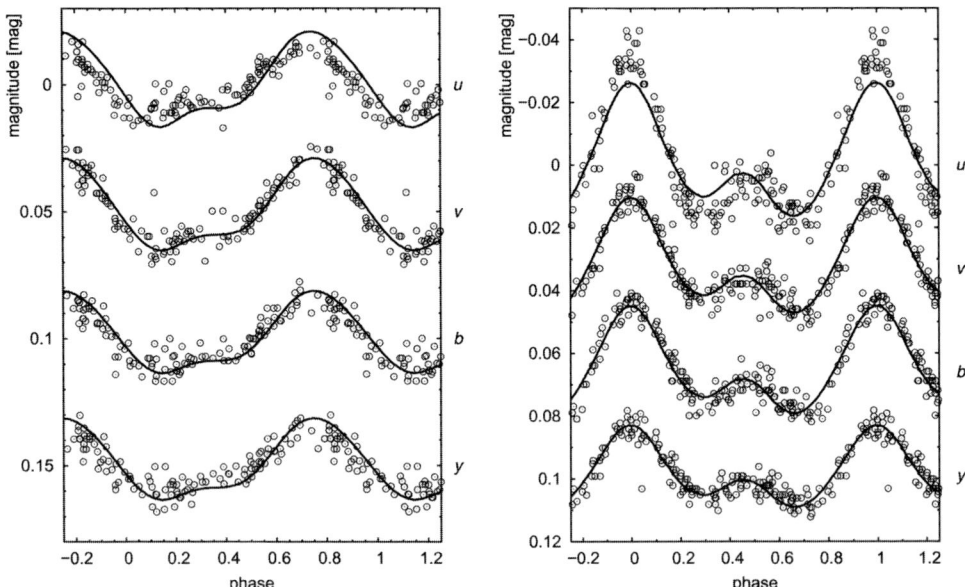

Figure 1. Left: Comparison of the predicted light variations of HD 37776 (calculated using the silicon and helium surface distribution according to Khokhlova et al. (2000)) and the observed one (Adelman & Pyper 1985, Adelman 1997b) in different colours. Right: Comparison of the predicted light variations of HR 7224 (calculated using the iron and silicon surface distribution according to Lehmann et al. (2007)) and the observed one (Adelman 1997a) in different colours.

3. Conclusions

We conclude that the observed light variations of HR 7224 and HD 37776 can be explained as a result of inhomogeneous distribution of helium, silicon and iron on the surface of these stars. This causes the redistribution of the flux from the ultraviolet to the visible part of the spectrum, and, due to the stellar rotation, the light variations.

A very good agreement of observed and simulated light variations serves as an independent test of modern model stellar atmospheres. The successful comparison of predicted and observed light curves would not be possible without using the calculations of radiative ionization cross-section done by the Opacity Project team Seaton et al. (1992).

Acknowledgements

Grants GA ČR 205/08/0003, 205/06/0217, and VEGA 2/6036/6.

References

Adelman, S. J. 1997a, *A&AS* 122, 249
Adelman, S. J. 1997b, *A&AS* 125, 65
Adelman, S. J. & Pyper, D. M., 1985, *A&AS* 62, 279
Groote, D. & Kaufmann, J. P. 1981, in: Upper Main Sequence CP stars, 23d Liège p. 435
Hubeny, I. & Lanz, T. 1995, *ApJ* 439, 875
Khokhlova, V. L., Vasilchenko, D. V., Stepanov, V. V., & Romanyuk, I. I., 2000, *AstL* 26, 177
Krtička, J., Mikulášek, Z., Zverko, J., Žižňovský, J. 2007, *A&A* 470, 1089
Krtička, J., Mikulášek, Z., Zverko, J., Žižňovský, J., & Zvěřina, P. 2008, A&A, in preparation
Lanz, T. & Hubeny, I. 2007, *ApJS* 169, 83
Lehmann, H., Tkachenko, A., Fraga, L., Tsymbal, V., & Mkrtichian, D. E. 2007, *A&A* 471, 941
Seaton, M. J., Zeippen, C. J., Tully, J. A., et al. 1992, *Rev. Mexicana Astron. Astrofis.* 23, 19

Binary Evolutionary Models

Z. Han[1] & Ph. Podsiadlowski[2]

[1] National Astronomical Observatories/Yunnan Observatory, Kunming, 65011, P.R. China
email: zhanwenhan@hotmail.com

[2] Department of Physics, University of Oxford, Keble Road, Oxford OX1 3RH, UK
email: podsi@astro.ox.ac.uk

Abstract. In this talk, we present the general principles of binary evolution and give two examples. The first example is the formation of subdwarf B stars (sdBs) and their application to the long-standing problem of ultraviolet excess (also known as UV-upturn) in elliptical galaxies. The second is for the progenitors of type Ia supernovae (SNe Ia). We discuss the main binary interactions, i.e., stable Roche lobe overflow (RLOF) and common envelope (CE) evolution, and show evolutionary channels leading to the formation of various binary-related objects. In the first example, we show that the binary model of sdB stars of Han *et al.* (2002, 2003) can reproduce field sdB stars and their counterparts, extreme horizontal branch (EHB) stars, in globular clusters. By applying the binary model to the study of evolutionary population synthesis, we have obtained an "a priori" model for the UV-upturn of elliptical galaxies and showed that the UV-upturn is most likely resulted from binary interactions. This has major implications for understanding the evolution of the UV excess and elliptical galaxies in general. In the second example, we introduce the single degenerate channel and the double degenerate channel for the progenitors of SNe Ia. We give the birth rates and delay time distributions for each channel and the distributions of companion stars at the moment of SN explosion for the single degenerate channel, which would help to search for the remnant companion stars observationally.

Keywords. binaries: close, galaxies: elliptical and lenticular, cD, stars: evolution, subdwarfs, supernovae: general, white dwarfs, ultraviolet: galaxies

---

1. General Principles of Binary Evolution

About half of the stars are in binaries and binary evolution plays a crucial role in the formation of many interesting objects, such as Algols, FK Comae (FK Com) stars, cataclysmic variables (CVs), planetary nebulae (PNe), barium (Ba) stars, CH stars, type Ia supernovae (SNe Ia), AM Canum Venaticorum (AM CVn) stars, low mass X-ray binaries (LMXB), high mass X-ray binaries (HMXB), symbiotic (Sym) stars, blue stragglers (BSs), pulsars, subdwarf B (sdB) stars, double degenerates (DDs) etc. Binary evolution is also important in the study of evolutionary population synthesis, which is a power tool in the study of galaxies.

A binary system (of low/intermediate mass) has two components: the primary (the initially more massive one) and the secondary. As the binary evolves, the primary expands and may fill its Roche lobe on the Hertzsprung gap or on the giant branch. Roche lobe overflow (RLOF) begins and the primary's envelope mass transfers to the secondary. Given the mass ratio q of primary to secondary less than a critical value q_c (Hjellming & Webbink, 1987; Webbink, 1988; Han & Webbink, 1999; Han *et al.*, 2002) at the onset of the mass transfer, where q_c mainly depends on the entropy profile of the primary's envelope and the angular momentum loss from the system, the mass transfer is stable, leading to a wide white dwarf (WD) binary. Given the mass ratio q larger than q_c, the mass transfer is dynamically unstable, leading to the formation of a common envelope

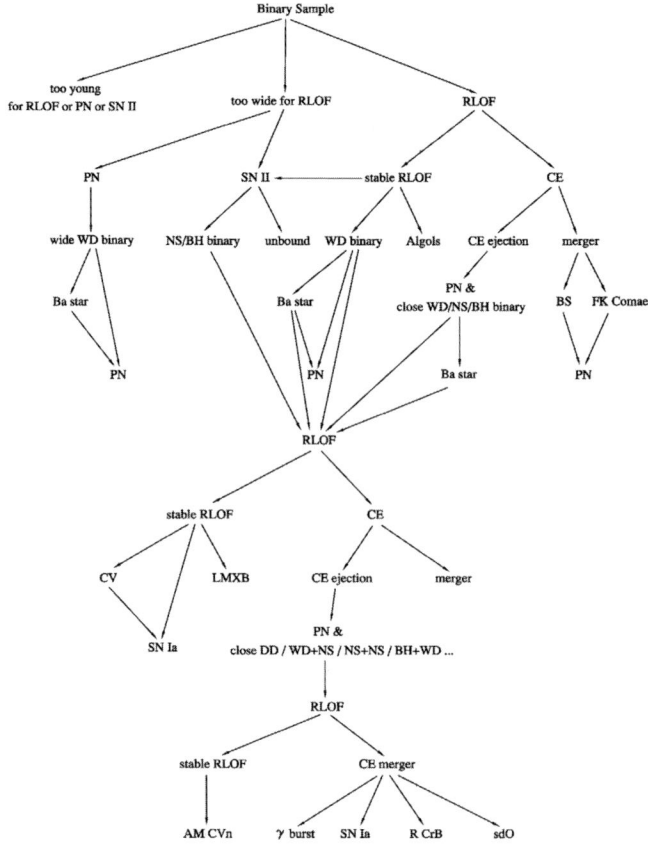

Figure 1. A simplified version of flow chart of binary evolution. See paragraph 1 of Section 1 of the text for the acronyms.

(CE; Paczyński, 1976). The CE engulfs the core of the primary and the secondary, and does not co-rotate with the embedded binary. The friction between the CE and the embedded binary makes the orbit decay, and a large amount of orbital energy released is deposited into the CE. If the CE can be ejected, a close WD binary forms, otherwise a fast rotating merger is resulted. For a WD binary system, the secondary continues to evolve and may experience mass transfer. Similar to the process described above, the mass transfer may lead to the formation of a CE and the CE ejection produces a double degenerate.

Figure 1 is a flow chart of binary evolution. It is by no means comprehensive, but it does show evolutionary channels leading to various objects for a binary system with given conditions.

A good stellar evolution theory should give and predict the statistical properties of a stellar population as well as the properties of individual stars or binaries. Binary population synthesis (BPS) is to evolve a large number of stars (including binaries) in order to investigate statistical properties of stars and check evolutionary mechanisms for different types of stars. In a BPS study, we first generate a binary sample (10 million binaries), then evolve the sample according to stellar evolution model grids and taking into account binary interactions, and we obtain different types of binary-related objects, which can be directly compared to observations.

In addition to the grids of stellar evolutionary models, we adopt the following input in BPS simulations:

(1) A constant star-formation rate is taken over the last 15 Gyr for a field population, or alternatively, a single star burst for a globular cluster or an elliptical galaxy.

(2) The initial mass function of Miller & Scalo (1979) is adopted.

(3) We mainly adopt a constant initial mass ratio distribution $n(q')$, where $q' = 1/q$ is the ratio of secondary to primary.

(4) We take the distribution of separations to be constant in $\log a$ for wide binaries, where a is the orbital separation. The adopted distribution gives that $\sim 50\%$ of stellar systems are binary systems with orbital periods less than 100 yr.

The main model parameters in BPS are mass transfer efficiency α_{RLOF} for the first stable RLOF (where the accretor is a main sequence star), CE ejection efficiency α_{CE} and thermal contribution α_{th} for CE evolution. The mass transfer efficiency α_{RLOF} is the fraction of the envelope mass that is transferred onto the secondary rather than is ejected from the system, where we assume that the matter lost from the system carries away the specific angular momentum of the system. A typical value for α_{RLOF} is 0.5. The CE ejection efficiency α_{CE} is the fraction of the released orbital energy used to overcome the binding energy of the envelope during the spiral-in process of a CE. The thermal contribution α_{th} defines the fraction of the internal energy of thermodynamics (including recombination energy as well as the thermal energy) contributing to the binding energy of the CE. A CE is ejected if

$$\alpha_{\mathrm{CE}} \Delta E_{\mathrm{orb}} \geqslant E_{\mathrm{gr}} - \alpha_{\mathrm{th}} E_{\mathrm{th}}, \qquad (1.1)$$

where ΔE_{orb} is the orbital energy released during the spiral-in process, E_{gr} the gravitational binding energy of the CE, E_{th} the internal energy of the CE. Both E_{gr} and E_{th} are calculated from detailed stellar models and therefore the prescription here is different from the λ prescription but appear to be more physical. We refer the reader to Han, Podsiadlowski & Eggleton (1994), Dewi & Tauris (2000) and Podsiadlowski, Rappaport & Han (2003) for the details. The inclusion of the internal energy in ejecting the CE seems to be the most plausible in the explanation of both long-period and short-period binaries containing compact objects (Han et al., 1995a; 1995b; Dewi & Tauris, 2000). Webbink (2007) has made a physical investigation on both the energetics of CE evolution and the angular momentum prescription (i.e. the γ prescription, Nelemans et al., 2000), and convincingly showed the necessity of recombination energy term for common envelope evolution. Previous studies, e.g., Han, Eggleton, Podsiadlowski & Tout (1995), Han, Podsiadlowski & Eggleton (1995), Han (1998), Han et al. (2002), Han et al. (2003), Han & Podsiadlowski (2004), have showed that both α_{th} and α_{th} are close to one.

2. The binary model for subdwarf B stars and the UV-upturn of elliptical galaxies

Subdwarf B (sdB) stars† are core helium-burning stars with very thin hydrogen envelope (Heber, 1986). They are important in many aspects of astrophysics, e.g., stellar evolution, distance indicators, Galactic structure, and the long-standing problem of far-ultraviolet excess in early-type galaxies (Kilkenny et al., 1997; Green, Schmidt & Liebert, 1986; Han, Podsiadlowski & Lynas-Gray, 2007).

† In this paper, we collectively refer to helium-core-burning stars with thin hydrogen envelopes as sdB stars, even if some of them may in reality be sdO or sdOB stars

Maxted *et al.* (2001) showed that the majority of field sdB stars are in binaries, and this has posed a serious challenge to stellar evolution theory. Han *et al.* (2002; 2003) proposed a binary model for their formation. In the model, there are three types of formation channels for sdB stars: stable RLOF for sdB binaries with long orbital periods, CE ejection for sdB binaries with short orbital periods, and the merger of helium WDs to form single EHB stars. In the stable RLOF channel, the mass donor fills its Roche lobe near the tip of the first giant branch and experiences a stable mass transfer, and its envelope is striped off by the RLOF, and the naked helium core (with thin hydrogen envelope) get ignited to produce a sdB stars. In the CE ejection channel, the mass donor also fills its Roche lobe near the tip of the first giant branch to have a dynamically unstable mass transfer leading to the formation of a CE. The CE ejection leaves a naked helium core (with thin hydrogen envelope) and the naked helium core is ignited to produce a sdB star. In the CE channel, the donor star needs to fill its Roche lobe closer to the tip of the first giant branch, or, in other words, the minimum core mass required for the donor star at the onset of mass transfer is larger than that in the stable RLOF channel to produce a sdB star. This is simply because that the time scale of the CE evolution is much shorter than that of the stable RLOF and the core does not grow by much in the CE process. In the merger channel, a close helium WD pair coalesces due to angular momentum loss via gravitational wave radiation. The binary model of Han *et al.* (2002; 2003) has successfully explained the main observational characteristics of field sdB stars: their distributions in the orbital period-minimum companion mass diagram, and in the effective temperature-surface gravity diagram; their distributions of orbital period and mass function; their binary fraction and the fraction of sdB binaries with WD companions; their birth rates; and their space density. The model is indeed a step forward and is widely used in the study of sdB stars (O'Tool, Heber & Benjamin, 2004).

Moni Bidin *et al.* (2006), Moni Bidin, Catelan & Altmann (2008) have done radial-velocity surveys for extreme horizontal branch (EHB) stars, the counterparts of field sdB stars, in globular clusters. They found that there is a remarkable lack of close binary systems in EHB stars. This is surprising as compared to the high binary fraction in field sdB stars. They speculated that there may exist a binary fraction-age relation for sdB stars. Han (2008) showed that such a relation does exist and the binary model of Han *et al.* (2002; 2003) can reproduce the EHB stars in globular clusters, in particular, the low binary fraction of the EHB stars. The main reason for the low binary fraction is that the stars in a globular cluster are all old, and the envelopes of donor stars in the CE channel are loosely bound, leading to wide EHB binaries rather than close ones.

One of the first major discoveries soon after the advent of UV astronomy was the discovery of an excess of light in the far-ultraviolet (far-UV) in elliptical galaxies (see the review by O'Connell, 1999). This came as a complete surprise since elliptical galaxies were supposed to be entirely composed of old, red stars and not to contain any young stars that radiate in the UV. Since then it has become clear that the far-UV excess (or upturn) is not a sign of active contemporary star formation, but is caused by an older population of helium-burning stars or their descendants with a characteristic surface temperature of 25,000 K (Ferguson *et al.*, 1991).

The origin of this population of hot, blue stars in an otherwise red population has, however, remained a major mystery. As we described above, the binary model of Han *et al.* (2002; 2003) reproduces Galactic hot subdwarfs (synonymous with sdB stars in this paper). The key feature of the channels in the model is that they provide the missing physical mechanism for ejecting the envelope and for producing a hot subdwarf. Moreover, since it is known that these hot subdwarfs provide an important source of far-UV light in our own Galaxy, it is not only reasonable to assume that they will also

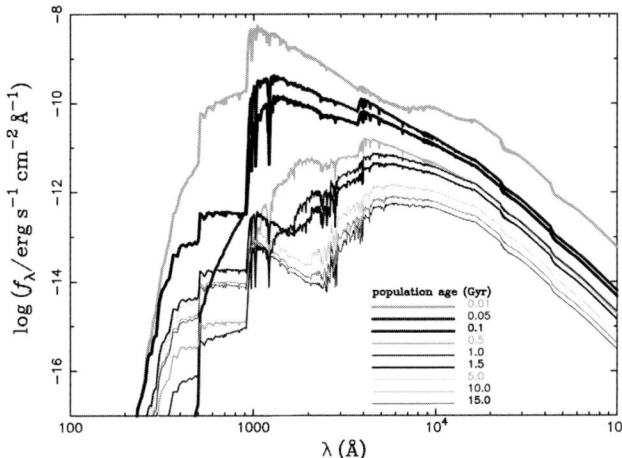

Figure 2. The evolution of the rest-frame intrinsic spectral energy distribution (SED) for a simulated galaxy in which all stars formed at the same time, i.e. a simple stellar population (SSP). The stellar population (including binaries) has a mass of $10^{11} M_\odot$ and the galaxy is assumed to be at a distance of 10 Mpc. The figure is for the standard simulation set (with $\alpha_{\rm CE} = \alpha_{\rm th} = 0.75$ and $\alpha_{\rm RLOF} = 0.5$) in Han, Podsiadlowski & Lynas-Gray (2007) and no offset is applied to the SEDs. Note that the line sections of between 500 Å and 900 Å for population ages of 1.5 and 5.0 Gyr overlap.

contribute significantly to the far-UV in elliptical galaxies, but is in fact expected. It would, therefore, be "a priori" to apply the Han et al. model to the study of the UV-upturn problem.

To quantify the importance of the effects of binary interactions on the spectral appearance of elliptical galaxies, we have performed the first population synthesis study of galaxies that includes binary evolution (see also Bruzual & Charlot, 1994; Worthy, 1994; Zhang, Li & Han, 2005). It is based on a binary population synthesis model of Han et al. (2002; 2003) that has been calibrated to reproduce the short-period hot subdwarf binaries in our own Galaxy that make up the majority of Galactic hot subdwarfs (Maxted et al., 2001). The population synthesis model follows the detailed time evolution of both single and binary stars, including all binary interactions, and is capable of simulating galaxies of arbitrary complexity, provided that the star-formation history is specified. To obtain galaxy colours and spectra, we have calculated detailed grids of spectra for hot subdwarfs using the ATLAS9 (Kurucz, 1992) stellar atmosphere code, which calculates plane-parallel atmospheres in local thermodynamic equilibrium. For the spectra and colours of single stars with hydrogen-rich envelopes, we use the comprehensive BaSeL library of theoretical stellar spectra (Lejeune, Cuisinier & Buser, 1997; 1998).

Figure 2 shows our simulated evolution of the spectral energy distribution (SED) of a galaxy in which all the stars formed at the same time. The total mass of the stellar population (including binaries) is $10^{11} M_\odot$ and the galaxy is taken to be at a distance of 10 Mpc. At early times, the far-UV flux is entirely caused by the contribution from young stars. Hot subdwarfs from the various binary evolution channels become important after about 1.1 Gyr, which corresponds to the evolutionary timescale of a $2 M_\odot$ star, and soon start to dominate completely. After a few Gyr the far-UV SED no longer changes appreciably relative to the visual flux. One immediate implication of this is that the model predicts that the magnitude of the UV excess $(1550 - V)$, defined as the relative ratio of the flux in the V band to the far-UV flux (Burstein et al., 1988), should not

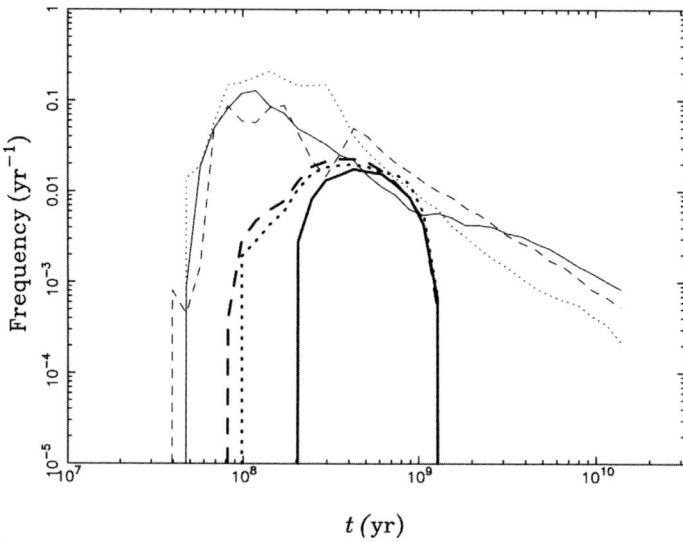

Figure 3. The evolution of birthrates of SNe Ia for a single star burst of $10^{11} M_\odot$ of solar metallicity. Solid, dashed and dotted lines are for $\alpha_{CE} = \alpha_{th} = 1.0$, 0.75, 0.5, respectively ($\alpha_{RLOF} = 0.5$). Thin lines are for the double degenerate channel, and thick lines are for the single degenerate channel.

evolve significantly with look-back time or redshift. Indeed, this is exactly what seems to have been found in recent observations (Brown *et al.*, 2003; Rich *et al.*, 2005).

We found that our binary model can naturally explain many observations of early-type galaxies in spite of its simplicity, and UV-upturn is expected to be universal (from dwarf to giant ellipticals; see Lisker & Han, 2008). The model also predicts that the magnitude of UV-upturn does not depend much on metallicity or redshift. We refer the reader to Han, Podsiadlowski & Lynas-Gray (2007) for the details.

3. Progenitors of Type Ia supernovae

Recent progress in cosmology is largely due to the use of Type Ia supernovae (SNe Ia) as a *calibrated* distance indicator (Riess *et al.*, 1998; Perlmutter *et al.*, 1999). The nature of their progenitors is still unclear, raising doubts as to the calibration which is purely empirical and based on nearby SN Ia sample. The SNe Ia are believed to be thermonuclear explosions of carbon-oxygen (CO) WDs. Observational characteristics of SNe Ia imply that the explosion occurs when a CO WD reaches the Chandrasekhar limit. There are mainly two channels to create Chandrasekhar-mass CO WDs: *the single degenerate channel*, where the CO WD accretes mass from a non-degenerate companion (Hachisu, Kato & Nomoto, 1999a; Han & Podsiadlowski, 2004), and *the double degenerate channel*, where two CO WDs with a total mass larger than the Chandrasekhar mass coalesce (Iben & Tutukov, 1984; Webbink & Iben, 1987)†.

Employing Eggleton's stellar evolution code (Eggleton, 1971; 1972; 1973; Han, Podsiadlowski & Eggleton, 1994; Pols *et al.*, 1995) and adopting the prescription of Hachisu *et al.* (1999) for the accretion efficiency of a CO WD, Han & Podsiadlowski (2004) carried out detailed binary evolution calculations for about 2300 close CO WD binaries, and mapped out the initial parameters in the orbital period-secondary mass

† Note, however, that in this case it is quite likely that the merger product experiences core collapse rather than a thermonuclear explosion(Nomoto & Iben, 1985).

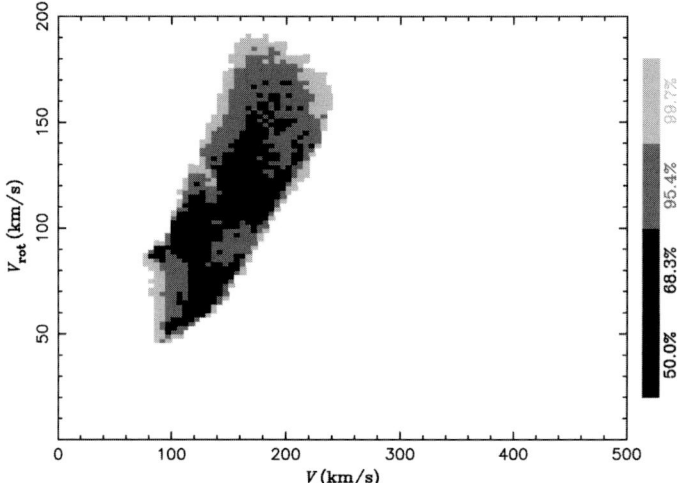

Figure 4. Snapshot probability distribution of companion stars in the plane of $(V_{\rm orb}, V_{\rm rot})$ at the current epoch, where $V_{\rm orb}$ is the orbital velocity and $V_{\rm rot}$ the equatorial rotational velocity for the companion stars at the moment of SN explosion. The probability decreases from inner regions to outer regions. Regions, from inside to outside with corresponding gradational grey scale in the legend (from bottom to top), together with the inner regions, contain 50.0%, 68.3%, 95.4%, and 99.7% of all the systems, respectively. The model adopts a constant star formation rate over the last 15 Gyr and $\alpha_{\rm CE} = \alpha_{\rm th} = 0.75$ ($\alpha_{\rm RLOF} = 0.5$). For a similar model but with $\alpha_{\rm CE} = \alpha_{\rm th} = 1.0$, the distribution is similar, but the upper edge of 190 km/s moves down to 170 km/s.

plane (for a range of WD masses) which lead to a SN Ia. They have implemented these results in a binary population synthesis (BPS) study to obtain the birth rates for SNe Ia for a constant star formation rate. The Galactic birth rate is lower than (but comparable to) that inferred observationally. They have also obtained the evolution of birth rates with time for a single star burst. We see from Fig. 3 that the time delay of SN Ia explosion from star burst is ~ 0.1 to ~ 1 Gyr for the single degenerate channel. The birth rates from the double degenerate channel reach to peaks at ~ 0.1 Gyr and decays with age (approximately $\propto t^{-1}$). Meng, Chen & Han (2008) did similar investigations, but for 10 metallicities in order to investigate the metallicity effect.

For the single degenerate model, the remnant companion star after SN explosion would be a fast rotator and have a high space velocity. Fig. 4 is the distribution of companion star in the plane of orbital velocity-rotational velocity at the moment of SN explosion, as derived from a BPS study with the implementation of the results of Han & Podsiadlowski (2004). Note, however, the ejecta of SN explosion would impact the companion and the companion obtains a kick velocity and the total velocity may be higher by up to 10% (Meng, Chen & Han, 2007). See Han (2008) for more distributions of companion stars.

Acknowledgements

This work was in part supported by the Natural Science Foundation of China under Grant Nos 10433030, 10521001 and 2007CB815406.

References

Brown, T. M. *et al.* 2003, *ApJ*, 584, L69
Burstein, D. *et al.* 1988, *ApJ*, 328, 440
Bruzual, A. G. & Charlot, S. 1993, *ApJ*, 405, 538

Dewi, J. D. M. & Tauris, T. M. 2000, *A&A* 360, 1043
Ferguson, H. C. et al. 1991, *ApJ*, 382, L69
Green, R. F., Schmidt, M., & Liebert, J. 1986, *ApJS*, 61, 305
Eggleton, P. P. 1971, *MNRAS*, 151, 351
Eggleton, P. P. 1972, *MNRAS*, 156, 361
Eggleton, P. P. 1973, *MNRAS*, 163, 179
Hachisu, I., Kato, M., & Nomoto, K. 1999a, *ApJ*, 522, 487
Hachisu, I., Kato, M., Nomoto, K., & Umeda, H. 1999b, *ApJ*, 519, 314
Han Z. 1998, *MNRAS*, 296, 1019
Han Z. 2008a, *ApJ*, 677, L109
Han Z. 2008b, *A&A*, in press (astro-ph/0804.4535)
Han, Z., Eggleton, P. P., Podsiadlowski, Ph., & Tout, C. A. 1995b, *MNRAS* 277, 1443
Han Z. & Podsiadlowski Ph. 2004, *MNRAS*, 350, 1301
Han, Z., Podsiadlowski, Ph., & Eggleton, P. P. 1994, *MNRAS* 270, 121
Han, Z., Podsiadlowski, Ph., & Eggleton, P. P. 1995a, *MNRAS* 272, 800
Han, Z., Podsiadlowski, Ph., & Lynas-Gray, A. E. 2007, *MNRAS*, 380, 1098
Han, Z., Podsiadlowski, Ph., Maxted, P. F. L., Marsh, T. R., & Ivanova, N. 2002, *MNRAS*, 336, 449
Han, Z., Podsiadlowski, Ph., Maxted, P. F. L., & Marsh, T. R. 2003, *MNRAS*, 341, 669
Han, Z. & Webbink, R.F. 1999, *A&A*, 349, L17
Heber, U. 1986, *A&A*, 155, 33
Hjellming, M. S. & Webbink, R. F. 1987, *ApJ*, 318, 794
Iben, I.Jr. & Tutukov, A. V. 1984, *ApJS*, 54, 335
Kilkenny, D., Koen, C., O'Donoghue, D., & Stobie, R. S. 1997, *MNRAS*, 285, 640
Kurucz, R. L. 1992, in: B. Barbuy & A. Renzini (eds.), *The Stellar Populations of Galaxies*, IAU Symp. (Dordrecht: Kluwer), vol. 149, p. 225
Lejeune, T., Cuisinier, F., & Buser, R. 1997, *A&AS*, 125, 229
Lejeune, T., Cuisinier, F., & Buser, R. 1998, *A&AS*, 130, 65
Lisker, T. & Han, Z. 2008 *ApJ*, in press (astro-ph/0803.2512)
Maxted, P. F. L., Heber, U., Marsh, T. R., & North, R. C. 2001, *MNRAS*, 326, 1391
Meng, X., Chen, X., & Han, Z. 2007 *PASJ*, 59, 835
Meng, X., Chen, X., & Han, Z. 2008 *MNRAS*, submitted (astro-ph/0802.2471)
Miller, G. E. & Scalo, J. M. 1979, *ApJS* 41, 513
Moni Bidin, C., Catelan, M., & Altmann, M. 2008, *A&A*, 480, L1
Moni Bidin, C., Moehler, S., Piotto, G., et al. 2006, *A&A*, 451, 499
Nelemans G., Verbunt, F., Yungelson, L. R., & Portegies Zwart, S. F. 2000, *A&A* 360, 1011
Nomoto, K. & Iben, I.Jr. 1985, *ApJ*, 297, 531
O'Connell, R. W. 1999, *ARA&A*, 37, 603
O'Tool, S. J., Heber, U., & Benjamin, R. A. 2004, *A&A*, 422, 1053
Paczyński, B. 1976, in: P. P. Eggleton, S. Mitton & J. Whelan (eds.), *Structure and Evolution of Close Binaries*, IAU Symp. (Dordrecht: Kluwer), vol. 73, p. 75
Perlmutter, S., et al. 1999, *ApJ*, 517, 565
Podsiadlowski, Ph., Rappaport, S., & Han, Z. 2003, *MNRAS* 341, 385
Pols, O. R., Tout, C. A., Eggleton, P. P. & Han, Z. 1995, *MNRAS*, 274, 964
Rich, R. M. et al. 2005, *ApJ*, 619, L107
Riess, A., et al. 1998 *AJ*, 116, 1009
Webbink, R. F. 1988, in: J. Mikołajewska, M. Friedjung, S. J. Kenyon, & R. Viotti (eds.), *The Symbiotic Phenomenon* (Dordrecht: Kluwer), p. 311
Webbink, R. F., 2007, in: EF. Milone, D. A. Leahy & D. W. Hobill (eds.), *Short Period Binary Stars*, (Springer), in press (astro-ph/0704.0280)
Webbink, R. F. & Iben, I.Jr. 1988, in: A. G. D. Philipp, D. S. Hayes & J. W. Liebert (eds.), *IAU Colloq. No. 95* (Davis Press: Schenectady), p. 445
Worthey, G. 1994, *ApJS*, 95, 107
Zhang, F., Li, L., & Han, Z. 2005, *MNRAS*, 364, 503

Discussion

BELCZYNSKI: Comment: Not only common envelope (CE) efficiency is very uncertain, but the very treatment of CE is highly uncertain. Question: How do you make single degenerate scenario of SNIa work? Since, it is known that it is rather hard to accumulate hydrogen and significantly increase white dwarf mass.

HAN: Indeed, common envelope evolution is the least understood process in binary evolution. The evolution is parameterized and calibrated with observations, which gives acceptable results. For the mass growth of white dwarf, I adopted the accumulation efficiency of Hachisu's. The mass transfer rates need to be in the right range for the WD to increase in its mass.

CHRISTENSEN-DALSGAARD: Comment: Many SdB stars are observed to pulsate. Thus we may constrain their structure from observed frequencies. Question: To what extent does the structure of the SdB star depend on the formation channel? That determines an ability to distinguish them from astroseismology.

HAN: The properties of SdB stars originated from different channels are quite different. SdB stars from stable RLOF channel have thick hydrogen envelopes, the SdB stars from common envelope ejection channel have thin envelopes. The merger channel produces SdB stars with no or extremely thin hydrogen envelopes, and the mass range of the SdB stars is also larger.

VINK: You mentioned most SdBs are in binaries, which holds for the field. For clusters such as NGC 6752, Moni Bidin *et al.* (2006) looked for close binaries in a sample of 51 objects, but found none. Are these all mergers?

HAN: Yes. There exists a relation between SdB binary fraction and population age, for which I have submitted a Letter to A&A. For an old stellar population, SdB stars from the merger channel dominate.

WANG: Do you care about the age of the binary components? For example if the secondary is a pre-main-sequence stars.

HAN: Both components have the same age, but the equivalent age of each component may be different. The mass gainer is rejuvenated and its equivalent age is smaller.

YI: Regarding SdB production, what is the source of the biggest uncertainty?

HAN: The biggest uncertainties are from common envelope ejection efficiency and the initial distribution of mass ratios. However, these can be calibrated b comparing theoretical results with observations of many binary related objects. We just used those calibrated parameters to produce SdB population and then applied to the UV-upturn problem in an 'a priori' way, and the UV-upturn is explained naturally.

Conference photograph

The speaker, Z. Han (left), is chatting with A. Sadowski during a coffee time.

The role of binary stars in stellar population synthesis

Zhongmu Li[1,2] and Zhanwen Han[1]

[1] National Astronomical Observatories/Yunnan Observatory, the Chinese Academy of Sciences, Kunming, 650011, China

[2] Graduate University of the Chinese Academy of Sciences
email: zhongmu.li@gmail.com

Abstract. More than about 50% stars are in binaries, but the effects of binary evolution were not taken into account in most previous stellar population synthesis studies. In fact, binaries can affect the integrated peculiarities such as spectral energy distributions (SEDs), colours, and line-strength indices of populations. With the effects of binary stars taken into account, some new results for stellar population studies will be shown. We discuss how binaries affect the colours and Lick indices of simple stellar populations, and the measurement of stellar ages and metallicities.

Keywords. Galaxies: stellar content, galaxies: evolution, galaxies: formation

1. Introduction

Evolutionary stellar population synthesis is a powerful method to study the stellar contents, and then the star formation histories of galaxies. It is widely used in astrophysics studies and is being updated day by day. Although most works (e.g., Thomas et al. 2005, Li et al. 2006) use spectral methods such as SED and line-index methods, photometric methods are also used by many works (e.g., Li et al. 2007, Li & Han 2007a, Li & Han 2007b) as it is shown that some colours can disentangle the well-known stellar age–metallicity degeneracy (e.g., Worthey 1994). However, most works take single-star stellar population (ssSSP) models (e.g., Fioc & Rocca-Volmerange 1997, Bruzual & Charlot 2003, Vazdekis et al. 2003, and Maraston 2005). This is actually different from the real case of galaxies and star clusters. Many works show that galaxies and star clusters contain a lot of binary stars, and the binary fraction seems larger than 50%. This suggests that binaries may play an important role in stellar population synthesis studies, and the role of binary stars should be taken into account. In fact, binary stars seem very important for stellar population synthesis studies, as they can reproduce the special stars, e.g., blue stragglers (e.g., Li & Han 2007c), and the UV-upturn of elliptical galaxies (e.g., Han et al. 2007). Binaries can also change the determinations (using both line-index or photometric methods) of the ages and metallicities of stellar populations. They are also important for galaxy formation and evolution studies. We introduce our work on the study of the role of binary stars in stellar population synthesis studies.

2. Stellar population model and methods used in the work

A rapid stellar population synthesis (RPS) model (Li & Han 2007c) is used in the work, because there is no more appropriate model. The model takes two different initial mass functions, i.e., the ones of Salpeter (1955) and Chabrier (2003), and it models both ssSSPs and binary-star stellar populations (bsSSPs) on the basis of a simple and easily used statistical isochrone database (see Li & Han 2007c), in which the rapid stellar evolution code of Hurley et al. (2002) (hereafter Hurley code) is used to evolve both single and binary stars. Most binary interactions are considered when using the Hurley code to evolve stars of bsSSPs. Except the metallicity, masses of two components, separation of two components, and eccentricity of each binary, other parameters are taken by the code naturally. As a whole, the systemic errors in the stellar population synthesis model is about 6%. For the work, 50% (the typical fraction in the Galaxy) binaries are assumed in each bsSSP. In order to investigate the effects caused only by binaries, we compare the integrated peculiarities of an ssSSPs to those of a bsSSP that has the same stellar age and metallicity as the ssSSP. The stellar-population parameters (age and metallicity) determined using theoretical ssSSPs are also compared with those determined by bsSSPs. This can help us to understand the differences between the estimates of stellar-population parameters when taking different stellar population models.

3. Main results

In our work, the effects of binaries on the colours and line-strength indices of stellar populations are tested, respectively. The effects on the estimates of two stellar-population parameters, i.e., age and metallicity, are also investigated. The main results are as follows.

1) When studying the effects of binaries on the colours of populations, it is found that binaries make the colours of populations bluer, compared to those of ssSSPs. In Fig. 1, the evolution of $(B-V)$ colour is shown, for both ssSSPs and bsSSPs. The results for other colours are not shown here, because other colours are shown similar results as $(B-V)$. When we try to find the reasons, special stars, especially the blue stragglers in stellar populations are found to be very important. They can significantly change the colours of populations as they are very luminous and blue, and contribute a lot to the light of stellar populations.

2) Binary stars can affect the line strength indices of stellar populations clearly. In Figs. 2 and 3, as examples, we compare the evolution of an age-sensitive index, Hβ, and a metallicity-sensitive, Mgb. We can see that binaries make Hβ index larger and Mgb index smaller when comparing to ssSSPs. In fact, binaries can also make other age-sensitive indices such as H$_{\delta A}$ and H$_{\gamma F}$ larger, while making other metallicity-sensitive indices such as iron indices smaller. The binary effects are mainly caused by the special stars generated by binary evolution, but the changes in surface element abundance that results from binary interactions can also contribute to them. As a whole, it suggests that ssSSP models can only measure stellar-population parameters different from the real ones because of the presence of binaries in galaxies and star clusters.

3) Because binaries change the integrated peculiarities of populations compared to ssSSPs, they affect the determination of stellar-population parameters. Our results show that ssSSPs determine less ages when taking a line-index method (using Hβ and [MgFe] indices), and determine lower metallicities (0.003 on average) when taking a photometric method (using $U-R$ and $R-K$ colours) for populations containing binaries. When taking two line-strength indices for work, the difference between stellar ages determined via bsSSPs and ssSSPs increases with increasing age or decreasing metallicity of populations. It can be as large as 6 Gyr for populations with ages near 15 Gyr and the metallicity of 0.004. A 3-D relation of the difference between ages determined by bsSSPs and ssSSPs

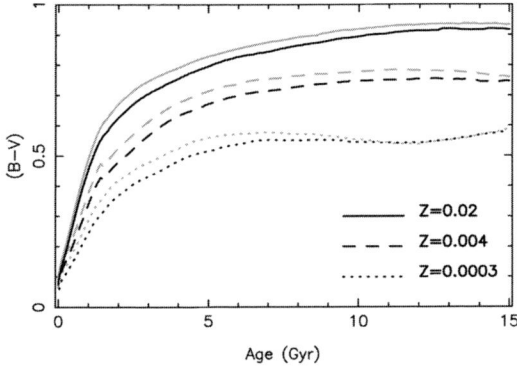

Figure 1. Comparisons of the evolution of $(B-V)$ colour of bsSSPs and ssSSPs. Solid, dashed, and dotted lines are for metallicities of 0.0003, 0.004, and 0.02, respectively. Black and gray lines are for bsSSPs and ssSSPs, respectively.

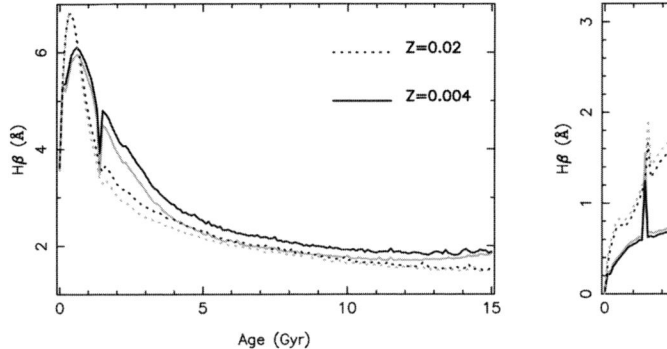

Figure 2. Comparisons of the evolution of Hβ index of bsSSPs and ssSSPs. Solid and dotted lines are for metallicities of 0.004 and 0.02, respectively. Black and gray lines are for bsSSPs and ssSSPs, respectively.

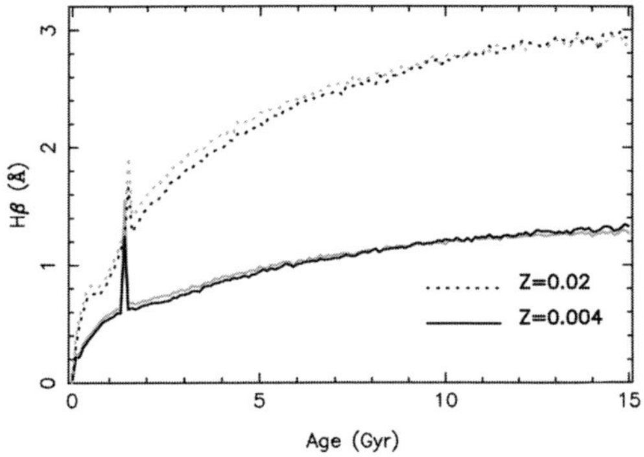

Figure 3. Comparisons of the evolution of Mgb index of bsSSPs and ssSSPs. Solid and dotted lines are for metallicities of 0.004 and 0.02, respectively. Black and gray lines are for bsSSPs and ssSSPs, respectively.

$(t_b - t_s)$, stellar age, and metallicity is shown in Fig. 4. In addition, it shows that ssSSP models and bsSSPs model can give similar results for stellar age determination.

4. Conclusion and disscussion

Binary stars play an important role in stellar population synthesis studies. They can make populations bluer, with larger age-sensitive indices and less metallicity-sensitive indices, for populations with different initial mass functions. When using bsSSP models instead of ssSSP models, larger stellar ages and metallicities will be obtained, via line-index and photometric methods, respectively. Although the results shown give us an image of the role of binary stars, it is far from well understanding the question. In fact, binary stars and new star formations have similar effects on stellar population studies. This makes it more difficult to get the accurate star formation histories of galaxies. In addition, the binary fraction in different galaxies or star clusters may be different. It seems necessary to measure the binary fraction of galaxies and then build stellar population models for different galaxies and star clusters.

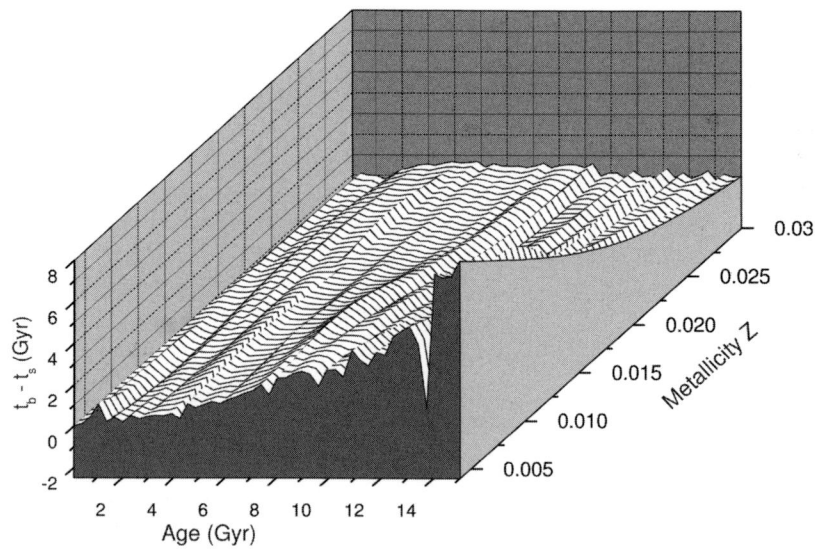

Figure 4. Comparisons of the isochrones of a pair of solar-metallicity ($Z = 0.02$) bsSSP and ssSSP. Black points show the isochrone of the ssSSP, and gray points show the isochrone of the bsSSP. Age and metallicity are the real parameters of populations, and for a population, "$t_b - t_s$" is obtained by subtract the age measured using ssSSPs from that measured using bsSSPs.

Acknowledgements

This work is supported by the Chinese National Science Foundation (Grant Nos. 10433030, 10521001 and 2007CB815406) and the Youth Foundation of Knowledge Innovation Project of the Chinese Academy of Sciences (07ACX51001).

References

Bruzual, G. & Charlot, S. 2003, MNRAS 344, 1000
Chabrier, G. 2003, ApJ 586, L133
Fioc, M. & Rocca-Volmerange, B. 1997, A&A 326, 950
Han, Z., Podsiadlowski, P., & Lynas-Gray, A. E. 2007, MNRAS 380, 1098
Hurley, J. R., Tout, C. A., & Pols O. R. 2002, MNRAS 329, 897
Li, Z. & Han, Z. 2007a, MNRAS 385, 1270
Li, Z. & Han, Z. 2007b, A&A 471, 795
Li, Z. & Han, Z. 2007c, MNRAS, in press, arXiv:0708.1204
Li, Z., Zhang, F., & Han, Z. 2006, ChJA&A 6, 669
Li, Z., Han, Z. & Zhang, F. 2007, A&A 464, 853
Maraston, C. 2005, MNRAS 362, 799
Salpeter, E. E. 1955, ApJ 121, 161
Thomas, D., Maraston, C., Bender, R., & Mendes de Oliveira, C. 2005, ApJ 621, 673
Vazdekis, A., Cenarro, A. J., Gorgas, J., Cardiel, N., & Peletier, R. F. 2003, MNRAS 340, 1317
Worthey, G. 1994, ApJS 95, 107

Discussion

F. HERWIG: Could you explain what causes the spreading at the turn-off in the binary model compared to the single-star simulation. Is it just the super-position of slightly photometrically different binary components? Why does it extend further to the blue than the single-star isochrone?

Z.M. LI: I am not clear about the reason, but this is real. The results shown perhaps are affected by the discrimination of observation. I am not sure about this as I did not test it. I will check and give answers in the future. Thanks.

H.-G. LUDWIG: What do we know about the binary fraction in external (perhaps Local Group) galaxies?

Z.M. LI: I think the binary fraction in different galaxies and star clusters is different and it is necessary to study further. The 50% fraction is very a widely used value.

C. BELCZYNSKI: You predict a large change in age estimate for old stellar populations (low mass stars). How big is this effect for very young populations ($\sim$ 4-5 Myr) containing massive stars (2-40 $M_\odot$)?

Z.M. LI: Not too much for young populations, as binaries contribute less to the light of young populations, relatively.

L. DENG: (1) Where are the photometric binaries in the CMD of M67? (2) How do you deal with the atmospheric models for remnants, such as Blue Stragglers if they are still binaries?

Z.M. LI: (1) I have no clear idea about this, because the data taken by our work was obtained by other people. This is actually not very important for our work. It needs to give more detailed investigation. (2) I did not take this point into account. It is very important to give further studies.

(F. KUPKA: More a comment than a question: since you're using photometric indices in the IR to determine stellar properties, one problem we have in that wavelength region is the lack of atomic data to calculate reliable line blanketing and thus fluxes and finally, colours, from model atmospheres. As long as you only consider low Z values that may not matter. But to compute IR photometric indices for, say, a G or K type giant, we really need better atomic data.

Z.M. LI: It is absolutely right. Thank you.

E. GLEBBEEK: Remark: You point out binaries are important to consider when comparing CMDs of star clusters. They are important for another reason as well: they affect the cluster dynamics. Question: What sort of dynamical model did you use for the simulations you showed?

Z.M. LI: There is no dynamical model. The stellar evolution models shown are for stars evolving in isolation.

N. Langer: Is the difference between your results for single stars and binaries not partly smaller than the error bar in your result for binaries due to uncertainties in binary evolution and population synthesis?

Z.M. Li: The difference is larger than the typical error in the Lick-index method, but it seems less than the uncertainty of photometric study. But the observational uncertainty depends on surveys. The changes caused by binaries are typically less than systematic errors of binary evolution because Hurly code gives rough evolution for stars.

Rotational mixing in close binaries

S. E. de Mink[1], M. Cantiello[1], N. Langer[1], S.-Ch. Yoon[2], I. Brott[1], E. Glebbeek[1], M. Verkoulen[1], and O. R. Pols[1]

[1] Astronomical Institute Utrecht, Princetonplein 5, 3584 CC Utrecht, The Netherlands
[2] Dep. of Astronomy & Astrophysics, Univ. of California, Santa Cruz, CA95064, USA
email: S.E.deMink@uu.nl, M.Cantiello@uu.nl, N.Langer@uu.nl

Abstract. Rotational mixing a very important but uncertain process in the evolution of massive stars. We propose to use close binaries to test its efficiency. Based on rotating single stellar models we predict nitrogen surface enhancements for tidally locked binaries. Furthermore we demonstrate the possibility of a new evolutionary scenario for very massive ($M > 40 M_\odot$) close ($P < 3$ days) binaries: Case M, in which mixing is so efficient that the stars evolve quasi-chemically homogeneously, stay compact and avoid any Roche-lobe overflow, leading to very close (double) WR binaries.

Keywords. stars: rotation, binaries: close, stars: Wolf-Rayet, stars: abundances

1. Introduction

Rotation plays an important role in the evolution of massive stars: it causes deformation of the star due to the centrifugal force, it interplays with stellar mass loss, and it can induce instabilities leading to internal mixing of the star. Mixing induced by rotation has been successful in explaining the ratio of red to blue super giants and it has been

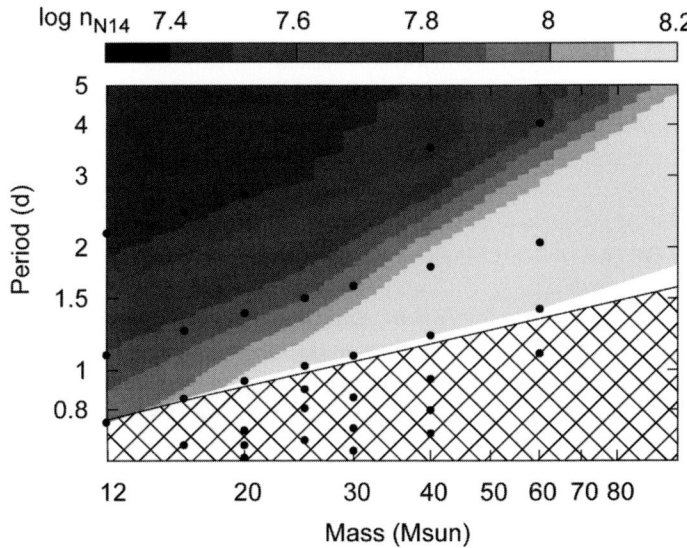

Figure 1. The nitrogen surface abundance is plotted in color shading for rotating single stellar models (black dots) at a metallicity of Z=0.004 with different masses and initial spin periods at the end of their main sequence evolution. In a tidally locked binary the spin period corresponds to the orbital period. The hashed region is excluded for binaries as the two stars are in contact at zero age.

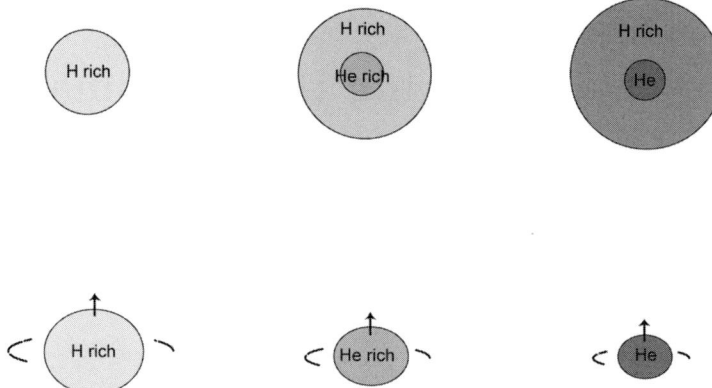

Figure 2. Cartoon representation the evolution of a core hydrogen burning star. A slow rotor will form a core-envelope structure (top row). Fast rotators will evolve quasi chemically homogeneously, and they will stay compact gradually becoming a WR star (bottom row).

invoked to explain the helium and nitrogen enhancements observed in OB stars (Heger & Langer 2000; Maeder & Meynet 2000, and references therein).

Although extensive literature exists on the subject, the efficiency of rotationally induced mixing is still very uncertain. The VLT-FLAMES survey of massive stars (Evans *et al.* 2005), which resulted in rotational velocities and surface abundances of about one thousand O and early B stars provided a major step forward. However it raised more questions than it answered regarding rotational mixing (Hunter *et al.* 2007). This motivated us to formulate potential observational tests to constrain the corresponding uncertain physical parameters. For this purpose we focus on tidally locked binaries.

The advantage of using binaries is that the major stellar parameters, such as the masses, radii and effective temperatures, can be accurately determined (e.g. Hilditch *et al.* 2005) and also, if high resolution spectra are available, the surface abundances. This enables us to test our stellar models directly against well understood systems (see also de Mink *et al.* 2007). A second advantage of using close binaries is that the tidal forces synchronize the spin period of the stars with the orbital period, such that $P_{\rm spin} = P_{\rm orbit}$. This enables us to determine the rotation rate of the stars much more accurately than in the case of single stars, where it can only be estimated from $v \sin i$, derived from spectral fitting. The inclination i of th rotation axis is generally not known for a particular single star.

In this contribution we use of rotating single stellar models (as published by Yoon *et al.* (2006), to which we refer for details) to demonstrate two predictions for detached tidally locked binaries. In Section 2 we discuss the surface enhancement which can be expected in close binaries. In Section 3 we discuss a new evolutionary binary scenario for the most massive close binaries, case M, in which rotational mixing is so efficient that the stars stay compact and avoid any Roche lobe overflow.

2. Surface abundances

The faster the initial rotation of a star of a given mass, the more efficient is rotational mixing. For example a 20 solar mass single star rotating with an initial equatorial rotational velocity of 180 km/s (corresponding to a spin period of about 1.5 days) enhances

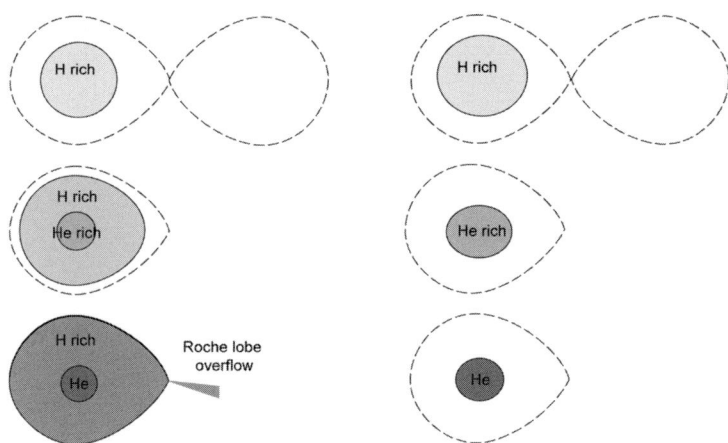

Figure 3. Cartoon of the evolution of normal star inside a Roche lobe (left column) and a quasi-chemically homogeneous evolving star which avoids Roche-lobe overflow: *Case M* (right column).

its surface nitrogen abundance by about 0.5 dex over the course of its main sequence evolution.

Figure 1 shows the surface N abundance at the end of core H burning for the grid of single stellar models published by Yoon *et al.* (2006). In a tidally locked binary the spin period corresponds to the orbital period. The figure therefore shows the maximum surface abundances which we can expect for close detached binaries. For example the sample of OB type binaries with orbital periods ranging from 1-5 days by Hilditch *et al.* (2005) should show enhanced N abundances by up to 0.4 dex. Currently no spectra are available of these systems with high enough resolution to determine the surface abundances. In more massive systems the enhancements will be even larger. In principle even one well-determined binary systems could serve as a strong test case for rotational mixing. In fact, even if in close binaries mixing is enhanced by processes such as tides or irradiation, these observations can be used to set an upper limit to the efficiency of rotational mixing.

3. Chemically homogeneous evolution in binaries

In rapidly rotating stars mixing can be so efficient that the stars fail to form the usual core/envelope structure. These stars stay compact during core H burning and gradually become WR stars (Maeder 1987; Yoon & Langer 2005), illustrated by the cartoon in Figure 2. This type of evolution has been suggested to lead to the formation of long GRB progenitors (e.g. Yoon & Langer 2005; Cantiello *et al.* 2007).

Figure 4 shows that for tidally locked binary systems there is a small range in the parameter space where chemically homogeneous evolution can occur in synchronously rotating binaries. This implies that in such close massive binaries the stars will stay compact and avoid Roche-lobe overflow completely (see the cartoon in Figure 3).

To demonstrate that this situation, predicted on the basis of single stellar models, can actually occur in binary evolution models we performed some preliminary calculations. Figure 5 shows the evolutionary track of a non-rotating 100 $M_\odot$ star at a metallicity of $Z = 10^{-5}$ together with four tracks corresponding to the evolution of a 100 $M_\odot$ star in a close binary with an equal mass companion, in orbits with initial periods of 1.7, 1.4, 1.2 and 1.15 days, respectively.

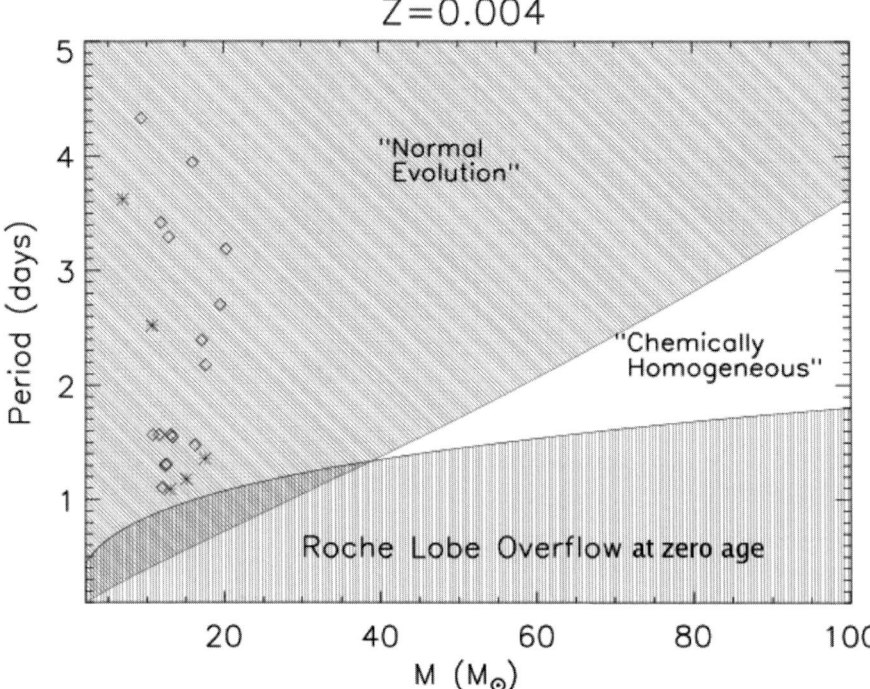

Figure 4. Parameter space for binary systems in which chemically homogeneous evolution can occur. The black hashed region is excluded as binaries do not fit in such close orbits at zero age. The symbols show observed systems in the Small Magellanic Cloud (Hilditch *et al.* 2005).

According to non-rotating models such close systems would fill their Roche lobe during their core hydrogen burning and start so called Case A mass transfer (Kippenhahn & Weigert 1967). The stars in the 1.7 day and the 1.4 day systems stay more compact than the corresponding non-rotating star, due to efficient rotational mixing. However, they cannot avoid Roche lobe overflow before the end of their main sequence evolution.

Although one may intuitively expect that an even closer system would fill its Roche lobe earlier, the primary in the 1.2 day orbit system stays compact enough to avoid mass transfer during the main sequence. It becomes brighter at almost constant radius until central hydrogen exhaustion. Then it contracts until the small amount of H that is still left in the outer layers ignites. The star expands during H shell burning and fills its Roche lobe. In the slightly more compact system with an initial orbital period of 1.15 days, mixing is so efficient that at the end of hydrogen burning both stars are basically pure helium stars, and Roche lobe overflow is completely avoided.

4. Conclusion

Close detached binaries are potentially strong test cases to constrain the uncertain physics of rotational mixing. Furthermore, we show that, contrary to expectation, the very closest massive binaries could avoid mass transfer altogether. In addition to the classic binary cases A, B and C (Kippenhahn & Weigert 1967; Lauterborn 1970), we find a new evolutionary scenario for very close massive binaries, which we named Case M, where the letter M refers to the importance of mixing. In this case both stars are so efficiently mixed, that they remain compact and avoid Roche-lobe overflow during the main sequence, and probably beyond.

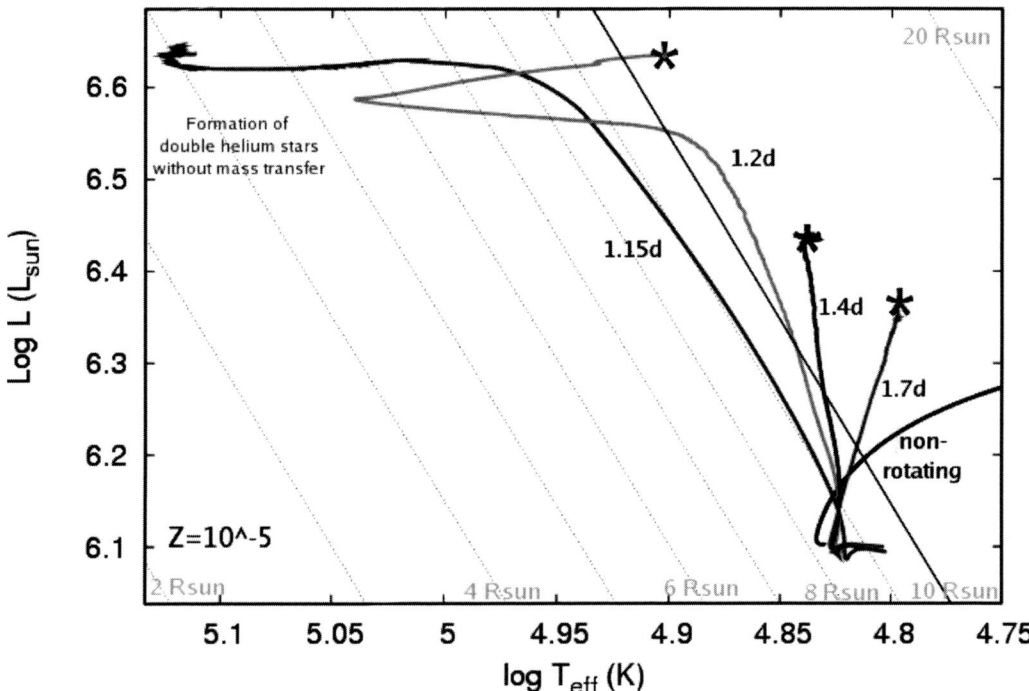

Figure 5. Evolution of stars of 100 M$_\odot$ in the HR diagram, showing the effects of rotational mixing in binaries. The diagonal lines are lines of constant radii. The black line shows a non-rotating 100M$_\odot$ single star for reference. The colored lines show the evolution of 100M$_\odot$ stars which orbit around a 98 M$_\odot$ mass companion with an initial orbital period of 1.7d (purple), 1.4d (blue), 1.2d (green), until the onset of Roche-lobe overflow (asterisk symbol). The red curve corresponds to an initial period of 1.15d. This system ends as two massive helium star in a close orbit.

According to this evolutionary scenario, double helium star systems in very close orbits can be made. Perhaps WR20a, a binary consisting of two detached core hydrogen burning stars of about 80 solar masses in a 3.6 day orbit (Bonanos et al. 2004), is an example of this type of evolution. It remains to be investigated whether this evolutionary scenario can lead to the formation of two long GRB progenitors.

References

Bonanos, A. Z., Stanek, K. Z., Udalski, A., et al. 2004, *Astrophys. J. L.*, 611, L33
Cantiello, M., Yoon, S.-C., Langer, N., & Livio, M. 2007, *Astron. Astrophys.*, 465, L29
de Mink, S. E., Pols, O. R., & Hilditch, R. W. 2007, *Astron. Astrophys.*, 467, 1181
Evans, C. J., Smartt, S. J., Lee, J.-K., et al. 2005, *Astron. Astrophys.*, 437, 467
Heger, A. & Langer, N. 2000, *Astrophys. J.*, 544, 1016
Hilditch, R. W., Howarth, I. D., & Harries, T. J. 2005, *Mon. Not. Roy. Astron. Soc.*, 357, 304
Hunter, I., Dufton, P. L., Smartt, S. J., et al. 2007, *Astron. Astrophys.*, 466, 277
Kippenhahn, R. & Weigert, A. 1967, *Zeitschrift fur Astrophysik*, 65, 251
Lauterborn, D. 1970, *Astron. Astrophys.*, 7, 150
Maeder, A. 1987, *Astron. Astrophys.*, 178, 159
Maeder, A. & Meynet, G. 2000, *Astron. Astrophys.*, 361, 159
Yoon, S.-C. & Langer, N. 2005, *Astron. Astrophys.*, 443, 643
Yoon, S.-C., Langer, N., & Norman, C. 2006, *Astron. Astrophys.*, 460, 199

Discussion

DENG: I am very interested in your case M, as they are always blue and bright. Can we call them "super-blue-stragglers"? Is there any observational evidence for that?

DE MINK: Good point. Probably WR20a belongs to the cluster Westerlund 2. Its components could be "super-blue-stragglers" with respect to the cluster. However, the cluster is so young that it may be hard to see whether the stars of WR20a are significantly bluer than the other cluster members. We are currently investigating which other observed systems are potential case M in candidates.

HAN: Is there any observational evidence for close BH/NS binaries with low-mass companions? Such binaries would be difficult to form from other channels, such as common-envelope evolution.

DE MINK: That is a good idea and could be an interesting application. I think it deserves further investigation.

BELCZYNSKI: (1) For an example if you take a ($80 M_\odot + 60 M_\odot$ binary) – do you know if these stars are synchronized? Any observations? (2) Is it possible that these stars overfill their Roche Lobes, before they get synchronized? (i.e., they evolve very rapidly and may expand rather fast.)

DE MINK: (1) The time scale for synchronization is very short compared to the main sequence life time, even for these vary massive stars, so we assume it is tidally locked. High resolution spectra of this system are available, but to my knowledge they have not been used to determine the rotational velocity of the stars directly. It would be interesting as a test for the efficiency of the tides in such an extreme system like WR20a. (2) Apart from there pre-main sequence evolution, which is very uncertain, I would expect that, if the stars filled their Roche Lobes before, they would still fill their Roche Lobes today, which they don't.

HERWIG: Could this also work in low-mass binaries to lead to close double-degenerate systems – maybe at very low Z?

DE MINK: I do not think so. Rotational mixing becomes less efficient in lower mass stars. Therefore, the window in the binary parameter space, in which we expect case M to occur, closes for masses below 40 $M_\odot$. If mixing is more efficient in binaries we might be able to go to lower masses. Using the models by Yoon *et al.* (2006), which span a metallicity range from $Z=10^{-5}$ to 0.004, we find that the "window" for homogeneous evolution shifts to smaller periods but not to smaller masses.

MEYNET: Are there results obtained with models including magnetic fields, i.e., with the dynamo theory proposed by H. Spruit?

DE MINK: Yes. We plan to investigate the effects of magnetic fields in more detail, but qualitatively we find the following: if we "switch off" transport of angular momentum and chemical species by magnetic fields, the star will rotate more differentially and shear mixing becomes more important. It would be interesting to see under what conditions chemically homogeneous evolution occurs in your models and what this implies for tidally locked binaries.

Close binary evolution and blue straggler formation

P. Lu[1,2]†, L. Deng[1]

[1] National Astronomical Observatories, Chinese Academy of Sciences, Beijing 100012, P.R. China
email: lupin@bao.ac.cn

[2] Graduate University of Chinese Academy of Sciences, Beijing, 100049, P.R. China

Abstract. In order to discuss the contribution of mass transfer in primordial close binaries to the blue straggler population in young clusters, we use Eggleton's stellar evolution code to simulate a grid of case A binary evolutionary models with the initial donor mass $2.0 - 8.0\ M_\odot$ and mass ratio $0.1 - 0.9$. The models cover the whole case A binaries that will experience mass transfer between 30.0 Myr to 1.0 Gyr. Based on such detailed models, we present a simulation to compare with the *HST* observation of young cluster NGC 1831 which can be fit with an isochrone of $\log(\text{age}) = 8.65$. The results show very few blue stragglers could be produced by case A binary evolution. There must be some other mechanisms for blue straggler formation in young clusters.

Keywords. Stars: blue stragglers, stars: binaries: close, Galaxy: open clusters and associations: individual (NGC 1831).

1. Introduction

Blue stragglers (BSs) are a kind of very special stars that lie on the extension of main sequence, in the area which is bluer and more luminous than the main sequence turn off (MSTO) in color magnitude diagram (CMD). Since the first observation of BSs in globular cluster M3 (Sandage 1953), they have been widely observed in all types of stellar systems by now. As the brightest and bluest "main sequence stars", they make a considerable contribution to the blue side of a stellar system's integrated spectrum energy distribution (ISED) (Deng *et al.* 1999, Schiavon *et al.* 2004, Xin & Deng 2005).

Based on single stellar evolution theory, stars massive than the MSTO should already have evolved off the main sequence. BSs seem to be the stars which have longer life time for their mass or they formed later than the others in the cluster. Both the two reasons can not explain the existence of BSs very well. Several mechanisms have been proposed to explain BS formation and now it is widely accepted that BSs are formed in binaries by directly collision or primordial binary evolution.

Direct collision is usually related to dense environment in the cores of star clusters (Fregeau *et al.* 2004). It is supposed to be an important scnario of BSs formation in the dense cluster center. We name the BSs produced under this way dynamical blue stragglers (DBSs). Primordial binaries evolution is another mechanisms of BSs formation. Primordial binaries means the binaries that already exist when the cluster formed. It is highly affected by initial binary fraction. Most dynamical interactions in dense cluster cores tend to destroy binaries, so this mechanism is thought to be important in a sparse

† Present address: 20A Datun Road, Chaoyang District, Beijing, China(100012)

environment (Mathys 1991). We call them primordial BSs (PBSs). The two mechanisms work at the same time in the evolution of stellar systems.

In 1964, McCrea proposed that BSs could be formed as the remnants of mass exchange in close binary systems. The massive star in the binary evolves faster, and mass transfer begins after the primary fills up its Roche lobe envelope. Then, as the hydrogen-rich material of the primary's envelope expanded out of the Roche lobe and transfers to the surface of the secondary through inner Lagrangian point, the companion will gradually grow in mass and become bluer and more luminous. As a result of mass exchange, the secondary becomes the massive one in the binary and will have a longer life time remaining on main sequence.

Mass transfer scnario can be divided into 3 subtypes – case A, case B and case C - based on the evolution phase of the donor when mass transfer begins (Kippenhahn & Weigert 1968). They are related to the time mass transfer begins when the donor is still on main sequence with a hydrogen burning core, it has evolved off the main sequence to the red giant branch before helium core ignition, or later evolutionary phase.

In this paper, we only focus on the mass transfer scnario and we limit our models in case A. A very detailed modelling of case A binary evolution is carried out in this paper. In order to investigate the distribution of BSs formed via case A mass transfer, we present a Mento-Carlo simulation to compare with the Hubble Space Telescope (HST) observations of young clusters. Young clusters are chosen so as to minimum the effect of dynamical evolution of stellar systems to BS formation. The calculation of our models is described in Sect. 2. and the result of Mento-Carlo simulation is given in Sect. 3. Summary and conclusion is presented in the final section.

2. The model of primordial blue stragglers

We use Eggleton's stellar evolution code which has been updated (Han et al. 1994,2000 & Pols et al. 1995,1998) to compute close binary evolution. The opacity library radiative (Iglesias & Rogers 1996) and molecular opacities (Alexander & Ferguson 1994) is adopted in this code.

So as to simplify the calculation, we take some assumptions which are reasonable in these models. The binary systems are assumed to be conservative in mass and angular momentum for no obvious mass loss in intermediate mass stars. The initial eccentricity e = 0 is adopted in the calculation. Tidal evolution is neglected because it has little influence on circular orbital binaries. Magnetic braking and stellar spins is also ignored for simplicity.

In the calculation, Roche lobe overflow (RLOF) is considered as a boundary condition with the mass transfer rate

$$dm/dt = const. \times max[0, (R_{star}/R_{lobe} - 1)^3] \tag{2.1}$$

where R_{star} is the radius of the donor and R_{lobe} is the Roche lobe radius (Eggleton 1983). We adopt the const. = 500 $M_\odot$ yr^{-1} to keep a steady RLOF. Mass transfer history of the donor extracted from the calculation is recorded and then used as the input in a subsequent calculation of the companion. The material is assumed to be deposited on the surface of the secondary with zero falling velocity and homogeneously distributed over the outer layer. For the evolution of close binary, the formation of binary merger and subsequence evolution of a merger is not very certain so we manually terminate the code when binary systems contact or merge.

Figure 1 shows an example of case A binary evolution model with initial donor mass 2.9 $M_\odot$, companion 2.6 $M_\odot$, orbital separation 13.0 $R_\odot$ and solar composition [Z = 0.02,

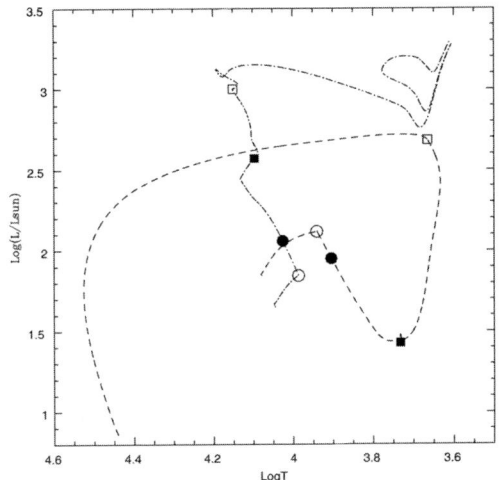 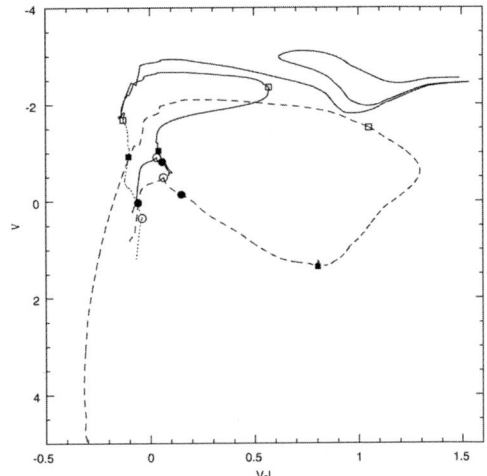

Figure 1. The evolutionary tracks of the two stars in our example binary. The dashed and dash-dotted lines are the evolutionary tracks of the donor and the accretor. The open circles, filled circles, filled squares and open squares on the tracks mean: mass transfer begins, mass ratio equals 1, the system is 446.7 Myr old, and mass transfer terminates.

Figure 2. The synthetic evolutionary track of the two components in our example binary system. The solid line is the synthetic evolutionary track of the two components. The dashed and dotted lines are the evolutionary tracks of the donor and the accretor respectively. All the marks means the same as in Fig. 1.

Table 1. Main results of the example binary ($2.9M_\odot+2.6M_\odot$)

Epoch	Age($10^9 yrs$)	$P(d)$	$a(R_\odot)$	Mass($M_\odot$)	lg($L/L_\odot$)	lgT_{eff}	X_C	Y_C	$\dot{M}(M_\odot yr^{-1})$
1	0.0000	2.3161	13.0000	2.9000	4.0781	1.8437	0.700	0.280	0.0
				2.6000	4.0433	1.6495	0.700	0.280	
2	0.3896	2.3161	13.0000	2.9000	3.9402	2.1169	0.089	0.892	0.0
				2.6000	3.9861	1.8452	0.336	0.645	
3	0.3906	2.2957	12.9237	2.7500	3.9040	1.9499	0.086	0.895	9.1768E-07
				2.7500	4.0256	2.0584	0.337	0.644	
4	0.4467	4.2559	19.5029	1.5640	3.7329	1.4268	0.015	0.966	2.7073E-09
				3.9207	4.0960	2.5719	0.249	0.732	
5	0.4761	117.476	178.1303	0.3993	3.6652	2.6840	0.000	0.981	0.0
				5.0852	4.1500	3.0004	0.112	0.869	

The columns are (1) model serial number, (2) age, (3) period, (4) orbital separation, (5) mass, (6) luminosity, (7) effective temperature, (8) hydrogen abundance in the core, (9) helium abundance in the core, (10) mass transfer rate.

$Y = 0.28$]. Roche lobe mass transfer of the donor begins at log(age) = 8.59 and then the secondary becomes more and more massive. At the meanwhile, it climbs up on the extension of main sequence and gradually become bluer and more luminous on HR diagram. Our comparison sample young cluster NGC 1831 can be fit with an isochrone of log(age) = 8.65, so we also mark the position of the two components at that age with filled square on Fig 1. When mass transfer terminate, the donor becomes a red giant with only 0.4 $M_\odot$. The helium core can not be ignited and it will quickly turn to a helium white dwarf. As a result of mass transfer, the secondary remains on main sequence with a hydrogen burning core. The parameters of the binaries at some epochs are listed in Table 1.

Usually, close binaries especially those belongs to case A cannot be visually resolved in observations, so we somehow need to get the synthetic spectrum energy distribution of the two components in a binary system. For the HST observation of NGC 1831, it is convenient to convert the magnitude in F555 and F814 band to standard V and I magnitude. So we also need to get the synthetic Johnson-Cousins V and I magnitude of all our models by convolving correspond filter response.

A standard spectrum library for evolutionary synthesis has been presented by Lejeune et al. (1997,1998) and they also obtained the colors and bolometric corrections, synthesized in the $(UBV)_J(RI)_C JKLL'M$ system. The library covers wide ranges of fundamental parameters: T_{eff}: 50000 K $\sim$ 2000 K, log$g$: 5.5 $\sim$ -1.02 and [M/H]: +1.0 $\sim$ -5.0. Given effective temperature, surface gravity and solar composition, we can easily get the Johnson-Cousins V and I magnitude of all our models by an interpolation. The evolutionary tracks of the example binary in CMD are shown in Fig. 2. The synthetic evolutionary track is also shown in this figure. The evolution history of synthetic color V-I is given in Fig. 3. Mass ratio becomes 1 right after mass transfer begins. After that, the synthetic color is dominated by the secondary and move slowly to the blue side. As a result of mass exchange, the synthetic color can remain in the blue region for a longer time. As the donor climbs up along the Hayashi line and expands it's envelope, the color will quickly evolve to the red side. Mass transfer keeps on going until the donor begins contract at the top of Hayashi line and then the secondary dominates the color again. The synthetic model will evolve to the very blue and luminous region and remain there for about 10 Myr and then follow the evolution of the companion.

Blue stragglers formed via mass transfer in old open clusters M67 has been studied (Tian et al. 2006). So we focus on case A binaries in young clusters. To simulate a continuous process in the evolution of young clusters, we computed a grid of close binary evolution models with an initial intermediate donor mass from 2.0 to 8.0 $M_\odot$. Mass ratio is from 0.1 to 0.9 with 0.1 interval and the orbital separation covers the whole case A. The models cover the whole case A binary evolution that will experience mass transfer between 30.0 Myr to 1.0 Gyr. A 'snapshot' of all these models in the CMD at log(age) = 8.65 are taken for the subsequent Mento-Carlo simulation.

3. Mento-Carlo simulations of the primordial BSs in NGC 1831

In order to investigate the distribution of BSs formed via close binary mass transfer in young clusters and compare with observations. We choose NGC 1831 as a sample to perform our Mento-Carlo simulation. The cluster can be fit with an isochrone of log(age) = 8.65 with solar composition.

3.1. Initial parameters of the cluster NGC 1831

To simulate a cluster for the comparison with observations, we adopt the initial parameters of the cluster shown below.

1) Initial mass function

We adopt the initial mass function given by Kroupa, Tout & Gilmore (1990,1991). They found a mass generating function following Eggleton (Eggleton, Fitchett & Tout 1989),

$$M(X) = 0.33[\frac{1}{(1-X)^{0.75} + 0.04(1-X)^{0.25}} - \frac{1}{1.04}(1-X)^2] \quad (3.1)$$

where X is a random number from 0 to 1 and M is the total mass of the two components in a binary system.

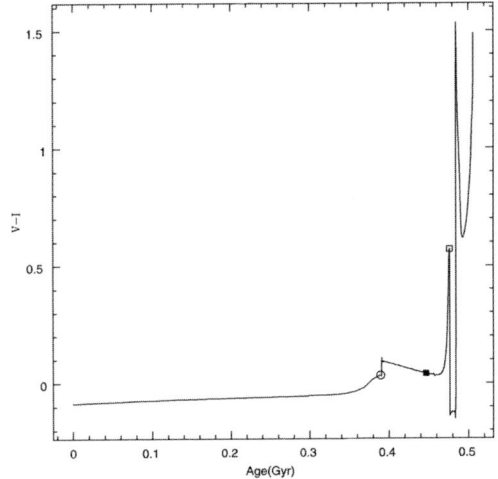 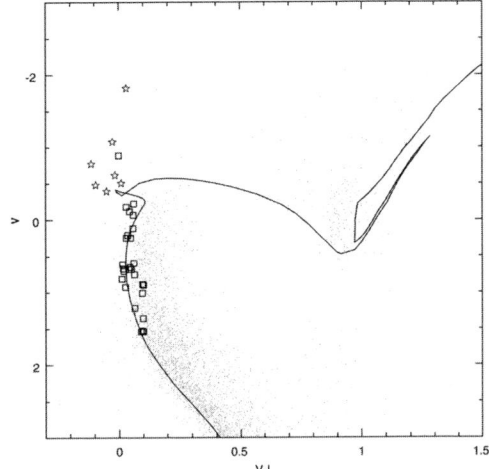

Figure 3. Evolution history of the synthetic color of the example binary system. The open circle, filled square and open square means: mass transfers begins, the system is 446.7 Myr old and mass transfer terminates.

Figure 4. The solid line is the 446.7 Myr theoretical isochrone. The open asterisks are the high possibility BS candidates and the open squares are the 28 binaries from our simulation.

2) Initial mass ratio

We adopt a uniform distribution of initial mass ratio (Hurley, Tout, Aarseth & Pols 2001).

$$1 > q > max[0.1(M(X) - 0.1), 0.02(M(X) - 50.0)] \quad (3.2)$$

where the M is the total mass of a binary from the initial mass function and limited to 0.1-50.0 $M_\odot$.

3) Initial orbital separation

A flat distribution of initial orbital separation from Pols & Marinus (1994) are used here.

$$\Gamma(\alpha) \propto \alpha^{-1}, \alpha_{max} > \alpha > \alpha_{min}, \alpha_{max}/\alpha_{min} \approx 5000 \quad (3.3)$$

where $\alpha = R/D$. R is the radius of stars and D is the orbital separation. α_{min} is the minimum size of the Roche lobe that just fits a zero-age star and $\alpha_{max}/\alpha_{min} \approx 5000$ corresponds to 100 AU orbital separation for a 2 $M_\odot$ primary which makes sure that about two thirds of all binaries are close and will interact at some evolutionary stage.

In the CMD of NGC 1831, we pay more attention to the region from MSTO to 2 mag below, which limits the initial donor mass 2.0 – 3.2 $M_\odot$. The initial binary fraction in this region is set to be 50%. Finally, we obtain 772 primordial binaries with initial donor mass 2.0 – 3.2 $M_\odot$ using the distribution of initial parameters listed above.

3.2. Results

After an interpolation in the precious 'snapshot' we have gotten, only 28 binaries can be interpolated and their positions in the CMD are shown in Fig 4 with open square. The points marked with asterisk are the high possibility candidate of BSs in the observation. Among the 28 binaries, only 1 is bluer and more luminous than MSTO thus could be a primordial blue straggler.

4. Summary and conclusion

The simulation shows very few BSs could be predicted by case A binary evolution in young clusters. There is a severe lack of BSs if we only consider mass transfer scnario in case A binaries. Based on the data of our models, the reason could be:

1. The distribution of initial parameters in simulation, especially the orbital separation. A flat distribution of initial orbital separation with an upper limit $\alpha_{max} \approx 5000 \alpha_{min}$ can only give about 10% case A binaries. More of them belong to case B and C.

2. Coalescence. The evolutions of contact binaries and mergers are very important for case A. During the evolution, case A binaries could easily contact or merge especially for those with higher donor mass, small or intermediate mass ratio and small orbital separation. So, at log(age) = 8.65, a significant portion of our close binary models have already contacted or merged.

Then as a result, finally, only 3%–4% of the total binaries can be interpolated thus even fewer blue stragglers can be produced via case A binary evolution. So, case A binary mass transfer might not be the main mechanism of BSs formation in young clusters. Coalescence and case B binary evolution needs to be considered or other mechanisms need to be introduced to the formation of blue stragglers in young clusters such as collision and so on.

Acknowledgements

We would like to thank Q. Liu for the data reduction of NGC 1831 and providing us the data. We also would like to acknowledge B. Tian for writing a document on how to use Eggleton's stellar evolution code.

References

Alexander, D. R. & Ferguson, J. W. 1994, *ApJ* 437, 879
Chen, X. F. & Han, Z. W. 2004, *MNRAS* 355, 1182
Deng, L., Chen, R., Liu, X. S., & Chen, J. S. 1999, *ApJ* 524, 824
Eggleton, P. P. 1971, *MNRAS* 151, 351
Eggleton, P. P. 1972, *MNRAS* 156, 361
Eggleton, P. P. 1973, *MNRAS* 163, 279
Han, Z., Podsiadlowski, Ph., & Eggleton, P. P. 1994, *MNRAS* 270, 121
Han, Z., Tout, C. A., & Eggleton, P. P. 2000, *MNRAS* 319, 215
Hurly, J. R., Tout, C. A., Aarseth, S. J., & Pols, O. R. 2001, *MNRAS* 323, 630
Hurly, J. R., Pols, O. R., Aarseth, S. J., & Tout, C. A. 2001, *MNRAS* 363, 293
Iglesias, C. A. & Rogers, F. J. 1996, *ApJ* 464, 943
Kroupa, P., Tout, C. A., & Gilmore, G. 1991, *MNRAS* 251, 293
Lejeune, Th., Cuisinier, F., & Buser, R. 1997, *A&AS* 125, 229
Lejeune, Th., Cuisinier, F., & Buser, R. 1998, *A&AS* 130, 65
McCrea, W. H. 1964, *A&A* 128, 147
Pols, O. R. & Marinus, M. 1994, *A&A* 288, 475
Pols, O. R., Tout, C. A., Eggleton, P. P., & Han, Z. 1995, *MNRAS* 274, 964
Pols, O. R., Schroder, K.-P., Hurley, J. R., Tout, C. A., & Eggleton, P. P. 1998, *MNRAS* 298, 525
Sandage, A. R. 1953, *AJ* 58, 61
Tian, B., Deng, L., Han, Z., & Zhang, X. B. 2006, *A&AS* 455, 247
Xin, Y. & Deng, L. 2005, *ApJ* 619, 824

Discussion

K. STEPEIN: If I understand correctly, your model predicts that BSs should be Algols?

P. LU: Yes, most BSs formed via case A should be Algols.

K. STEPEIN: A lack of a sufficient number of close binaries which can transform into BSs may result from neglecting angular momentum loss.

P. LU: More binaries could become close during the evolution by considering angular momentum loss, but at the meanwhile, the disadvantage is that more close binaries will merge by taking account of momentum loss. We can not tell exactly the subsequent evolution.

O. PLOS: An explanation for the small number of blue stragglers formed by case A mass transfer in your models, could be that most blue stragglers in young clusters are actually formed by case B mass transfer. In rather massive binaries the parameter space for case B is larger than for case A. This would also explain the lack of Algols among observed blue stragglers, because case B mass transfer produces wide, detached remnants.

J. CHRISTENSEN-DALSGAARD: Blue stragglers in old clusters can be in the Cepheid instability strip; in young clusters they may be β cephei stars. In these cases observations of oscillations could constrain the internal structure. Do the different formation scenarios yield different internal structures that we could tell apart with astero-seismology.

P. LU: In fact, BSs in our models are binaries. Their synthetic positions on CMD may in the Cepheid instability strip, but actually they may not be β cephei stars respectively. BSs formed via directly collision or merger may be different, but we still could not exactly describe the detailed internal structure.

The LOC team of the meeting: starting from the left, Yu Xin, Chunlin Tian, Yanping Wang, Guoqing Liu, Guohu Zhong, Xiaoshan Yun, Yi Hu, Yangping Luo, Pin Lu (the speaker), Changqing Luo, Qiang Liu.

The single degenerate channel for the progenitors of Type Ia supernovae

Xiangcun Meng[1,2], Xuefei Chen[1] and Zhanwen Han[1]

[1] National Astronomical Observatories/Yunnan Observatory, the Chinese Academy of Sciences, Kunming, 650011, China,
email: conson859@msn.com

[2] Graduate School of the Chinese Academy of Sciences

Abstract. We have carried out a detailed study of the single-degenerate channel for the progenitors of type Ia supernovae (SNe Ia). In the model, a carbon-oxygen white dwarf (CO WD) accretes hydrogen-rich material from an unevolved or a slightly evolved non-degenerate companion to increase its mass to Chandrasekhar mass limit. Incorporating the prescription of Hachisu et al. (1999a) for the accretion efficiency into Eggleton's stellar evolution code and assuming that the prescription is valid for all metallicities, we performed binary stellar evolution calculations for more than 25,000 close WD binary systems with various metallicities. The initial parameter spaces for SNe Ia are presented in an orbital period-secondary mass ($\log P_{\rm i}, M_2^{\rm i}$) plane for each Z.

Adopting the results above, we studied the birth rate of SNe Ia for various Z via binary population synthesis. From the study, we see that for a high Z, SNe Ia occur systemically earlier and the peak value of the birth rate is larger if a single starburst is assumed. The Galactic birth rate from the channel is lower than (but comparable to) that inferred from observations.

We also showed the distributions of the parameters of the binary systems at the moment of supernova explosion and the distributions of the properties of companions after supernova explosion. The former provides physics input to simulate the interaction between supernova ejecta and its companion, and the latter is helpful for searching the companions in supernova remnants.

Keywords. Binaries: close, stars: evolution, supernovae: general, white dwarf: metallicity

---

Although type Ia supernovae (SNe Ia) appear to be good cosmological distance indicators and have been applied successfully in determining cosmological parameters (e.g. Ω and Λ; Riess et al. 1998; Perlmutter et al. 1999), the exact nature of SNe Ia is still unclear, especially the progenitor model of SNe Ia (see the reviews by Hillebrandt & Niemeyer 2000; Leibundgut 2000). At present, the single-degenerate Chandrasekhar model (Whelan & Iben 1973; Nomoto, Thielemann & Yokoi 1984) is the most widely accepted model. In the model, a CO WD accretes hydrogen-rich material from its companion until its mass reaches a mass of $\sim 1.378 M_\odot$ (close to Chandrasekhar mass, Nomoto, Thielemann & Yokoi 1984), and then explodes as a SN Ia. The companion is probably a main sequence star or a slightly evolved star (WD+MS). The discovery of the potential companion of Tycho's supernova also verified the reliability of the WD + MS model (Ruiz-Lapuente et al. 2004; Ihara et al. 2007). The purpose of this paper is to study the progenitor model comprehensively and systematically.

1. Binary evolution calculation

We use the stellar evolution code of Eggleton (1971, 1972, 1973) to calculate the binary evolutions of WD+MS systems. The code has been updated with the latest input physics

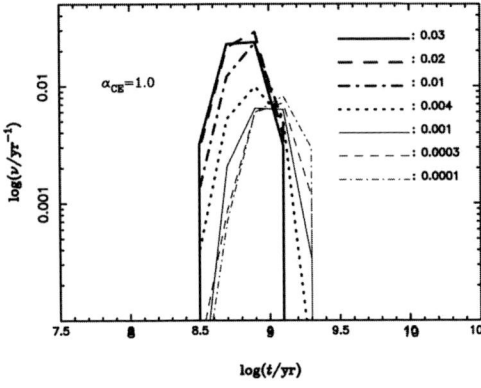 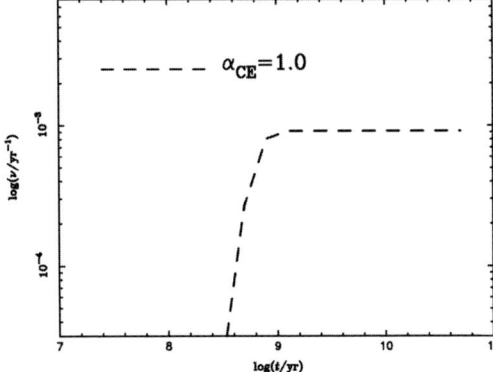

Figure 1. The evolution of the birth rate of SNe Ia for a single starburst of $10^{11} M_\odot$ for different metallicities with $\alpha_{CE} = 1.0$.

Figure 2. The evolution of the birth rate of SNe Ia for a constant star formation rate (Z=0.02, SFR=$5 M_\odot \text{yr}^{-1}$) with $\alpha_{CE} = 1.0$.

over the last three decades (Han, Podsiadlowski & Eggleton 1994; Pols et al. 1995, Pols et al. 1998). Roche lobe overflow (RLOF) is treated within the code described by Han et al. (2000). Ten metallicities are chosen here (i.e. $Z = 0.0001$, 0.0003, 0.001, 0.004, 0.01, 0.02, 0.03, 0.04, 0.05 and 0.06). The opacity tables for these metallicties are compiled by Chen & Tout (2007) from Iglesias & Rogers (1996) and Alexander & Ferguson (1994).

Instead of solving stellar structure equations of a WD, we adopt the prescription in Hachisu et al. (1999a) on the accretion of the hydrogen-rich material from its companion onto the WD based on an assumption of optically thick wind (Hachisu et al. 1996) (see Meng, Chen & Han. 2008 for the prescription in details).

We calculated more than 25,000 WD+MS binary systems with various metallicities, and obtained a large, dense model grid. We summarize the final outcomes of all the binary evolution calculations in the initial orbital period-secondary mass ($\log P^i$, M_2^i) planes (see Figs. 1 to 10 in Meng, Chen & Han. 2008). The contours for SNe Ia are shown by solid lines in these figures. From these figures, we see that the contour moves from the left lower region to the right upper region with metallicity Z in the ($\log P^i$, M_2^i) plane. This means that the progenitor systems for SNe Ia have a more massive companion and a longer orbital period for a high Z. We wrote our results into a FORTRAN code and the code can be downloaded on X. Meng's personal web site *http://www.ynao.ac.cn/~bps/download/xiangcunmeng.htm*.

2. The results of binary population synthesis

To investigate the birth rate of SNe Ia, we followed the evolution of 10^7 binaries for various Z with Hurley's rapid binary evolution code (Hurley et al. 2000, 2002). The results of grid calculations in section 1 are incorporated into the code. Since the code is valid just for $Z \leqslant 0.03$, only seven metallicities (i.e. $Z = 0.03$, 0.02, 0.01, 0.004, 0.001, 0.0003 and 0.0001) are examined here. The primordial binary samples are generated in a Monte Carlo way and a circular orbit is assumed (see Meng, Chen & Han. 2008 for details). The results are shown in Figs. 1 and 2. From Fig. 1, we see that most supernovae occur between 0.2 Gyr and 2 Gyr after star formation, and a high metallicity leads to a systematically earlier explosion time. The peak value of birth rate increases with metallicity Z. However, Fig 2 shows that the WD + MS channel can only account for about 1/3 of the Galactic SNe Ia observed(3-$4\times 10^{-3} \text{yr}^{-1}$, van den Bergh & Tammann 1991; Cappellaro & Turatto 1997). Therefore, there may be other channels or mechanisms contributing to SNe Ia.

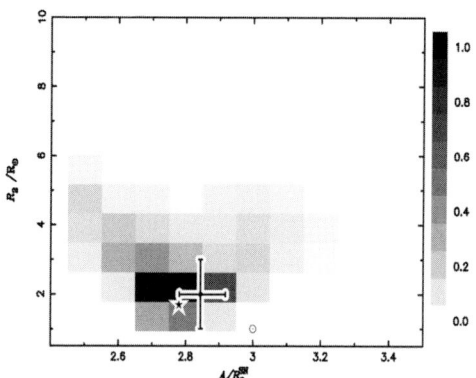
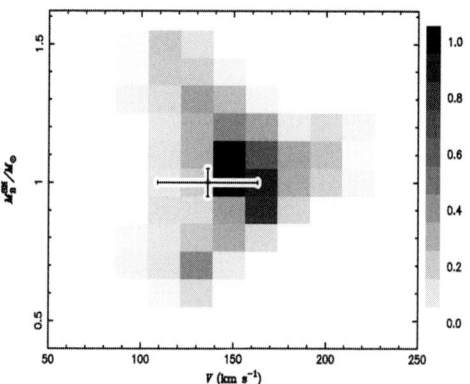

Figure 3. The distribution of the radii and the ratios of separations to radii of companions at the moment of supernova explosion for the case of $Z = 0.02$ and $\alpha_{CE} = 1.0$. Cross represents the potential candidate of the companion of Tycho's supernova, Tycho G (Ruiz-Lapuente *et al.* 2004; Branch 2004). The length of the bars of the cross represents observational error. Solar symbol and pentacle represent the main-sequence and subgiant companion model in Marietta *et al.* (2000), respectively.

Figure 4. The distribution of the masses and the space velocities of companions in SN Ia remnants for $Z = 0.02$ and $\alpha_{CE} = 1.0$. The cross represents the potential candidate of the companion of Tycho's supernova, Tycho G (Ruiz-Lapuente *et al.* 2004; Branch 2004), and the length of the bars of the cross represents observational errors.

We also showed the distributions of the parameters of the binary systems at the moment of supernova explosion (Fig. 3) and the properties of companions after supernova explosion (Fig. 4). The former may provide physics input when one simulates the interaction between supernova ejecta and its companion, and the latter may help to search for the companions in supernova remnants. From Fig. 4, we see that the properties of the potential companion of Tycho's supernova, Tycho G, are well consistent with our binary population synthesis results. We also noticed from Fig. 3 that the solar model used by Marietta *et al.* (2000) when they simulated the interaction between explosion ejecta and the companion is not a typical one (see the solar symbol in the figure).

A SN Ia remnant is not spherically symmetric due to the companion, and then the spectrum of the SN Ia may be polarized (Leonard & Filippenko 2005). Based on the numerical simulation of Marietta *et al.* (2000) and Kasen *et al.* (2004), we found that at least 75% of all SNe Ia may be detected by spectropolarimetry. At present, almost all SNe Ia, which are observed by spectropolarimetry, have various degrees of polarization signal (Leonard *et al.* 2005).

Since there exists mass-loss from a progenitor binary system by an optically thick wind, the material lost from the system may become circumstellar materials (CSM). The CSM may be the origin of the color excess of SNe Ia. Based on the single degenerate model with optically thick wind and using a simple analytic method, we reproduced the distribution of the color excess of SNe Ia obtained from observation if a wind velocity of 10 km/s is adopted. If the wind velocity is larger than 100 km/s, the reproduction is bad.

3. Conclusion

We systematically studied the single degenerate channel of SNe Ia, (i.e. WD+MS channel), and showed the parameter space leading to SNe Ia from this channel. Our work may provide help for the future studies of SNe Ia. We also provided the properties of companions after supernova explosion, which may help to search for the companion in SN Ia remnants. We found that at least 75% of all SNe Ia may be detected by spectropolarimetry, and then in the future, statistical studies about the spectropolarimetric observations of SNe Ia may verify the reliability of the WD + MS channel.

Acknowledgements

This work was in part supported by Natural Science Foundation of China under Grant Nos. 10433030, 10521001, 2007CB815406 and 10603013.

References

Alexander, D. R. & Ferguson, J. W. 1994, *ApJ* 437, 879
Branch, D. 2004, *Nature* 431, 1044
Cappellaro, E. & Turatto, M. 1997, *in Ruiz-Lapuente P., Cannal R., Isern J., eds, Thermonuclear Supernovae. Kluwer, Dordrecht* p. 77
Chen, X. & Tout, C. A. 2007, *ChJAA* 7, 2, 245
Eggleton, P. P. 1971, *MNRAS* 151, 351
Eggleton, P. P. 1972, *MNRAS* 156, 361
Eggleton, P. P. 1973, *MNRAS* 163, 279
Hachisu, I., Kato, M., & Nomoto, K. 1996, *ApJ* 470, L97
Hachisu, I., Kato, M., Nomoto, K., & Umeda, H. 1999a, *ApJ* 519, 314
Han, Z., Podsiadlowski, P., & Eggleton, P. P. 1994, *MNRAS* 270, 121
Han, Z., Tout, C. A., & Eggleton, P. P. 2000, *MNRAS* 319, 215
Han, Z. & Podsiadlowski, Ph. 2006, *MNRAS* 368, 1095
Hillebrandt, W. & Niemeyer J. C. 2000, *ARA&A* 38, 191
Hurley, J. R., Pols, O. R., & Tout, C. A. 2000, *MNRAS* 315, 543
Hurley, J. R., Tout, C. A., & Pols, O. R. 2002, *MNRAS* 329, 897
Iglesias, C. A. & Rogers, F. J. 1996, *ApJ* 464, 943
Ihara, Y., Ozaki, J., Doi, M., et al. 2007, *PASJ* 59, 811, arXiv: 0706.3259
Kasen, D., Nugent, P., Thomas, R. C., & Wang, L. 2004, *ApJ* 610, 876
Leibundgut, B. 2000, *A&ARv* 10, 179
Leonard, D. C. & Filippenko, A.V 2005, *in Turatto M. et al., eds, in 1604 - 2004, Supernovae as Cosmological Lighthouses, (San Francisco: ASP)*, (astro-ph/0409518)
Leonard, D. C., et al. 2005, *ApJ* 632, 450
Marietta, E., Burrows, A., & Fryxell, B. 2000, *ApJS* 128, 615
Meng, X., Chen, X., & Han, X. 2008, arXiv: 0802.2471
Nomoto, K., Thielemann, F-K., & Yokoi, K. 1984, *ApJ* 286, 644
Perlmutter, S., et al. 1999, *ApJ* 517, 565
Phillips, M. M. 1993, *ApJ* 413, L105
Pols, O. R., Tout, C. A., Eggleton, P. P., et al. 1995, *MNRAS* 274, 964
Pols, O. R., Schröder, K. P., Hurly, J. R., et al. 1998, *MNRAS* 298, 525
Reindl, B., Tammann, G. A., Sandage, A., et al. 2005, *ApJ* 624, 532
Riess, A. et al. 1998, *AJ* 116, 1009
Ruiz-Lapuente, P., et al. 2004, *Nature* 431, 1069
van den Bergh, S. & Tammann, G. A. 1991, *ARA&A* 29, 363
Whelan, J. & Iben, I. 1973, *ApJ* 186, 1007

Discussion

J. CHRISTENSEN-DALSGAARD: Is the Tycho supernova the only Type Ia case where the companion may have been identified?

X.C. MENG: At present there is not others. Even for Tycho G, it is only a potential candidate of the companion of Tycho's supernova.

Modelling the evolution and nucleosynthesis of carbon-enhanced metal-poor stars

O. R. Pols[1], R. G. Izzard[1]†, M. Lugaro[1] and S. E. de Mink[1]

[1]Sterrekundig Instituut Utrecht, P. O. Box 80000, NL-3584 TA Utrecht, The Netherlands
email: O.R.Pols@uu.nl, R.G.Izzard@uu.nl, M. A. Lugaro@uu.nl, S. E. deMink@uu.nl

Abstract. We present the results of binary population simulations of carbon-enhanced metal-poor (CEMP) stars. We show that nitrogen and fluorine are useful tracers of the origin of CEMP stars, and conclude that the observed paucity of very nitrogen-rich stars puts strong constraints on possible modifications of the initial mass function at low metallicity. The large number fraction of CEMP stars may instead require much more efficient dredge-up from low-metallicity asymptotic giant branch stars.

Keywords. Stars: AGB and post-AGB, stars: evolution, stars: binaries, stars: abundances, stars: mass function

1. Introduction

One of the most striking results of recent large surveys for very metal-poor stars in the Galactic halo is the large proportion of highly carbon-enriched objects among them. These carbon-enhanced metal-poor (CEMP) stars, usually defined as metal-poor stars with [C/Fe] > 1.0, make up at least 10 per cent and probably as much as 20–25 per cent of very metal-poor stars with [Fe/H] < −2 (Frebel *et al.* 2006; Lucatello *et al.* 2006).

The majority (about 80 per cent, according to Aoki *et al.* 2007) of CEMP stars are also enriched in Ba and other heavy elements produced by slow neutron captures (the *s*-process) in asymptotic giant branch (AGB) stars. For these so-called CEMP-s stars a likely scenario is pollution by mass transfer from a more massive AGB companion in a binary system, which has since become a white dwarf. Supporting evidence for this scenario comes from radial velocity monitoring, which suggests that all CEMP-s stars could statistically be binaries (Lucatello *et al.* 2005a). The remaining fraction of CEMP stars that do not show s-process enrichments (the CEMP-no stars) are typically more metal-poor than the CEMP-s stars and so far have shown no evidence for binarity (Aoki *et al.* 2007). These stars exhibit a variety of abundance patterns, and may have formed instead from material ejected from rapidly rotating massive stars (Meynet *et al.* 2006) or from faint core-collapse supernovae (Umeda & Nomoto 2005). Confusing such a clear distinction are a surprisingly large number of CEMP stars enhanced in both *s*-process and *r*-process (rapid neutron-capture) elements, whose origin is still quite unclear (Jonsell *et al.* 2006).

Within the mass transfer scenario, the large proportion of CEMP-s stars requires the existence of a sufficient number of binary systems with primary components that have undergone AGB nucleosynthesis. In recent studies (Lucatello *et al.* 2005b; Komiya *et al.* 2007) it has been argued that this requires a different initial mass function (IMF) at low metallicity, weighted towards intermediate-mass stars. If true, this in turn has important consequences for the chemical evolution of the halo and, by implication, of other galaxies.

† Present address: Institut d'Astronomie et d'Astrophysique, Université Libre de Bruxelles, CP226, Boulevard du triomphe, B-1050 Bruxelles, Belgium

However, the model calculations on which these estimates are based still contain many uncertainties regarding the evolution and nucleosynthesis of low-metallicity AGB stars, the efficiency of mass transfer, and the evolution of the surface abundances of the CEMP stars themselves.

In this contribution we explore the effect of some of these uncertainties on the number fraction of CEMP stars by means of a binary population synthesis study. Apart from carbon, we concentrate on nitrogen and fluorine enrichments as possible tracers of the origin of CEMP stars, the latter motivated by the recent discovery of fluorine in a CEMP star (see Sect. 2). In Sect. 3 we present the first results of our binary population synthesis simulations, and in Sect. 4 we give our conclusions.

2. Nitrogen and fluorine in CEMP stars

Apart from carbon, substantial enhancements of nitrogen with respect to iron are common among CEMP stars, typically with [C/N] > 0. Detailed AGB nucleosynthesis models of low initial mass ($< 2.5\,M_\odot$) produce carbon (by the 3α reaction during thermal pulses and subsequent convective dredge-up), but do not produce nitrogen because it is burned during the same helium shell flashes. On the other hand, AGB models of higher mass convert the dredged-up carbon effectively into nitrogen by CN-cycling at the bottom of the convective envelope (hot bottom burning, HBB). The surface abundances of these more massive AGB stars approach the CN-equilibrium ratio of [C/N] ≈ -2. Detailed evolution models of AGB stars (Karakas & Lattanzio 2007) indicate that HBB sets in at significantly lower mass at low metallicity (between 2.5 and $3\,M_\odot$ at [Fe/H] = -2.3) than at solar metallicity (around $5\,M_\odot$). One may thus expect a population of so-called nitrogen-enhanced metal-poor (NEMP) stars, with [C/N] < -0.5. Although a few examples of such stars are known, mostly at [Fe/H] < -2.9, they appear to be very rare (Johnson et al. 2007). As we show in Sect. 3 the number of NEMP stars sets an additional constraint on possible changes to to IMF at low metallicity.

Recently, Schuler et al. (2007) derived a super-solar fluorine abundance of [F/Fe] = $+2.9$ in the halo CEMP star HE 1305+0132. This is the most iron-deficient star, [Fe/H] = -2.5, for which the fluorine abundance has been measured. Enhancements of carbon and nitrogen are also measured ([C/Fe] = $+2.7$; [N/Fe] = $+1.6$), and Ba and Sr lines are seen in its spectrum (Goswami 2005), placing HE 1305+0132 in the group of CEMP-s stars.

Fluorine can be made in AGB stars as a by-product of the ^{14}N $+^4$He reaction under neutron-rich conditions during thermal pulses (Lugaro et al. 2004). Fluorine enhancements of up to 30 times solar have been measured among Galactic AGB stars (Jorissen et al. 1992), demonstrating that these stars can indeed produce fluorine efficiently. Detailed AGB nucleosynthesis models (Karakas & Lattanzio 2007) show that fluorine is produced and dredged to the surface alongside carbon in low-metallicity AGB stars with masses $< 3\,M_\odot$. At larger masses fluorine is destroyed by proton captures as a result of HBB.

In Fig. 1 we show the enhancements of F and C+N relative to hydrogen as observed in HE 1305+0132 and compare these to the abundance ratios in the material lost by AGB stars at [Fe/H] = -2.3 according to the Karakas & Lattanzio (2007) models. The figure shows that AGB stars with masses between 1.7 and $2.3\,M_\odot$ produce fluorine and carbon in the right amounts to account for the observed abundances, after accretion of the material by a low-mass companion and subsequent dilution in its envelope by a factor 6 to 9 (Lugaro et al. 2008). Assuming a mass of $0.8\,M_\odot$ and a convective envelope mass fraction of 60%, this implies that the star should have accreted 0.05–$0.12\,M_\odot$ from its companion, corresponding to 3–11% of the mass lost by the AGB star. These constraints

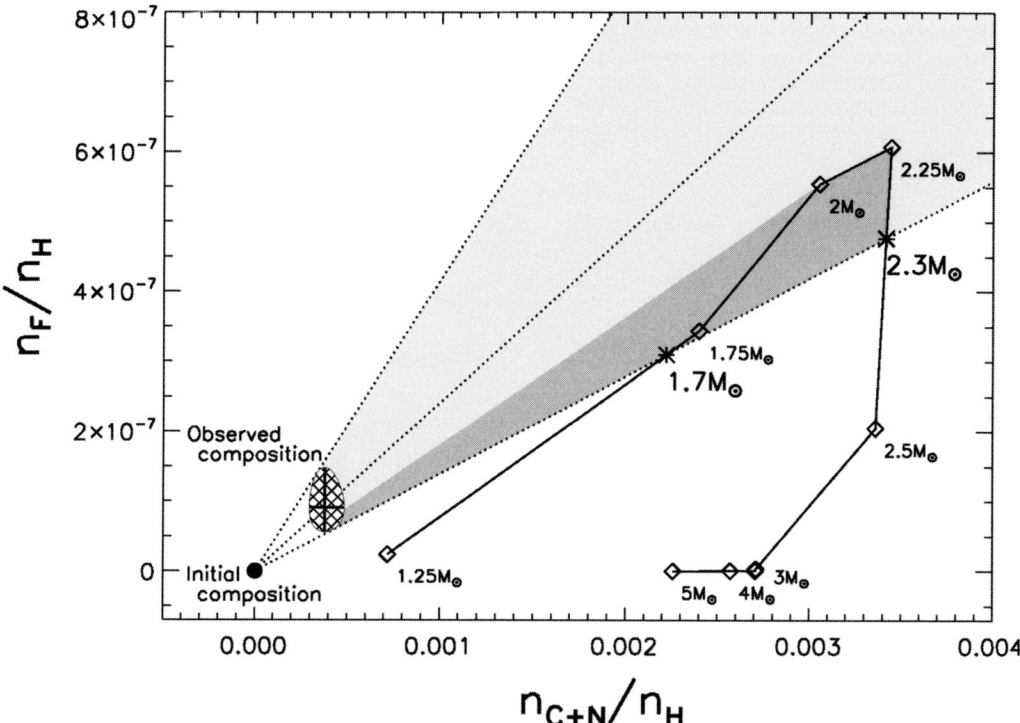

Figure 1. Abundances by number of F and C+N with respect to H as observed in HE 1305+0132 from Schuler *et al.* (2007; hatched ellipsoid showing 1σ errors) compared to those computed in the material lost in AGB star winds at [Fe/H] = -2.3, with labels indicating the initial masses. The light gray area is the region of the plot where the F and C+N abundances from the AGB companion should lie in order to match the observed abundances, after dilution. The darker gray area represents the region covered by the models with masses in the range 1.7–2.3 $M_\odot$ indicated by the asterisks.

fit well within the binary mass transfer scenario. Furthermore this result indicates the potential of using fluorine as a tracer for the origin of CEMP stars.

3. Binary population nucleosynthesis of CEMP stars

We have simulated populations of metal-poor halo stars in binary systems using the rapid synthetic binary nucleosynthesis code of Izzard *et al.* (2004) and Izzard *et al.* (2006). The code uses fits to detailed single-star evolution and nucleosynthesis models, in particular the AGB models of Karakas & Lattanzio (2007), and follows the surface abundances as a star evolves through dredge-up episodes. A prescription for hot bottom burning in massive AGB stars is included, calibrated against the same detailed AGB models. Binary evolution is followed according to the prescriptions of Hurley *et al.* (2002). We model mass transfer by stellar wind accretion, according to the Bondi & Hoyle (1944) prescription, and by Roche-lobe overflow (RLOF). In binaries with AGB primaries, RLOF is usually unstable and leads to the ejection of a common envelope without any further accretion onto the companion. In our default model we assume efficient thermohaline mixing, in accordance with the findings of Stancliffe *et al.* (2007).

In all models we use solar-scaled initial abundances according to Anders & Grevesse (1989) with $Z = 10^{-4}$, corresponding to [Fe/H] = -2.3, which is the lowest metallicity for which we have detailed AGB models. We select stars with ages between 10 and 13.7 Gyr

Table 1. Number fractions of CEMP and NEMP stars among halo stars at [Fe/H] = −2.3 and log g < 4.0 resulting from binary population synthesis, for different sets of physical ingredients in the evolution models. The last two columns give the fraction of CEMP-s stars and of fluorine-rich FEMP stars, relative to the number of CEMP stars.

model		f(CEMP)	f(NEMP)	$\frac{\text{NEMP}}{\text{CEMP}}$	$\frac{\text{CEMP-s}}{\text{CEMP}}$	$\frac{\text{FEMP}}{\text{CEMP}}$
1A	default physics	2.30 %	0.35 %	0.15	0.29	0.85
2A	no thermohaline mixing	4.20 %	0.52 %	0.12	...	...
3A	calibrated 3DUP	9.43 %	0.34 %	0.036	0.99	0.82
4A	no therm. mix. + calibr. 3DUP	14.96 %	0.47 %	0.031	0.75	0.87

(roughly corresponding to the age of the halo) and log g < 4.0 (thus including turnoff stars and (sub)giants but excluding unevolved main-sequence stars). Among this sample we designate as CEMP stars those with [C/Fe] > 1.0 and as NEMP stars those with [N/Fe] > 0.5 and [C/N] < −0.5, following the definition of Johnson *et al.* (2007). Note that these definitions partly overlap.

We can compare our model results with the statistics of the SAGA database of metal-poor stars (Suda *et al.* 2008). We selected 375 stars from the database in a metallicity range [Fe/H] = −2.3±0.5 and log g < 4.0. Of these, 296 have a C abundance measurement and 69 classify as CEMP stars, yielding a CEMP fraction of 18–23 %. Only one star classifies as a NEMP star, giving a very small nominal NEMP fraction of <0.3 %. If we consider an extended metallicity range −4 < [Fe/H] < −2 in order to improve the number statistics, we find 6 NEMP stars and a NEMP fraction of about 1.5 %. We conclude that CEMP stars outnumber NEMP stars by at least a factor 10.

In our default model (1A) we assume that the initial primary masses M_1 are distributed according to the solar neighbourhood IMF as derived by Kroupa *et al.* (1993), the initial periods come from a flat distribution in log P and the initial mass ratios from a flat distribution in $q = M_2/M_1$. In Table 1 we present the resulting number fractions of CEMP and NEMP stars (first row). This model clearly fails to account for the large observed CEMP fraction, although the small NEMP fraction is consistent with the observations. The last two columns give the number ratio of CEMP-s stars (defined as those CEMP stars with [Ba/Fe] > 0.5) to CEMP stars and the ratio of FEMP stars (defined as stars with [F/Fe] > 1.0) to CEMP stars. Although all our CEMP stars are also enriched in Ba, in this model the majority have [Ba/Fe] < 0.5.

In the next rows of Table 1 we vary some of the uncertain physical ingredients in our models, while keeping the input distributions the same. In model 2A we switch off thermohaline mixing, which has the effect of increasing the surface abundances of C, N and Ba in turnoff stars with shallow convection zones relative to our default model. The numbers of CEMP stars and NEMP stars correspondingly increase, although the CEMP fraction remains too low to be compatible with observations. In model 3A we vary some of the parameters relating to third dredge-up (3DUP), in particular we assume a smaller value (by up to 0.1 $M_\odot$) of the minimum core mass at which 3DUP occurs, and after dredge-up has started we assume efficient 3DUP (with $\lambda \geqslant 0.8$) regardless of how small the envelope mass is. In practice this means that almost all AGB stars with initial masses > 0.8 $M_\odot$ undergo efficient dredge-up. These modifications are motivated by similar changes needed to reproduce the carbon-star luminosity function of the Magellanic Clouds (Izzard & Tout 2004) and the abundances of Galactic post-AGB stars (Bonačić Marinović *et al.* 2007). This leads to an increase of the CEMP fraction to almost 10 %, much closer to but still short of the observed value. Model 4A is a combination of 2A and 3A and results in a CEMP fraction of 15 %. We note that in these latter two models

Table 2. Number fractions of CEMP and NEMP stars among halo stars at [Fe/H] = −2.3 and $\log g < 4.0$ as in Table 1, for the default physical ingredients while varying the input distributions.

model		f(CEMP)	f(NEMP)	$\frac{\text{NEMP}}{\text{CEMP}}$
1A	default $N(M_1, q, P)$	2.30 %	0.35 %	0.15
1B	$N(q, P)$ from Duquennoy & Mayor (1991)	3.50 %	0.71 %	0.20
1C	$N(M_1)$ from Miller & Scalo (1979)	3.15 %	0.62 %	0.20
1D	$N(M_1)$ from Lucatello et al. (2005b)	4.81 %	1.35 %	0.28
1E	$N(M_1)$ from Komiya et al. (2007)	13.47 %	26.61 %	1.98

nearly all CEMP stars are CEMP-s stars, in accordance with observations. We also note that in all models the majority (> 80 %) of CEMP stars are expected to be fluorine-rich, with [F/Fe] > 1.0.

In Table 2 we present CEMP and NEMP fractions of the default physical model while varying the initial distributions of binary parameters. In model 1B we assume mass ratios and periods drawn from the distributions derived by Duquennoy & Mayor (1991) for the local population of G dwarfs (i.e. a log-normal period distribution with a broad peak at 170 years and a q distribution with a broad peak at $M_2/M_1 = 0.23$). This leads to an increase by a factor of 1.5–2 in the number of CEMP and NEMP stars as the peak in the period distribution coincides with the period range in which mass transfer is effective.

Models 1C, 1D and 1E explore the effect of varying the initial mass function, by assuming the default q and P distributions in combination with a log-normal form of the IMF. The Miller & Scalo (1979) IMF also represents the solar neighbourhood but gives somewhat higher CEMP and NEMP fractions than the Kroupa et al. (1993) IMF. Model 1D assumes the IMF suggested by Lucatello et al. (2005b) as required to reproduce the large CEMP fraction (it has a median mass of $0.79\,M_\odot$, compared to $0.10\,M_\odot$ for the Miller & Scalo IMF). It results in a larger CEMP fraction but still falls short of the observed value. The discrepancy between our and Lucatello's results arises mainly because in our (default) models the initial primary mass and period range contributing to CEMP stars are smaller than they assumed. Model 1D also shows an increased NEMP fraction, the result of a larger weight of intermediate-mass stars (with $M > 2.7\,M_\odot$ undergoing HBB) in this IMF. This effect is much more extreme when we assume the IMF suggested by Komiya et al. (2007) which has a median mass of $10\,M_\odot$. Although it gives rise to a substantial CEMP fraction, the CEMP stars are outnumbered by NEMP stars by a factor of two. This is not compatible with the observed limits on the number fraction of NEMP stars.

4. Conclusions

The detection of a large fluorine overabundance in the CEMP star HE 1305+0132 is well explained within the AGB binary mass transfer scenario (Lugaro et al. 2008). On the other hand, models of rapidly rotating massive stars do not produce fluorine (Meynet et al. 2006). Therefore fluorine appears to be a useful discriminant between different scenarios proposed for the origin of CEMP stars. Our population synthesis results indicate that (rather independent of model assumptions) at least 80 % of CEMP stars formed by AGB mass transfer should be enriched in fluorine ([F/Fe] > 1.0), i.e. most CEMP-s stars should also be FEMP stars.

Our binary population synthesis models show that the paucity of NEMP stars among metal-poor halo stars is incompatible with a strongly modified IMF at low metallicity, heavily weighted towards intermediate-mass stars, as has been suggested by Komiya et al. (2007) in order to explain the high proportion of CEMP stars. Another possible explanation for the ubiquity of CEMP-s stars is that low-metallicity AGB stars undergo much more efficient dredge-up than shown by the detailed evolution models available to date.

Acknowledgements

We would like to thank Takuma Suda for making the SAGA database of metal-poor stars available to us in electronic form ahead of publication.

References

Anders, E. & Grevesse, N. 1989, *Geochim. Cosmochim. Acta* 53, 197
Aoki, W., Beers, T. C., Christlieb, N., Norris, J. E., Ryan, S. G., & Tsangarides, S. 2007, *ApJ* 655, 492
Bonačić Marinović, A., Izzard, R. G., Lugaro, M., & Pols, O. R. 2007, *A&A* 469, 1013
Bondi, H., & Hoyle, F. 1944, *MNRAS* 104, 273
Duquennoy, A., & Mayor, M. 1991, *A&A* 248, 485
Frebel, A., Christlieb, N., Norris, J. E., et al. 2006, *ApJ* 652, 1585
Goswami, A. 2005, *MNRAS* 359, 531
Hurley, J. R., Tout, C. A., & Pols, O. R. 2002, *MNRAS* 329, 897
Izzard, R. G., & Tout, C. A. 2004, *MNRAS* 350, L1
Izzard, R. G., Tout, C. A., Karakas, A. I., & Pols, O. R. 2004, *MNRAS* 350, 407
Izzard, R. G., Dray, L. M., Karakas, A. I., & Lugaro, M., & Tout, C. A. 2006, *A&A* 460, 565
Johnson, J. A., Herwig, F., Beers, T. C., & Christlieb, N. 2006, *ApJ* 658, 1203
Jonsell, K., Barklem, P. S., Gustafsson, B., Christlieb, N., Hill, V., Beers, T. C., & Holmberg, J. 2006, *ApJ* 451, 651
Jorissen, A., Smith, V. V., & Lambert, D.L. 1992, *A&A* 261, 164
Karakas, A. & Lattanzio, J. C. 2007, *PASA* 24, 103
Komiya, Y., Suda, T., Minaguchi, H., Shigeyama, T., Aoki, W., & Fujimoto, M. Y. 2007, *ApJ* 658, 367
Kroupa, P., Tout, C. A., & Gilmore, G. 1993, *MNRAS* 262, 545
Lucatello, S., Tsangarides, S., Beers, T. C., Carretta, E., Gratton, R. G., & Ryan, S. G. 2005, *ApJ* 625, 825
Lucatello, S., Gratton, R. G., Beers, T. C., & Carretta, E. 2005, *ApJ* 625, 833
Lucatello, S., Beers, T. C., Christlieb, N., Barklem, P. S., Rossi, S., Marsteller, B., Sivarani, T., & Lee, Y. S. 2006, *ApJ* 652, L37
Lugaro, M., Ugalde, C., Karakas, A. I., Görres, J., Wiescher, M., Lattanzio, J. C., & Cannon, R. C. 2004, *ApJ*, 615, 934
Lugaro, M., de Mink, S. E., Izzard, R. G., et al. 2008, *A&A* 484, L27
Meynet, G., Ekström, S., & Maeder, A. 2006, *A&A* 447, 623
Miller, G. E., & Scalo, J. M. 1979, *ApJS* 41, 513
Schuler, S. C., Cunha, K., Smith, V. V., Sivarani, T., Beers, T. C., & Lee, Y. S. 2007, *ApJ* 667, L81
Stancliffe, R. J., Glebbeek, E., Izzard, R. G., & Pols, O. R. 2007, *A&A* 464, L57
Suda, T., Katsuta, Y., Yamada, S., et al. 2008, *PASJ* 60, in press, arXiv:0806.3697
Umeda, H. & Nomoto, K. 2005, *ApJ* 619, 427

Discussion

C. CHARBONNEL: You spoke about observations of s-process elements in CEMP in general. Are there some precise data, regarding the low- and high-mass s-elements, that may help discriminate the progenitor (e.g., AGB vs massive stars)?

O. POLS: There are certainly many detailed abundance determinations of heavy elements in CEMP stars, and these have been tested against s-process models by several groups. In general, these observations agree with AGB models, although different assumptions must be made for different stars regarding the free parameters in these models (such as the ^{13}C pocket size), and there are still many open questions in this respect. Some CEMP stars show both an s-process and r-process signature, for these one may need pre-enrichment by massive stars and AGB mass transfer...

N. LANGER: Could the fluorine in CEMP stars also come from neutrino-induced nucleosynthesis from supernovae?

O. POLS: Indeed, supernovae can also be sources of fluorine although this has not been confirmed observationally to my knowledge. However, SNe typically produce more Oxygen than Carbon, contrary to what is seen in CEMP stars. One would need to get rid of the Oxygen, or suppose a different source for the carbon. AGB stars naturally explain both C and F enhancements simultaneously.

The delighted people, from left: O. Pols (the speaker), S. Campbell and S.E. de Mink.

A Spectroscopic Study of Blue Stragglers in M67

G. Q. Liu[1,2], L. Deng[1], M. Chávez[3] and E. Bertone[3]

[1]National Astronomical Observatories, Chinese Academy of Sciences,
Beijing 100012, P. R. China
email: lgq@bao.ac.cn, licai@bao.ac.cn

[2]Graduate University of Chinese Academy of Sciences,
Beijing 100049, P. R. China

[3]INAOE - Instituto Nacional de Astrofísica, Óptica y Electrónica,
Luis Enrique Erro 1 - 72840 Tonantzintla, Puebla, Mexico
email: mchavez@inaoep.mx, ebertone@inaoep.mx

Abstract. Spectrophotometric observations of the complete sample of twenty four blue stragglers (BSs) in the old galactic open cluster M67 (NGC2682) have been collected, using the Guillermo Haro Observatory in Cananea, Mexico. All the calibrated spectra were re-calibrated by the Beijing Arizona Taipei Connecticut (BATC) photometric system which includes fluxes in 11 photometric bands covering $\sim$3600–10000 Å. The goal of the current work is to provide observational constraints on spectral properties of BSs by determining the effective temperature ($T_{\rm eff}$) and surface gravity ($\log g$). The overall results, obtained by applying the flux fitting method, indicate that $T_{\rm eff}$ and surface gravities of BSs in M67 are fully compatible with those expected for main sequence stars.

Keywords. Stars: blue stragglers – Stars: Hertzsprung-Russell (HR) diagram – Galaxy: open clusters and associations: individual: M67

1. Introduction

BSs were first discovered in the globular cluster M3 by Sandage (1953). They lie above the turn off point of main sequence as a brighter and bluer extension of cluster's main sequence and appear to straggle away from their regular evolutionary processes, hence their name *blue stragglers*. Through decades of research since their discovery, blue stragglers are widely observed in almost all kinds of stellar systems, such as open clusters, galactic halo and even dwarf galaxies (Stryker 1993).

Nowadays, based on observational and theoretical studies, it is generally believed that the popular formation ways of blue stragglers are all correlated with interactions between stars, such as stellar collisions in high density and mass transfer in close binaries (Ahumada 1999; Bacon *et al.* 1996; Ferraro *et al.* 1997; Gilliland *et al.* 1992; Leonard 1989; Livio 1993; Ouellette & Pritchet 1998; Piotto *et al.* 1999; Stryker 1993; Tian *et al.* 2006). Due to their peculiar nature, it is very important to study BSs, because in a stellar population they are among the most massive and luminous stars, whose contribution to the integrated light of their host stellar systems could not be predicted by the standard theory of stellar evolution (Bressan *et al.* 1993). In fact, BSs greatly affect the spectral energy distribution (SED) of the whole population, particularly at ultraviolet and blue wavelengths (Deng *et al.* 1999; Manteiga *et al.* 1989; Xin *et al.* 2007, 2008).

However, in spite of many investigations, a conceivably definite explanation of the BS phenomenon has not been reached. Generally, they are historically regarded as core

Table 1. The blue straggler population of M67.

Name	R.A. (2000)	Dec. (2000)	ExpTime (s)	n
BS005	8:51:11.78	11:45:22.24	2400	4
BS018	8:52:10.75	11:44:06.07	1200	2
BS025	8:51:27.04	11:51:52.22	1200	3
BS029	8:51:48.65	11:49:15.36	2400	4
BS034	8:51:34.31	11:51:10.23	2400	4
BS038	8:51:32.61	11:48:52.02	1200	2
BS040	8:51:26.45	11:43:50.75	1200	2
BS043	8:51:14.37	11:45:00.70	2400	4
BS046	8:51:20.82	11:53:25.65	2400	4
BS047	8:51:03.52	11:45:02.68	1200	2
BS065	8:51:21.77	11:52:38.00	2400	4
BS093	8:51:32.57	11:50:40.42	1200	2
BS111	8:51:19.92	11:47:00.50	1500	2
BS115	8:51:37.72	11:37:03.54	1200	2
BS116	8:50:55.70	11:52:14.50	2400	4
BS126	8:49:21.49	12:04:23.00	1200	2
BS131	8:51:28.40	12:07:38.30	1200	2
BS139	8:51:39.24	11:50:03.66	1500	2
BS143	8:51:21.25	11:45:52.63	1200	2
BS182	8:51:15.47	11:47:31.74	1800	2
BS184	8:50:47.69	11:44:51.33	3300	4
BS185	8:51:28.17	11:49:27.06	3000	4
BS206	8:48:59.84	11:44:51.66	600	1
BS216	8:51:20.59	11:46:16.36	1500	2

hydrogen burning stars (Benz & Hills 1987, 1992) and it has been usually assumed that the spectral properties of BSs are compatible with those of main sequence stars at the same loci on the CMD. However, whether or not the spectral properties of BSs can be actually represented by main sequence objects, has not yet been fully investigated. Our goal is to provide observational constraints on spectral properties of BSs by determining the effective temperature and surface gravity.

2. Observations and reductions

Our working sample is 24 BSs in M67, which have nearly 100% membership probabilities as determined with both proper motion and radial velocity observations (Girard et al. 1989; Sanders 1977). The catalog is shown in Table 1, where we give, in columns (1) to (5), the BSs identification numbers from Fan et al. (1996), the equatorial coordinates, the integrated exposure times, and the number of spectra collected for each object. Observations were carried out during a three-night run in February 2005, using the 2.12 m telescope of the Guillermo Haro Observatory at Cananea, Mexico. The spectra were at a low-resolution of 3.2 Å per pixel with wavelength coverage roughly from 3600 to 6900 Å. For the data reduction we followed the standard procedures using IRAF packages. The flux calibrated spectra were subsequently re-calibrated by using photometric magnitudes in the Beijing-Arizona-Taipei-Connecticut (BATC) system (Deng et al. 1999). The BATC system includes 15 intermediate-band filters covering the interval 3500–10000 Å.

Figure 1 shows the observed spectral energy distribution (SED) of the BSs. The spectra are roughly ordered in a temperature sequence decreasing from top to bottom. Vertical dotted lines indicate the location of four major spectral features, namely, H_α (6562 Å), H_β (4860 Å), H_γ (4340 Å), and Ca_K (3933 Å).

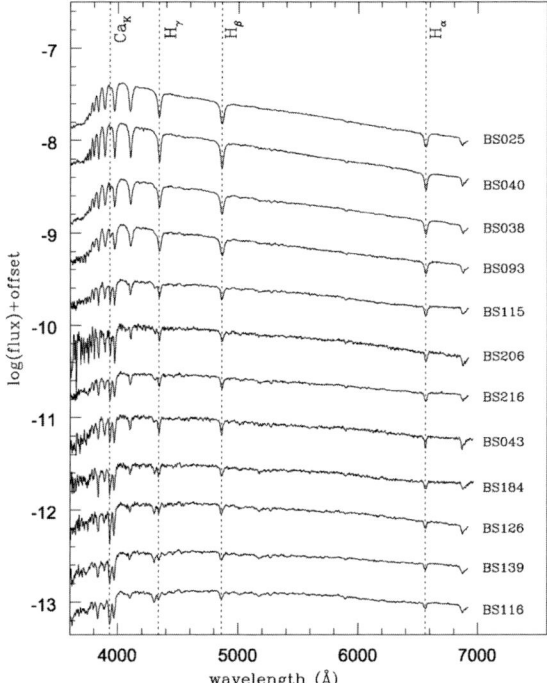

Figure 1. The relative flux-calibrated spectra of BSs. The spectra are roughly ordered in a temperature sequence decreasing from top to bottom.

3. Spectral fitting and analysis

We applied the flux fitting method to determine the effective temperature and surface gravity of our sample stars. For this task, we made use of three different spectral libraries, both observed and theoretical, namely, the Pickles (1998) empirical library, the Lejeune et al. (1997, 1998) data set (which is mainly based on Kurucz grid of low resolution fluxes), and the high resolution BLUERED theoretical library (Bertone et al. 2004, 2008). In all cases, we assumed a solar chemical composition for BSs (Bressan & Tautvaišiene 1996, Hobbs & Thorburn 1991). Each calibrated spectrum was compared to every entry from the spectral atlases. The algorithm we have applied finds the spectrum that produced the minimum standard deviation σ of the flux residuals, computed at each λ point.

Figure 2 presents the best fits for two representative BSs, BS038 and BS139. The smaller panels below each spectrum correspond to the residuals between observed and theoretical spectra, and dotted lines indicate the loci of 3σ values. The model spectra very well reproduce the observed ones, except at blue and ultraviolet wavelengths, where discrepancies can be ascribed to uncertainties associated to both observational and theoretical spectra.

The results for the full stellar sample are reported in Table 2, where columns (1)–(7) list, respectively, the object ID, the parameter pairs ($T_{\rm eff}$, log g) and the identification numbers of Sanders (1977), for easing cross identification. Note that effective temperatures derived from the data sets are in good agreement (the discrepancy is of the order of a few percent, apart from the BLUERED estimate of BS005, which is significantly lower). The same is true for the surface gravities, which closely match the luminosity classes, when comparing the results from the two low-resolution libraries; the log g

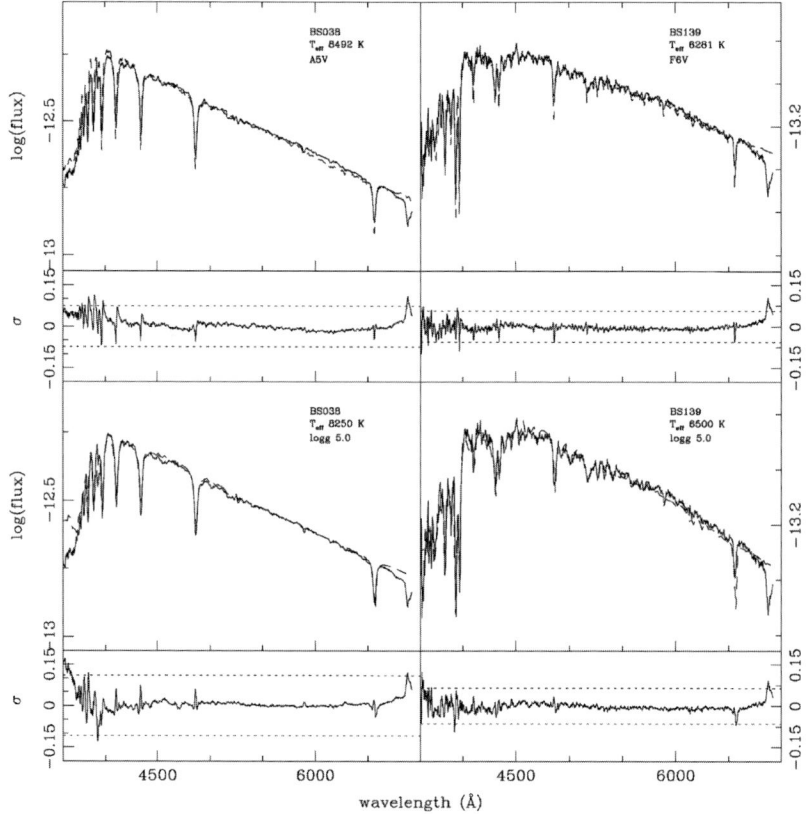

Figure 2. Best fits for two representative BSs, BS038 and BS139. Solid lines and dashed lines indicate, respectively, the observed spectra and the best fitting reference ones from the Pickles spectra (top panels) and the Lejeune *et al.* library (bottom panels).

estimates from BLUERED are sistematically a bit higher. The overall results indicate that BSSs in M67 span a wide range in $T_{\rm eff} \sim 5600 - 12600$ K and surface gravities that are fully compatible with those expected for main sequence objects ($\log g = 3.5 - 5.0$ dex).

There are eight binary candidates in our sample as revealed by other people previous observations. BS029 (S1267), BS043 (S975), BS047 (S752), BS111 (S997), and BS115 (S1195) were identified as spectroscopic binaries with long periods. BS034 (S1284) has been considered as a BS in its final stage of mass transfer with a white dwarf companion. BS046 (S1082) was detected to be a complex unusual eclipsing binary system, or even a triple system, whose SED could be explained by the sum of a close binary and another main sequence star. BS184 (S1036) is observed as a W UMa type binary with small amplitude light variations.

Even though the limited resolution of the observations prevents us from corroborating their binary nature, our analysis demonstrates that both spectral libraries provide equally good fits for single and binary objects.

4. Conclusions

Low resolution spectra of the complete sample of 24 blue stragglers in the old galactic open cluster M67 (NGC2682) has been collected. The whole data set was implemented

Table 2. Parameters derived from fittings.

Name[a]	BATC-Pickles T_{eff} (K)	Sp.Type	BATC-Lejeune T_{eff} (K)	$\log g$ (dex)	BATC-BLUERED T_{eff} (K)	$\log g$ (dex)	S^b
BS005	12589	B6IV	12625	5.00	11050	3.1	977
BS018	8790	A3V	8500	4.50	8500	4.0	1434
BS025	8492	A5V	8500	4.25	8950	5.0	1066
BS029*	8054	A7V	7813	4.38	8100	5.0	1267
BS034*	8054	A7III	7688	4.38	7900	5.0	1284
BS038	8054	A7III	7750	4.00	8050	5.0	1263
BS040	8790	A3V	8625	4.75	8450	4.2	968
BS043*	6469	F5V	6500	3.50	6700	4.7	975
BS046*	6776	F2V	6625	3.88	6850	4.7	1082
BS047*	7586	F0III	7250	4.00	7500	4.8	752
BS065	5636	G2V	5750	3.50	6000	4.2	1072
BS093	7586	F0III	7625	4.25	7800	5.0	1280
BS111*	6281	F6V	6500	3.50	6600	4.6	997
BS115*	6776	F2V	7000	4.25	7050	4.7	1195
BS116	6039	F8V	5938	3.75	6150	4.5	792
BS126	6281	F6V	6250	3.50	6450	4.8	277
BS131	6776	F2V	6875	4.75	6950	4.7	1273
BS139	6039	F8V	6000	3.50	6200	4.6	984
BS143	6039	F8V	6000	3.50	6100	4.1	1005
BS182	6281	F6V	6250	4.00	6500	4.7	751
BS184*	6281	F6V	6375	4.25	6500	4.6	1036
BS185	6531	F5V	6438	4.00	6700	4.7	145
BS206	6776	F2V	6750	4.50	6900	4.7	2204
BS216	6531	F5V	6500	4.50	6700	4.7	2226

NOTE: a, stellar identification from Fan et al. (1996); b, Sanders number (Sanders 1977); *, binary population.

for a comparison with three different stellar atlases aimed at determining the effective temperature and surface gravity. All objects have effective temperature and surface gravity values in agreement to the expected values for objects in the hydrogen-burning stage (see details in Liu et al. 2008). Therefore, in terms of spectral properties, blue stragglers can indeed be represented by empirical or theoretical data of main sequence stars independently of the distinctive formation mechanisms and different physical nature.

Limited by the spectral resolution of the current observational data set, it is not possible to assess binarity and the formation mechanism of the sample of BSs in M67. We anticipate that a detailed chemical abundance analysis at high resolution for the sample will show signatures of these dynamical and physical processes.

5

Acknowledgements

We would like to thank the National Science Foundation of China (NSFC) for support through grants 10573022, and the Ministry of Science and Technology of China through grant 2007CB815406. MC and EB would like to thank CONACyT through the grants SEP-2005-49231 and SEP-2004-47904.

References

Ahumada, J. A. 1999, *RMxAC* 8, 89
Bacon, D., Sigurdsson, S., & Davies, M. B. 1996, *MNRAS* 281, 830
Benz, W. & Hills, J. G. 1987, *ApJ* 323, 614
Benz, W. & Hills, J. G. 1992, *ApJ* 389, 546
Bertone, E., Buzzoni, A., Chávez, M., & Rodríguez-Merino, L. H. 2004, *AJ* 128, 829
Bertone, E., Buzzoni, A., Chávez, M., & Rodríguez-Merino, L. H. 2008, *A&A* accepted
Bressan, A., Fagotto, F., Bertelli, G., & Chiosi, C. 1993, *A&AS* 100, 647
Bressan, A. & Tautvaišiene, G. 1996, *Baltic Astron.* 5, 239
Deng, L., Chen, R., Liu, X. S., & Chen, J. S. 1999, *ApJ* 524, 824
Fan, X., Burstein, D., Chen, J. S., Zhu, J., Jiang, Z., Wu, H., Yan, H., Zheng, Z., Zhou, X., Fang, L.-Z. and 16 coauthors 1996, *AJ* 112, 628
Ferraro, F. R., Paltrinieri, B., Fusi Pecci, F., Cacciari, C., Dorman, B., Rood, R. T., Buonanno, R., Corsi, C. E., Burgarella, D., & Laget, M. 1997, *A&A* 324, 915
Gilliland, R. L. & Brown, T. M. 1992, *AJ* 103, 1945
Girard T. M., Grundy W. M., Lopez C. E., & van Altena W. F. 1989, *AJ* 98, 227
Hobbs, L. M. & Thorburn, J. A. 1991, *AJ* 102, 1070
Leonard, P. J. T. 1989, *AJ* 98, 217
Lejeune, Th., Cuisinier, F., & Buser, R. 1997, *A&AS* 125, 229
Lejeune, Th., Cuisinier, F., & Buser, R. 1998, *A&AS* 130, 65
Liu, G. Q., Deng, L., Chávez, M., & Bertone, E. *et al.* 2008, *MNRAS*, submitted
Livio, M. 1993, *ASPC* 53, 3
Manteiga, M., Martinez R. C., & Pickles, A. J. 1989, *ApSS* 156, 169
Ouellette, J. A., & Pritchet, C. J. 1998, *AJ* 115, 2539
Pickles, A. J. 1998, *PASP* 110, 863
Piotto, G., Zoccali, M., King, I. R., Djorgovski, S. G., Sosin, C., Dorman, B., Rich, R. M., & Meylan, G. 1999, *AJ* 117, 264
Sandage, A. R. 1953, *AJ* 58, 61
Sanders, W. L. 1977, *A&AS* 27, 89
Stryker, L. L. 1993, *PASP* 105, 1081
Tian, B., Deng, L., Han, Z., & Zhang, X. B. 2006, *A&A* 455, 247
Xin, Y., Deng, L., & Han, Z. 2007, *ApJ* 660, 319
Xin, Y., Deng, L., de Grijs, R., *et al.* 2008, *MNRAS* 384, 410

Discussion

FERNANDES: Why do you think that the database of Lejeune (that was built to "normal" stars) can be used for BS stars?

LIU: The formation mechanism and evolutionary state of BSs are not clear, limited by few spectral observations. BSs' loci on CMD (above turn off point and as an extension of main sequence) indicate that the spectral properties of BSs could be similar with normal main sequence stars. So, it could be worthy to test spectral properties of BS with Lejeune spectra.

LANGER: Could you constrain the He-abundance from the colours derived from your observations?

LIU: We could not do this based on current low resolution observation. We anticipate that a detailed chemical abundance including He-abundance analysis at high resolution for the sample will be good for this purpose.

STEPIEN: Could you see any signatures of binarity among the observed stars?

LIU: Limited by the spectral resolution of the current observational data, no firm conclusion of binarity can be draw.

POLS: Can you obtain any constraints on the abundances (e.g. the He abundance) from your spectra? To have such information for a complete sample of blue stragglers could be very interesting.

LIU: Yes, I also think the abundance information of BSs is interesting. But our observations are made in low resolution which does not allow abundance determination.

The speaker, G. Q. Liu (right), is fighting back on N. Langer's challenge.

The Art of Modelling Stars in the 21st Century
Proceedings IAU Symposium No. 252, 2008
L. Deng & K.L. Chan, eds.

© 2008 International Astronomical Union
doi:10.1017/S1743921308023302

The Missing Population of Be+Black Hole X-Ray Binaries

Aleksander Sądowski[1], J. Ziółkowski[1], K. Belczyński[2] and T. Bulik[3]

[1]Copernicus Astronomical Center, ul. Bartycka 18, 00-716 Warsaw, Poland
email: jz@camk.edu.pl

[2]Department of Astronomy, New Mexico State University,
Frenger Mall, Las Cruces, NM 88003, USA
email: kbelczyn@lanl.gov

[3]Astronomical Observatory of Warsaw University, Al. Ujazdowskie 4,
00-478 Warsaw, Poland
email: tb@astrouw.edu.pl

Abstract. At present, 117 Be/neutron star (Be/NS) X-ray binaries (XRBs) are known in the Galaxy and the Magellanic Clouds, but not a single Be/black hole (Be/BH) binary was found so far. We carried out the calculations of stellar population synthesis to investigate the case of the apparently missing population of Be/BH XRBs. According to our calculations, the main reason of this disparity is the fact that within the orbital period range where Be XRBs are found (∼10 to ∼300 days), these systems are formed predominantly with a NS component. The systems with a BH component are formed predominantly with much longer orbital periods and they are not easy to detect.

Keywords. X-rays: binaries, (stars:) binaries: close, stars: evolution, stars: emission-line, Be

1. Introduction

The binary systems composed of a Be star and a neutron star (Be/NS type systems) form the most numerous class of X-ray binaries (XRBs) in our Galaxy. These systems consist of a NS orbiting a Be type star on a rather wide (orbital periods in the range of ∼10 to ∼300 days), frequently excentric, orbit. NS has a strong magnetic field and, in vast majority of cases, is observed as an X-ray pulsar (with the spin periods in the range of 34 ms to about 6000 s). The Be component is deep inside its Roche lobe and the mass accretion on a NS is occuring through the interaction of a NS with the excretion disc around Be component.

At present, 117 Be/NS type XRBs are known in the Galaxy and the Magellanic Clouds (which is almost a half of the total number of the known NS XRBs). Other classes of XRBs are less numerous: we know 87 X-ray bursters (which are also NS XRBs) and 48 X-ray pulsars not associated with a Be type companion (which form still other classes of NS XRBs: 35 of these NSs are associated with a supergiant type companion and 13 with a low mass companion). In addition, we know 58 black hole candidate (BHC) systems (among them 24 confirmed BH systems). However, not a single BHC binary containing a Be type component (Be/BH binary) was found so far.

This disparity (117 Be/NS type systems out of 252 known NS XRBs vs. not a single Be/BH type system among 58 known BH XRBs) called the attention of the researchers already for some time. Zhang et al. (2004) noted that, according to stellar population synthesis calculations by Podsiadlowski *et al.* (2003), BH binaries are formed predominantly with relatively short orbital periods ($P_{orb} < 10$ days). If this is the case, then, according to Zhang *et al.*, the excretion disc truncation mechanism (Artymowicz & Lubow, 1994) might be so efficient, that the accretion rate is very low and the system remains dormant (and therefore invisible) for almost all the time. One should note, however, that Podsiadlowski *et al.* considered, essentially, BH systems with Roche lobe filling secondaries, which definitely is not the case of Be XRBs. Therefore, their results are not relevant for the case of Be/BH XRBs.

We carried out the calculations of stellar population synthesis to investigate the case of the apparently missing population of Be/BH XRBs using the Star Track code described by Belczyński, Kalogera & Bulik (2002) and Belczyński *et al.* (2008).

2. Definition of a Be XRB

The most characteristic observational property of Be stars distinguishing them from other B stars is the presence of excretion discs producing the characteristic emission lines. The underlying cause of the presence of this disc is, in turn, rapid rotation. In the context of XRBs, the presence of an excretion disc is crucial, because it permits the relatively efficient accretion on the compact companion, even in the case of large orbital separation. It is not clear how Be stars achieved their fast rotation (although different hypothesis like rapid rotation at birth or spin-up due to binary mass transfer are advanced see e.g. McSwain & Gies, 2005). The fraction of Be stars among all B stars is similar for single stars and for those in binary systems (one quarter to one third).

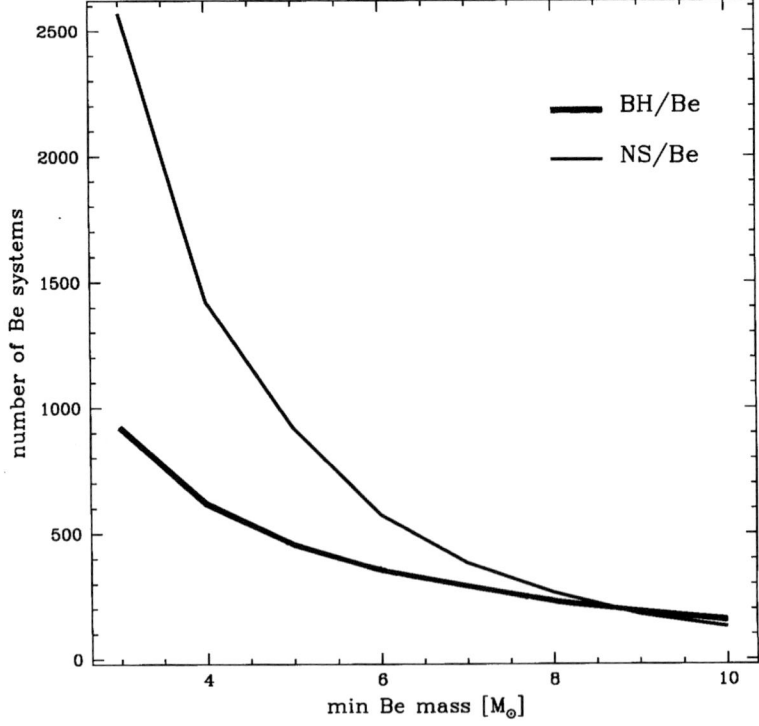

Figure 1. The expected numbers of Be/NS (thin line) and Be/BH (thick line) binaries as functions of the assumed minimal mass of a Be component.

For the purpose of our calculations, we we assumed, for simplicity, that one quarter of all B stars are always Be stars and that these stars are always efficient mass donors, independently of the size of the binary orbit (as is, in fact, observed in Be/NS XRBs). Therefore, according to our definition, a Be XRB is a system composed of a compact object (NS or BH) and a main sequence B star (and we apply a factor 0.25 to the number of such systems, to account for the fact that not every B star is a Be star).

3. Preliminary Results

Fig. 1 shows that, when we count the total expected numbers of Be/NS and Be/BH binaries, these numbers should be, roughly comparable. The estimated masses of observed Be stars cover the range from $\sim$2.3 $M_\odot$ (Lejeune & Schaerer, 2001) to $\sim$25 $M_\odot$ (McSwain & Gies, 2005). Therefore, if we assume $3M_\odot$ as a reasonable lower limit for the mass of a Be component, then the Be/NS systems should outnumber Be/BH systems only by a factor of about 2.5.

The reason for the observed large disparity becomes obvious, when we look at Fig. 2. According to our calculations, the distribution of the orbital periods is completely different for Be/NS and Be/BH systems. Within the orbital period range where Be XRBs are found ($\sim$10 to $\sim$300 days), Be systems are formed predominantly with a NS component. The ratio of the expected number of Be/NS systems to the expected number of Be/BH systems is, for this orbital period range, larger than 50. The systems with a BH component are formed predominantly with much longer orbital periods. Such systems are very difficult to detect, both due to very long orbital periods and due to, probably, very low luminosities (the accretion at such large orbital separations must be very inefficient).

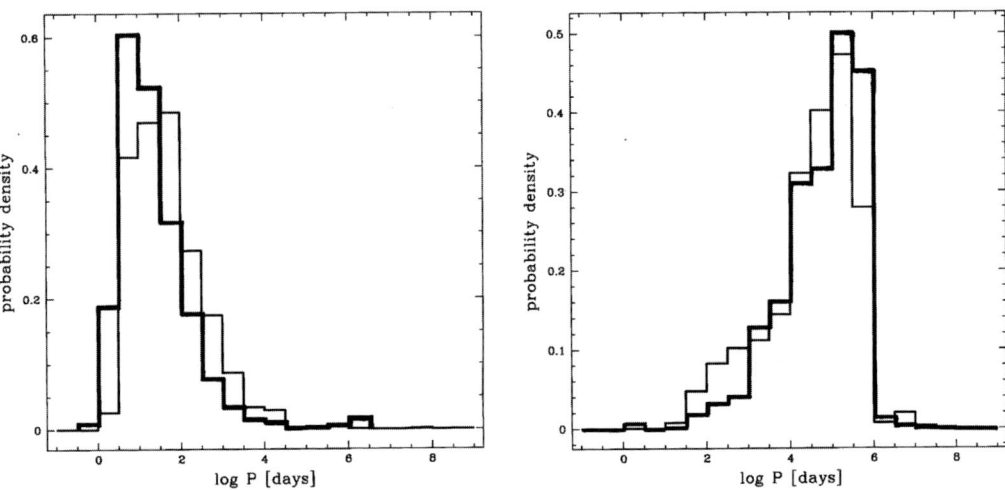

Figure 2. The expected distributions of orbital periods of Be/NS (left panel and Be/BH (right panel) binary systems. Thick lines correspond to the minimum mass of a Be component equal 3 $M_\odot$ and the thin lines to the minimum mass equal 8 $M_\odot$.

We should stress, that the results presented above are only very preliminary results. We plan to carry out further calculations. In particular, we shall try to explain the physical reasons for so different orbital periods distributions for the systems with NSs and with BHs.

We should also stress that, while our results provide a major factor explaining the observed disparity in the numbers of Be/NS and Be/BH systems, this might be not the only factor. Another possible factor may be related to the previous evolution of a Be star. If, indeed, a B star must be a member of a binary system and undergo a mass transfer in order to become a Be star, then one can imagine that the systems composed of a Be star and a relatively less massive companion (which collapses to a NS) remain bound, while those composed of a Be star and a relatively more massive companion (which collapses to a BH) are disrupted in the process of supernova explosion.

Acknowledgements

This work was partially supported by the polish MNiSW grant NN203065933 (2007–2010).

References

Artymowicz, P. & Lubow, S. H. 1994, *ApJ* 421, 651
Belczyński, K., Kalogera, V., & Bulik, T. 2002, *ApJ* 572, 407
Belczyński, K., Kalogera, V., Rasio, F.A., Taam, R.E., Zezas, A., Bulik, T., Maccarone, T.J., & Ivanova, N. 2008, *ApJ Suppl* (in press), also astro-ph/051181
Lejeune, T. & Schaerer, D. 2001, *A&A* 366, 538
McSwain, M.V. & Gies, D.R. 2005, *ApJ Suppl* 161, 118
Podsiadlowski, Ph., Rappaport, S., & Han, Z. 2003, *MNRAS* 341, 385
Zhang, F., Li, X.-D., & Wang, Z.-R. 2004, *ApJ* 603, 663

Discussion

PODSIADLOWSKI: I did not quite understand the evolution that produces a black hole with a ~ 3 $M_\odot$ companion and an orbital period of 400 days. Such systems should have experienced a common envelope phase.

BELCZYŃSKI: Indeed such systems experience a common envelope (CE) phase. Progenitor system consists of a massive star (~ 40 $M_\odot$) and a companion may have several solar masses. A system is initially on a very wide orbit (2000–4000 $R_\odot$). When the primary evolves two important (for the CE survival) things are happening: (i) core of the massive primary grows in mass (nuclear burning). (ii) mass of the primary envelope decreases in mass (wind mass loss). If the orbit is wide enough (as it is for progenitors of BHs with low mass companions) the primary fills its Roche lobe when substantial part of its envelope is gone. The following CE phase tightens the orbit by factors 10-100. Usually, it is expected that such a CE phase leads to a merger of two stars, since it is argued that there is not enough binary orbital energy in the system (due to the very low mass of the companion star) to unbind the massive envelope of the primary. However, in some cases enough of the envelope is lost that the orbital decay during the CE phase does not lead to a merger. The exposed helium core of the primary follows through subsequent evolution and forms a black hole. The massive black hole with low mass companion system is formed. Such a scenario requires that the stellar winds of massive stars are lowered by factor of about 2 (so that the wind mass loss from the massive primary does not prevent Roche lobe overflow and the following CE phase), the assumption that is well within the observational mass loss estimates available for massive stars. The example of such an evolution is given in full detail in Belczynski & Bulik 2002, ApJ Lett. 574, L147. The presented example was used to explain self-consistently the formation of microquasar

GRS 1915+105, a binary system hosting the most massive known stellar black hole in our Galaxy.

LANGER: Which physics principle explains the key feature in your analysis, namely that BH binaries have much larger periods than NS binaries?

BELCZYŃSKI: In general, the black hole binaries originate from the (initially) wider systems for a simple reason that stars that form black holes are more massive (and therefore larger in size) than neutron star progenitors. To avoid the merger during one of the subsequent Roche lobe overflow phases the binary systems that form black holes with companions (and not the mergers) are wider. However, for the specific subpopulations of binary systems discussed in our paper we are still investigating the detailed evolutionary channels and we will present the detailed analysis and the full answer to this question in our journal submission.

The speaker, A. Saldowski, is presenting his work.

The Art of Modeling Stars in the 21st Century
Proceedings IAU Symposium No. 252, 2008
L. Deng & K.L. Chan, eds.

© 2008 International Astronomical Union
doi:10.1017/S1743921308023326

Stellar radii from long-baseline interferometry

Pierre Kervella

LESIA, Observatoire de Paris, CNRS UMR 8109, UPMC, Université Paris Diderot,
5 Place Jules Janssen, 92195 Meudon, France
email: Pierre.Kervella@obspm.fr

Abstract. Long baseline interferometers now measure the angular diameters of nearby stars with sub-percent accuracy. They can be translated in photospheric radii when the parallax is known, thus creating a novel and powerful constraint for stellar models. I present applications of interferometric radius measurements to the modeling of main sequence stars. Over the last few years, we obtained accurate measurements of the linear radius of many of the nearest stars: Procyon A, 61 Cyg A & B, α Cen A & B, Sirius A, Proxima... Firstly, I describe the example of our modeling of Procyon A (F5IV-V) with the CESAM code, constrained using spectrophotometry, the linear radius, and asteroseismic frequencies. I also present our recent results on the low-mass 61 Cyg system (K5V+K7V), for which asteroseismic frequencies have not been detected yet.

Keywords. techniques: interferometric, stars: individual (Procyon, 61 Cyg), stars: evolution

1. Introduction

Long-baseline interferometry now routinely provides stellar angular diameters with sub-percent accuracy. These measurements are independent from the classical observables obtained by photometry or spectroscopy, and therefore represent a very valuable addition to constrain the stellar evolution and structure models. After a short description of the principle of interferometric angular size measurements (Sect. 2), I present in Sect. 3 the results we obtained on the benchmark star Procyon A. This star is specially interesting as a number of seismic oscillation frequencies have been detected by spectroscopy, and I discuss the synergy between the interferometric radius measurement and asteroseismology. Section 4 is then dedicated to the discussion of our most recent results on the low-mass binary star 61 Cyg.

2. Visibilities and angular diameters: a brief interferometry primer

This article is not intended to give an extensive tutorial on optical long-baseline interferometry, but rather explain its general principle with simplified concepts. For the interested reader, several excellent introductions to interferometry are available either in publications (Lawson *et al.* 2000; Perrin & Malbet 2003; Malbet & Perrin 2007) or through specialized web sites (OLBIN 2008; JMMC 2008).

An interferometer can be defined as a non-connex pupil telescope, i.e. a telescope whose pupil is split in a number of separate pieces. This simple definition can be used to better understand the angular resolution of an interferometer from its point spread function (PSF). Let's first consider the PSF of a single-dish telescope equiped with a primary mirror of diameter $D = 10$ m, no central obscuration and no atmospheric perturbation (or with a perfect adaptive optics system). The Rayleigh resolution criterion states that such a telescope is able to separate details that are $\theta_{\rm tel} = 1.22 \, \lambda/D = 55$ milliarcseconds

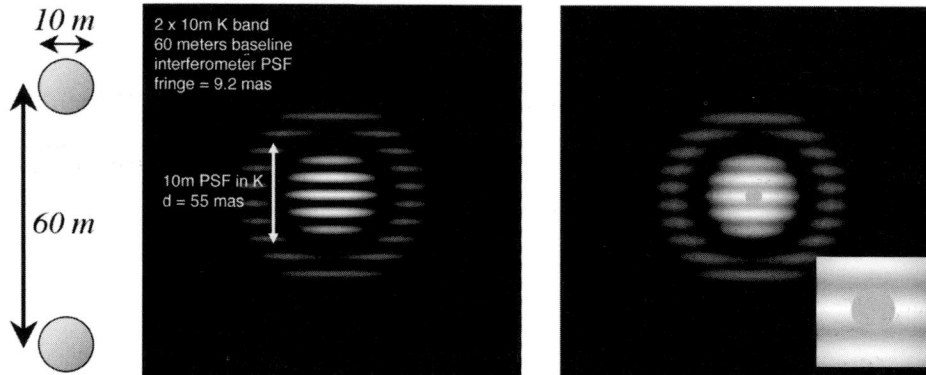

Figure 1. *Left:* Pupil of an interferometer of two 10 m telescopes separated by 60 m. *Middle:* Point Spread Function of this interferometer, when operated in the K band. The fringe spacing is 9.2 mas. *Right:* Image of α Cen A ($\theta \approx 8.5$ mas) formed by this interferometer, with the superimposed stellar disk. The fringes have a lower contrast than on the left, as the star is partially resolved. As shown in the insert on the right, the light from the stellar disk "leaks" in the dark fringes.

(mas) apart on the sky, in the K band ($\lambda = 2.2\,\mu$m). As a comparison, the largest solar-type star angularly is α Cen A, with $\theta \approx 8.5$ mas, i.e. more than six times smaller than the resolution limit of the telescope. Now let's consider the PSF of an interferometer of two 10 m telescopes separated by a baseline of 60 m (Fig. 1, left). As this setup is a particular case of the classical Young's experiment, the diffraction pattern of this instrument is made of a series interference fringes superimposed on the PSF of an individual telescope. Thanks to the long baseline, the spacing of the interference fringes is 6 times smaller than the single telescope PSF along the direction of the baseline (Fig. 1, middle), and corresponds to a resolution of 9.2 mas. Although this value is still a bit larger than α Cen A, it becomes comparable, and the image formed by the interferometer (Fig. 1, right) is now significantly different from the PSF. The finite angular size of α Cen A causes a reduction of the contrast of the fringes, that can be measured with great accuracy. There is a direct relation between the visibility of the fringes and the angular size of the star (Zernike-Van Cittert theorem), that allows to retrieve the angular size of the star. The best interferometric instruments currently available can provide contrast measurements with a relative precision of $\approx 0.1\%$, and angular diameters to $\pm 0.1\%$ or better. Intuitively, the reason for the change of the contrast with the increasing resolution is that the size of the star becomes larger than the spacing of a fringe, and its flux starts to "leak" in the dark fringes flanking the bright central fringe.

Naturally, interferometry does not provide direct *radius* measurements. In order to retrieve the linear photospheric size of a star, one needs its parallax. Thanks to the *Hipparcos* catalogue (ESA 1997), most nearby stars have high accuracy parallaxes. However, one should keep in mind that their accuracy can still be limiting to compute the radius, particularly when the angular diameter is known with very high accuracy ($\lesssim 1\%$).

3. Procyon A: the synergy of interferometry and asteroseismology

Procyon A is among the brightest stars in the sky and is easily visible to the naked eye. This made it an ideal target for a number of spectro-photometric calibration works. It is also a visual binary star classified F5IV-V, with a white dwarf (WD) companion orbiting the main component in 40 years. The influence of this massive companion on the

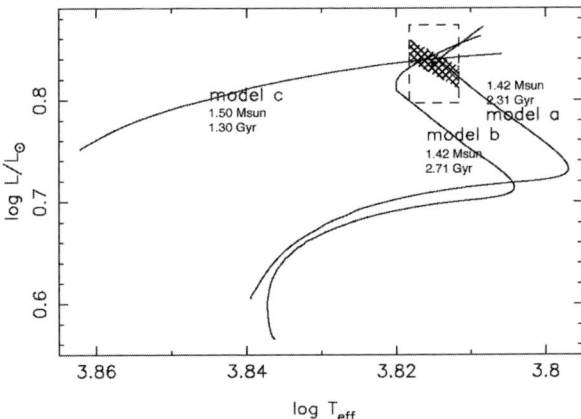

Figure 2. *Left:* Evolutionary tracks of three models of Procyon in the Hertzsprung-Russell diagram. *Right:* Corresponding model parameters (Kervella et al. (2004).

apparent motion of Procyon was discovered by Bessel (1844), and Girard et al. (2000) obtained astrometric masses of 1.497 ± 0.037 $M_\odot$ and 0.602 ± 0.015 $M_\odot$, respectively for Procyon A and B. It has also been an asteroseismic target for a decade and Martic et al. (1999; 2001) measured a large frequency spacing of 54-55 μHz.

In Kervella et al. (2004), we presented several models of Procyon computed with the CESAM code (Morel 1997; Morel & Lebreton 2007). Because numerous new studies and observational constraints (like the direct diameter) exist today, we re-examined the status of Procyon. Our models were constrained using the spectroscopic effective temperature and the linear diameter value that we derived from our observations with the VINCI/VLTI instrument (Kervella et al. 2000; 2003a). Figure 2 (left) shows the evolutionary tracks of three models that converge into the uncertainty box in the HR diagram. The corresponding parameters are listed in the table of Fig. 2 (right). The added value of interferometry is clearly visible on this diagram, as the surface of the diagonal shaded area, set by the constraint from the interferometric radius, is much smaller than the classical $L - T_{\rm eff}$ uncertainty box represented as a rectangle. More generally, the radius constraint always appears as a diagonal zone in the classical $L - T_{\rm eff}$ HR diagram with logarithmic scales, as the photospheric radius is linked to the these two quantities by Stefan-Boltzmann's law: $L = 4\pi R^2 \sigma T_{\rm eff}^4$. From this law, a 10× increase in the radius of a star causes a 100× increase in luminosity (at constant $T_{\rm eff}$) or a $\sqrt{10}\times$ decrease in $T_{\rm eff}$ (at constant L).

Although our model c converges in the radius-limited uncertainty domain of the HR diagram (Fig. 2, left), the associated large frequency spacing of 56.4 μHz is too large compared to the values measured by Martic et al. ($\overline{\Delta\nu_0} \sim 54 - 55\,\mu$Hz). Moreover, the corresponding model age of 1.3 Gyr is too young for the cooling time of the white dwarf Procyon B. Provencal et al. (2002) found that its progenitor ended its lifetime 1.7 ± 0.1 Gyr ago, an age incompatible with the 1.3 Gyr of model c and thus with the 1.50 $M_\odot$ mass of Procyon A from Girard et al. (2000). By reducing the mass of Procyon to 1.42 $M_\odot$ (models a and b), we could reproduce both the observed radius and seismic large frequency spacing accurately. For our prefered model a of Procyon A, we derive an age of 2 314 Myr. Subtracting the cooling age of the WD companion to our determination of the age of Procyon A leads to a lifetime of ≈ 600 Myr for the progenitor of Procyon B. This indicates that the mass of the progenitor is approximately 2.5 $M_\odot$. Such a star is expected to evolve towards a $\approx 0.6\,M_\odot$ white dwarf, just as Procyon B, thus strengthening our coherent picture of this stellar system. Two years after the publication of our work in 2004, Gatewood & Han (2006) redetermined the astrometric mass of Procyon, and found $M = 1.43 \pm 0.03 M_\odot$, in excellent agreement with our "interfero-seismic" modeling.

An extensive review of the remarkable synergy between interferometry and asteroseismology is given by Cunha et al. (2007). Kjeldsen & Bedding (2003) showed that the large frequency spacing $\overline{\Delta\nu_0}$ is proportionnal to the square root of the density of the star: $\overline{\Delta\nu_0} \sim 134.9\sqrt{(m/M_\odot)/(R_\star/R_\odot)^3}$ [μHz]. From this expression, it is clear that the interferometric radius, combined with $\overline{\Delta\nu_0}$ brings a strong constraint on the mass of the star. But interferometry is not only useful for measuring stellar radii: it can also efficiently constrain the stellar atmosphere structure from limb darkening measurements. The interested reader can refer to Aufdenberg et al. (2005) for a comparison of the limb darkening predictions of several models of the atmosphere of Procyon with interferometric visibility measurements obtained over a broad range of wavelengths.

4. 61 Cyg A & B: a low-mass nearby binary system

The cool dwarfs 61 Cyg A and B are the nearest stars in the northern hemisphere. They are a visual binary pair with a very long orbital period (≈ 700 yrs). In 1838, 61 Cyg became the first star whose distance from Earth was estimated accurately (Bessel 1838), shortly before Procyon's, and it is now known with an exquisite accuracy. Its proper motion of more than $5''$ per year, first determined by Piazzi in the XVIIIth century, makes it one of the fastest moving stars in terms of apparent displacement. Although some of this motion comes from the proximity of 61 Cyg to us, the pair is also moving fast into space relative to the Sun, at 108 km/s, indicating that 61 Cyg is not a member of the thin disk of our Galaxy. The proximity of 61 Cyg makes it a northern analog of the numerical modeling benchmark α Cen. The spectral types of its two members (K5V and K7V) ideally complement our previous studies of α Cen A & B (G2V+K1V; Kervella et al. 2003b; Bigot et al. 2006). The masses of 61 Cyg A & B are controversial at approximately 0.74 and 0.46 $M_\odot$ (Gorshanov et al. 2006) or 0.67 and 0.59 $M_\odot$ (Walker et al. 1995). With effective temperatures of about 4 400 and 4 000 K, they shine at luminosities of only 0.15 and 0.08 $L_\odot$. There is no confirmed planet around them, although indications exist that 61 Cyg B could host a giant planetary companion (Gorshanov et al. 2006). The abundances of heavy chemical elements have been determined (Luck & Heiter 2005, 2006) in these stars which are found slightly metal poor (≈ -0.2 dex), so a priori older than the Sun but belonging to the galactic disk.

The measured CHARA/FLUOR visibilities translate into the following angular diameters: $\theta_{\text{LD}}(61\,\text{Cyg A}) = 1.775 \pm 0.013$ mas, $\theta_{\text{LD}}(61\,\text{Cyg B}) = 1.581 \pm 0.022$ mas. The limb darkening (LD) models were taken from Claret (2000) for the K band. From the combination of these angular diameters and the trigonometric parallaxes taken from van Altena et al. (1995) for 61 Cyg A and the *Hipparcos* catalogue (ESA 1997) for 61 Cyg B ($\pi_A = 286.9 \pm 1.1$ mas., $\pi_B = 285.4 \pm 0.7$ mas), we derive the following photospheric linear radii: $R(61\,\text{Cyg A}) = 0.665 \pm 0.005\ R_\odot$, $R(61\,\text{Cyg B}) = 0.595 \pm 0.008\ R_\odot$. The relative uncertainties on the radii are therefore $\pm 0.8\%$ and $\pm 1.4\%$. Thanks to the high precision of the parallaxes (0.38% and 0.25%), the radius accuracy is limited in this case by the precision of the angular diameter measurements.

Using the available classical constraints (photometry, spectroscopy) and the radii, we computed a series of CESAM2k models (Morel & Lebreton 2007). We selected as the most plausible models those satisfying first the luminosity and radius constraints and second the effective temperature constraint. The models of 61 Cyg A and B converge simultaneously to the radii-limited uncertainty boxes for masses of 0.69 and 0.61 $M_\odot$

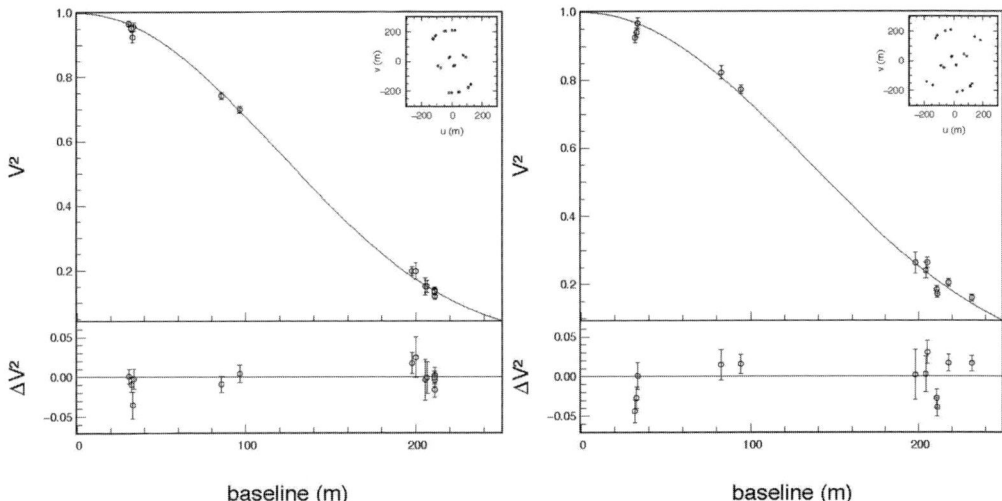

Figure 3. Visibility data from CHARA/FLUOR interferometric measurements (open circles) and the adjusted limb darkened disk model visibility curves (solid curves) for 61 Cyg A (left) and B (right) from Kervella *et al.* (2008).

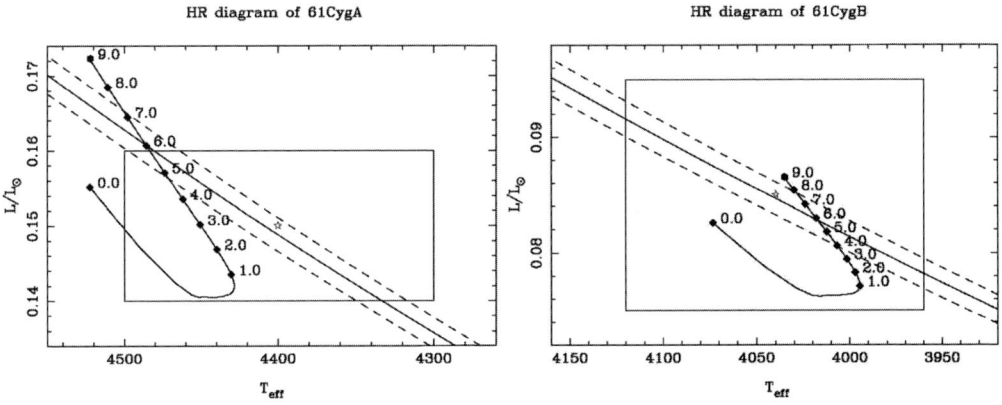

Figure 4. CESAM2k evolutionary tracks of 61 Cyg A (left) and B (right). The labels indicate the age in Gyr relatively to the ZAMS. The rectangular box represents the classical $L - T_{\text{eff}}$ error box, and the diagonal lines represent the radius and its uncertainty.

and an age of 6.0 ± 1.0 Gyr (Fig. 4). However, it appears difficult to go beyond the present modeling of the binary system. The main reason is that the masses are presently not constrained sufficiently well by the long period astrometric orbit, and no seismic frequencies have been detected yet in this system. The detection of such oscillations would bring precious constraints to stellar structure models in the cool, low-mass part of the HR diagram.

5. Conclusion

We presented two examples of stellar evolution modeling constrained by interferometric radius measurements: Procyon A and 61 Cyg A & B. In both cases, the radius reduces spectacularly the error boxes in the HR diagram, and thus brings a strong constraint on the evolutionary status of the stars. But, as shown by our Procyon modeling and Cunha *et al.* (2007), the best results are obtained when asteroseismic frequencies are available.

In this case, the radius and frequencies combine to constrain the mass of the system very tightly (Creevey *et al.* 2007). Several hundred stars are accessible to high accuracy radius measurements to 3% or better by interferometry, and even more will soon be within reach of the new generation of interferometric facilities. So one can expect new and exciting results in this field in a near future.

Acknowledgements

This research took advantage of the SIMBAD and VIZIER databases at the CDS, Strasbourg (France), and NASA's Astrophysics Data System Bibliographic Services.

References

Aufdenberg, J. P., Ludwig, H.-G., & Kervella, P. 2005, ApJ, 633, 424
Bessel, F. W. 1838, MNRAS, 4, 152
Bessel, F. W. 1844, MNRAS, 6, 136
Bigot, L., Kervella, P., Thévenin, F., & Ségransan, D. 2006, A&A, 446, 635
Claret, A., 2000, A&A, 363, 1081
Creevey, O. L., Monteiro, M. J. P. F. G., Metcalfe, T. S., *et al.* 2007, ApJ, 659, 616
Cunha M. S., Aerts C., Christensen-Dalsgaard J., Baglin A., *et al.* 2007, A&ARv, 14, 217
ESA 1997, The Hipparcos and Tycho Catalogues, ESA SP-1200
Gatewood, G. & Han, I. 2006, AJ, 131, 1015
Girard, T. M., Wu, H., Lee, J. T., *et al.* 2000, AJ, 119, 2428
Gorshanov, D. L., Shakht, N. A., & Kisselev, A. A. 2006, Ap, 49, 386
JMMC, Jean-Marie Mariotti Center web site, *http://www.mariotti.fr/*
Kervella, P., Coudé du Foresto, V., Glindemann, A. & Hofmann, R. 2000, SPIE, 4006, 31
Kervella, P., Gitton, Ph., Ségransan, D., *et al.* 2003a, SPIE, 4838, 858
Kervella, P., Thévenin, F., & Ségransan, D., *et al.* 2003b, A&A, 404, 1087
Kervella, P., Thévenin F., Morel, P., *et al.* 2004, A&A, 413, 251
Kervella, P., Mérand, A., Pichon, B., *et al.* 2008, A&A, submitted
Kjeldsen, H., & Bedding, T. 1995, A&A, 293, 87
Lawson, P. (ed.) 2000, Course Notes from the 1999 Michelson Summer School, JPL Pub. 00-09
Luck, R. E., & Heiter, U. 2005, AJ, 129, 1063
Luck, R. E., & Heiter, U. 2006, AJ, 131, 3069
Malbet, F., & Perrin, G. (eds.) 2007, New Astronomy Reviews, 51, 8-9, 563, Proceedings of the EuroSummer School "Observation and Data Reduction with with the VLT Interferometer" (*http://vltischool.obs.ujf-grenoble.fr*)
Martic, M., Schmitt, J., Lebrun, J., *et al.* 1999, A&A, 351, 993
Martic, M., Lebrun, J.-C., Schmit, J., Appourchaux, Th., & Bertaux, J.-L. 2001, in SOHO 10 / GONG 2000 Workshop, A. Wilson ed., ESA SP-464, p 431
Morel, P. 1997, A&AS, 124, 597
Morel, P., & Lebreton, Y. 2007, Ap&SS, doi:10.1007/s10509-007-9663-9
OLBIN, Optical Long Baseline Interferometry News web site, *http://olbin.jpl.nasa.gov/*
Perrin, G., & Malbet, F. (eds.) 2003, EAS Publications Series, 6, Proceedings of the EuroWinter School "Observing with the VLTI", *http://www.mariotti.fr/obsvlti/obsvlti-book.html*
Provencal, J. L., Shipman, H. L., Koester D., Wesemael, F. & Bergeron, P. 2002, ApJ, 568, 324
Van Altena, W. F., Lee, J. T., & Hoffleit, E. D. 1995, The General Catalogue of Trigonometric Stellar Parallaxes, Fourth Edition, Yale University Observatory
Walker, G. A. H., Walker, A. R., Irwin, A. W. *et al.* 1995, Icarus, 116, 359

Discussion

WOITKE: Your analysis works best for "clean stars". One should be aware that K-band interferometry is quite sensitive to e.g. close circumstellar material, hot dust, etc., which can lead to misleading conclusions, in particular for red giants.

KERVELLA: (1) Dust, e.g., in disks, emits at longer wavelengths and is usually not a problem in the K-band. (2) This is a very good point, however, specifically for dusty envelopes around evolved stars, that can have a significant impact on the radius estimation.

STEPIEN: Is it true that all existing models of low mass main sequence stars show systematically smaller radii than observed?

KERVELLA: I am not an expert in this field of stellar modeling, but it is clear that the adjustment of some parameters related in particular to the treatment of convection would possibly improve the agreement with the models. Such models certainty are possible, but I am not aware of them.

LUDWIG: In view of the systematic differences of the stellar radii at low T_{eff} between the measurements and theoretical mass-radius relationship: Have the stellar structure models been calculated assuming a constant-presumably solar calibrated-mixing length parameter? Remark on interferometric measurements: For constraining stellar atmosphere models visibilities at optical (or even shorter!) wavelength are particularly valuable.

KERVELLA: (1) Yes, as far as I can tell, the models were computed assuming a constant value of the MLT parameter. (2) Indeed! Measurements of the limb darkening at visible wavelength would be very useful. But due to turbulence, that is stronger in the visible, these measurements are significantly more difficult to obtain there in the infrared.

FERNANDES: I wonder if the claimed disagreement between models and observations in the mass-radius diagram (at the low mass regime) is mainly related with activity more than convection.

KERVELLA: This is a very good point. The low mass dwarfs with spectral types K and M tend to show rather strong magnetic activity (flares in particular), that can possibly bias the T_{eff} and L estimates. However, interferometric measurements are relatively insensitive to these perturbations.

The speaker, P. Kavella, is presenting his work.

The Y^2 Isochrones Getting an Extra Dimension

Sukyoung K. Yi, Yong-Cheol Kim[1], Pierre Demarque[2], Young-Wook Lee[1], Sang-Il Han[1] and Do Gyun Kim[1]

[1] Yonsei University, Department of Astronomy, Seoul 120-749, Korea
email: yi@yonsei.ac.kr

[2] Yale University, Department of Astronomy, New Haven 06520-8101, CT, USA

Abstract. The Yonsei-Yale Isochrones have been widely used since its birth in 2001. We announce a major upgrade mainly making varieties of helium values available. The recent works on the globular clusters with extreme helium abundances have called for such a need. The new version of the Y^2 Isochrones are available for $[\alpha/Fe] = 0$ through 0.6, $\Delta Y/\Delta Z = 1.5$ through 3.0, and extreme helium abundances (Y = normal 0.05, 0.1, 0.15, 0.2), and for 11 metallicity grids, with full capability of interpolation. The database will be powerful for making population models. Besides, the accuracy of the models on the lower main sequence has been substantially improved. We illustrate the major upgrades and demonstrate the power of the new grids.

1. Introduction

Since Demarque and Larson (1964) used the concept of isochrone to derive the age of NGC 188, isochrones have been widely utilized in stellar population studies. This purely theoretical achievement helps setting one of the most important constraints on cosmology, because the age of the universe is a distinctive product of any cosmological model.

The Yonsei-Yale (Y^2) Isochrones (Yi et al. 2001) have been popular mainly for the following reasons. First, it reproduces the shapes of the observed color-magnitude diagrams (CMDs) of star clusters very well, which is important for cluster age estimation. Second, it covers a wide parameter space (11 metallicity grids, 3 alpha enhancement grids), which is particularly useful for population synthesis. Lastly, it is available for two options of the temperature-color transformation schemes. The Green et al. transformation appears superior in the metal-poor regime, but in the metal-rich regime Lejeune et al. calibration seems to work better (see Yi et al. 2001 for discussion).

2. Lower Main Sequence

The Y^2 isochrones, like most others, however had shortcomings. Most notable was the inaccuracy of its low-mass main sequence models. As (implicitly) demonstrated in Yi et al. (2001, see their Figure 10), their low-mass models are much redder than fainter than observation. This problem is particularly visible when the super-accurate photometry data have been obtained by space telescopes. Figure 1 shows the recent HST data of the globular cluster NGC 6387 (Richer et al. 2008). The data are for the proper-motioned confirmed member stars only. The incredible clearness of the main sequence down to the brown dwarf limit is unprecedented. Such a robust data provides a testbed for stellar evolution theory while posing a threat as well. We compare our updated isochrones to the observed HRD in Figure 1. The models are not yet the final, well-calibrated ones, but

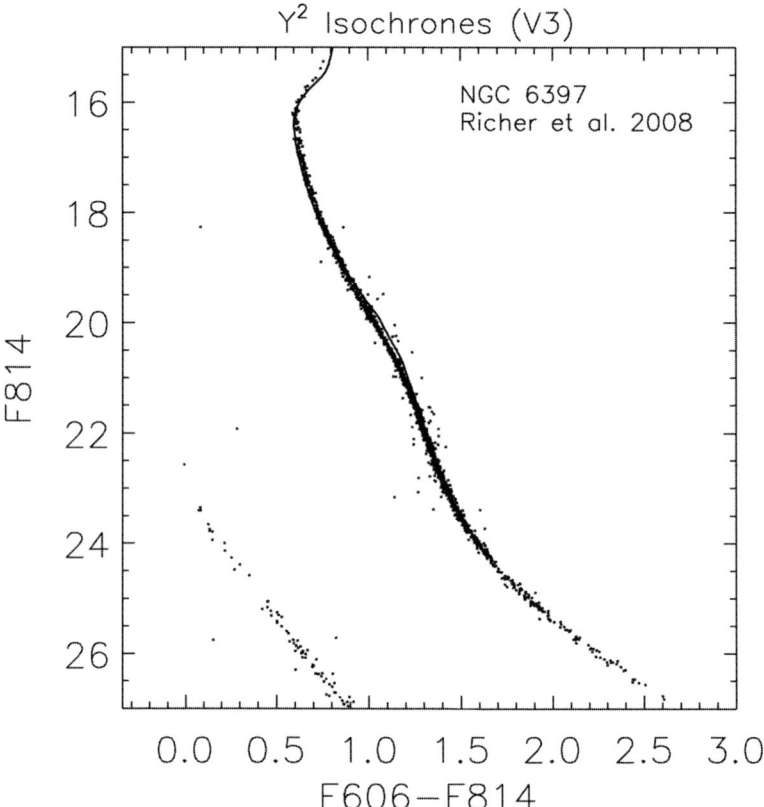

Figure 1. The HST observed HRD of NGC 6397 from Richer *et al.* (2008) compared with our new isochrone that has been improved for the accuracy on the lower main sequence.

simply based on the theoretical Kurucz spectral library. Yet, the match is surprisingly good. The delicate curvatures are reasonably reproduced. It should be noted that our models reach down to 0.2 solar mass, and so missing the lower end of the main sequence observed.

A very important remark to make is that such a good match is not possible when simple filter conversion formulae (from HST filters to Johnson VI) are used. In this sense, it is critical to use the isochrones specifically computed for the filter systems used for the observation.

3. Helium abundances

A remarkable new information has recently emerged on the chemical compositions of some stars. Observations for the colour-magnitude diagrams on globular clusters ω Cen and NGC 2808 revealed multiple populations. The most massive globular cluster ω Cen for example is now known to have up to five different metallicities both for the main sequence and the red giant branch (Bedin *et al.* 2004). Most shockingly, the bluest main sequence is found spectroscopically to be more metal-rich (Piotto *et al.* 2005) which implies an extremely high helium abundance of $Y \approx 0.4$. Interestingly, Lee *et al.* (2005b) noted that such extreme helium stellar populations would evolve into the extremely hot part of the HB explaining the hitherto mysterious origin for the EHB stars of ω Cen. Lee *et al.* claims that the same phenomenon is seen in NGC 2808 as well. Figure 2 shows a

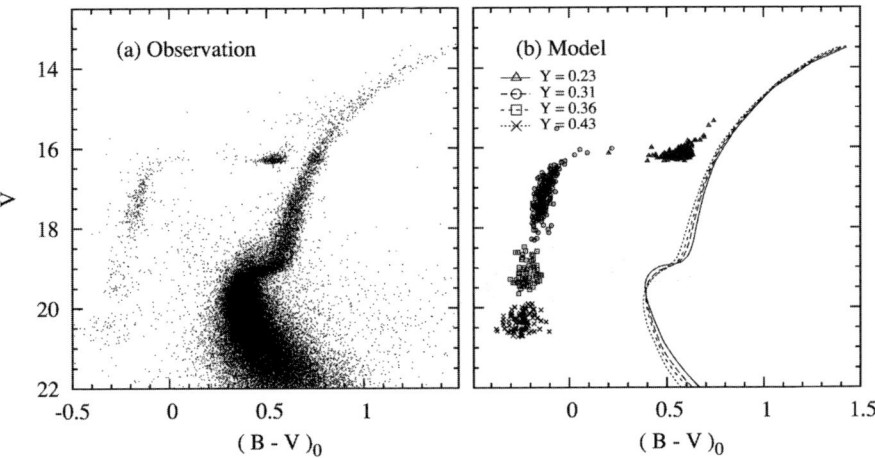

Figure 2. The observed and modeled colour-magnitude diagrams of the globular cluster NGC 2808. *left*: The cluster shows an exceptionally wide distribution of horizontal branch stars. *right*: It can be precisely reproduced by theory for example by assuming a large range of helium abundance. Original version from Lee *et al.* (2005b) and kindly provided by Chul Chung.

comparison between the observed HRD and our synthetic HRD using our latest stellar models. One can notice the wide scatter on the observed main sequence and an incredibly spread-out HB. NGC2808 has been a challenge to stellar evolution theory in this sense. Now with a new hypothesis of extreme helium abundances, we were able to reproduce both MS and HB spreads rather easily. Similar phenomena have been found amongst M87 globular clusters, too (Kaviraj *et al.* 2007).

Such a high helium abundance could in fact be more mysterious than the origin of the EHB stars itself, hence became a hot topic. The high value of helium abundance ($Y \approx 0.4$) seems particularly impossible when it is combined with its low metallicity empirically constrained ($Z \approx 0.002$–0.003). This leads to $\Delta Y/\Delta Z \approx 70$ which is extremely unlikely from the galactic chemical enrichment point of view unless some exotic situation is at work, such as the chemical inhomogeneity in the proto-galactic cloud enriched by first stars (Choi & Yi 2007; 2008). Inspired by this wave, we extend our models towards non-standard values of helium abundance.

4. Prospects

The sample fits we show in this paper demonstrate the power of the up-to-date stellar models in understanding (or at least reproducing) the details of stellar evolution. Whether such models are truly realistic will be answered only when we test the underlying hypotheses against more targets. Recent compilations of HST data on Galactic globular clusters by the Ata Sarajedini group and others are particularly exciting for providing such opportunities.

Our new isochrone database will likely be released in early 2008 in a published form. It will contains all the core hydrogen burning stages but core helium burning stages will be available soon after that. We expect it to be a powerful database to investigate various topics related with stellar populations with. Such examples include the origin of helium-rich populations, the UV upturn in massive spheroids, galactic halo formation, etc. Meanwhile, the availability of new hyper-accurate space photometry data will be critical to test and further elaborate stellar models beyond yesterday's imagination.

Acknowledgements

We thank Harvey Richer and Ivan King for sending us their new HST photometry data on NGC 6397. We are grateful to Chul Chung for regenerating the NGC 2808 model HRD that suits the format of this paper. This research has been supported by Korea Research Foundation (SKY).

References

Bedin, L. R., Piotto, G., Anderson, J., Cassisi, S., King, I. R., Momany, Y., & Carraro, G. 2004, ApJ, 605, 125
Demarque, P. & Larson, R. B. 1964, ApJ, 140, 544
Choi, E. & Yi, S. K. 2007, MNRAS, 375, L1
Choi, E. & Yi, S. K. 2008, MNRAS, 386, 1332
Kaviraj, S., Sohn, S. T., O'Connell, R. W., Yoon, S.-J, Lee, Y.-W., & Yi, S. K. 2007, MNRAS, 377, 987
Lee, Y.-W. et al. 2005, ApJ, 621, L57
Piotto, G. et al. 2005, ApJ, 621, 777
Richer, H. B. et al. 2008, AJ, 135, 2141
Yi, S., Demarque, P., Kim, Y.-C., Ree, C. H., Lejeune, Th., & Barnes, S. 2001, ApJS, 136, 417

Discussion

Z. HAN: You used an extreme helium enrichment to explain the CMD of ω Centauri. I was wondering whether you have any other pieces of evidence for the enrichment?

S.K. YI: The extreme helium abundance is derived indirectly from the CMD fitting and spectroscopic metal measurements. There is no other "direct" evidence as far as I know.

L. DENG: What is the main difference between your Y^2 isochrone and other isochrones e.g., Padova, Geneva?

S.K. YI: Y^2 isochrones do not cover large-mass stars of M>5M$_\odot$ but focus on making precision models for low-mass stars, compared to others. We also attempt to cover a wide parameter space so that they can be useful for the studies of unknown populations. There are other difference in the input physics as well.

G. MEYNET: (1) Will you provide isochrons with the solar abundance given by Asplund et al. (2005)? (2) Can you say a few words about that He–rich stars in ω Cen?

S.K. YI: (1) maybe in the future. (2) ω Cen is reported to have extreme–helium population of Y$\sim$0.4, $\triangle Y/\triangle Z \sim 70$. Candidate origins include SNe, AGB stars, and rotating massive stars. We think (Choi & Yi 2007, 2008) AGB stars are very unlikely under the current theoretical assumptions. Rotating massive stars seem plausible, but it requires an explanation why only some globular clusters show this, and not galaxies.

Blue Stragglers from Primordial Binary Evolution

Xuefei Chen and Zhanwen Han

Yunnan observatory/National Astronomy Observatories, CAS
email: xuefeichen717@hotmail.com

Abstract. Binaries exist in all clusters and much evidence suggests that close-binary evolution makes an important contribution to the blue straggler population, at least in some clusters as well as in the field. Here we present different channels to blue stragglers from primordial binary evolution and examine their contributions to the integrated spectral energy distribution of the host clusters in theory via binary population synthesis.

Keywords. binary:close, stars: blue stragglers, stars: evolution

1. Introduction

Blue stragglers (BSs) were first found by Sandage in 1953, and many mechanisms have been provided to explain them ever since. These stars exist in all populations and their existence shows an incomplete understanding of stellar evolution and perhaps also of star formation within clusters. These strange objects lie above the main-sequence turnoff region in color-magnitude diagrams, and may significantly affect the integrated spectral energy distributions (ISEDs) of their host clusters by contributing excess spectral energy in the blue and ultraviolet (Xin & Deng 2005). It is becoming clearer that several mechanisms are responsible for the blue-straggler phenomenon. Generally the mechanisms may be divided into two types: one is from single stars and the other is from binaries. Both of them have advantages and limitations (see the review by Stryker 1993). It is likely that more than one mechanism occurs even within the same cluster to produce BSs, and primordial binaries are important contributors in some clusters as well as in the field.

There are three different channels leading to BS formation from primordial binary evolution. Channel A: mass transfer. During stable Roche lobe overflow (RLOF), the accretor goes upward along the main sequence, if it is a main sequence star, and becomes a BS when it is more massive than the turnoff of the host cluster. Channel B: contact binary. Also during stable RLOF, the accretor is also likely to fill its Roche lobe and the system becomes a contact binary. The contact binary eventually coalesces as a single main-sequence star, if both of the two components are on the main sequence. So the remnant may be a BS if it is more massive than the turnoff. Channel C: dynamically unstable RLOF between two main-sequence components. In this case, a common envelope (CE) is formed and the binary merges into a single star if CE cannot be ejected. The merger is a BS if its mass is beyond the turn-off mass of the host cluster. All binary evolutionary cases (i.e. cases A, B and C) can produce BSs from mass transfer but in various period range. Binary coalescence from case A evolution is a popular hypothesis for single BSs.

2. Binary Population Synthesis

We employ the binary population synthesis code originally developed by Han in 1994, which has been updated regularly ever since. The main input of the code is a grid

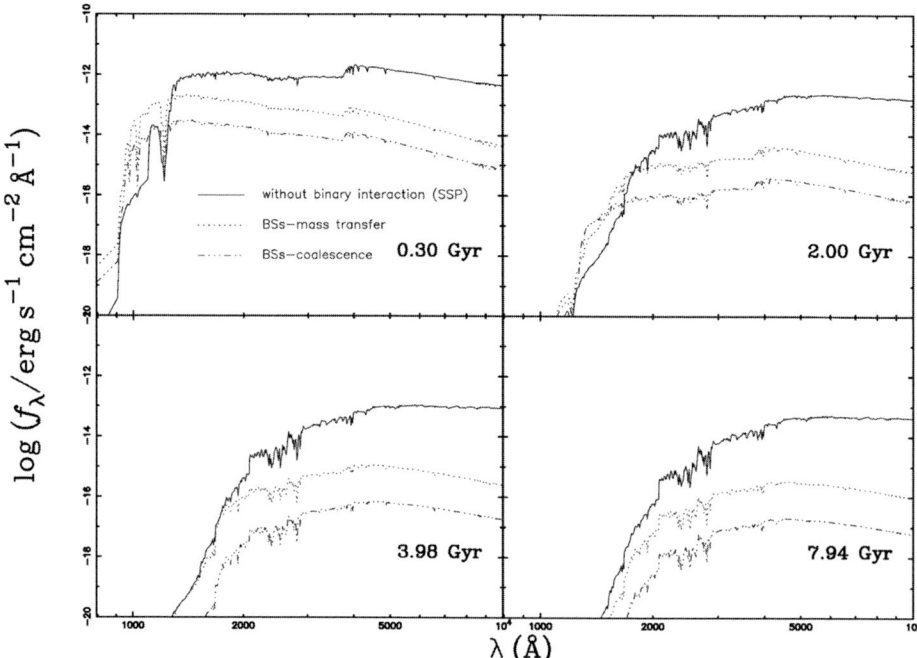

Figure 1. Integrated rest-frame intrinsic SEDs for a stellar population (including binaries) with a mass of $10^{10} M_\odot$ at a distance of 10Mpc. We see that BSs resulting from primordial binary evolution are dominant contributors to the ISED in UV and blue bands between 0.3 and 2.0 Gyr. The BSs and SSP have comparable energy in UV and blue bands between 2 and 4 Gyr.

of stellar evolution, which is calculated from Eggleton's stellar evolution code. Here a Population I grid ($Z = 0.02$) is adopted, and single stars are evolved via interpolations in the model grid. We adopt the critical mass ratio for stable RLOF $q_c = 3.2$ when the primary is on the main sequence or during Hertzsprung gap. If the mass donor is on the FGB or AGB, we use $q_c = 1.5$. For case A, we assume that RLOF is always conservative while various values of mass transfer efficiency are adopted for other cases. The mass lost from the system takes away a specific angular momentum in units of the specific angular momentum of the system. In order to obtain the colours and the ISED of the populations produced in our simulations, we use the latest version of the comprehensive BaSeL library of theoretical stellar spectra (Lejeune et al. 1997, 1998), which gives the colours and ISEDs of stars with a wide range of Z, $\log g$ and T_eff. Figure 1 shows the preliminary results from binary population synthesis.

Acknowledgements

This work was in part supported by the Natural Science Foundation of China under Grant Nos. 10603013, 10433030, 10521001 and 2007CB815406.

References

Han Z., Podsiadlowski Ph., & Eggleton P. P., 1994, MNRAS, 270, 121
Lejeune T., Cuisinier F., & Buesel R., 1997, A&AS, 125, 229
Lejeune T., Cuisinier F., & Buesel R., 1998, A&AS, 130, 65
Stryker L. L., 1993, PASP, 105, 1081
Xin Y. & Deng L., 2005, ApJ, 619, 824

A Model for Adiabatic Mass-loss

Hongwei. Ge[1,3], R. F. Webbink[2] and Z. Han[1]

[1]National Astronomical Observatories/Yunnan Observatory, Chinese Academy of Sciences, Kunming 650011, China; email: hongwei.ge@gmail.com
[2]Department of Astronomy, University of Illinois, 1002 W. Green St. Urbana, IL 61801, USA
[3]Graduate University of Chinese Academy of Sciences, Beijing 100039, China

Abstract. We describe our work on the development and application of a stellar structure code to compute model sequences representing donor stars in interacting binaries subject to rapid (adiabatic) mass-loss. The donor star is assumed to remain in hydrostatic equilibrium, but no heat flow is allowed. These sequences can be used to define bifurcation sequences in close binary evolution, and to circumscribe possible survivors of common envelope evolution.

Keywords. stars: evolution, stars: interiors, stars: mass loss, binaries: close

1. Introduction

Mass transfer is the defining feature of close binary evolution. That mass transfer typically occurs on one of three characteristic time scales. If the tidal (Roche) lobe of the donor star expands sufficiently rapidly in response to mass transfer to permit that star to remain in thermal equilibrium, mass transfer proceeds on a nuclear time scale. At the opposite extreme, mass transfer may be so rapid that the hydrostatic expansion of the donor star drives it beyond its Roche lobe, even in the absence of thermal relaxation in its interior; in this case, mass transfer proceeds on a dynamical time scale, and typically results in common envelope evolution or binary merger. In the intermediate case, donor stars stable against dynamical mass transfer, but unable to remain in thermal equilibrium, mass transfer proceeds on a thermal time scale.

We investigate the conditions for dynamical mass transfer in binary stars by constructing model sequences subject to adiabatic mass loss, that is, sequences in which the entropy (and composition) profiles are determined by the initial model of the sequence, and held fixed throughout the sequence. The initial model may be selected from any point in an evolutionary sequence for the donor star. Idealized mass loss sequences of this sort, treating donor stars as simple, compound, or centrally-condensed polytropes, were published by Hjellming & Webbink (1987); detailed adiabatic sequences, employing full stellar models, were first calculated by Hjellming (1989a,b). Our goal is not only to explore threshold conditions for common envelope evolution in greater detail, but also to use our results to limit possible survivors of common envelope evolution, relevant to, e.g. SNe Ia, hot subdwarfs (Han et al., 2002).

2. Adiabatic Mass-loss

In the limit that mass loss from the donor star in an interacting binary is rapid compared to the thermal time scale of that star, the asymptotic response of that donor is adiabatic. Heat flow is negligible, but, to a good approximation, the donor remains in hydrostatic equilibrium (except very near the L_1 point). We can be treat this situation in terms of mass loss from a single star.

We construct an evolutionary sequence of single-star models of a given mass and metallicity. Any model along this sequence may be selected as the starting model for an adiabatic mass loss sequence; that model would correspond to such a donor as it first fills its Roche lobe. The profiles in specific entropy, $s(m, X)$, and composition, $X(m)$ are held fixed along an adiabatic sequence as mass is removed from the surface. Only one other thermodynamic constraint is then needed to specify completely the state of the gas at any point in the interior of a model. That constraint is provided by the local pressure, $P(m)$, through the equation of hydrostatic equilibrium. It is sufficient, then, to solve only the equations for mass continuity and hydrostatic equilibrium, subject to the usual boundary conditions on central mass and surface optical depth, to generate an adiabatic mass loss model. The local density and temperature through the interior of the model are implicitly defined in terms of the local specific entropy, pressure, and composition.

The model solution can also be used to reconstruct the luminosity profile through the interior of a mass-losing model. Given the local state of the gas, we write the entropy gradient (2.1) in terms of the pressure and composition gradients:

$$\mathrm{d}s/\mathrm{d}m = c_{\mathrm{P}}(\nabla - \nabla_{\mathrm{a}})\,\mathrm{d}\ln P/\mathrm{d}m + \partial s/\partial X \cdot \mathrm{d}X/\mathrm{d}m. \qquad (2.1)$$

During adiabatic mass loss, $|\mathrm{d}\ln P/\mathrm{d}m|$ increases at a each mass layer, as that layer is brought toward the stellar surface, and pressure decreases. The ambient temperature gradient, ∇, is therefore driven toward the adiabatic temperature gradient, ∇_{a}, since $\mathrm{d}s/\mathrm{d}m$ and $\mathrm{d}X/\mathrm{d}m$ are fixed. In radiative zones, the stellar luminosity at mass coordinate m then follows directly from $\nabla_{\mathrm{r}} = \nabla$. In convective zones, it is necessary to invert mixing-length theory to calculate separately the contributions of radiative and convective energy fluxes, but this is straightforward algebraically. The luminosity profile provides valuable insight into the thermal relaxation of a donor star under rapid mass loss.

3. Computing Method

Our adiabatic mass loss code is written in FORTRAN95. It is based on Eggleton's (1971, 1972, 1973) stellar evolution code, updated as described by Han *et al.* (1994) and Pols *et al.* (1995, 1998). Initial stellar models are calculated using Evolve ZAMS (Paxton 2004), which is derived from Eggleton's code.

Acknowledgements

RFW acknowledges support from US National Science Foundation grant AST 0406726 to the University of Illinois; ZH acknowledges support from NSFC grant No. 10521001.

References

Eggleton, P. P., 1971, *MNRAS* 151, 351
Eggleton, P. P., 1972, *MNRAS* 156, 361
Eggleton, P. P., 1973, *MNRAS* 163, 179
Han, Z., Podsiadlowski, P., & Eggleton, P. P. 1994, *MNRAS* 270, 121
Han, Z., Podsiadlowski, P., Maxted, P. F. L., Marsh, T. R., & Ivanova, N. 2002, *MNRAS* 336, 449
Hjellming, M. S. 1989a, *Space Sci. Revs* 50, 155
Hjellming, M. S. 1989b, Ph.D. Thesis, University of Illinois at Urbana-Champaign
Hjellming, M. S. & Webbink, R. F. 1987, *ApJ* 318, 794
Paxton, B. 2004, *PASP* 116, 699
Pols, O. R., Tout, C. A., Eggleton, P. P., & Han, Z. 1995 *MNRAS* 274, 964
Pols, O. R., Schröder, K.-P., Hurley, J. R., Tout, C. A., & Eggleton, P.P. 1998 *MNRAS* 298, 525

ns in the 21st Century
The history of KZ Hya and its unseen companions

S. Y. Jiang

National Astronomical Observatories, Chinese Academy of Sciences, Beijing 100012, P.R. China

Abstract. KZ Hya is a short-period high amplitude metal pool population II pulsating variable. Its spectral type is B9-A7 III/IV. Its average effective temperature is 7640K. But its mass is only 0.97 solar mass. From normal stellar evolution and H-R diagram, we can not get such a solar mass star at post main sequence stage with so high effective temperature and so early type spectra. We observe this star since 1984 till now, 23years past. Finally we prove it is inside a binary with at least 2 unseen companions. The most massive companion has mass larger than 0.76 solar mass, mostly may be 0.99 to 3.99 solar mass. That means this companion must be a massive white dwarf. The distance between tow companions is about 10 AU. If the companion is white dwarf, this binary are fairly inside the nebula. This system is very old, older than 7.59 billion years. The nebula should be already diluted to very low density so that we cant see the nebula directly. As its spectra type is B9III/VI at some time of maximum light and the visual absolute magnitude is 2.78, about 2 magnitudes higher than our sun. We can image that at the end of AGB stage of the companion, the strong fast winds from hot central core push away the outer atmosphere of KZ Hya. Later KZ Hya absorbed a part of Helium rich material from the companion. This will cause hydrogen content X decrease from 0.75 to about 0.62. Then KZ Hya looks like a hot post main sequence star

Keywords. Stars: variable: other, (stars:) binaries: general, stars: white dwarfs

1. Introduction

KZ Hya = HD 94033 = SAO 179271 (V = 9.498?10.243) is one of the 13 known field SX Phe stars. Its pulsating period is 0.05951d. Due to high velocity and deficiency in metals (Fe/H = −2.40, Z = 0.0001), KZ Hya is clearly belong to Population II. Its spectral type is B9-A7 III/IV. Its average effective temperature is 7640K. But its mass is only 0.97 solar mass. ¿From normal stellar evolution and H-R diagram, we can not get such a solar mass star at post main sequence stage with so high effective temperature and so early type spectra. Yang *et al.* (1985) get a periodic variation of the O-C values of its maximum light time with period about 9 years and a binary hypothesis was proposed. Liu *et al.* (1991) confirm this hypothesis. Fu *et al.* (2008) renewed the binary solution and suggest its unseen companion is a massive white dwarf with mass larger than 0.86 solar mass. Here we try to use a binary evolution model to solve this problem.

2. The evolution of unseen companion

For single star, to get a white dwarf with mass between 1.399 to 0.999 solar mass for Z = 0.0001, its zero age main sequence mass must be between 6.0 to 3.0 solar mass. Its Hertzsprung Gap Time is 61 to 269 MY. Mass still is 6.0 to 3.0 solar mass. The core helium burning time is 62 to 271.3 MY, mass change to 5.999 to 2.999 solar mass. The first AGB time is 69 to 310.1 MY, mass change to 5.96 to 2.97 solar mass. The second AGDB time is 69.1 to 310.6 MY, with mass 5.933 to 2.94 solar mass. Then arrive to

C/O white dwarf stage, the age is 70 to 312.7 MY, with mass of 1.399 to 0.999 solar mass. Now the age is 7590 MY, the planetary nebula is already disappeared and the core white dwarf must be cooled to about 8000 K with 10 magnitude fainter than KZ Hya.in V band, so we cant see it in normal way.

3. The evolution of KZ Hya

Suppose its zero age main sequence mass is 0.80 solar mass with Z = 0.0001. When the unseen companion evolves to C/O white dwarf, large amount of Helium rich masses lost out by fast wind. Parts of them will absorbed by KZ Hya and the distance between companions enlarged to about 10 AU as today we observed. So the mass of KZ Hya become todays 0.97 solar mass. This will change the X content from 0.75 to about 0.62. As a main sequence star with X as low as 0.62, its evolution is quite special, the surface effective temperature is much higher than normal solar mass star and its radius also larger to about 1.51 solar radius. This is what we expected to explain the special observed parameters of KZ Hya.

4. The binary evolution

Really, for a binary, the evolution is much difference with single star. If we have a zero age binary with masses of 6.5 and 0.80 solar masses respectively. The semi-major axes of main component is 3.3 au 703–704 solar radii , the ellipticity e is 0.495 to 0.5. After 70.5 MY, the main component will fill in its Roche lobe, with mass about 3 solar masses. Then evolve to about 1 solar mass carbon white dwarf and 0.97 solar masses post main sequence star with period of about 10500 days or about 29 years and semi-major axis of about 5 au. The real solution is very sensitive on the input parameter of masses, ellipticity and semi-major axis. A small change can get solution with stage of fill in Roche lobe or without fill in Roche lobe stage and very different final components masses and orbital period. The calculation only can get circle orbit but our observations shown e = 0.29 should be caused by the orbital inclination angle i is not 90 degrees.

5. Conclusion

Now we get result that KZ Hya is evolved from a old binary system Its zero age mass only 0.8 solar mass. Later KZ Hya absorbed 0.17 solar mass of helium from its second time red giant companion and changed its X content from 0.75 to 0.62 and it surface temperature increase to about 7600K in average.

References

Yang, X., Jiang, S., & Guo, Z. 1985, *Chinese Astronomy and Astrophysics*, 9, 324
Liu, Y., Jiang, S., & Cao, M. 1991, *IBVS*, 3606
Fu, J. N., Khokhuntod, P., Rodriguez, E., Boonyarak, C., Marak, K., Lopez-Gonzalez, M. J., Zhu, L. Y., Qian, S. B., & Jiang, S. Y. , 2008, *Astronomical Journal* 135, 1958

The Art of Modelling Stars in the 21st Century
Proceedings IAU Symposium No. 252, 2008
L. Deng & K. L. Chan, eds.

Structure and evolution of W UMa-type systems

Lifang Li[1], Fenghui Zhang[1], Zhanwen Han[1], Dengkai Jiang[1,2] and Tianyu Jiang[1,2]

[1]National Astronomical Observatories Yunnan Observatory, Chinese Academy of Sciences,
P. O. Box 110, Kunming, Yunnan Province, 650011, P.R. China
email: gssephd@public.km.yn.cn .or. llf@ynao.ac.cn

[2]Graduate University of Chinese Academy Sciences, Beijing 100039, China

Abstract. We summarize and discuss our recent works on the structure and evolution of low-mass W UMa-type contact binary stars. Three conclusions are given as followings: (1) The energy transfer is taken place in the radiative region of common envelope of W UMa systems; (2) The magnetic activity level of W UMa systems is weaker than that of non-contact binaries or rapid-rotating single stars; (3) The evolutionary outcome of W UMa systems might be the rapid-rotating single stars, and an average lifetime is derived to be about 7 Gyr for W UMa systems.

Keywords. binaries: close, stars: evolution, stars: rotation, stars: activity.

1. Introduction

W UMa binaries are very common eclipsing binaries in solar neighbourhood and are classified into A- and W-type systems. A-subtype W UMa stars show primary minima resulting from the eclipse of the larger, more massive component, whereas the opposite is true for W-subtype ones. The pioneer works on the structure and evolution of contact binaries were carried out by Lucy (1976) and Robertson & Eggleton (1977).This issue has been recently investigated by many authors (Li, Han & Zhang 2004,2005; Yakut & Eggleton 2005, etc). Serious uncertainties on evolution of contact binaries concern the energy transfer (Kahler 2002) and the evolutionary outcome. In this paper, we summarize our recent works on the structure and evolution of W UMa contact binaries.

2. The models

It is not yet clear where and how the energy is transferred in W UMa systems. Li, Han & Zhang (2004) constructed a conservative model to investigate the energy transfer in W UMa systems using Eggleton's code (Eggleton 1971). The results are shown in Figure 1. It is seen in Fig. 1 that the model exhibits the observational properties of A-subtype systems if energy transfer occurs in the base or the whole of common envelope, and that the model shows the observational properties of W-subtype systems if the energy transfer occurs in the outermost layers. This suggests that the energy transfer in W UMa systems might be taken place in the radiative region of common envelope.

Li, Han & Zhang (2005) investigated non-conservative evolution of low-mass W UMa contact binaries with angular momentum (AM) loss owing to magnetic stellar wind (MSW) (Hurley et al 2002). The time dependence of some quantities are shown in Figure 2. As seen from figure 2 that W UMa contact binaries would merge into a fast-rotating single star owing to Darwin's instability (Eggleton & Kiseleva-Eggleton 2001) after about

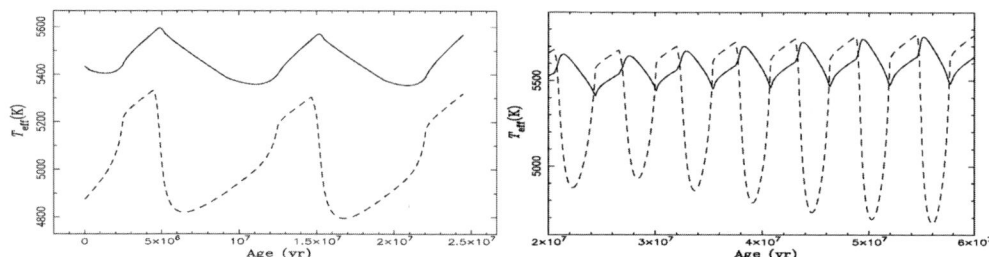

Figure 1. The time dependence of the effective temperature of the primary (solid line) and the secondary (dashed line). Left panel shows the models in which the energy transfer is taken placed in the base or in the whole of the common envelope, and right panel shows the model in which the energy transfer is taken placed in the outmost layers of the common envelope.

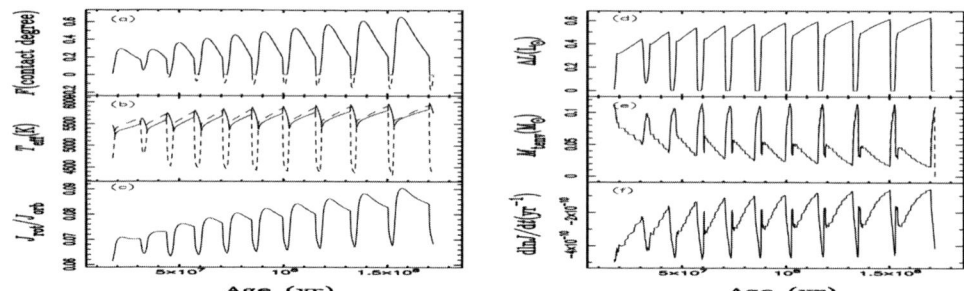

Figure 2. The time dependence of contact degree, F, the effective temperature of the primary (solid line) and the secondary (dashed line), together with the ratio of the rotating angular momentum to the orbital angular momentum, the luminosity transfer rate, the total mass contained in convective envelopes ($M_{\rm tenv}$), and the rate of angular momentum loss (Li, Han & Zhang 2005).

1200 cycles (about 7Gyr), and that the rate of AM loss in contact evolution is much lower than that in non-contact stage, which is very consistent with the result predicted by observations (Stepien *et al.* 2001, and references therein) and a polytropic model of Rucinski (1992). This might be caused by less total mass contained in the convective envelopes of contact binaries, comparing with that of non-contact binaries.

Acknowledgements

This work was partly supported by the Chinese Natural Science Foundation (10673029, 10773026 and 10433030), and by the Yunnan Natural Science Foundation (2007A113M and 2005A0035Q).

References

Eggleton, P. P., 1971, *MNRAS*, 151, 351
Eggleton, P. P. & Kiseleva-Eggleton, L., 2001, *ApJ*, 562, 1012
Hurley, J. R., Tout, C. A., & Pols, O. R., 2002, *MNRAS*, 329, 897
Kähler, H., 2002, *A&A*, 395,899
Li, L., Han, Z., & Zhang, F., 2005, *MNRAS*, 351, 137
Li, L., Han, Z. & Zhang, F., 2005, *MNRAS*, 360, 272
Lucy, B., 1976, *ApJ*, 205, 208
Robertson, J. A. & Eggleton, P. P., 1977, *MNRAS*, 179, 359
Rucinski, S. M., 1992, *AJ*, 103, 960
Stępień, K., Schmitt, H. M. M., & Voges, W., 2001, *A&A*, 370, 157
Yakut, K. & Eggleton, P. P., 2005, *ApJ*, 629, 1055

A Spectroscopic Study of Barium Stars

G. Q. Liu[1,2], Y. C. Liang[1] and L. Deng[1]

[1] National Astronomical Observatories, Chinese Academy of Sciences,
Beijing 100012, P. R. China
email: lgq@bao.ac.cn, ycliang@bao.ac.cn

[2] Graduate University of Chinese Academy of Sciences,
Beijing 100049, P. R. China

Abstract. We present an analysis of eight barium stars, providing their atmospheric parameters ($T_{\rm eff}$, log g, [Fe/H], ξ_t) and chemical abundances, based on the high signal-to-noise ratio and high resolution Echelle spectra. The s-process elements Y, Zr, Ba, La, Eu show obvious overabundance relative to the Sun. And Na, Mg, Al, Si, Ca, Sc, Ti, V, Cr, Mn, Ni show comparable abundances to the Solar ones. The results of theoretical model of wind accretion for binary systems can explain the observed abundance patterns of the neutron capture process elements in these Ba stars, which means that their overabundant heavy-elements could be caused by accreting the ejecta of AGB stars, the progenitors of the present white dwarf companions in the binary systems.

Keywords. Stars: abundances — Stars: atmospheres — Stars: chemically peculiar — Stars: evolution — binaries: spectroscopic

1. Introduction

Barium stars appear as a distinct group of chemically peculiar red giants, which show enhanced features of Ba II, Sr II, CH, CN and sometimes C_2 lines. These Ba stars belong to binary systems and could have accreted the matter ejected by their companions (the former AGB stars and now evolved into white dwarfs) about 1×10^6 years ago through wind accretion, disk accretion, or common envelope ejection. The elements heavier than iron are synthesized in the interior of AGB stars through the slow neutron capture process (s-process) (Liang *et al.* 2000). At present, there is a large sample of Ba stars with measurements of orbital elements, absolute magnitudes and kinematics. However, the corresponding heavy-element abundances have not been obtained from high resolution observations. Therefore, we observed the high resolution and high signal-to-noise ratio spectra of Ba stars to obtain their chemical abundances, using the Coudé Echelle Spectrograph of National Astronomical Observatories (NAOC) mounted on the 2.16 m telescope at Xinglong station (Xinglong, P. R. China).

Table 1. Atmospheric parameters of sample stars and heavy element abundances.

HD	$T_{\rm eff}$	log g	ξ_t	[Fe/H]	[Y/Fe]	[Zr/Fe]	[Ba/Fe]	[La/Fe]	[Eu/Fe]
4395	5477	3.60	1.3	−0.16	0.48	0.44	0.53	—	0.43
180622	4391	2.24	1.5	0.21	0.45	0.42	0.45	0.62	0.52
201657	4284	2.17	1.7	−0.31	0.62	0.67	1.21	1.43	0.63
201824	4552	1.67	1.5	−0.40	0.46	0.60	1.27	1.27	0.66
210946	4577	2.42	1.6	−0.22	0.33	0.68	0.81	0.91	0.21
211594	4490	2.44	1.6	−0.23	0.86	0.71	1.26	1.24	0.60
216219	5553	3.64	1.4	−0.34	—	0.87	1.05	1.10	0.49
223617	4501	2.27	1.5	−0.10	0.50	0.65	0.85	1.04	0.53

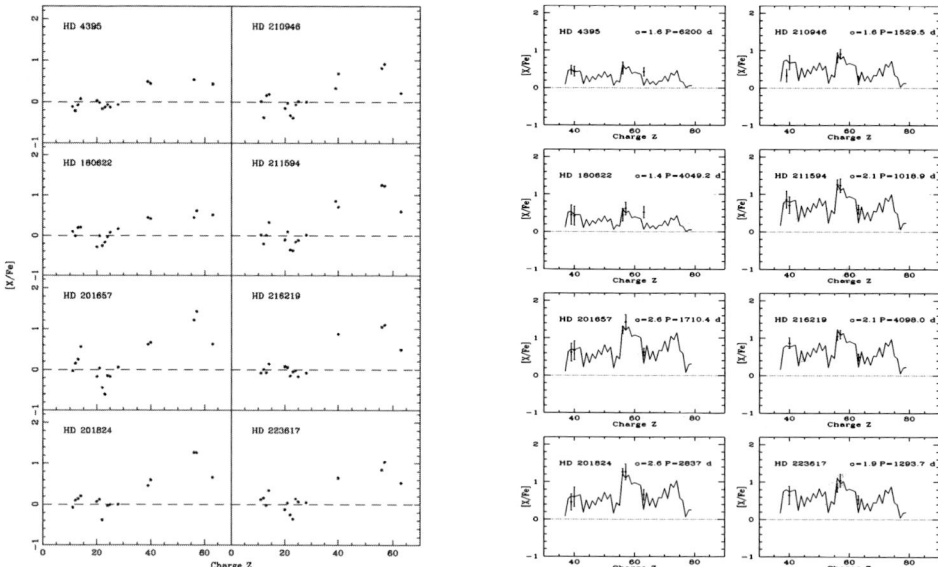

Figure 1. The abundance patterns of sample stars.

Figure 2. This figure shows the comparisons between theoretical abundances and the observed abundances of the sample stars.

2. Results and Conclusions

We obtain the chemical abundances of eight barium stars based on the input atmospheric parameters (Table 1) and the measured EWs of the absorption lines (see details in Liu et al. 2008). The neutron capture process elements Y, Zr, Ba, La, Eu show obvious overabundance relative to the Sun (Figure 1), for example, their [Ba/Fe] values are from 0.45 to 1.27 (Table 1). Other elements, including Na, Mg, Al, Si, Ca, Sc, Ti, V, Cr, Mn, Ni, show comparable abundances to the Solar ones, and their [Fe/H] cover a range from −0.40 to 0.21, which means they belong to disk stars.

The wind accretion model was applied to predict the theoretical heavy element abundances of Ba stars in binary systems, and then compare these theoretical predicts with the observed abundance patterns of our sample stars as shown in Figure 2 (see details in Liang et al. 2003). The predicted results by the model can explain well the observed abundance patterns of s-process elements in sample stars with orbital period longer than 1500 (or 1600) days, which is consistent with the suggestions of Jorissen et al. (1998). It is interesting to notice that the heavy element abundance patterns of two sample stars with $P > 1000$ days, HD 211594 and HD 223617, can also been explained by wind accretion.

Acknowledgements

This work was supported by the Natural Science Foundation of China (NSFC) Foundation under No.10403006 and the National Basic Research Program of China (973 Program) No.2007CB815404, 2007CB815406.

References

Jorissen A., Van Eck S., Mayor M., & Udry, S., 1998, A&A, 332, 877
Liang Y. C., Zhao G., & Zhang B., 2000, A&A, 363, 555
Liang Y. C., Zhao G., Chen Y. Q., Qiu H. M., & Zhang B., 2003, A&A, 397, 257
Liu G. Q., Liang Y. C., Deng L., 2008, ChJAA, be accepted

Evolutionary scenario for W UMa-type stars

K. Stępień[1] and K. Gazeas[2]

[1] Warsaw University Observatory, email: kst@astrouw.edu.pl
[2] Harvard-Smithsonian Center for Astrophysics, email: kgaze@physics.auth.gr

Abstract. An alternative to TRO model of a W UMa-type star is presented in which the binary is past mass exchange with mass ratio reversal. The secondary is hydrogen depleted and both components are in thermal equilibrium. Evolution in contact is driven by orbital angular momentum loss and mass transfer from the secondary to primary component, similarly as it is observed in Algols. Temperature equalization of both components results from an assumed energy transfer by a large scale flow encircling the whole system in the common envelope.

Keywords. stars: activity, binaries: close, stars: evolution, stars: late type

---

W UMa-type stars are believed to originate from close detached binaries which lose angular momentum (AM) via a magnetized wind until a primary overflows its critical Roche lobe and transfers mass to the secondary. Details of the mass transfer (MT) and further evolution of the binary is a matter of debate. Lucy (1976) developed a model with a secondary swollen due to the transfer of a limited amount of mass ($\sim 0.1 - 0.2 M_\odot$). A contact binary is formed but it is out of thermal equilibrium and undergoes thermal relaxation oscillations (TRO, see also Eggleton & Kiseleva-Eggleton 2002). Its further evolution is driven by an increasing luminosity of the evolving primary, which forces the secular MT from the secondary. An alternative model has been developed by Stępień (2004, 2006) who observed that the time scale of orbital AM loss of a cool close binary with an initial period equal to a couple of days is the same as the evolutionary time scale of its primary. Due to this coincidence, the primary reaches its Roche lobe when it is close to, or just beyond TAMS. The model assumes that MT following the overflow of the Roche lobe by the primary leads to a common envelope phase during which mass ratio reversal takes place, much the same way as in Algols. Depending on the initial parameters and details of evolution in the common envelope phase, the binary emerges from that phase either directly as a contact binary or as a very short-period Algol transformed subsequently into the contact binary after the surplus AM is lost.

Neglecting spin AM of both components, compared to orbital AM, the AM loss rate of a close binary by a magnetized wind is given by

$$\frac{dH_{\rm orb}}{dt} = -4.9 \times 10^{41}(R_1^2 M_1 + R_2^2 M_2)/P_{\rm orb}.$$

Here AM is in cgs units, period in days, masses and radii in solar units and time in years. The formula is based on semi-empirically determined AML rate of single, cool stars. The supersaturation effect is allowed for by assuming $P_{\rm orb} \equiv 0.4$ days for periods shorter than 0.4 days.

Together with AM mass is also carried away by magnetized winds. Mass loss rate is often treated as a free parameter, particularly in close binary studies (Eggleton & Kiseleva-Eggleton 2002). We assume, however, that the mass loss rate of each component is the same as of a single star with the same basic characteristics. Mass loss rates of single active stars were recently determined from empirical data by Wood et al. (2002). According to their results, active stars lose mass at a constant rate per surface area, so

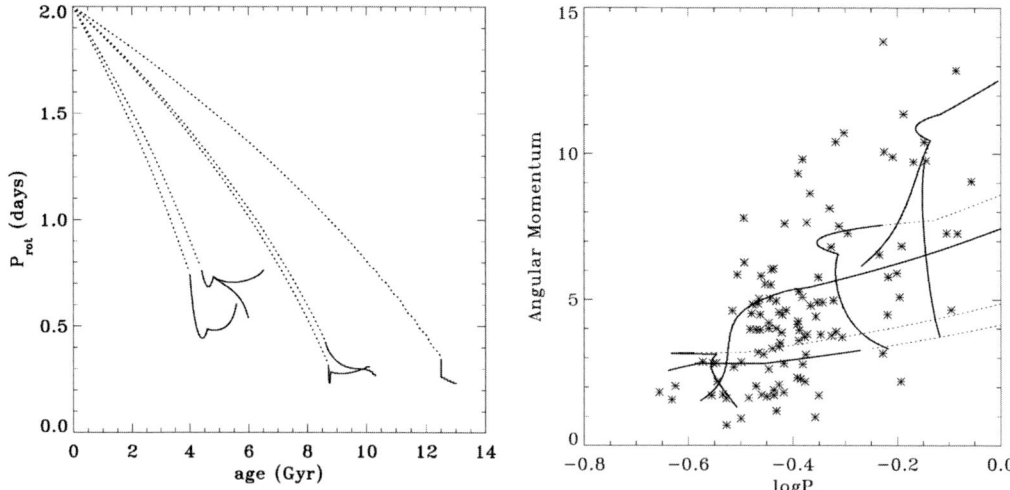

Figure 1. Computed time variations of orbital period of binaries with masses (from left to right): 1.3+1.1, 1.3+1.0, 1.0+0.8, 1.0+0.5 and 0.9+0.45 $M_\odot$ (left) and AM variations of the same binaries compared to the observed values of over 100 W UMa-type stars. Dotted lines denote detached stage and solid lines the contact phase.

that a parametric formula can be used for each component: $\dot{M}_{1,2} = -10^{-11} R_{1,2}^2$, where mass loss rates are in $M_\odot$/year, radii in solar units and numbers denote the respective components.

Once a contact binary is formed, energy transfer from the primary to secondary takes place in a way described e.g. by Kähler (2004). He argued that a large scale flow takes place between the inner and outer Roche lobe. The flow is separated from the convective envelope of the secondary by a radiative layer so that, contrary to what Lucy (1976) assumed, the convective layers of both components do not have identical adiabatic constants.

During evolution in contact a self-regulating mechanism develops in which AM loss makes orbit shrink but the secular MT from the secondary widens it. As a result, MT rate adapts itself to the AM loss rate so that contact between both components is maintained. Model calculations show that the resulting average MT rates are about $3 - 4 \times 10^{-10} M_\odot$/year and typical lifetimes in contact are about 1-1.5 Gyr. Coalescence of both components ends the contact phase. Detailed evolutionary calculations of several sample binaries were obtained by Gazeas & Stępień (2008). Fig. 1 presents period evolution of those binaries starting from the (assumed) initial value of 2 days (left) and comparison of the predicted AM variations with the observed values (right).

References

Eggleton, P. P. & Kiseleva-Eggleton, L. 2002, *ApJ* 575, 461
Gazeas, K. & Stępień, K. 2008, astro-ph0803.0212
Kähler, H. 2004, *A&A* 414, 317
Lucy, L. B. 1976, *ApJ* 205, 208
Stępień, K. 2004, in: A.K. Dupree & A.O. Benz (eds.) *Stars as Suns: Activity, Evolution and Planets*, IAU Symp. No. 219 (San Francisco: Astr. Soc. of Pacific), p. 967
Stępień, K. 2006, *AcA* 56, 199
Wood, B. E., Müller, H. R., Zank, G. P., & Linsky, J. L. 2002, *ApJ* 574, 412

The eclipsing binary IU Per and its intrinsic oscillations

X. B. Zhang†

National Astronomical Observatories, Chinese Academy of Sciences, Beijing, 100012, China
email: xzhang@bao.ac.cn;

Abstract. The results of a long-term time-series photometry of the short-period eclipsing binary IU Per are reported. The observation confirms the intrinsic δ Scuti-like pulsation of the star as discovered by previous authors. A photometric solution for the binary system was carried out with the new data. Based on which, the pure oscillation light variations from the mass-accreting primary component were extracted. A Fourier analysis reveals four pulsation modes. Combining with the photometric solution, a preliminary mode identification was given.

Keywords. binaries: eclipsing – stars: oscillations – stars: individual: IU Per.

1. Introduction

Pulsating stars in eclipsing binary systems are of peculiar interests for the theoretical study of stellar structure and evolution since binarity of these stars provides more useful information of the components. The existence of stellar pulsation in eclipsing binary systems has been noted early in 1970s. However, there are only about 40 of such samples have been found up to date.

The less-studied eclipsing binary IU Per is very probable a new sample consisting of a d Scuti-like pulsating component as reported by Kim *et al.* (2005). With a period of 0.8570257 day (Kreiner, 2004), it is one of the pulsating eclipsing system with the shortest orbital among all the samples. To check the pulsation nature of the binary system, we have observed the star photometrically for long time. It presents here the preliminary results of the observations.

2. Observations

All the observations were carried out at the Xinglong Station of NAOC. The fast photometry in 2006 was made with a 4-channel photoelectric photometer mounted on the 85-cm reflector. A single narrow band b filter was used. Good data were collected on 3 nights. The V-band CCD photometry of IU Per was performed from Nov.03 to 15, 2007 by using the APM 50cm telescope. The data were taken with a 1320?1320 CCD camera which provides a field size of 20.2 arcmin with a scale of 0.92 arcsec/pixel.. Exposure times were set from 10 to 15 seconds according to the weather conditions. Useful data were collected on 9 nights. A total number of 8400 frames were obtained. Adopting two stars, GSC0008-743 and GSC0008-949, as comparison and check stars, respectively, the differential photometry was extracted. The photometry precision is generally better than 0.005 mag. Fig. 1 presents an example of the time-series photometry. Inspecting the light curves. The short-term light variations in addition to the eclipsing light changes can be clearly seen. It confirms the discovery of Kim *et al.* (2005)

† Present address: National Astronomical Observatories., 20A Datun Road, Beijing, 10012, China.

Table 1. Results of the Fourier analysis

f_i	Frequency (c/d)	Amplitude (mmag)	S/N	Q (×100)	Mode
f_1	43.1314±0.0011	3.08±0.07	11.26	1.194	6H
f_2	47.7959±0.0017	1.94±0.07	8.16		3×F
f_3	15.9031±0.0031	1.06±0.07	3.81	3.239	F
f_4	40.9849±0.0036	0.98±0.07	3.23	1.257	5H

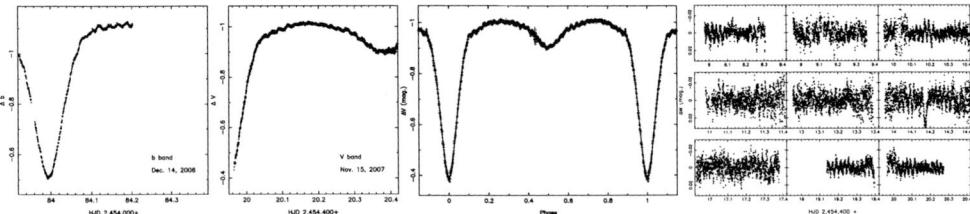

Figure 1. a, Real-time light curves of IU Per; b, The phased V light curve along with the theoretical fittings; c, The pulsating light variations extracted from the CCD data

3. Light curve analysis

A total of 7 new epochs of light minima were detected. With which, the orbital period of the system was renewed as 0.857025 days. The phased V-band light curve was shown in Fig. 1 along with a theoretical fitting based on a photometric solution by using the 2003 version of the W-D code. The light curve synthesis yields a semi-detached configuration for the system with the secondary filling its own Roche lobe. The mass ratio of the system is derived to be about 0.27. According to the behavior of light variations at the eclipses, it is identified that the intrinsic oscillations are very probably from the mass-accreting primary component.

Based on the photometric solution, the theoretical time-series light curves for each star were computed. Following $l_{obs.} = l_1 f_{pul.} + l_2$, the intrinsic pulsation light variations from the primary star were then extracted as shown also in Fig. 1. A power spectral analysis was performed with the code Period04 to study the pulsation nature. It reveals four reliable pulsating frequenciess. The results are given in Table 1. From the results of photometric solution, the mean density of the primary star was computed as $\rho_1/\rho_\odot = 0.265$. With this value, a prilimary mode identitication was given.

4. Conclusions

The δ Scuti-like intrinsic oscillations of the eclipsing binary IU Per were confirmed. The star was proved to be a semi-detached algol system. With a new method, the intrinsic oscillation light variations from the mass-accreting component were extracted and the pulsating modes were identified. It suggests that IU Per could be a new member of the oscillating EA stars of peculiar interest for its short orbital period and the multi-periodic oscillations.

Reference

Kim, S. L., Lee, J. W., Koo, J. R., Kang, Y. B., & Mkrtichian, D. E., 2005, IBVS, No. 5629

The duration properties of Swift Gamma-Ray Bursts

Z.-B. Zhang[1,2] and C.-S. Choi[1]

[1] International Center for Astrophysics, Korea Astronomy and Space Science Institute,
36-1 Hwaam, Yusong, Daejon 305-348, South Korea;
email: z-b-zhang@163.com

[2] Yunnan Observatory, National Astronomical Observatories, Chinese Academy of Sciences,
P. O. Box 110, Kunming 650011, China

Abstract. We report the systematic analysis of the durations for Swift gamma-ray bursts (GRBs) and compare the results with those of pre-Swift data. We show that the durations of Swift bursts also have two log-normal distributions that are clearly divided at $T_{90} = 2$ s. Their intrinsic durations also show a bimodal distribution but shift systematically toward the smaller value compared with the observed one. This study confirms the spectra of short GRBs are in general harder than the long GRBs and shows that this trend becomes weak in the source frame.

Keywords. gamma-rays: bursts – gamma rays: theory

1. Introduction

Based on an analysis of durations using initial BASTE data, Kouveliotou et al. (1993) divided gamma-ray bursts (GRBs) into two classes, i.e., long GRBs (LGRBs) with $T_{90} > 2$ s and short GRBs (SGRBs) with $T_{90} < 2$ s. Koshut et al. (1996) pointed out that the observed duration distribution may vary with instruments. It is therefore necessary to investigate if there exists a new GRB class and/or what physical factors produce such properties (Gehrels et al. (2004)). In this study, we focus on the relevant issues using the updated Swift data.

To study the intrinsic properties of GRBs, we selected six data sets, namely s1 - s6. As of 2007 July 1, Swift has detected 75 LGRBs (s1) and 20 SGRBs (s2) with known duration and redshift. 44 GRBs from s1 also have available E_p values and constitute our sample s3. In s2, only 11 sources have a measured E_p and are employed to build the sample s4. For the remaining 9 bursts in s2 without the measured redshifts, we assigned a redshift value of $z = 0.5$ to the 9 bursts, approaching the median redshift of $z = 0.4$, as assumed by Norris and Bonnell (2006). In our fifth sample set, s5, we include 48 pre-Swift LGRBs whose z and T_{90} are already measured, in which 18 sources, less than half ($\sim$38%) of the 48 pre-Swift bursts, are detected by the BATSE mission and constitute our sample s6. The detailed sample selection can be referred to Zhang & Choi (2008).

2. Results

2.1. Observed T_{90} distribution

Right panel in Figure 1 shows the T_{90} distribution for the 95 Swift GRBs, which include s1 and s2 samples. The best fit with a two-lognormal function gives the center values ($T_{90,p1} = 0.28$ s and $T_{90,p2} = 42.83$ s) and the widths ($w_1 = 19.05$ s and $w_2 = 18.20$ s) with the reduced Chi-square $\chi^2/dof = 0.67$, which are roughly consistent with those calculated from the BATSE data (McBreen et al. 1994; Meegan et al. 1996; Paciesas et al. 1999; Horváth 2002; Nakar 2007). The superposed function has a minimum around 2 s as found by Kouveliotou et al. (1993), indicating that the Swift sources are also

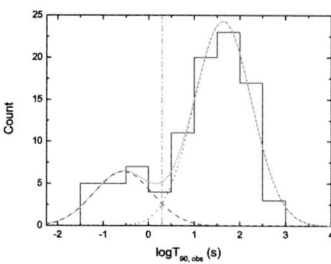 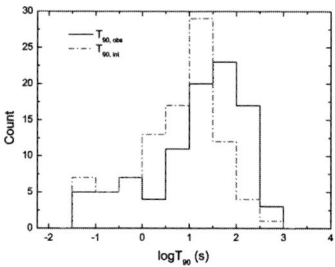

Figure 1. *Left panel:* shows the bimodal distribution of durations for the 95 GRBs (s1 and s2; histogram) and two-component log-normal fit to the data (solid line). The GRBs are divided into two classes at $T_{90} \simeq 2$ s (vertical line): LGRBs (dotted line) and SGRBs (dashed line). *Right panel* shows a comparison between the observed (solid line the same as Fig. 1) and the intrinsic (dashed-dotted line) T_{90} distributions.

divided into two classes, SGRBs and LGRBs, although the Swift is more sensitive to long soft bursts than the BATSE (Band 2006; Gehrels et al. 2007).

2.2. Intrinsic T_{90} distribution

The increasing of the fraction of GRB sources with higher redshift may lead to an evident discrepancy of the intrinsic T_{90} distributions between Swift and pre-Swift. The transformation of T_{90} from observer frame to source frame is generally expressed as $T_{90,int} = T_{90}/(1+z)^{\omega}$, in which $\omega = 1$ has been used, corresponding the case of no energy stretching. As shown in Figure 1 (right panel), the $T_{90,int}$ has a bimodal distribution and is significantly shifted toward shorter durations than the observed one. The best fit with a two-lognormal function gives two centers ($T_{90,p1} = 0.13$ s, $T_{90,p2} = 12.30$ s) and two widths ($w_1 = 10.96$ s and $w_2 = 17.38$ s) with $\chi^2/\text{dof} = 0.92$, indicating that the distribution of $T_{90,int}$ is indeed bimodal but systematically narrower and shifted towards low values of duration in comparison to the observed one.

3. Conclusions

We find that Swift T_{90} have two-lognormal distributions divided clearly at $T_{90} \simeq 2$ s. This implies that the classification in terms of duration is unchanged from pre-Swift to the Swift era. Swift $T_{90,int}$ also show a bimodal distribution but shifted systematically toward the smaller value and the distribution exhibits a narrower width relative to the observed one. In addition, the trend of LGRBs with a relatively softer spectrum largely weakens in the source frame, relative to SGRBs.

References

Balázs, L. G., et al. 2003, *A&A*, 401, 129
Band, D. L. 2006, *ApJ*, 644, 378
Gehrels, N. et al. 2004, *ApJ*, 611, 1005
Gehrels, N., Cannizzo, J. K., & Norris, J. P. 2007, *NJPh*, 9, 37
Horváth, I. 2002, *A&A*, 392, 791
Kouveliotou, C., et al. 1993, *ApJ* 413, L101
Koshut, T. M., et al. 1996, *ApJ*, 463, 570
McBreen, B., Hurley, K. J., Long, R. et al. 1994, *MNRAS*, 271, 662
Meegan, C. A., et al. 1996, *ApJS*, 106, 65
Nakar, E. 2007, *PhR*, 442, 166
Norris, J. P. & Bonnell, J. T. 2006, *ApJ*, 643, 266
Paciesas, W., Meegan, C., Pendleton, G., et al. 1999, *ApJS*, 122, 465
Zhang, Z. B. & Choi, C. S. 2007, *A&A*, 484, 293

The energy transfer in W UMa binary stars

Dengkai Jiang[1,2], Jiangcheng Wang[1], Zhanwen Han[1]
Tianyu Jiang[1,2], and Lifang Li[1]

[1] National Astronomical Observatories/Yunnan Observatory, Chinese Academy of Sciences,
Kunming 650011, China
email: jiangdengkai@hotmail.com

[2] Graduate University of Chinese Academy Sciences, Beijing 100039, China

Abstract. The properties of W UMa binary stars are studied based on the well-determined physical parameters of 132 W UMa systems. It is found that the energy transfer rate has a maximum value at q∼0.58. The relation between the energy transfer rate and the temperature deviation is also investigated, and the temperature of the secondary component is related to the energy transfer rate.

Keywords. stars: evolution—(stars:) binaries: close—(stars:) binaries: general

1. Introduction

W UMa binary stars are very common eclipsing variables in which both components are overflowing their inner Roche lobe surfaces. The surface temperatures of two components of W UMa binary stars are nearly equal, though the masses of two components are often greatly different. The over-luminosity of the secondaries is the result of the energy transfer between the two components within a convective envelope (Lucy 1968). Mochnacki (1981) suggested that the energy transfer rate of W UMa binary star depends on the mass ratio. Li, Han & Zhang (2004) discussed the region of energy transfer in the common envelope of W UMa binary stars, and showed that the energy transfer may take place in the radiative region of common envelope. On the base of the well-determined physical parameters of 132 W UMa binary stars collected from the literature, the energy transfer of W UMa binary stars has been analyzed.

2. The outcome of W UMa binary stars

The energy transfer rate can be calculated by

$$U = \frac{st^4 - q^\alpha}{1 + q^\alpha} \qquad (2.1)$$

where $s = S_2/S_1$, $t = T_{2e}/T_{1e}$; $S_{1,2}$, $T_{1,2e}$ are the surface area and the effective temperature of the components, respectively (Mochnacki 1981). The distribution of the energy transfer rate vs the mass ratio of W UMa binary stars is shown in the left panel of Figure 1. It is seen that the energy transfer rate of W-types is greater than that of A-types (Li et al 2008). The energy transfer rate of W UMa binary stars in fact increases with increasing mass ratio if $q \sim 0.6$ as Wang (1994). But when the q is greater than this value, the energy transfer rate is decreased quickly. We can obtain a polynomial relation between the energy transfer rate and mass ratio by the least-squares fitting as the following,

$$U = -3.634q^3 + 2.899q^2 + 0.309q + 0.0769 \qquad (2.2)$$

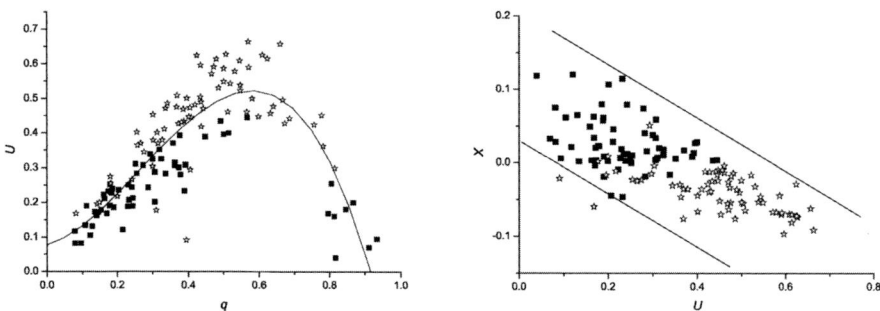

Figure 1. *Left panel*: The energy transfer rate as a function of mass ratio of W UMa binary stars. The solid line represents the Polynomial fit. *Right panel*: Relation of the energy transfer rate and the temperature deviation. Full squares:A-types; open stars:W-types.

according to equation (2.2), the energy transfer rate of W UMa binary stars peaks at a mass ratio of about 0.581.

The temperature deviation is defined as

$$\chi = \frac{T_{1e} - T_{2e}}{T_{2e}} \qquad (2.3)$$

where T_{1e} and T_{2e} are the effective temperatures of the primary and the secondary, respectively. The relation between the energy transfer rate and the temperature deviation is shown in the right panel of Figure 1. The W UMa binary stars seem to load in a strip limited by two solid lines, and there is a tendency for a decreasing energy transfer rate with increasing temperature deviation. It is clear that the temperature deviation is related to energy transfer rate. This indicates that the temperature of the secondary component increases with increasing energy transfer rate, or even exceed the temperature of the primary after the energy transfer rate is high enough.

3. Summary

We present the energy transfer of W UMa binary stars, and show the relation between the energy transfer rate and the mass ratio, together with the relation between the energy transfer rate and the relative temperature deviation. This study indicates that there is a peak of the energy transfer rate at q = 0.58. The temperature of the secondary component is correlated with the energy transfer rate.

Acknowledgements

This work was partly supported by the Chinese Natural Science Foundation (10673029, 10773026, 10433030 and 10521001), and by the Yunnan Natural Science Foundation (2007A113M and 2005A0035Q).

References

Li, L., Zhang F., & Han Z., 2004, *MNRAS*, 351, 137
Li, L., Zhang, F., Han, Z., Jiang, D., & Jiang, T., 2008, *MNRAS*, inpress
Lucy, L. B., 1968, *ApJ*, 151, 1123
Mochnacki S., 1981, *A&A*, 245, 650
Wang, J. M., 1994, *AAnn Astropys.pJ*, 434, 277

The Effect of Binary Interactions in Infrared Passbands

F. Zhang, L. Li and Z. Han

National Astronomical Observatories/Yunnan Observatory, Chinese Academy of Sciences, PO Box 110, Kunming, Yunnan Province 650011, China
email: gssephd@public.km.yn.cn or zhangfh@ynao.ac.cn

Abstract. We present the integrated J, H, K, L, M and N magnitudes and the colours involving infrared bands, for an extensive set of instantaneous-burst binary stellar populations (BSPs) by using evolutionary population synthesis (EPS). By comparing the results for BSPs *WITH* and *WITHOUT* binary interactions we show that the inclusion of binary interactions makes the magnitudes of populations larger (fainter) and the integrated colours smaller (bluer) for $\tau \geqslant 1\,\mathrm{Gyr}$. Also, we compare our model magnitudes and colours with those of Bruzual & Charlot (2003, hereafter BC03) and Maraston (2005, hereafter M05). At last, we compare these model broad colours with Magellanic Clouds globular clusters (GCs) and Milky Way GCs. In $(V-R)$–[Fe/H] and $(V-I)$–[Fe/H] diagrams it seems that our models match the observations better than those of BC03 and M05.

Keywords. infrared: general, binaries: general, stars: evolution, galaxies: clusters: general.

Introduction In previous paper (Zhang *et al.* 2005) we took into account binary interactions (BIs) in evolutionary population synthesis (EPS) models, presented the integrated $U-B$, $B-V$, $V-R$ and $V-I$ colours of binary stellar populations (BSPs), while did not give the infrared magnitudes and colours because larger fluctuations exist. However, these results in infrared passbands are very important in EPS models because the infrared light can reflect the metallicity of populations and the visible/infrared colours are the candidates of breaking the degeneration between age and metallicity.

Results We present the infrared integrated magnitudes and colours for BSPs. The ages of BSPs are in the range 1-15 Gyr, the metallicities are in the range $0.0001 - 0.03$.

In Fig. 1 we present the bolometric magnitude M_{BOL}, K magnitude, $B-V$ and $V-K$ colours at $Z = 0.02, 0.004$ and 0.0001 for BSPs *WITH* and *WITHOUT* BIs, the results of Bruzual & Charlot (2003, hereafter BC03) using Salpeter (1955, hereafter BC03-S) and Chabrier (2003, hereafter BC03-C) IMFs, the results of Maraston (2005, hereafter M05) using Salpeter (1955, hereafter M05-S) and Kroupa (2001, hereafter M05-K) IMFs, and the recent $V-K$ of Bruzual (2007, hereafter B07) at solar metallicity. By comparison we see that **(i)** the magnitudes and colours of BSPs *WITH* BIs are greater (fainter) and smaller (bluer) than those *WITHOUT* BIs, respectively. **(ii)** The magnitudes of BC03-S, BC03-C, M05-S and M05-K are greater than ours. **(iii)** The shape of the evolutionary curves of these colours is significantly different. In Fig. 2 we compare the model colours with Magellanic Clouds globular clusters (GCs) with the type of Searle, Wilkinson & Bagnuolo(1980, hereafter SWB) in the range of 3-7 and the young star clusters in the merger remnant galaxy NGC 7252 in $(B-V)$.vs.$(U-B)$ and $(B-V)$.vs.$(V-K)$ diagrams. It shows that the BC03 and our models agree with the observations in $(B-V)$.vs.$(U-B)$ diagram; while in $(B-V)$.vs.$(V-K)$ diagram larger discrepancies exist among models. In Fig. 3 we compare the model broad colours with Milky Way GCs in colour-metallicity

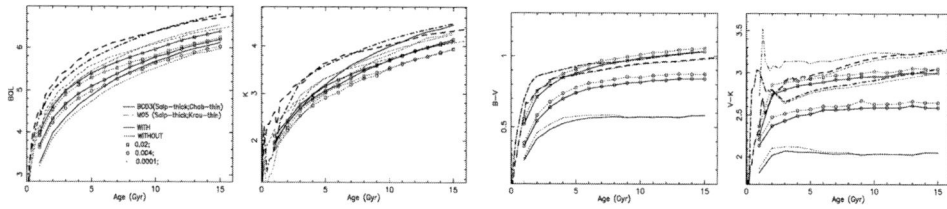

Figure 1. The bolometric magnitude ($M_{\rm BOL}$), K magnitude, $B-V$ and $V-K$ colours for BSPs *WITH* and *WITHOUT* BIs at $Z = 0.02, 0.04$ and 0.0001. Also shown are the results of BC03-S, BC03-C, M05-S and M05-K at solar metallicity. In $V-K$ diagram the recent results of B07 are shown (dash-dot-dot-dot).

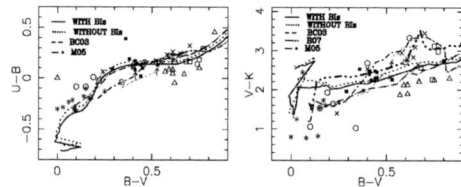

Figure 2. $B-V$ versus $U-B$, $B-V$ versus $V-K$ colours of star clusters. The different symbols represent Magellanic Clouds GCs with the SWB type in the range 3-7. Solid rectangles show the young star clusters in the merger remnant galaxy NGC 7252. The full and dotted lines show the evolution of BSPs *WITH* and *WITHOUT* BIs at $Z = 0.01$, the ages of BSPs are greater than a few Myr. The dashed lines are the BC03-S (thick) and BC03-C (thin) models at $Z = 0.008$, the dot-dash lines are the M05-S (thick) and M05-K (thin) models at $Z = 0.01$ with age $\log \tau > 8$ yr, respectively. In $V-K$ diagram the values of B07 at solar metallicity are also shown (dash-dot-dot-dot).

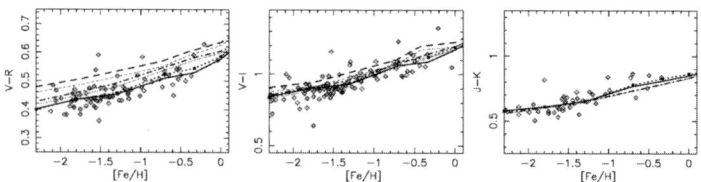

Figure 3. Comparison of colours with data of Milky Way GCs. The lines represent the same models as in Fig. 2. All models have the same age $\tau = 13\,\mathrm{Gyr}$

diagrams. It shows that our models match the observations better than M05 and BC03 in $(V-R)$-[Fe/H] and $(V-I)$-[Fe/H] diagrams.

Acknowledgments This work was funded by the Chinese Natural Science Foundation (Grant Nos 10773026, 10673029, 10433030 & 10521001) and by Yunnan Natural Science Foundation (Grant Nos 2005A0035Q & 2007A113M).

References

Bruzual, G. A. 2007, *IAUs, Stellar populations as Building Blocks of Galaxies* 241, 125 (**B07**)
Bruzual, G. A. & Charlot, S. 2003, *MNRAS* 344, 1000 (**BC03**)
Chabrier, G. 2003, *PASP* 115, 763
Kroupa, P. 2001, *MNRAS* 322, 231
Maraston, C. 2005, *MNRAS* 362, 799 (**M05**)
Salpeter, E. E. 1955, *ApJ* 121, 161
Searale, L., Wilkinson, A., & Bagnuolo, W. G. 1980, *ApJ* 239, 803
Zhang, F., Han, Z., Li, L., & Hurley, J. R. 2005, *MNRAS* 357, 1088

Secular period decreasing of 17 detached chromospherically active binaries

C.Q.Luo[1]†, Y.P. Luo[1], X.B. Zhang[2], L.C. Deng,[2] Z.Q. Luo[1], S.Z. Yang[1]

[1] Institute of Theoretical Physics, China West Normal University, IN Nanchong Sichuan, 637002,China
email: changqingluo@126.com; luoyangping789@163.com; zqluo@tom.com; szyangcwnu@126.com

[2] National Astronomical Observatories, Chinese Academy of Sciences, IN Beijing, 100012, China
email: xzhang@bao.ac.cn; licai@bao.ac.cn.

Abstract. The long-term orbital period changes of detached chromospheric active binaries were surveyed. 17 of such systems are found to be undergoing secular period decreasing with the rates (dP/dt) of -3.05×10^{-9} to -3.77×10^{-5} days per year. The longer the orbital period, the more rapidly the period decreases. Following Stepien (1995), the period decreasing rate due to the angular momentum loss (AML) caused by magnetic wind is computed for each system. A comparison shows that the observed dP/dt's are obviously higher than that of the theoretical predictions by 1-3 orders of magnitude. It suggests that the magnetic wind is not likely the determinant mechanism driving the AML in close binaries.

Keywords. Detached active binaries, period decreases, magnetic wind.

1. Introduction

Angular momentum loss (AML) plays an important role in the evolution of binary stars. AML via magnetized wind is usually thought as the main mechanism for late-type close binaries. Applying the observational results of spin-down of single stars, Stepien (1995) had deduced a formula to compute the AML via magnetic wind in close binary systems. With which he had given a good interpretation on the frequencies and formation time-scale of contact systems evolved from detached binaries. However, a comparison study of this formula to the virtual AML observed in detached late-type binaries is still lack. We report here a period study of active detached binaries. The results are discussed comparison with Stepien's model.

2. The working sample and data

A large number of detached late-type binaries which have definite physical parameters (masses and radii) were compiled as the preliminary candidates. Based on the data collected by the database of *BBSAG*, *AAVSO* and *VAR.ASTRO*, the behavior of period changes of all these candidates were surveyed. 17 systems, which have sufficient observed times of light minima and present strict long-term orbital period decreases, were finally included in the program sample. The basic data of the 17 stars were given in Table 1. The times of minima used for period study are mainly taken from the two database mentioned above.

† Present address: National Astronomical Observatories, A20 Datun Road,Chaoyang District,Beijing, 10012,China.

Table 1. Physical parameters of the six CABS and orbital period decrease rates

Stars	M_1 ($M_\odot$)	M_2 ($M_\odot$)	R_1 ($R_\odot$)	R_2 ($R_\odot$)	Orbital Period (days)	dP/dt(Obs) (days/yr)	dP/dt(Com) (days/yr)
RT And	1.230	0.910	0.92	1.26	0.62892820	-5.15×10^{-8}	-3.38×10^{-10}
WY Cnc	1.071	0.530	0.58	0.93	0.82936767	-8.66×10^{-8}	-1.81×10^{-10}
MM Her	1.270	1.199	2.89	1.56	7.96032045	-7.96×10^{-7}	-6.08×10^{-10}
RT Lac	1.482	0.600	4.81	4.41	5.07392578	-1.29×10^{-6}	-3.88×10^{-9}
UV Psc	0.991	0.758	0.83	1.11	0.86104648	-3.18×10^{-8}	-2.75×10^{-9}
BH Vir	1.020	1.001	1.11	1.25	0.81687110	-1.07×10^{-8}	-3.78×10^{-10}
AR Lac	1.260	1.130	2.72	1.52	1.98315308	-1.20×10^{-6}	-9.02×10^{-10}
AR Mon	2.613	0.800	10.8	14.2	21.0710678	-2.96×10^{-5}	-1.22×10^{-8}
CM Dra	0.231	0.213	0.23	0.25	1.26838898	-4.57×10^{-7}	-3.70×10^{-11}
ER Vul	1.099	1.051	1.08	1.11	0.69809484	-3.05×10^{-9}	-3.28×10^{-10}
RS Cvn	1.440	1.380	4.00	1.99	4.79767597	-2.84×10^{-6}	-1.21×10^{-9}
RU Cnc	1.470	1.460	4.90	1.90	10.1729706	-1.91×10^{-6}	-1.26×10^{-9}
SS Cam	1.832	1.748	6.40	2.20	4.82353386	-2.67×10^{-6}	-2.38×10^{-9}
SZ Psc	1.620	1.241	5.10	1.50	3.96571971	-5.70×10^{-6}	-2.02×10^{-8}
V471 Tau	0.760	0.740	0.83	0.01	0.52118310	-1.38×10^{-9}	-1.34×10^{-10}
VV Mon	1.498	1.412	6.20	1.80	6.05033577	-2.39×10^{-6}	-2.33×10^{-9}
Z Her	1.554	1.310	2.73	1.85	3.99279951	-1.59×10^{-7}	-7.15×10^{-10}

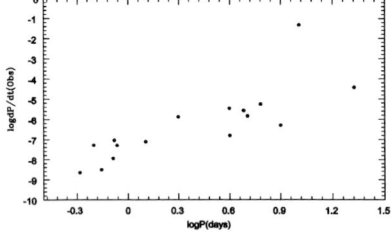

Figure 1. The observed P-dP/dt relation.

Figure 2. A comparison of the period changing rate between observations and calculations.

3. Results and Conclusions

The results of the period analysis for the 17 systems are given in Table 1. The predicted period changing rate with respect to the magnetic wind computed for each star is also given. Figure 1 represents a relation of the observed dP/dt vs. P. It shows that the period decreasing rate seems to be decrease with time, i.e., the longer the period is, the more rapidly the period decreases. This may be caused by some instabilities due to the strong interactions between the components among the short-period systems. A comparison between the observed period decreasing rates and the calculated values is shown in Figure 2. It shows that the observed dP/dt are obviously higher than the calculated values by 1-3 orders of magnitude. It is suggested that the magnetic wind seems not to be the only factor for AML.

Acknowledgements

This work is supported by the Natural Science Foundation of China through grants 10778719, 10573022 & 10773015.

Hydrodynamic Processes in Massive Stars

Casey A. Meakin[1,2,3]

[1] Astronomy Department, University of Arizona, Tucson, AZ 85721, USA
[2] Astronomy and Astrophysics Center, University of Chicago, Chicago, IL 60637, USA
[3] Joint Institute for Nuclear Astrophysics, University of Chicago, Chicago, IL, 60637, USA
email: casey.meakin@gmail.com

Abstract. The hydrodynamic processes operating within stellar interiors are far richer than represented by the best stellar evolution model available. Although it is now widely understood, through astrophysical simulation and relevant terrestrial experiment, that many of the basic assumptions which underlie our treatments of stellar evolution are flawed, we lack a suitable, comprehensive replacement. This is due to a deficiency in our fundamental understanding of the transport and mixing properties of a turbulent, reactive, magnetized plasma; a deficiency in knowledge which stems from the richness and variety of solutions which characterize the inherently non-linear set of governing equations. The exponential increase in availability of computing resources, however, is ushering in a new era of understanding complex hydrodynamic flows; and although this field is still in its formative stages, the sophistication already achieved is leading to a dramatic paradigm shift in how we model astrophysical fluid dynamics. We highlight here some recent results from a series of multi-dimensional stellar interior calculations which are part of a program designed to improve our one-dimensional treatment of massive star evolution and stellar evolution in general.

Keywords. Hydrodynamics, convection, stars: evolution, methods: numerical

1. The Challenge at Hand

Massive stars play a central role in a variety of astrophysical contexts:
(*a*) Nucleosynthetic yields and galactic chemical evolution
(*b*) Black hole and neutron star formation rates; supernova and gamma ray burst rates
(*c*) "feedback" with the ISM and IGM through winds, ionizing photons, and explosions
(*d*) Supernova theory through progenitor evolution and global asymmetries due to convection and rotation

Each of these diverse topics depends deeply on our ability to correctly model the life and death of an individual massive star. But the evolution of a massive star relies critically on the correct treatment of the hydrodynamic transport processes which are operating throughout the stellar plasma. The difficulty encountered in modeling the same transport properties even in the far less exotic conditions present in the Earth's atmosphere and oceans is a stark reminder of the challenges faced by the stellar evolutionist.

1.1. Moore's Law

"Make no little plans. They have no magic to stir men's blood and probably will not themselves be realized." —**Daniel Burnham, American architect and urban planner**.

While computational tools and numerical experiments are beginning to provide profound new insights into the nature of turbulent flows, the astrophysical conditions encountered in stellar interiors reminds us that we have a long way to go before we can fully resolve such flows.

For instance, consider the ratio between the largest length scale present in a stellar convection zone and the smallest scales on which velocity fluctuations can occur before being smoother out by viscous forces. In a turbulent medium the smallest scales are connected to the largest scales through a cascade of energy. The ratio between largest and smallest scales can be related to the Reynolds number of the flow if we adopt the Kolmogorov energy spectrum (Kolmogorov (1941), Kolmogorov (1962)), so that $l_{max}/l_{min} \sim \text{Re}^{3/4}$ (see Boris (2007)). An often cited example is that for the conditions present in the solar convection zone, where we find something like $\text{Re} \sim 10^{10}$, so that $l_{max}/l_{min} \sim 3 \times 10^7$. Therefore, to model a cubical region which contains all of the relevant scales from the largest eddy to the viscous damping scale we would required our calculation to contain along the lines of $N \sim 10^{22}$ computational cells. For comparison, the largest turbulence calculations carried out to date have $N = (2048)^3 \sim 10^{10}$ (e.g., Kritsuk et al. (2007)) to $N = (4096)^3 \sim 0.7 \times 10^{11}$ (on the Earth Simulator) computational cells. Therefore, we need an increase in computing resources by a factor of $\sim 10^{12}$.

Moore's law states that the computational resources available for a given cost doubles every 18 months,

$$\log_2(\text{flops}/\$) = \text{time}/(18 \text{ months}). \qquad (1.1)$$

How long we must wait until a fully resolved simulation of stellar turbulence is possible at a funding level comparable to current computational astrophysics levels? If Moore's law holds, we will be able to afford a computing cluster which is a faster by a factor of 10^{12} in $t \approx 18\text{months} \times \log_2 10^{12} \approx 60$ years.†

Computers have just surpassed the petaFLOPS barrier, performing one thousand trillion (10^{15}) floating point operations per second (FLOPS). At this computing speed, it would take ~ 4 months to compute just one floating point operation per cell in our fully resolved 10^{22} zone stellar turbulence calculation. Although this remains a prohibitively expensive calculation, it is exciting that a hundred fold increase in speed has been achieved since 2002 (when the 10 terraFLOPS mark was passed), just 6 years ago‡. (This is equivalent to a Moore's law doubling period of only ~ 11 months.) This example illustrates the telescoping nature of technological advance summarized by Moore's law, and provides a truly visceral sense of realism about our earlier estimate that a fully resolved stellar turbulence calculation will be feasible in only 60 years.

Do these considerations lead us to the conclusion that it is premature to perform stellar convection simulations, and suggest that we should instead wait until adequate computational resources are available? There are several grounds on which to reject this line of reasoning. (1) Significant development is needed in software and data management strategies which is arguably best approached by pushing our present resources to their limits. (2) Analyzing and designing numerical experiments is also a developing art, and we are still learning how to query data in order to inspire and test new theoretical ideas; a creative process which is also best approached by getting our hands dirty. (3) It is possible that many of the resulting flow features captured by our incompletely resolved numerical experiments are nevertheless robust because of an inherently universal property of turbulent flow; the "turbulent cascade". We briefly discuss this topic in the next subsection.

---

† Some of the implications of management strategies when considering such large calculations in the context of Moore's law are examined in Gottbrath et al. (1999).
‡ See http://www.top500.org for a summary of the fastest computing systems in operation as well as historical data.

1.2. The ILES Approach

The large eddy simulation (LES) approach entails explicitly modeling the largest eddies in a turbulent flow and using a turbulence model to incorporate the mixing, dissipation, and dynamical consequences of the smaller, unresolved scales. One of the motivating factors behind this approach is that the majority of the kinetic energy in the flow is contained in the largest scales. Another, relates to how the largest and smallest scales of motion couple in a turbulent flow through an inertial range cascade which. As observed by Boris (2007): "The physically important aspects of the fluid dynamics of turbulent flow can be notably insensitive to the small-scale details of how it is computed."

This latter observation underlies a somewhat recent shift in perspective about how to model turbulence and carry out LES simulations. In particular, there has been a shift away from developing sophisticated subgrid turbulence models, and instead taking advantage of the insensitivity of the large scale motions on the detailed properties of the smallest scale motions where dissipation occurs. Instead, the basic physically motivated numerical algorithms used in modern hydrodynamics methods ensure that (1) conservation, (2) monotonicity, (3) causality, and (3) locality are built into the solutions (Boris (2007)). This approach has been dubbed Implicit Large Eddy Simulation (ILES).

2. New Resources, New Tools

A number of groups have begun to model stellar interiors and atmospheres in multi-dimensions using simulation codes designed for massively parallel processing environments; platforms which employ thousands to hundreds of thousands of microprocessors simultaneously on a single calculation. Developing the software infrastructure necessary to perform numerical simulations on modern equipment is just as important as the advances in hardware manufacturing which Moore's law describes, particularly in light of the changing face of parallel processing architectures. Multi-core processors (cell processors) and hardware heterogeneity is likely to play a prominent role in the future (Turner (2007)). These changes in computing architecture depend on advances in multi-threading programming capabilities. Software and information management technologies are also needed to effectively manipulate data sets which will soon exceed a petabyte (1 PB = 10^3 TB = 10^6 GB). Post processing data is already beginning to use a significant fraction of the total number of FLOPS required for a computational project.

Vast parallelism favors numerical schemes which have a high degree of communication locality. Anelastic (Gough (1969)) and implicit methods (including low-Mach number solvers e.g. Almgren *et al.* (2006), Lin *et al.* (2006)) require solutions to elliptic equations which are burdened by *global* communication at each time step, an operation which is not ideally suited to large parallel platforms. The computational advantages that these approximate methods have traditionally held over explicit schemes are being lost in the new era of massively parallel computing. The good news is that the enormous increases in computing resources is making fully compressible, multi-physics, explicit solutions accessible for an increasingly rich set of astrophysical problems. Modelers are now able to incorporate a higher level of realism into their models, including magnetic fields, realistic equations of state, sophisticated radiation transport schemes, multi-species flow, and combustion physics (i.e., nuclear reactions). While some concerns have been raised concerning the use of fully compressible solvers for low Mach number flows (Schneider *et al.* (1999)), it isn't clear that these short comings are afflicting present simulations of stellar convection. Direct comparison between fully compressible, anelastic methods, and analytic results for very low mach number flow ($M < 10^{-3}$; see Meakin & Arnett

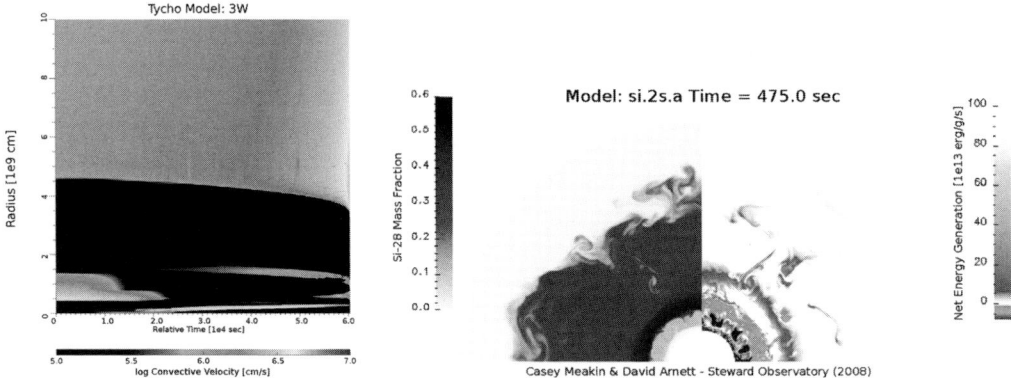

Figure 1. Pre-supernova silicon burning hydrodynamics: (left) The radial dependence of the convective velocity is shown as a function of time for a one-dimensional 23 $\mathcal{M}_\odot$ stellar model as it approaches core collapse, which commences at the very end of the time-sequence shown. The innermost convection zone is due to silicon burning, and a transition from core to shell burning can be seen. The overlying convection zones are driven by oxygen, neon, and carbon burning shells. This model was evolved with the TYCHO stellar evolution code. (right) This snapshot shows the distribution of ^{28}Si and net energy generation for a two dimensional hydrodynamic simulation of the TYCHO model ∼1000 s before core collapse. Silicon, oxygen, neon, and carbon are burning in concentric shells progressively further away from the iron-rich core which will soon undergo gravitational collapse. The outer boundary of the oxygen burning convection zone is strongly perturbed by the convective motions which eventually mixes the carbon, neon and oxygen burning shells together prior to core collapse (Meakin 2006b).

(2007b)) suggest a promising outlook for fully compressible solvers such as the piecewise parabolic method (PPM) of Colella & Woodward (1984).

3. New Data

The nuclear timescale $\tau_{nuc} = X/\dot{X}$, for fuel X which characterize the different evolutionary phases in a stars life are generally many orders of magnitude larger than the advective timescale $\tau_{adv} = L/v$, for fluid with a speed v traversing a region of size L. These disparate timescales make computing the entirety of a stars life in three dimensions prohibitively expensive in the foreseeable future. However, the condition $\tau_{nuc} \gg \tau_{adv}$ allows us to separate the problem so that we can study a snapshot of the evolution, and use this snapshot to formulate a theory of stellar hydrodynamics which we can then incorporate into a 1D stellar evolution code. During the later burning phases and just prior to core collapse in massive stars, we find $\tau_{nuc} \sim \tau_{adv}$. Under these circumstances, the timescale for nuclear evolution becomes small enough to simulate directly.

3.1. *Pre Core-collapse Silicon Shell Burning and Symmetry Breaking*

We have begun a program of multi-dimensional stellar interior modeling which tackles both the quasi-steady and dynamic evolution. Some preliminary work on simulating the reactive hydrodynamic flow associated with pre-core collapse silicon burning in a shell which surrounds an iron core is described in Meakin & Arnett (2006) and Meakin & Arnett (2007a) (see Figure 1). An interesting discovery is the strong interaction between the turbulent convection and the intervening stably stratified layers. Stable layers are significantly distorted by the convective motions, allow for coupling between different burning zones through waves excited in the stable layers (wave cavities), and significant amounts of material is entrained from the convective boundaries into the burning zones. These effects lead to large asymmetries as core collapse is approached which could play an important role in seeding instabilities and affecting the outcome of core collapse and the subsequent supernova explosion.

3.2. *Quasi-Steady Oxygen Shell Burning*

The neutrino cooled oxygen shell burning epoch is an ideal evolutionary phase to study the physics of quasi-steady state stellar convection. The acceleration of this burning stage due to neutrino cooling reduces the ratio between the thermal and hydrodynamic timescales, hence easing the burden of obtaining a relaxed model (see e.g. Arnett (1996), Ch. 10). Recently, we have extended oxygen shell burning simulations to include significantly larger computational domains, longer evolutionary timescales, and 3D flow (Meakin (2006), Meakin & Arnett (2007b), Meakin & Arnett (2007c)). A snapshot of the turbulent flow within an oxygen burning shell is presented in Figure 2. While we find a statistically converged, smooth, quasi-steady state, it is characterized on smaller timescales by significant intermittency and fluctuations. This is illustrated in Figure 3, in which we present both a time averaged radial profile and a space-time diagram showing the evolution of the buoyancy work in a convective oxygen burning shell.

4. Processes and Theory

It is important to bear in mind that numerical simulations of stellar convection are not complete and faithful representations of the actual flows present within a stellar interior. What simulation does provide is a fully non-linear solution with a large number of degrees of freedom which is constrained by an ever more realistic astrophysical context (equation of state, background structure and source terms, better nuclear energetics, etc). These solutions provide the theorist with (1) insight into the fundamental processes which might be operating in a stellar interior, and (2) estimates for the amplitudes and length scales present in the flow which drive instabilities on smaller, unresolved scales. The data from these numerical experiments which inspire theoretical ideas, must ultimately be augmented by a richer, and broader theory of basic processes.

4.1. *Reynolds Averaged Equations*

We develop a kinetic energy (KE) equation in Meakin & Arnett (2007c) by decomposing the velocity $\mathbf{u}$, density ρ, and pressure p fields into mean and fluctuating components $\varphi = \varphi_0 + \varphi'$, employing the hydrostatic equilibrium condition, and performing averages,

$$\partial_t \overline{\langle \rho E_K \rangle} + \nabla \cdot \overline{\langle \rho E_K \mathbf{u_0} \rangle} = -\nabla \cdot \overline{\langle \mathbf{F_p} + \mathbf{F_K} \rangle} + \overline{\langle p' \nabla \cdot \mathbf{u}' \rangle} + \overline{\langle \mathbf{W_b} \rangle} - \varepsilon_K \qquad (4.1)$$

where E_K is the kinetic energy per gram, $\mathbf{W_b}$ is the buoyancy work term, ε_K is the viscous dissipation of kinetic energy, $p' \nabla \cdot \mathbf{u}'$ represents the compressional work done by turbulent fluctuations, and $\mathbf{F_K}$ and $\mathbf{F_p}$ are kinetic energy and pressure-correlation fluxes. A complimentary equation for the internal energy can may be developed (see Meakin & Arnett (2007c), Arnett et al. (2008)).

One of the primary aims in turbulence (and stellar convection) research is to develop physical models for the various terms in these equations, such as the dissipation and flux terms. The reliability of these model terms is only as good as the physical assumptions on which they are based. Often, one is forced to resort to mathematically motivated, ad hoc or phenomenologically based closure models to develop a working theory, and these "theories" are often replete with adjustable parameters which absorb our ignorance about various flow properties. For instance, a commonly used model for the kinetic energy flux is to assume (e.g., Stellingwerf (1982), Kuhfuss (1986)),

$$\mathbf{F_K} \propto -\nabla(E_K) \qquad (4.2)$$

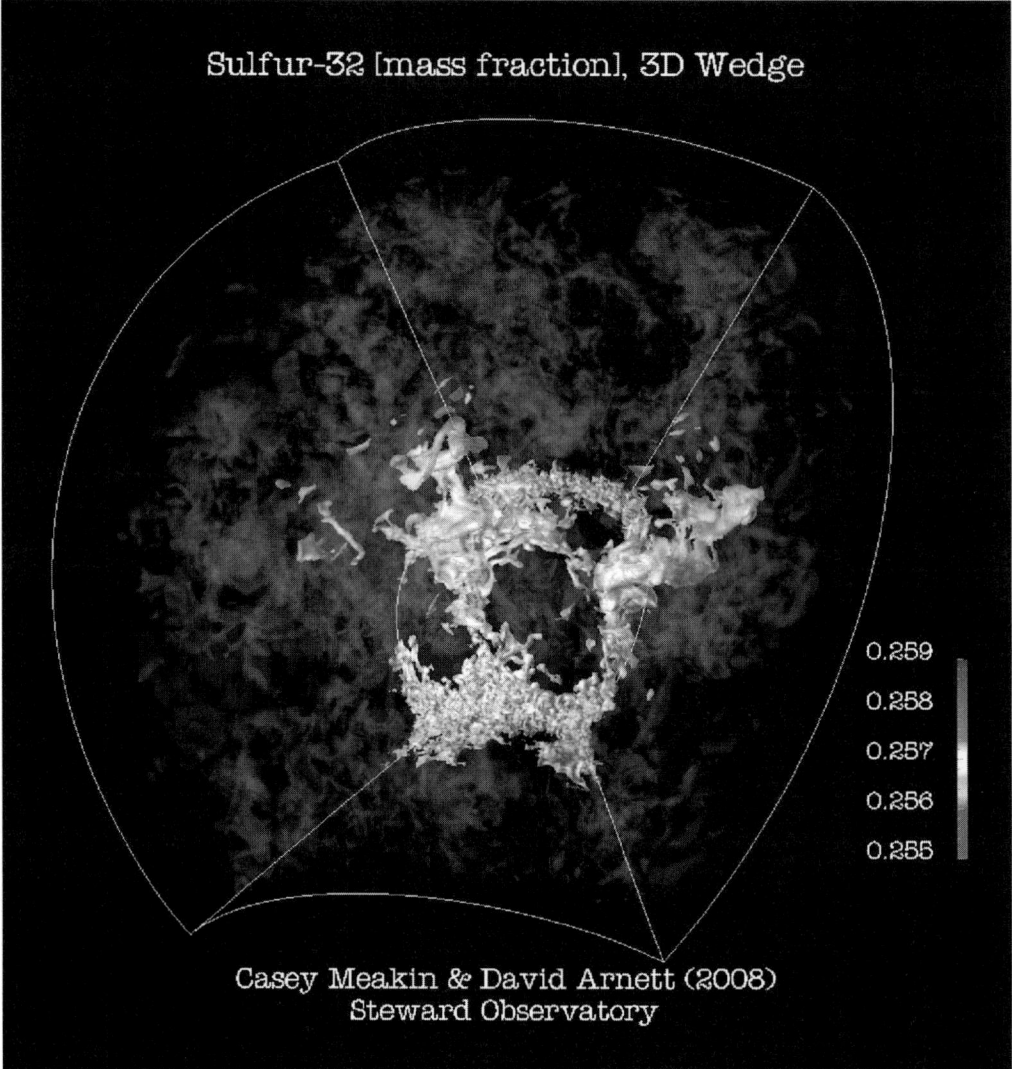

Figure 2. The turbulent flow field in a deep (4 pressure scale height) oxygen burning shell is shown for a computation with a large angular domain ($120° \times 120°$), the boundaries of which are shown outlined by white lines. The domain, which is described by spherical polar coordinates, is oriented so that the polar direction is roughly in the up-down direction, and the azimuthal direction is oriented roughly in the left-right direction. The mass fraction of ^{32}S is visualized in order to give a sense of the topology and the complex, multi-scale, turbulent nature of the flow. Material with a high mass fraction of ^{32}S is being entrained into the turbulent oxygen burning convective shell from the underlying silicon and sulfur rich core. The computational domain contains 17 million cells. Evolving the flow for 5 convective turnover times requires $\sim$1 million cpu-hours on a computing cluster equipped with quad Intel Xeon EM64T 2.8GHz processor cores. (Data from Meakin & Arnett, 2008 in preparation.)

which is sometimes referred to as the down gradient approximation (DGA). Although this model is contradicted by experiment, fundamental theory, and numerical simulation (see e.g., Pope (2000)) it remains the cornerstone of many modern turbulence theories which are used in stellar evolution modeling.

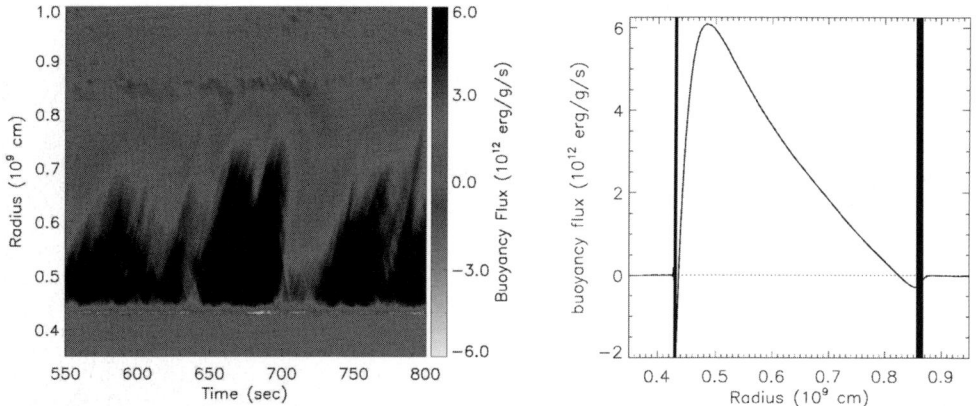

Figure 3. Buoyancy flux. (left) Time-series diagrams and (right) time-averaged radial profiles are shown for 3D oxygen shell burning model described in Meakin & Arnett (2007c).

One avenue for moving beyond simplified turbulence models and closures such as DGA is to draw upon the physical intuition garnered from (1) ever more realistic numerical simulation, (2) better laboratory flow visualization techniques, and (3) cross pollination between the different fluid dynamics sub-discplines (e.g., oceanography, astrophysics, laboratory combustion, etc).

4.2. Dissipation and the Mixing Length

The dissipation of the piecewise parabolic method (PPM) acts on the smallest scales. The numerical dissipation characteristics of this hydrodynamics algorithm compares remarkably well to other approaches used to model turbulence (see Benzi et al. (2007), Boris (2007)). The good energy conservation properties of the finite volume, conservative PPM scheme allows us to infer the dissipation rate of kinetic energy in our convection simulations, which is shown in Figure 4(left). Guided by the dependence of dissipation on the kinetic energy scale of the flow in homogeneous turbulence, we posit that the dissipation in the convective shell can be written as,

$$\varepsilon_K = v_t^3/L \quad (4.3)$$

where v_t is the rms turbulent velocity fluctuation at a given radius, and L is a "damping length". Dissipation calculated according to this expression is shown in Figure 4 by the thin line and compares remarkably well to the inferred damping rate, strongly supporting our ansatz about the nature of the dissipation. The damping length L, which represents the largest eddy in the system, is comparable to the depth of the convective shell. Solar surface convection simulations (R.Stein, private communication) also show this type of dissipation, though the dissipation length L is about four pressure scale heights and not the entire depth of the convection zone, but still significantly larger than a pressure scale height.

In Arnett et al. (2008) we consider the implications that this form of dissipation has for the mixing length theory (MLT) of convection. In particular, we show that if the dissipation length L scales with the depth of the convection zone, the near balance between dissipation and buoyancy driving,

$$\overline{\langle \rho' g v_t \rangle} = \varepsilon_K \approx v_t^3/L \quad (4.4)$$

implies that the mixing length parameter α is a function of the depth of the convection

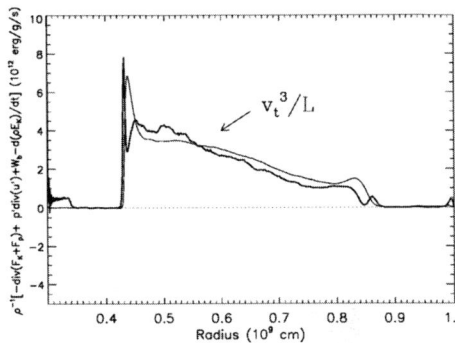
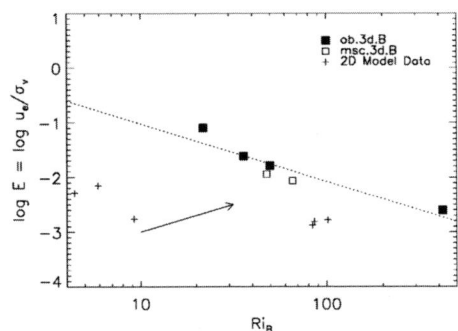

Figure 4. Derived flow properties for a numerical simulation of turbulent oxygen shell burning in a 23 $\mathcal{M}_\odot$ stellar model. (left) The inferred turbulent kinetic energy dissipation rate (thick line) is compared to a dissipation rate calculated using the local rms turbulent velocity v_t according to $\varepsilon_K = v_t^3/L$ with "damping length" L. The rms turbulence velocity is based on the velocity variance across the horizontal plane and two convective turnover times for a fixed radius. The damping length represents the largest eddy in the simulation, which is found to be comparable to the depth of the convection zone in this model. (right) The normalized entrainment rate (boundary layer migration rate in units of the rms turbulence velocity at the boundary) is plotted against the bulk Richardson number, Ri_B. The dashed line shows the best-fit power law to the 3D data. The 2D entrainment rates fall everywhere below the 3D trend due to the incompatible turbulence mixing properties between 2D and 3D flow (see Meakin & Arnett (2007c)).

zone,

$$\alpha^2 \propto L/H_p. \quad (4.5)$$

While this result depends on various measured properties of the specific flow at hand (such as the correlation coefficient between temperature and velocity fluctuations, $\alpha_E = \langle T'v'\rangle/(T'_{\rm rms}v'_{\rm rms})$) comparing a wide range of diverse simulations suggest some universality (e.g., $\alpha_E \approx 0.7$ for a both oxygen shell burning, solar surface, and ideal box convection simulations; see Meakin & Arnett (2007c), Arnett et al. (2008)). That α is not a universal constant appears to be a robust result.

4.3. Entrainment and Buoyancy Flux

The interaction of a turbulent convective region with a bounding stably stratified layer is a long standing challenge to stellar interior modelers. Various phenomenological models have been formulated to treat the mixing that takes place at this interface, but can generally be classified as either (1) a ballistic picture in which eddies penetrate the stable layers until buoyancy breaking halts their motions (Zahn (1991)); (2) a diffusive type process operating within the stable layers which mixes material from the convection zone into the surroundings; or (3) an instantaneous mixing in a region of a fixed, parametrized size. While these prescriptions for mixing have been able to solve various astrophysical quandaries (such as cluster color-magnitude diagram fitting), they are not based on robust, self-consistent physical models and contain parameters which are not grounded in more basic physical considerations and must be calibrated. The universality of these parameters is therefore under question, and like the α in MLT are likely not universal constants.

A more detailed and rigorous analysis of the dynamics taking place at a turbulent boundary layer has been considered by both the geophysical and laboratory fluid dynamics communities, and much progress has been made in elucidating fundamental processes which mediate the mixing rates at these boundaries. One of the primary indicators of boundary layer dynamics is the bulk Richardson number (Fernando (1991)),

$$Ri_B = \frac{bL}{v_t^2} \qquad (4.6)$$

for buoyancy jump b, outer scale L, and rms turbulent velocity v_t. For large values of Ri_B, the boundary layers are strongly stratified compared to the strength of the turbulence and mixing proceeds slowly and the boundary remains relatively undistorted. Boundaries with small values of Ri_B are strongly distorted by the turbulence, which is attended by more rapid mixing rates. Entrainment rates, defined as a boundary layer migration speed u normalized by the rms turbulent velocity scale, is often found to be well characterized by a simple power law dependence "entrainment law",

$$E = \frac{u}{v_t} = A Ri_B^{-n} \qquad (4.7)$$

where A and n are constants fitted to experimental and simulation data. These observations are connected to the underlying hydrodynamic processes through the buoyancy evolution of the boundary layers (see e.g., §7 in Meakin & Arnett (2007c)). An interesting result is that this same power law dependence also holds for the astrophysical convection simulations analyzed in Meakin & Arnett (2007c) (see right panel in Figure 4).

The evolution of buoyancy is related to the "buoyancy flux" through,

$$\partial_t b = -\nabla(q) \qquad (4.8)$$

where $q = \rho' v' g / \rho_0$ which is related to the buoyancy work term in the Reynolds averaged KE equation above by $q = \rho_0 \mathbf{W_b}$. This conservation law for buoyancy describes the exchange between the kinetic energy in turbulence and the potential energy of stratification. A fundamental theory of mixing at convective boundaries will model these terms (some progress is being made; see Fernando & Hunt (1997), McGrath et al. (1997)). While the time and horizontally averaged profiles of the buoyancy flux is smooth, a spatio-temporal decomposition reveals that the smooth profile arises from a highly dynamic underlying behavior (Figure 3).

Acknowledgements

I would like to thank the IAU for a travel grant that made my attendance at this meeting possible. This work is supported in part at the University of Arizona by the National Science Foundation under Grant 0708871 and by NASA under Grant NNX08AH19G. This work was also supported in part at the University of Chicago by the National Science Foundation under Grant PHY 02-16783 for the Frontier Center "Joint Institute for Nuclear Astrophysics" (JINA).

References

Almgren, A. S., Bell, J. B., Rendleman, C. A., & Zingale, M. 2006, ApJ, 637, 922
Arnett, D. 1996, Supernovae and Nucleosynthesis: An Investigation of the History of Matter, from the Big Bang to the Present, by D. Arnett. Princeton: Princeton University Press, 1996
Arnett, D., Meakin, C. A., & Young, P. A., 2008, ApJ, submitted
Benzi, R., Biferale, L., Fisher, R., Kadanoff, L., Lamb, D., & Toschi, F. 2008, Physical Review Letters, 100, 234503
Boris, J., 2007, in Implicit Large Eddy Simulations, ed. F. F. Grinstein, L. G. Margolin, & W. J. Rider, Cambridge University Press, p. 9
Colella, P. & Woodward, P. R. 1984, Journal of Computational Physics, 54, 174
Fernando, H. J. S. 1991, Annual Review of Fluid Mechanics, 23, 455

Fernando, H. J. S. & Hunt, J. C. R. 1997, Journal of Fluid Mechanics, 347, 197
Gough, D. O. 1969, Journal of Atmospheric Sciences, 26, 448
Gottbrath, C, Bailin, J, Meakin, C. A., Thompson, T, & Charfman, J. J. 1999, ArXiv Astrophysics e-prints, arXiv:astro-ph/9912202
Kippenhahn, R., & Weigert, A. 1990, Stellar Structure and Evolution, XVI, 468 pp. 192 figs.. Springer-Verlag Berlin Heidelberg New York. Also A stronomy and Astrophysics Library,
Kolmogorov, A. N., 1941, Dokl. Akad. Nauk SSSR, 30, 299
Kolmogorov, A. N.,1962, J. Fluid Mech., 13, 82
Kritsuk, A. G., Norman, M. L., Padoan, P., & Wagner, R. 2007, ApJ, 665, 416
Kuhfuss, R. 1986, A&A, 160, 116
Lin, D. J., Bayliss, A., & Taam, R. E. 2006, ApJ, 653, 545
McGrath, J. L., Fernando, H. J. S., & Hunt, J. C. R. 1997, Journal of Fluid Mechanics, 347, 235
Meakin, C. A. 2006, Ph.D. Thesis, University of Arizona
Meakin, C. A. & Arnett, D. 2007c, ApJ, 667, 448
Meakin, C. A. & Arnett, D. 2007b, ApJ, 665, 690
Meakin, C. A. & Arnett, D. 2007a, IAU Symposium, 239, 296
Meakin, C. A. & Arnett, D. 2006, ApJL, 637, L53
Pope, S. B. 2000, Turbulent Flows, by Stephen B. Pope, pp. 806. ISBN 0521591252. Cambridge, UK: Cambridge University Press, September 2000
Schneider, T., Botta, N., Geratz, K. J., & Klein, R. 1999, Journal of Computational Physics, 155, 248
Stellingwerf, R. F. 1982, ApJ, 262, 330
Turner, J. A. 2007, LANL Rep. LA-UR-07-1037 (Los Alamos: LANL)
Zahn, J.-P. 1991, A&A, 252, 179

Discussion

LUDWIG: Remark: You pointed out correctly that there are different "mixing-length" representing different flow properties. The mixing-length parameters related to convective envelopes of late-type stars provides a measure of the entropy jump. This value cannot be directly compared to the values you derive, which represent other flow features.

MEAKIN: I agree. But I would still like to emphasize that if one were interested in studying correlation and trends in the mixing length parameters it would be very interesting to consider these trends with structural properties of the star, such as the depth of the convection zone.

LANGER: In 1D stellar models, most convective boundaries are characterized by steep (infinite) gradients in mean molecular weight. Would one then not mostly end up in your "stable" regime of environment, minimizing the whole effect?

MEAKIN: A survey of the properties of convection zone boundaries across the H-R diagram needs to be carried out. Based on the few phases of evolution that I've studied, I suspect the boundaries are much more hydrodynamically active than most people give them credit for.

PALACIOS: Remark: Concerning the problem of rotation in the solar convective zone and on how this is dealt with in 1D stellar evolution codes, we only consider (in 1D modeling) the radial rotation which is indeed almost uniform in the convective envelope of the Sun. Of course, we can not reproduce the latitudinal (conical) rotation with 1D evolution.

MEAKIN: Thank you.

MATHIS: Remark: There is two connects concerning the 1D modeling presented in the first slide:

– Firstly, the shellular rotation approximation is only admitted in the stellar radiation zones while the rotation is almost solid in convective zone.

– Secondly, in stellar evolution codes such as the Geneva, the STAREVOL and the CESAM codes, some processes such meridional circulation advection or internal gravity waves transport are treated rigorously in 2D, the average on an isobar being then taken to calculate the net affect on the mean angular velocity.

The speaker, C. Meakin, is presenting his work.

Shear Driven Turbulence and Coherent Structures in Solar Surface Simulations

F. Kupka

Max-Planck-Institute for Astrophysics, Karl-Schwarzschild Str. 1, D-85748 Garching, Germany
email: fk@mpa-garching.mpg.de

Abstract. Numerical simulations of convection near the solar surface are now advanced enough to reproduce both a large set of observational data and provide tests for convection models. We discuss the role of coherent structures in models of solar p-mode excitation, for which the analysis of numerical simulations has provided key inputs in the modelling. The robustness of these simulations is shown by a comparison illustrating the influence of boundary conditions on ensemble averaged quantities. In a concluding example advanced high resolution simulations are shown to resolve the onset of shear driven turbulence generated by up- and downflow structures.

Keywords. Convection, turbulence, Sun: interior, Sun: granulation, Sun: helioseismology

1. Some astrophysical constraints imposed by solar convection

A brief look on observations of solar surface convection by means of imaging spectroscopy and narrow band photometry (Title 2007, e.g.) shows the well-known pattern of bright patches (granules) embedded in a network of dark regions (intergranular lanes). This structure dominates the solar surface outside regions of enhanced magnetic activity. Analysis of spectroscopic data and corresponding numerical simulations of these structures has demonstrated that granules are upflow regions of gas which is hotter than its surrounding networks which in turn are created by regions of cold downflows. Since the granules cover a larger fraction of the surface than the intergranular network, mass conservation requires the upflows to have lower average velocities than their downflow counterparts (for a review see Spruit, Nordlund & Title 1990).

At current instrumental resolutions of about 100 km (Scharmer, Gudiksen, Kiselman, et al. 2002; Title 2007; Kosugi, Matsuzaki, Sakao, et al. 2007) the solar surface granulation looks like a laminar flow, as is also found in numerical simulations at somewhat higher resolution of 20 km to 50 km (see Stein & Nordlund 1998; Robinson, Demarque, Li, et al. 2003; Wedemeyer, Freytag, Steffen, et al. 2004; Vögler, Shelyag, Schüssler, et al. 2005; Steffen 2007). Are these surface layers representative for the entire solar convection zone? Not necessarily. Granules can cool off by radiating into space, while hot flows inside the Sun are surrounded by optically thick fluid, which increases the cooling time scale. More importantly, the solar surface layers are highly stratified. The local (pressure) scale height $H_p = P/(\rho g)$ increases by more than two orders of magnitudes from the top to the bottom of the solar convection zone, because the pressure P increases much faster towards the interior than the product of density ρ and (local) gravitational acceleration g. As can be seen from classical and current solar structure models published in textbooks (Stix 1989 or Weiss, Hillebrandt, Thomas, et al. 2004, e.g.), a density contrast of $\sim 625,000 : 1$ and a temperature contrast of $\sim 350 : 1$ is found between the bottom and the top of the solar convection zone. Since g changes by less than a factor of 2 (decreasing towards the top of the convection zone), which coincidentally is compensated by a similar change in mean molecular weight μ (increasing towards the top due to hydrogen becoming neutral), H_p is $\sim 50,000$ km near the bottom of the solar convection zone while it is ~ 150 km near its top. Similar holds for surface convection zones in cool main sequence stars other than the Sun. Finally, the flow speed as measured in Mach numbers is about 0.3 near the surface and as low as 10^{-4} close to the bottom of the solar convection zone (cf. also Stix 1989). Consequently, granules experience a very strong decompression in the solar photosphere which has no counterpart in the solar interior.

Taking mass conservation into account and assuming that horizontal flows are mostly subsonic one can estimate (Stein & Nordlund 1998, e.g.) that horizontal scales L are limited by $L \lesssim 4H_p/\mathrm{Ma}$. Here, Ma is the vertical flow Mach number (< 1), the vertical flow velocity in units of local sound speed. Thus, at the solar surface the granule diameters $D \sim L$ are less than about 2000 km, in agreement with observations which put D in the range of 1200 km (see Chap. 1.3 and 6.3 in Stix 1989). Consequently, current simulations of surface granulation cannot include the entire solar convection zone and have to be restricted to a "box-in-a-star" approximation (see also Freytag, Steffen & Dorch 2002). The other way round a simulation of the bottom of the solar convection zone cannot account for surface granulation, since it necessarily has to be global, if the expected energy carrying scales (of order H_p) should be represented in the simulation. For stars with expanded atmospheres the situation is more in favour of global simulations, since H_p at the stellar surface becomes large, as it scales with $1/g$, which makes a "star-in-a-box" approach feasible for simulations of their surface convection even with current high performance computers (see again Freytag, Steffen & Dorch 2002). Tests based on spectroscopy and helioseismology have given numerical simulations of solar and stellar surface convection strong observational support (see the above references and further observational tests as presented, for instance in Rosenthal, Christensen-Dalsgaard, Nordlund, *et al.* 1999 and in Asplund, Nordlund, Trampedach, *et al.* 2000) and thus corroborated the "box-in-a-star" approach for the case of (geometrically) thin stellar atmospheres.

2. Numerical simulations of solar and stellar surface convection

Numerical simulations of solar and stellar surface convection compute volume averages of the basic variables ρ, $\rho \boldsymbol{u}$, and ρe, the densities of mass, momentum, and total energy (with the latter defined as the sum of kinetic and thermal energy). Hence, in a terminology borrowed from engineering sciences, they are large eddy simulations (LES). On their numerical grid LES represent the large spatial scales where most of the kinetic energy is carried. For stellar surface convection the scales on which radiative cooling occurs have to be included, too, which can be achieved with current computational resources. Length scales in the flow smaller than the mesh width of the grid are accounted for by some *subgrid scale model* (or other approximations), while scales larger than the box size (grid extent) assumed in the simulation are handled by the boundary conditions. The simulations start from "typical" initial conditions and proceed through a relaxation phase until quasi-stationarity of key variables is found, such as time averaged total energy flux and root mean square velocities as a function of depth. For a discussion of relaxation of such simulations see Chan & Sofia (1986) and Kupka (2008). From that point onwards a statistical interpretation of the simulation results can be performed. Time integrations of horizontal averages are used to generate ensemble averages which can be compared to quantities used in stellar structure and evolution computations and to stellar observations (which are averages over time and, usually, over the stellar disk, too). Other tools include graphical visualization of the flow, particularly of three-dimensional features, by means of volume rendering and animations to study their development in time.

Unresolved scales cannot be avoided in numerical simulations of stellar convection, as non-linear interactions caused by advection processes become very large in stellar environments and hence dominate over viscous processes. This is expressed by a large Reynolds number Re = UL/ν, which compares length scales and velocities L and U of the energy carrying scales of the flow (for instance, the solar granules and their average velocity) with the kinematic viscosity ν. For solar granules, $L \sim D$, $U(L) \sim 2\ldots 3$ km s^{-1}, and $\nu \sim 1740$ cm^2 s^{-1}, where we take $U(L)$ from spectroscopy and numerical simulations (cf. Asplund, Nordlund, Trampedach, et al. 2000) and ν from Cowley (1990) (Tables 1 and 2 for a layer at the surface with T = 5660 K, with values for $\mu = \nu\rho$ based on Edmonds 1957). Thus, Re $\sim 10^{10}$ and advection completely dominates over viscous friction. This justifies modelling solar surface convection as a nearly inviscid flow. Assuming a Kolmogorov scaling for length scales $l \ll L$ one can estimate dissipation by viscous friction to occur at l_d where $L/l_d \sim \text{Re}^{3/4}$ (cf. Lesieur 1997), thus $L/l_d \sim 10^{7.5}$. Hence, on an equispaced spatial grid $N > 10^{22}$ points are required to resolve l_d. This is neither affordable, nor necessary, since most of the observed physics takes place at scales $l \sim L$. The assumption behind choosing a particular grid size $h \gg l_d$ is based on the expectation that the dynamically important processes take place for $l > h$ and thus, averages over scales $l \leqslant h$ can be represented by hyperviscosity, or non-linear numerical viscosity, or turbulent viscosity, combined with various treatments of shocks (for a discussion of choosing those scales in the context of stellar convection simulations see Kupka 2008).

3. Convection models and coherent structures

Coherent structures are defined as spatial regions that at a given time show some organization with respect to any quantity related to the flow (Lesieur 1997). The basic concept and its role in turbulent flows is already described in Townsend (1956) and even earlier references can be found in the literature (see also the reviews by Cantwell 1990 and Narasimha 1990). The up- and downflow pattern evident in convective flows in general and in stellar surface convection in particular is just one more example. It is instructive to compare the analysis of Stein & Nordlund (1998) with respect to the distribution of hot and cold areas in up- and downflows with measurements of convection in the atmosphere of the Earth, the planetary boundary layer, in Hartmann, Kottmeier & Raasch (1997) and Hartmann, et al. (1999). Clearly, the role of up- and downflows is reversed in the sense that upflows cover a smaller area in the planetary boundary layer than the downflows (which is usually explained by driving convection through heating at the bottom instead of cooling at the top). Moreover, the averages for the solar surface have to be taken over horizontal areas of equal optical Rosseland depth instead of fixed geometrical height to avoid that strong large scale fluctuations introduced by the granules themselves skew the statistical distribution under investigation. Taking that into account the distributions of velocity and temperature fluctuations around their horizontal mean look quite similar. Could that be used in convection models developed for both systems?

In Kupka & Muthsam (2007b) fluxes of velocity and temperature fluctuations as obtained from various convection models were tested with numerical simulations for cases of both inefficient convection (where the radiative flux still transports most of the energy, as, for instance, in surface convection in A-stars of intermediate temperature) and efficient convection. It was noted that for the case of efficient convection only a closure model explicitly accounting for the skewness of both velocity and temperature fluctuations gave satisfactory results when compared to the simulations (see also Kupka 2007). Previous models had relied on a diffusion approach (Xiong 1978) for third order moments (ensemble averages of products of fluctuations of the basic field variables relative to their mean), the quasi-normal approximation for fourth order moments (Canuto 1992, Canuto 1993), or used the eddy damping approach to improve over the latter (Canuto, Cheng & Howard 2001). None of these models explicitly accounts for the coherent, large scale nature of convective flows. This was concluded to be one of the major shortcomings in the previous models, when applied to the case of deep zones of efficient convection (Kupka & Muthsam 2007a-c and Kupka 2007).

In Gryanik & Hartmann (2002) and Gryanik, et al. (2005) it was suggested that the spatial asymmetry in the distribution of temperature and velocity fields inside a convection zone should be accounted for through a model which interpolates between the two cases of large and small skewness. In the first case, the model yields the same result as a two-scale mass-flux model (Gryanik & Hartmann 2002) which consists of averaging separately the velocity and temperature fields over columns defined by the locations where the difference between each field and its horizontal average changes sign. Fluctuations within the up- or downflow, or among different columns defined this way, are neglected in that limit. In the small skewness limit the model recovers the quasi-normal approximation. A completely different approach to account for the non-Gaussian nature of convective flows was suggested by Cheng, Canuto & Howard (2005), where the different terms appearing in the dynamical equations for third order moments (Canuto 1993, Canuto, Cheng & Howard 2001) are used to model the contributions neglected by the quasi-normal approach to fourth order moments. As explained in Canuto (2007), this yields a Reynolds stress model with properties very similar to plume models and traditional versions of mass-flux models. The model was found to be in much better agreement with observations of the planetary boundary layer than its predecessors.

Kupka & Robinson (2007) showed that the model of Gryanik & Hartmann (2002) and of Gryanik, et al. (2005) provides improvements by up to an order of magnitude compared to the quasi-normal approximation in consistency tests with numerical simulations for surface convection in the Sun and in a K dwarf. They corroborated the results previously found for the planetary boundary layer and later on confirmed for the case of convection in the ocean (Losch 2004). Encouraged by this wide range of applicability the model was extended in Belkacem, et al. (2006a) to account for the turbulent fluctuations within the downdrafts by a plume model. In this variant the model was applied in computations of excitation rates of solar p-modes (Belkacem, et al. 2006b). Mode excitation occurs due to shear stresses and entropy fluctuations (cf. also Samadi & Goupil 2001), thus the power injected into a p-mode depends quadratically on the vertical velocity field, and as a consequence of mass conservation a skewed vertical velocity field provides a higher mode excitation rate than its unskewed counterpart. The new model for p-mode excitation rates provides an improvement by an order of magnitude over previous models (local convection models without turbulent pressure and assuming a quasi-normal distribution of fourth order moments with zero skewness of velocity and temperature fields). Model predictions now agree with observations to within measurement uncertainties. In Samadi, et al. (2008) it was shown that the model also recovers the p-mode excitation rates derived for α Cen A. Although observations for this star are not yet accurate enough to distinguish between the inclusion of skewness and its neglect in the models, it was clearly shown that convection models must account for turbulent pressure to agree with observations and in addition that both skewness and a deviation of temporal fourth order correlations from a normal distribution can be probed with sufficiently accurate observations from asteroseismology. It should be pointed out here that probing a convection model with p-mode excitation rates is a much more stringent test than using just the p-mode frequencies, since the latter can be recovered with different temperature-pressure profiles of the surface layers of a solar (or stellar) model. The classical requirement of recovering the solar radius of the present Sun is an even weaker test, since integral quantities can be reproduced with many different internal thermal and hydrostatical structure profiles, as long as there is an adjustable parameter in the convection model and, say, an unknown helium content (to match the solar luminosity).

4. Comparison of numerical simulations among each other

How reliable are numerical simulations of solar surface convection for probing convection models? In addition to testing the simulations with observations (cf. Sect. 1), it is instructive to compare the predictions for statistical correlations by the different codes for various resolutions and boundary conditions. This could corroborate the numerical methods used in the computations of the flow and elucidate possible weaknesses of the simulations. It would demonstrate to what extent the flow field is influenced by the boundary conditions and the domain size chosen. The role of non-grey radiative transfer and the influence of different numerical methods and viscosity models on the large scale, coherently structured flow could be studied, too. In the following we discuss such a comparison with results that have kindly been provided by M. Steffen (CO^{5}BOLD code, Freytag, Steffen & Dorch 2002) and F.J. Robinson (CKS code, Chan & Sofia 1996, Kim & Chan 1998). The CO^{5}BOLD code is based on a Roe-scheme and combines a non-linear numerical viscosity with a model for subgrid-scale viscosity (Smagorinsky-Lilly model, Smagorinsky 1963). Equation of state and opacity in this code are taken from standard model atmosphere codes (Kurucz 1979, Gustafsson, et al. 1975, and more recent versions). The CKS code combines a conservative scheme with the same explicit subgrid-scale viscosity as used in the CO^{5}BOLD code, but additionally that viscosity term is boosted around shocks in the flow to handle the latter. Opacities and equation of state are the same which are commonly used in stellar structure and evolution models (Iglesias & Rogers 1996, Rogers, Swenson & Iglesias 1996, Alexander & Ferguson 1994). We compare results from two different runs with each code. All runs are based on the "box-in-a-star" approach with a Cartesian geometry, periodic horizontal boundary conditions, and a constant vertical gravitational acceleration set equal to the solar one. For the simulations with the CO^{5}BOLD code, the vertical boundary conditions are open (Steffen 2007 and priv. comm.). The high resolution case features a 5-bin non-grey radiative transfer, a grid of $400 \times 400 \times 165$ points with a constant horizontal resolution of 28 km and a variable vertical one of 12 km to 28 km. The resulting simulation box has a volume of $11.2 \times 11.2 \times 3.1$ Mm^3. The deep simulation case assumes grey radiative transfer, a grid of $200 \times 200 \times 250$ points with a constant horizontal resolution of 56 km and a constant vertical one of 21 km. The simulation box has a volume of $11.2 \times 11.2 \times 5.2$ Mm^3 in this case. For both simulations with the CKS code, closed vertical boundary conditions are imposed (free slip, with a constant energy input flux corresponding to the solar effective temperature). Radiative transfer is treated in a grey, 3D Eddington approximation. The first simulation is "model D" from Robinson, Demarque, Li, et al. (2003), also used in Kupka & Robinson (2007). It assumes a grid with $58 \times 58 \times 170$ points with a constant resolution of 50 km horizontally and 17.6 km vertically which result in a box volume of $2.9 \times 2.9 \times 3$ Mm^3. The second simulation (F.J. Robinson 2007, priv. comm.) has $117 \times 117 \times 190$ points with a constant resolution of 35 km horizontally and 15 km vertically which result in a box volume of $4.1 \times 4.1 \times 2.835$ Mm^3. Most importantly, the vertical location of the simulation box relative to the optical surface (with $\tau_{\mathrm{ross}} \approx 1$) has been shifted further upwards by about 150 km compared to the "model D"

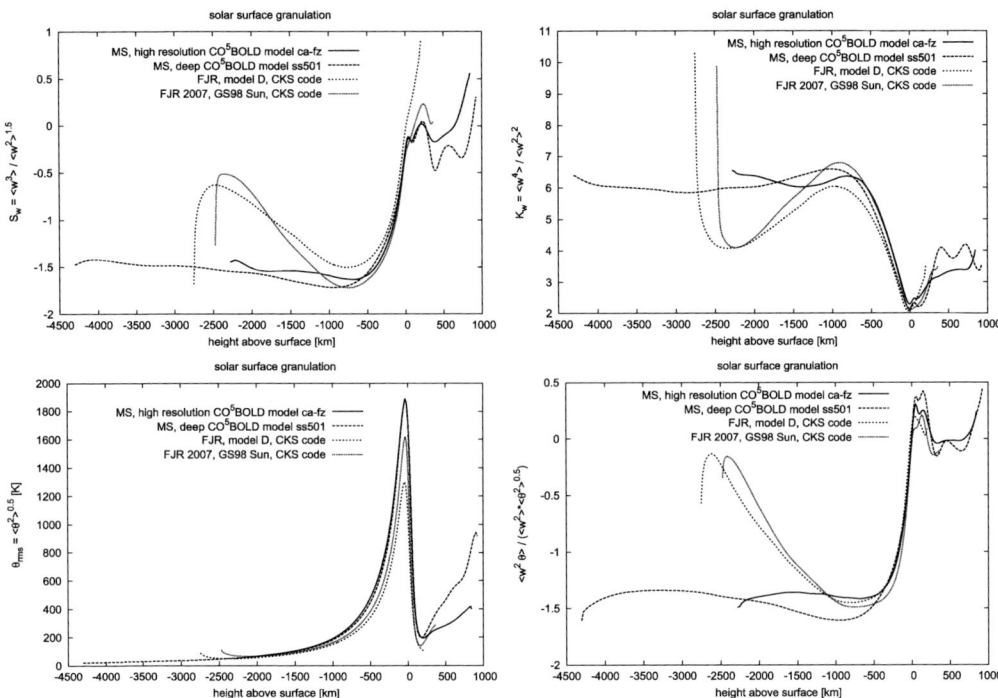

Figure 1. Comparison of simulations with open boundary conditions (data kindly provided by M. Steffen and denoted MS) and with closed boundary conditions (data kindly provided by F.J. Robinson and denoted FJR). Skewness and kurtosis of vertical velocity fluctuations (S_w, K_w), the root mean square of temperature fluctuations (in [K]), and a normalised cross-correlation, $\overline{w^2\theta}/(\overline{w^2}\,\overline{\theta^2})^{0.5}$ are shown.

simulation. This significantly reduces the influence of the upper boundary, as follows from the comparison of various ensemble averages from all four simulation runs presented in Fig. 1. The data are given as a function of height (with a zero-point chosen to be the solar surface at $\tau_{\mathrm{ross}} \sim 1$ in "model D" while data from other simulations are plotted with a constant vertical shift applied to their own measure of height such that just below the surface the average temperature agrees among the simulations shown). The quantities have been computed from a time average of the horizontal averages of the local fluctuations of velocity and temperature (which were taken relative to their horizontal average at each point in time included in the averaging). The influence of closed vertical boundary conditions extend to between 1.5 and 2.5 pressure scale heights, as discussed in Kupka (2008). Since the boxes considered are sufficiently deep, all four simulations agree quite well between the surface at 0 km to about 1000 km further below. This range includes the entire superadiabatic layer and the top of the quasi-adiabatic interior of the solar convection zone. Compared to K_w and $\overline{w^2\theta}/(\overline{w^2}\,\overline{\theta^2})^{0.5}$ the skewness of vertical velocity, S_w, is more sensitive to the (upper) vertical boundary condition. Pushing that boundary sufficiently upwards, as done in the more recent model computed with the CKS code (FJR 2007), significantly improves the comparison with the CO^5BOLD simulations, which both have open upper boundary conditions (located even further upwards). Moving the closed boundary further upwards also allows for larger temperature fluctuations which again are in better agreement with the simulations with open boundary conditions.

5. Simulations of fully turbulent solar convection

At the bottom of the solar photosphere Re $\sim 10^{10}$, as mentioned in Sect. 1, while the Prandtl number Pr, the ratio of kinematic viscosity and thermometric conductivity, remains of order $\sim 10^{-9}$. Interestingly, the product of both, the Peclet number Pe = Re·Pr, is hence of order 10. Consequently, it is possible to model radiative cooling at the solar surface realistically (without resorting to artificially enhanced conductivities, as is necessary for convection simulations of the lower part of the solar convection zone, where Pe exceeds 10^6). Since up- and downflows in solar surface convection move relative to each other at almost sonic speed, one would expect large shear stresses to occur and a large acoustic flux to be created in the layers underneath the visible surface, which are no longer affected by strong radiative cooling. What resolution would be required to resolve the turbulence generated by the Kelvin-Helmholtz shear instability? If we estimate the energy carrying scale to be of the order of the granule diameter $L \sim D$ and assume a Kolomogorov scaling (cf. Sect. 2), we obtain that the grid size h should be of the order of 4 km to achieve an *effective Reynolds number* of about 2300 (Kupka 2008). In that case the shear driven turbulence is at least partially resolved on the computational grid. This kind of scaling is different in 2D (Kupka 2008), because conservation of vorticity along streamlines (cf. Lesieur 1997) reduces the number of internal degrees of freedom in the flow, and permits access to a different dynamical range in comparison with a 3D simulation. Thus, if properties related to turbulence in the flow are of interest in a simulation, 3D simulations are necessary not only quantitatively but also qualitatively.

To reduce the influence of periodic horizontal boundary conditions and in particular to minimize spurious "self-interactions" of granules in a "box-in-a-star" simulation it is advisable to consider a simulation volume that is large enough to hold on average up to half a dozen up- and downflow structures along any given horizontal direction. At a resolution of 5 km this would require about 1500^3 grid points. An alternative to such an expensive calculation is the use of local grid refinement. This permits to consider a more limited domain which only contains perhaps up to three up- and downflow structures along any given horizontal direction at the desired resolution. This high resolution domain is embedded in a larger simulation box with a lower spatial resolution. The grid refinement efficiently reduces the influence of boundary conditions except for those due to more laminar flow structures which occasionally enter the high resolution domain. An example for this strategy can be found in Muthsam, Löw-Baselli, Obertscheider, *et al.* (2007) who performed 2D simulations of solar surface convection at resolutions of better than 3 km based on this approach with the ANTARES simulation code. The simulations very clearly revealed Kelvin-Helmholtz instabilities due to shear between up- and downflows which in turn created all the flow structures expected for a turbulent flow in two spatial dimensions. Remarkably, the solar photospheric layers remained smooth, apart from waves and shock fronts occasionally entering from below. Since the code uses (among other options) a high resolution (5th order) WENO scheme (Liu, Osher & Chan 1994), it does not have to rely on artificial diffusion schemes once a threshold resolution (somewhat larger than 50 km horizontally in the solar case) is exceeded (Muthsam, Löw-Baselli, Obertscheider, *et al.* 2008). Thus, the onset of shear driven turbulence can be observed at lower resolution than with previously used methods (see also Kupka 2008).

What happens in such simulations in 3D? In Fig. 2 we show results from a simulation at $7.1 \times 9.8 \times 9.8$ km^3 resolution (first co-ordinate denotes the vertical direction) of $259 \times 425 \times 393$ cells (and thus a volume of $2.8 \times 4.2 \times 3.8$ Mm^3) embedded in a region with a resolution twice lower vertically and four times lower horizontally. It fills a volume of $2.8 \times 11.2 \times 11.2$ Mm^3 and is described in more detail in Muthsam, Löw-Baselli,

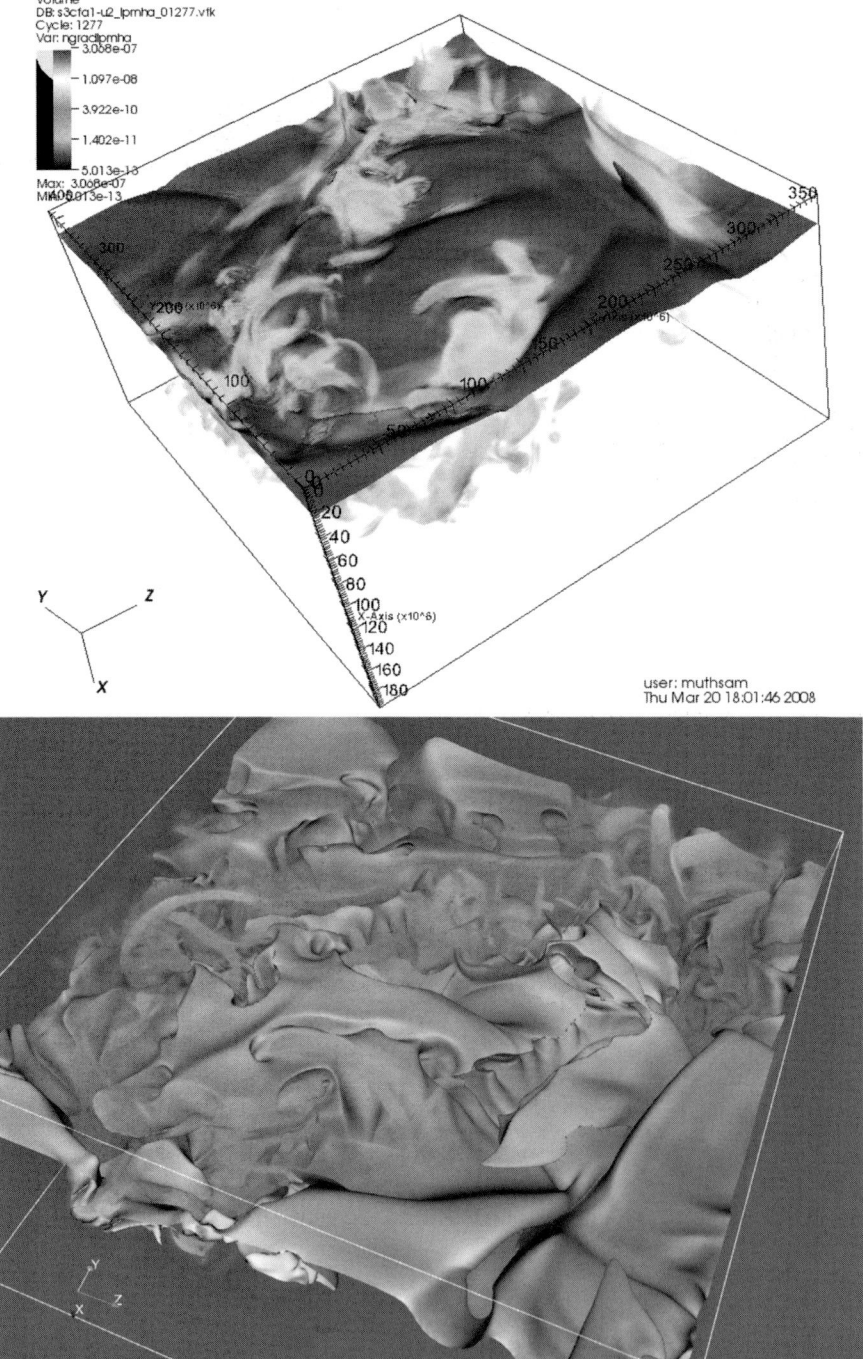

Figure 2. 3D high resolution simulations of shear driven turbulence in solar surface convection. The upper panel shows a snapshot of an isosurface at $T = 6000$ K with a volume rendering of strong local pressure fluctuations. The lower panel shows an isosurface at $T = 8000$ K at the same point in time (note the different viewing angle; volume rendering is less discernible).

Obertscheider, *et al.* (2008). The figure shows a decaying granule and its surroundings at two different depth levels: at the visible surface (upper panel) and underneath. At visible layers the isosurface still looks quit smooth and quasi-laminar. Below those surface layers, however, the simulation appears highly turbulent (lower panel). As pointed out in Muthsam, Löw-Baselli, Obertscheider, *et al.* (2008), at a resolution of close to 10 km, similar to the highest resolution case of Vögler & Schüssler (2007), the numerical scheme is still essential to track the onset of turbulence, which is less prominent and delayed to occur at higher resolution with a numerical method which requires a higher viscosity.

6. Summary and outlook

Numerical simulations have confirmed the role of coherent structures in astrophysical systems which had previously been inferred from observational data. They are now advanced enough to probe the advection properties of non-local convection models used to study solar p-mode excitation. The latter have shown that it is necessary to explicitly account for organised motion in convection modelling. In turn, direct comparisons of statistical properties among different numerical simulations allow quantifying the influence of boundary conditions and other physical simplifications and demonstrate the strengths and limitations of individual simulation codes. Recent simulations have shown how turbulence can be present in a flow despite being shrouded from direct observations, as is the case in solar surface convection. Future improvements in numerical techniques and computing capabilities will allow resolving shear driven turbulence for an even larger number of astrophysical problems and help in developing a more refined physical picture of turbulent flows in astrophysics. The simulations will continue to complement observations by providing another source of test data for convection models which in turn can be further refined as required by their expected continuous use in calculations of stellar structure and evolution for some time to come.

Acknowledgements

This work is partially supported by the DFG subproject KU 1954/3-1 of SPP 1276/1. The author is grateful to M. Steffen and F. J. Robinson for providing the data used for Fig. 1 and to H. J. Muthsam for providing the colour images used to produce Fig. 2.

References

Alexander, D. R. & Ferguson, J. W. 1994, *ApJ* 437, 879
Asplund, M., Nordlund, Å., Trampedach, R., Allende Prieto, C., & Stein, R. F. 2000, *A&A* 359, 729
Belkacem, K., Samadi, R., Goupil, M. J., & Kupka, F. 2006a, *A&A* 460, 173
Belkacem, K., Samadi, R., Goupil, M. J., Kupka, F., & Baudin, F. 2006b, *A&A* 460, 183
Cantwell, B. 1990, in J. L. Lumley (ed.), *Whither Turbulence ? Turbulence at the Crossroads*, Springer Lecture Notes in Physics, (Berlin: Springer-Verlag), vol. 357, p. 97
Canuto, V. M. 1992, *ApJ* 392, 218
Canuto, V. M. 1993, *ApJ* 416, 331
Canuto, V. M. 2007, in F. Kupka, I. W. Roxburgh & K. L. Chan (eds.), *Proc. IAU S 239*, (Cambridge: Camb. Univ. Press), p. 19
Canuto, V. M., Cheng, Y., & Howard, A. 2001, *J. Atmos. Sci.* 58, 1169
Chan, K. L. & Sofia, S. 1986, *ApJ* 307, 222
Chan, K. L. & Sofia, S. 1996, *ApJ* 466, 372
Cheng, Y., Canuto, V. M., & Howard, A. M. 2005, *J. Atmos. Sci.* 62, 2189
Cowley, C. R. 1990, *ApJ* 348, 328

Edmonds, F. N. 1957, *ApJ* 125, 535
Freytag, B., Steffen, M., & Dorch, B. 2002, *Astron. Nachrichten* 323, 213
Gryanik, V. M. & Hartmann, J. 2002, *J. Atmos. Sci.* 59, 2729
Gryanik, V. M., Hartmann, J., Raasch, S., & Schröter, M. 2005, *J. Atmos. Sci.* 62, 2632
Gustafsson, B., Bell, R. A., Eriksson, K., & Nordlund, Å. 1975, *A&A* 42, 407
Hartmann, J., Kottmeier, C., & Raasch, S. 1997, *Bound.-Layer Meteor.* 84, 45
Hartmann, J., et al. 1999, *Polar Research Rep.*, (Bremerhaven: Alfred Wegener Institute for Polar and Marine Sciences), vol. 305, p. 81
Iglesias, C. A. & Rogers, F. J. 1996, *ApJ* 464, 943
Kim, Y.-C. & Chan, K. L. 1998, *ApJ (Letters)* 496, L121
Kosugi, T., Matsuzaki, K., Sakao, T., Shimizu, T., Sone, Y., Tachikawa, S., Hashimoto, T., Minesugi, K., Ohnishi, A., Yamada, T., Tsuneta, S., Hara, H., Ichimoto, K., Suematsu, Y., Shimojo, M., Watanabe, T., Shimada, S., Davis, J. M., Hill, L. D., Owens, J. K., Title, A. M., Culhane, J. L., Harra, L. K., Doschek, G. A., & Golub, L. 2007, *Solar Phys.* 243, 3
Kupka, F. 2007, in: F. Kupka, I. W. Roxburgh & K. L. Chan (eds.), *Proc. IAU S 239*, (Cambridge: Camb. Univ. Press), p. 92
Kupka, F. 2008, in: W. Hillebrandt, F. Kupka (eds.), *Interdisciplinary Aspects of Turbulence*, Springer Lecture Notes in Physics, (Berlin: Springer-Verlag), in print
Kupka, F. & Muthsam, H. J. 2007a, in: F. Kupka, I. W. Roxburgh & K. L. Chan (eds.), *Proc. IAU S 239*, (Cambridge: Camb. Univ. Press), p. 80
Kupka, F. & Muthsam, H. J. 2007b, in: F. Kupka, I. W. Roxburgh & K. L. Chan (eds.), *Proc. IAU S 239*, (Cambridge: Camb. Univ. Press), p. 83
Kupka, F. & Muthsam, H. J. 2007c, in: F. Kupka, I. W. Roxburgh & K. L. Chan (eds.), *Proc. IAU S 239*, (Cambridge: Camb. Univ. Press), p. 86
Kupka, F. & Robinson, F. 2007, *MNRAS* 374, 305
Kurucz, R. L. 1979, *ApJS* 40, 1
Lesieur, M. 1997, *Turbulence in Fluids, 3rd ed.*, (Dordrecht: Kluwer Academic Publishers)
Liu, X., Osher, S., & Chan, T. 1994, *J. Comput. Phys.* 115, 200
Losch, M. 2004, *Geophys. Res. Lett.* 31, L23301
Muthsam, H. J., Löw-Baselli, B., Obertscheider, Chr., Langer, M., Lenz, P., & Kupka, F. 2007, *MNRAS* 380, 1335
Muthsam, H. J., Löw-Baselli, B., Obertscheider, Chr., Langer, M., Lenz, P., & Kupka, F. 2008, to be submitted
Narasimha, R. 1990, in J. L. Lumley (ed.), *Whither Turbulence ? Turbulence at the Crossroads*, Springer Lecture Notes in Physics, (Berlin: Springer-Verlag), vol. 357, p. 13
Robinson, F. J., Demarque, P., Li, L. H., Sofia, S., Kim, Y.-C., Chan, K. L., & Guenther, D. B. 2003, *MNRAS*, 340, 923
Rogers, F. J., Swenson, F. J., & Iglesias, C. A. 1996, *ApJ* 456, 902
Rosenthal, C. S., Christensen-Dalsgaard, J., Nordlund, Å., Stein, R. F., & Trampedach, R. 1999, *A&A* 351, 689
Samadi, R. & Goupil, M.-J. 2001, *A&A* 370, 136
Samadi, R., Belkacem, K., Goupil, M. J., Dupret, M.-A., & Kupka, F. 2008, *A&A*, in print
Scharmer, G. B., Gudiksen, B. V., Kiselman, D., Löfdahl, M. G., & Rouppe van der Voort, L. H. M. 2002, *Nature* 420, 151
Smagorinsky, J. 1963, *Mon. Weather Rev.* 91, 99
Spruit, H. C., Nordlund, Å., & Title, A. M. 1990, *Annu. Rev. Astron. Astrophys.* 28, 263
Steffen, M. 2007, in: F. Kupka, I. W. Roxburgh, & K. L. Chan (eds.), *Proc. IAU S 239*, (Cambridge: Camb. Univ. Press), p. 36
Stein, R. F. & Nordlund, Å. 1998, *ApJ* 499, 914
Stix, M. 1989, *The Sun* (Berlin: Springer-Verlag)
Title, A. M. 2007, in: K.A. van der Hucht (ed.), *Highlights of Astronomy*, (Cambridge: Camb. Univ. Press), vol. 14, p. 30
Townsend, A. A. 1956, *The structure of turbulent shear flow* (Cambridge: Camb. Univ. Press)
Vögler, A., Shelyag, S., Schüssler, M., Cattaneo, F., Emonet, T., & Linde, T. 2005, *A&A* 429, 335

Vögler, A. & Schüssler, M. 2007, *A&A*, 465, L43
Wedemeyer, S., Freytag, B., Steffen, M., Ludwig, H.-G., & Holweger, H. 2004, *A&A* 414, 1121
Weiss, A., Hillebrandt, W., Thomas, H.-C., & Ritter, H. 2004, *Cox & Giuli's Principles of Stellar Structure. Extended Second Edition*, (Cambridge: Cambridge Scientific Publ.)
Xiong, D. R. 1978, *Chinese Astronomy* 2, 118

Discussion

K. STEPIEN: Have you tried to calculate acoustic energy flux heating the chromosphere?

F. KUPKA: Not yet, but we plan to do that and are looking forward to it.

P. WOITKE: Concerning the "switch" between diffusion limits and bin-Z-dependent radiative transfer: (1) Why do you do that? (2) Do we have to worry about it? (3) What is your numerical recipe for this "switch"?

F. KUPKA: (1) Saving computational time. (2) The bin-method is only good around $\tau=1$ where it is constructed for. Applying it to too deep layers leads actually to errors. (3) We apply a simple geometric criterion, namely a specified height, about half of the computational domain is diffusive transfer.

(H.-G. LUDWIG: How far do we have to go in resolution in order to reach the regime where the turbulence properties in the models became really self-similar (or obey e.g., a Kolmogorov scaling law)?

F. KUPKA: I would say until we resolve shear driven turbulence so that its dynamics is captured on the computational grid, a Kolmogoro spectrum showing up at small scales should be a good indicator. We are currently performing simulations at even higher resolution to obtain further clues on that.

The speaker, F. Kupka (right), is relaxing in a sofa together with V. Canuto during coffee break.

Analysing the Contributions in Moment Equations of Reynolds Stress Models of Convection with Numerical Simulations

F. Kupka[1], H. J. Muthsam[2]

[1]Max-Planck-Institute for Astrophysics, Karl-Schwarzschild Str. 1, D-85748 Garching, Germany
email: fk@mpa-garching.mpg.de

[2]Institute of Mathematics, University of Vienna, Nordbergstraße 15, A-1090 Vienna, Austria,
email: herbert.muthsam@univie.ac.at

Abstract. We discuss how 3D numerical simulations can be used to analyse the different contributions within dynamical equations of non-local Reynolds stress models of convection.

Keywords. Convection, turbulence, stars: interiors

1. Models, numerical simulations, model cases, and discussion

Reynolds stress models for turbulent convection require to solve dynamical equations for the ensemble averages of velocity and temperature fields and their lower order moments. Because of the non-linearity of the underlying hydrodynamical equations these equations are part of an unclosed hierarchy which requires additional assumptions (closure approximations) to formulate predictive models. In Kupka & Muthsam (2007a-c) and Kupka (2007) a study of a set of moment equations proposed by Canuto & Dubovikov (1998) (combined with results from Canuto 1992, 1993) was presented. They extended earlier tests by Kupka (1999), which had been based on 3D numerical simulations of compressible convection by Muthsam *et al.* (1995, 1999), to the case of deep convection zones. We present further results using a term-by-term analysis of individual contributions in the dynamical equations. The simulations are briefly described in Kupka & Muthsam (2007a), and resolve all scales down to the dissipation range for a given, constant Prandtl number Pr, assume idealised microphysics, a perfect gas with $\gamma = 5/3$, prescribed radiative conductivities, and a cartesian geometry with a constant, downwards pointing gravitational acceleration. Horizontal boundary conditions are periodic, vertical ones are closed and stress-free with a constant energy flux imposed at the bottom and a constant temperature at the top. Radiative transfer is treated in the diffusion approximation. As in Kupka & Muthsam (2007a) we consider a thin zone of inefficient convection embedded in stably stratified layers ('model 3J', 72×50^2 grid points, Pr=1) along side simulations of a thicker zone with more efficient convection embedded in likewise manner ('model 155X' with Pr=0.1 and a resolutions of 160×140^2 points, first component vertical). Solutions of Reynolds stress models are obtained with a modified version of the code of Kupka (1999). The model equations are those suggested in Canuto & Dubovikov (1998) (CD98) with some extensions taken from Canuto (1992, 1993, 1997) and Canuto *et al.* (2001). For the time scales τ_θ and $\tau_{p\theta}$ the high Peclet number (Pe) limit of CD98 was taken, since their low Pe number limits are not applicable to moderate Prandtl numbers.

As shown in Fig. 1 for model '3J' the simulations at least broadly agree with the most complete model (compressibility terms and residuals are not shown, the unstable zone is

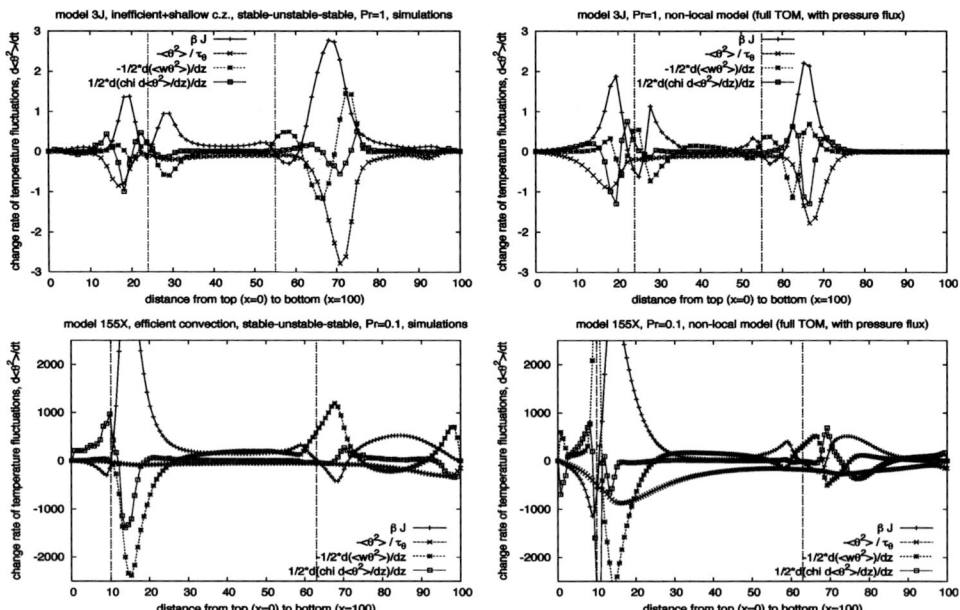

Figure 1. Model terms for the change rate of temperature fluctuations as found in simulations (left panels) and the model (right panels) for case '3J' (top row) and '155X' (bottom row).

located between the two vertical lines) on the overall shape and the contribution of the main terms in the dynamical equation for temperature fluctuations. Similar is found for the other equations of the model which helps to explain its success for shallow convection zones of A-type stars (Kupka & Montgomery 2002). But for deep, efficient convection ('155X') the discrepancies are much larger, both in terms of the shape and the size of the individual terms of the dynamical equation investigated here.

Acknowledgements

The work of F. Kupka is supported by the DFG subproject KU 1954/3-1 of SPP 1276/1. H.J. Muthsam acknowledges support from FwF projects P17024 and P18224.

References

Canuto, V. M. 1992, *ApJ* 392, 218
Canuto, V. M. 1993, *ApJ* 416, 331
Canuto, V. M. 1997, *ApJ* 482, 827
Canuto, V. M. & Dubovikov, M. S. 1998, *ApJ* 493, 834 (CD98)
Canuto, V. M., Cheng, Y., & Howard, A. 2001, *J. Atmos. Sci.* 58, 1169
Kupka, F. 1999, *ApJ* (Letters) 526, L45
Kupka, F. 2007, in: F. Kupka, I. W. Roxburgh, & K.L. Chan (eds.), *Proc. IAU S 239*, (Cambridge: Camb. Univ. Press), p. 92
Kupka, F. & Montgomery, M. H. 2002, *MNRAS* 330, L6
Kupka, F. & Muthsam, H. J. 2007a, in: F. Kupka, I. W. Roxburgh & K. L. Chan (eds.), *Proc. IAU S 239*, (Cambridge: Camb. Univ. Press), p. 80
Kupka, F. & Muthsam, H. J. 2007b, in: F. Kupka, I. W. Roxburgh & K. L. Chan (eds.), *Proc. IAU S 239*, (Cambridge: Camb. Univ. Press), p. 83
Kupka, F. & Muthsam, H. J. 2007c, in: F. Kupka, I. W. Roxburgh & K. L. Chan (eds.), *Proc. IAU S 239*, (Cambridge: Camb. Univ. Press), p. 86
Muthsam, H. J., Göb, W., Kupka, F., Liebich, W., & Zöchling, J. 1995, *A&A* 293, 127
Muthsam, H. J., Göb, W., Kupka, F., & Liebich, W. 1999, *New Astronomy* 4, 405

SONG – Stellar Observations Network Group

Frank Grundahl[1], J. Christensen-Dalsgaard[1], H. Kjeldsen[1], S. Frandsen[1], T. Arentoft[1], P. Kjaergaard [2] and U. G. Jørgensen[2]

[1]Danish AsteroSeismology Centre, Department of Physics and Astronomy, University of Aarhus, DK-8000 Aarhus C, Denmark
email: fgj@phys.au.dk

[2]Niels Bohr Institute, Copenhagen University, Juliane Maries Vej 30, DK-2100 Copenhagen, Denmark

Abstract. Several areas of stellar observations depend critically on nearly continuous observations of individual objects over very extended periods. Important examples are investigations of stellar oscillations to carry out asteroseismology, and the search for extra-solar planets. To meet this requirement we are establishing the SONG network, consisting of 8 sites with a 1-meter-class telescope with a suitable geographical distribution. These will be optimized for asteroseismology based on Doppler-velocity observations and the characterization of extra-solar planets with photometry, using gravitational microlensing. Funding has been obtained towards the construction of the prototype SONG telescope which will be set up on Tenerife, with first light expected in 2011. The full network will be established in parallel with the tests of the prototype and is planned to be operational in 2014.

Keywords. instrumentation: spectrographs, techniques high angular resolution, stars: oscillations, stars: planetary systems

---

1. Asteroseismology

During the past $\sim$25 years helioseismology has proven itself as an extremely valuable tool for studying the detailed properties of the solar interior; it is, however, only during the past $\sim$5 years that asteroseismology – seismology of stars other than the Sun – has become possible. This is due to the large progress in measuring precise stellar radial velocities achieved by the groups searching for extra-solar planets via the radial velocity method. Solar-like oscillations in stars manifest themselves as low-amplitude radial velocity variations due to the surface motions imposed by the oscillations.

In order to measure stellar oscillation frequencies to high precision and to separate closely spaced pairs of frequencies it is necessary to obtain long and continuous radial-velocity time series. At present it is very difficult to obtain time strings longer than around two weeks. This is the motivation for building a network of telescopes.

For asteroseismology SONG will be able to observe solar-like oscillations in stars brighter than $V = 6$ and thus obtain their oscillation spectra. Measuring the oscillations using radial velocities has a very substantial advantage over intensity observations since the background signal (noise) from the star is much lower (Grundahl *et al.* 2007).

To measure the oscillations the stars will be observed with a high-resolution spectrograph equipped with an iodine cell for velocity reference (Butler *et al.* 1996). For bright stars this is expected to reach precisions of 1m/s per minute of observation. A prediction of the expected velocity precision is shown in figure 1.

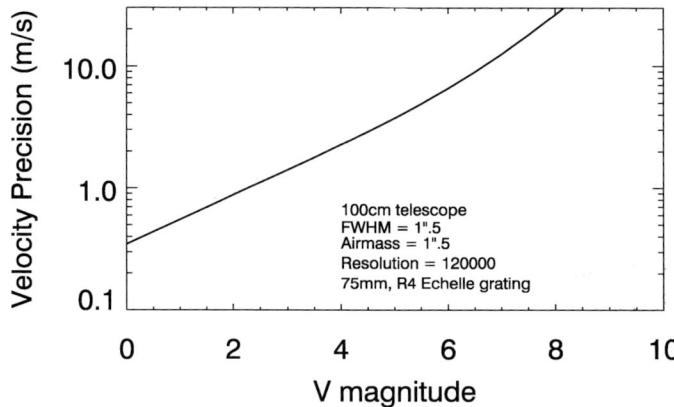

Figure 1. The radial velocity presision as a function of stellar magnitude for a 1m SONG telescope. The calculation is for a slowly rotating star resembling α Cen A.

2. Microlensing planet search

The main science driver for our exoplanet search is to be able to characterize the population of habitable exoplanets in our Galaxy. A planet is habitable if it is large enough to maintain an atmosphere, small enough to not transform itself into a gas planet, and in the right orbital distance from its star to house liquid water at its surface.

SONG will detect small planets using both the microlensing technique and Doppler velocity method. Recently the microlensing method has demonstrated the capability to detect low-mass planets (Beaulieau, et al. 2006; Bennett, et al. 2008), and for SONG it is expected that it will be possible to detect low-mass planets over a wider range of orbital periods than for the Doppler method (for further details, see Jørgensen 2008). The combination of long time series and high-resolution photometry and spectroscopy allows SONG to identify exoplanets in orbits all the way from far inside Mercury's orbit, to Neptune-like orbits and beyond. SONG will be able to detect habitable exoplanets around all types of stars, and for many orbital radii SONG may be able to detect planets as small as Mars (Jørgensen 2008).

The design of SONG is made such that within 5 years of observation we will be able to either characterize the Earth-like exoplanet population, or (in case of non-detection) will be able to put a 5% upper limit on the existence of Earth mass exoplanets around the weighted-average population of stars in our Galaxy.

References

Beaulieu et al. 2006, Nature 439, 437
Bennett et al. 2008, ApJ, in the press (arXiv:0806.0225)
Butler, R. P., Marcy, G. W., Williams, E., McCarthy, C., Dosanjh, P., & Vogt, S. S. 1996, PASP, 108, 500
Grundahl, F., Kjeldsen, H., Christensen-Dalsgaard, J., Arentoft, T., & Frandsen, S. 2007, CoAst 150, 300
Jørgensen, U. G. 2008, Phys. Scr., T129, in the press

Conference summary

Norbert Langer

Utrecht University

To summarize 54 mostly excellent and innovative talks, plus 57 interesting posters, is an impossible task, which I will not even try. This the more as the focus of this meeting was extraordinarily broad. We discussed many different processes in stars, from mixing to pulsations and mass transfer. And we discussed the whole spectrum of stellar types, up an down the main sequence, including the Sun, and into many branches of evolved states of single and binary stars. As we are all working on more or less particular niches in the field of stellar physics, this meant an extraordinary learning experience for most of us. Indeed, this conference offered a stellar physics course at the highest level, which can not be obtained in any other way.

Of course, there is a price to everything. In contrast to more specialized meetings, we could cover basically none of the topics which we discussed thoroughly. Furthermore, we had to leave out many interesting subjects of stellar physics entirely. The retained mix of depths and topics was, however, very interesting for most of us, and the (intended) theoretical inclination of most talks helped us to discuss in a common language, and with mutual interest. In fact this mix turned out to be extremely stimulating, and helped many of us to develop new ideas on things related to our fields of specialization.

In the following, I will go through the conference ordered by topic, and convey my subjective view of what struck me as most relevant. For simplicity, I will not mention any names, as the reader my easily find the details of the mentioned issues in these proceedings.

1. Micro-physics

It was quite impressive to see that the micro-physical functions which are required for any theoretical consideration of stellar structure, stellar atmospheres or stellar pulsations are being more and more refined and tested, and include more and more physical components.

We have seen that stellar opacities are now available for a very wide range of parameters, i.e. temperatures, densities and chemical compositions. In fact, stellar modelers are running out of excuses to *not* update their input opacities to the most appropriate data. Also molecules and dust are incorporated in modern opacity compilations, where local thermodynamic equilibrium is assumed to compute their occurrence. This really is a sufficient prescription for a vast spectrum of applications, while a discussion of the extra dimension of time dependent molecule and dust formation and destruction, which is important for many fields as well, has not been mentioned much at our meeting.

For the important issue of the equation of state (EOS), where two groups have made large progress in the last ten or so years, we heard that we have the choice between imperfect results from a mathematical exact approach, and results from a more pragmatic approach which may capture more of the relevant physics. It was fascinating to see how the adiabatic part of the Solar convection zone can be used as a laboratory to test the accuracy whatever EOS is used in a solar model, and it was reassuring to hear that the EOS is no major source of uncertainty for the modeling of non-degenerate stars, except if it gets to the level of accuracy required for studies of stellar pulsations.

The third topic under the heading of this section is the initial composition of stars, which relates most notably to the chemical composition of our Sun. It may, at first glance, be surprising that for our Sun, which is so close and bright, the surface composition is still uncertain. While the difficulty to determine the Solar helium abundance is acknowledged since a long time, it turns out that measuring even the most abundant metal in the Sun, oxygen, is also far from trivial. While radiation-hydrodynamic modeling of the solar atmosphere, as now routinely performed by several groups, provides a method of abundance determination which is in principle parameter-free, some parameters which describe unknown micro-physical processes — like the strength of an unknown nickel blend to an oxygen line, or collisional excitation of hydrogen — slip back in through the backdoor. We have seen furthermore that presumably simple issues as an equivalent width measurement add to the ambiguity of the Solar oxygen determination. Still, there was agreement at this meeting that there is less oxygen in the Sun than we thought about five years ago — which is good, as it makes the Sun more normal compared to the chemistry in the Solar neighborhood, and which is bad, as it destroys the nice agreement between observed and modeled solar oscillations (cf. below).

2. Mixing in stars

In the stellar interior, a wealth of different (magneto-) hydrodynamic instabilities may occur and lead to the transport of heat, chemical elements or angular momentum. Mixing! We heard many talks on this issue! And there was generally no consensus on the efficiency and consequences of any of these instabilities, while the interaction of more than one of them has not even been addressed give very few exceptions.

Clearly, the most important process in this context is convection, and it was impressive to see how much progress there has been obtained at all levels of sophistication! We have heard about new analytical formulations which include the key physical ingredients of convection, namely buoyancy, shear and vorticity, and which not only describe the transport of heat but also of angular momentum, and which will supersede the Mixing Length Theory eventually. We also heard about innovative methods to describe convection in 1-dimensional grid based calculations which still capture the asymmetry of the upflow and the downdrafts observed in multi-dimensional calculations, and which is thought to have important applications for the modeling of Cepheid pulsations. And finally, we have seen results from ground-braking 2D- and 3D-hydro modeling, which is nowadays applied to many different phases of stellar evolution. From those models we have seen that the interaction of convection and rotation may lead to a variety of rotation laws, which are mostly far from the simple case of rigid rotation. And we could see that convection driven by helium core- or shell flashes, even though hindered by mean molecular weight barriers, may propagate through gravity waves, as well as through inherently multi-dimensional entrainment processes.

In relation to convection, we have seen at lease seven talks which dealt with convective overshooting, where it was quite noticeable that the terminology in the community is not yet coherent. Terms like penetration, entrainment and wave propagation have been used in this context. And while a description of overshooting which is valid in general terms seems rather difficult right now, there was consensus on one result: overshooting is *diverse*, i.e. one can not expect the same efficiency in different situations. And while quite clearly the effects of overshooting may play a role all the way to the bitter end of stellar evolution, many questions may have no satisfactory answer yet. E.g., is significant overshooting always present in hydro-simulations, and if so, why? Is it true that convection zones never have boundaries? Just think of oil floating on water.

We also had many talks on rotationally induced mixing in stars. This topic is also quite complex since rotation may trigger a whole bunch of different instabilities, depending on the conditions inside the star, i.e., the shear instability, Eddington-Sweet circulations, the baroclinic instability, to name just the most important ones. It was interesting to see that evidence for rotational mixing seems to be present throughout the stellar mass spectrum on the main sequence. However, at low mass, where rotational mixing seems required to explain why the surface composition of most A to F-type stars is not affected by, e.g., gravitational settling, only very shallow mixing is needed. At high mass, where nitrogen enrichment is observed in some main sequence stars, deep mixing is required to obtain it. Also, in low mass stars, some shear needs to be generated in order for rotational mixing to become efficient, while in massive stars shear may help, but the Eddington-Sweet circulations may be efficient even in rigid rotators. Rotationally induced mixing may also have an important role in post-main sequence stars, e.g. in the context of the cool bottom processing of low mass red giant stars, or during the s-process nucleosynthesis in thermally pulsing AGB stars.

However, all the mentioned processes are not enough. There is ample observational evidence for the need of further transport processes! The signatures of gravitational settling and radiative levitation in Am/Fm stars are overwhelming, and the settling of helium is likely important for understanding the Sun. These processes may further be important for the so called lithium dip on the lower main sequence, and it may help explain the high enrichment factors seen in the carbon-enhanced metal-poor (CEMP-) halo stars. Angular momentum transport through gravity waves has been put forward as an explanation of the slow rotation of the Solar core. This very complicated mechanism clearly deserves further studies, e.g. to find out whether it has any direct effect on chemical mixing. A future cooperation with oceanographers seems most helpful in this case, who already studied this process in much more detail than astronomers. Thermohaline mixing and semiconvection, both thermally driven instabilities, are also clearly important in stars. While the former appears to very well account for surface abundances in low mass red giants, the latter, again studied in detail in the ocean, is predicted to occur in stars of almost any mass. Finally, rotationally induced magnetic fields may lead to the transport of matter and angular momentum, where the second appears to be required to understand the spins of young neutron stars and white dwarfs.

The problem is of course that none of the available theoretical prescriptions of the mentioned processes allows to make quantitative predictions without involving uncertain parameters. There is little predictive power in explaining one given phenomenon with one given mixing process by calibrating the parameters involved so that the observed phenomenon is reproduced. I feel that we are still struggling to understand for many of the mentioned processes whether they are fundamentally important or not! In particular, we have seen various examples at this meeting where the *same* observation (the Solar core rotation, the lithium dip, increased main sequence core masses, ...) could be explained by combinations of *different* mixing processes! We clearly have difficulties to cope with the inventiveness of Nature which produces so many ways to mix a star! And it may be of key important for future progress, besides increasing our analytical and modeling efforts, to design key observations which can cleanly serve to test one instability and not simultaneously a few others.

3. Pulsations

The field of stellar pulsations is rapidly progressing, with new space missions providing unprecedented high-quality data, and analytical and numerical methods becoming evermore sophisticated. The unrivaled star in the field is of course our Sun, where incredibly accurate data (1000σ error bars!) provide tough constraints to stellar modelers. This led in fact to a crisis, as there seems to be no possible Solar model which fulfills the seismic constraints when the new, lower Solar oxygen abundance is used. The gravitational settling of helium is helpful, but is unlikely to be stronger than already applied in the current models. Perhaps, the oxygen abundance is not quite *as* low? In fact, based on the observed oscillation frequency splittings, we have seen that one can derive the opacity which is required to lead to a consistent Solar model, which leads to a required metallicity of $Z = 0.017$. Is then perhaps the neon abundance larger than we think? It would help, but there is no clear evidence for this. For the Sun, we furthermore await the first undisputed detection of gravity modes, which will certainly pin down the core properties of the Sun in great detail.

The observations of p-modes in the sun and Solar-type stars have been shown to provide a powerful method to age-date these stars. The Solar age comes out as 4.68 Gyr, which is not a perfect match but demonstrates the capabilities of the method, which will be soon applied to many bright stars in the sky. It was further exciting to see that oscillation data combined with interferometry can not only provide accurate stellar ages, but also masses and radii.

Of course, pulsations provide the most promising window to look inside the stars, and to this respect we have seen already, and will see many more exciting results. A nice example at this conference was provided by the analysis of pulsations in PG 1159 post-AGB stars, which showed that in principle one might measure their rate of evolution by determining the change of their oscillation frequency with time, which then might provide the strongest observational constraints on the neutrino cooling which constitutes the dominant energy loss mechanism in these stars.

4. Stellar atmospheres, winds, mass loss

Having our Sun as an example, it clearly provides the test case for the best radiation-hydrodynamic model atmospheres so far, which indeed can reproduce the Sun's atmosphere in very great detail. Atmosphere models of this kind are helpful to derive accurate surface abundances, as mentioned above. But even in these models, further improvements are desirable, i.e. on the level of the micro-physics input, but also regarding turbulence e.g. induced by shear.

For hot stars, the state of the art are 1D-model atmospheres, which now routinely include the elements from hydrogen to nickel in non-LTE, and which keep leading to extraordinary discoveries, e.g., the detection of a fluorine overabundance by a factor of 100 in PG 1159 stars. In luminous hot stars, stellar winds are important, which can now be modeled simultaneously with the stellar atmosphere. As these winds are driven by photon scattering in metal lines, a metallicity dependence is predicted and observed which has important consequences for the the evolution of massive stars. The theoretical challenge today comes from the evidence that hot star winds are not smooth but clumpy outflows. The origin as well as the effects of this clumpiness are not yet well understood.

The atmospheres and winds of red giants and supergiants are even more complex than those of their hot counterparts. The self-consistent modeling is very difficult, and the wind driving in 1D models is only recovered for carbon-rich AGB stars with effective temperatures below 3200 K. The observed large wind anisotropy and the dynamic dust formation processes need to be considered to obtain more realistic AGB winds. On the other hand, we have seen that one may use statistical properties of observed samples of AGB stars to severely constrain the AGB mass loss rates.

And finally we have seen that we need models for what happens beyond the edge of the star, i.e. in its circumstellar environment, to be able to address the effects of feedback of star formation on the dynamical and chemical evolution of galaxies. All stars eject a significant fraction of their mass during their lives, which transports chemical elements, but also energy and momentum into their surroundings. While the low mass stars produce planetary nebulae, we learned that the definition of such a thing is a non-trivial matter. We were surprised by evidence for more than 90% of all planetary nebulae to be bipolar in the wider sense, making spherical planetary nebulae exceptional. The cause of that seems still unclear, except for indications that the shaping process already starts in the proto-planetary nebula phase. For the massive stars, it was shown that their wind bubbles and circumstellar shells may play a key role in the picture of self-regulated star formation. It turns out that multi-D simulations of these structures are necessary in order to correctly predict their energy deposition rate in the circumstellar medium, which may be as low as 1%.

5. Evolution

We heard many talks on stellar evolution modeling, which made clear that this task needs all of the above. The right micro-physics input is essential, and internal mixing needs to be considered. Pulsations may be needed to understand the mass loss process, or provide hard observational constraints to stellar models. And the atmosphere models provide the chemical composition for the onset of the evolution as well as for advanced stages, and the mass loss rates for the hot stars.

Moving up the main sequence, we have seen that we need more helium-rich stellar models of K and M dwarfs to understand the surprising different stellar populations in Galactic globular clusters. For the Galactic G dwarfs, understanding the distribution of observed lithium abundances is still a challenge; we have heard two very different explanation of the lithium gap. In the A and F type regime, we understand that we may need rotational mixing to keep the chemically normal stars as such, and we may require to consider mass loss to prevent the chemically anomalous Am/Fm stars from getting too extreme anomalies due to radiative levitation and gravitational settling. Finally, we may need rotational mixing to explain nitrogen enhancements in OB stars, and anisotropic mass loss may help to keep them rapidly rotating, which may lead to quasi-homogeneous evolution at low metallicity.

In the late evolutionary phases of stellar evolution, not only the uncertain processes discussed above play a role, but any mistakes in describing the earlier evolution may show effects as well. If then we extrapolate our knowledge to the low metallicity universe where we have little observational constraints, it may not be too surprising that our understanding becomes poorer. This may concern the so called super-AGB stars, which undergo core-carbon burning and thermal pulsing, and of which the most massive ones may undergo electron-capture induced core collapse and supernova explosions. This type of stars, which are largely neglected in most comprehensive compilations of stellar evolution models, occupy perhaps a very small initial mass range at solar metallicity; at low metallicity, however, they might be more abundant than any other supernova progenitor type. This rich site for nucleosynthesis is still largely unexplored. And perhaps equally uncertain is the role of very low metallicity intermediate mass stars in the pollution process of the carbon-enhanced metal-poor (CEMP-) halo stars. While binary models have often been used to explain some of their properties, we have seen that perhaps (dual) self-pollution can produce some of them as well.

And also for the massive stars, despite the local populations of main sequence stars are not well understood, much attention is give to their possible role in the early universe and at very low metallicity. This is motivated in part by an observed bias of long-duration gamma-ray bursts toward low metallicities, which may involve rotationally induced rotation as explanation. And at reduced mass loss rates, pair-creation supernovae appear possible at low metallicity, which is interesting in the context of very bright local supernovae, as well as for the potentially very massive stars of the first and second generation in the early universe.

6. Binaries

If it needed one more degree of physical complexity and modeling uncertainty: stellar binaries provide it. And to make the issue more severe: binaries are abundant! So they must play a role in any observed stellar population – and we have seen some examples of this – which means we have to try and model them!

Many classes of stars may exist only because there are binaries, and a prominent one which was discussed in this meeting are the sdB stars. Binary evolution models can explain their properties well, even though, of course, there are parameters involved. And for binaries to also affect the photometric properties of stellar populations, the explanation of the UV-upturn in the SED of elliptical galaxies being due to sdB stars may be one of the best examples. We have also seen that binaries may produce blue stragglers in star clusters, as well as the CEMP stars which were mentioned previously, where, again, fluorine may be the best discriminator between various possible progenitor evolution paths.

But some things get actually simpler in binaries, compared to single stars; E.g., the measurement of stellar masses. And it may also be that binary stars may provide the best constraints for uncertain mixing processes. We have seen that it is possible to put strong constrains on convective overshooting in main sequence stars by comparing models to suitable unevolved binaries. And in contrast to single stars, one thing seems rather clear in unevolved close binaries: the stars involved are not the product of a binary merger – which was shown to relate to 10% of all stars! – or Roche-lobe overflow. It was thus proposed that massive close binaries are the best test case for rotational mixing in single stars.

On the other extreme of evolution, in the supernova regime, binaries again play a clear and substantial role. We heard that the progenitor evolution of half of all core collapse supernovae is affected by a binary companion. And that can exert strong effects. E.g., even if single stars were to produce black holes above, say, $25\,M_\odot$, the mass donors in close binaries might produce neutron stars up to about $50\,M_\odot$. May be electron-capture induced supernovae (see above) occur preferentially in binaries, and explain X-ray binaries with low-eccentricity. And binaries are needed to explain Type Ia supernovae, short-duration gamma-ray bursts, but perhaps also long-duration gamma-ray bursts and even magnetars. So it is clearly worth investing into understanding binary evolution, and the observed populations of evolved binaries, like the Be-X-ray binaries, have been shown to provide essential clues on the uncertainties involved in the modeling process.

7. Final remarks

This meeting provided us with excellent reviews and contributions showing where we stand in our understanding of the essential processes acting inside stars, and in the art of modeling their evolution as single and binary stars. It was very useful to have no limit on the type of star we could discuss during this meeting, which lead to the identification of many cross links which could not have occurred in a more specialized meeting. This conference, and this proceedings book, will be a very useful tool to guide us into new directions for future research.

Acknowledgements

On behalf of all participants, I want to thank all the members of the LOC, and of the SOC foremost Licai Deng, for organizing such an interesting and smooth meeting in this beautiful setting.

A snapshot of the meeting. K. Stepien (standing) is asking the speaker a question.

Author Index

Arentoft, T. – 465
Aret, A. – **41**
Asplund, M. – **13**

Baraffe, I. – 121
Barnes, S. – 117
Belczyński, K. – 399
Bertone, E. – 391
Bi, S. L. – 2–Z5, **243**
Brott, I. – 365
Brun, A. S. – 3–T5, 255
Bulik, T. – 399

Caffau, E. – **35**, 75
Cai, T. – **45**
Campbell, S. W. – **235**
Canuto, V. M. – **67**
Cantiello, M. – **103**, 365
Chan, K. L. – **43**
Chang, Y. L. – **111**
Charbonnel, C. – **135**, **245**
Chávez, M. – 391
Chen, D. –**269**
Chen, L.² –157
Chen, L.² –345
Chen, X. – **247**
Chen, X. F. – 5–T7, **417**
Cheng, Y. – 67
Chevreton, M. – 157
Choi, C.-S. – 431
Christensen-Dalsgaard, J. – **135**, 465
Cui, D. N. – 113, 339
Cui, W. Y. – **113**

Däppen, W. – **27**
de Grijs, R. – 121
de Koter, A. – 283
de Mink, S. E. – **365**, –383
Deng, L. – 2–T2, **83**, 119, 121, 131, 371, 391, 425, 437
Dolez, N. – 157
Dotter, A. – 1

Ekström, S. – 317

Fan, W. S. – 265
Ferguson, W. J. – **1**
Frandsen, S. – 465
Fu, J.-N. – 157

Gai, N. – 115, 243
Gazeas, K. – 427

Ge, H. W. – **419**
Gehren, T. – 127
Geng, Y. Y. – **339**
Glebbeek, E. – 365
Grundahl, F. – **465**
Georgy, C. – 217
Gough, D. O. – 149
Guo, J. H. – **341**

Han, Z. – **349**, 359, 379, 417, 419, 423, 433, 435
Hensler, G. – **309**
Hirschi, R. – 317
Houdek, G. – **149**
Hu, Y. – 121
Huang, Y. F. – 343

Izzard, R. G. –383

Jiang, B. W. – 263, 267
Jiang, D. K. –5–Z4, **433**
Jiang, S. Y. – **421**
Jiang, T. Y. – 423, 433

Kim, S.-L. – 157
Kim, Y.-C. – **117**
Kawaler, S. – 215
Kervella, P. – **405**
Kjaergaard, P. – 465
Kjeldsen, H. – 465
Koesterke, L. – 223
Kovetz, A. – **245**, 261
Krtička, J. – **283**, **347**
Kruk, J. W. – 223
Kwok, S. – **197**
Kučinskas, A. – 75
Kubát, J. – 283
Kupka, F. – **451**, **463**

Laner, N. **467**
Lattanzio, J. C. – 235
Li, J. – **265**
Li, L. F. – **423**, 433, 435
Li, M. – 123
Li, Y. – 133
Li, Z. M. – **359**
Liang, Y. C. – **119**, 131, 425
Liu, G. Q. – **391**, **425**
Liu, Q. – **121**
Liu, Y – 111
Lugaro, M. – 383

Lu, P. – 343, **371**
Lu, Y. – 345, **343**
Ludwig, H.-G. – 35, **75**
Luo, C. Q. – **437**
Luo, Y. P. – 437
Luo, Z. Q. – 437

Maeder, A. – 317
Mathis, S. – **255**
Mao, D. – 27
Meakin, C. A. – **439**
Meng, X. C. – **379**
Meynet, G. – **317**
Michaud, G. – 289
Mikulášek, Z. – 347
Mocák, M. – **215**
Muijres, L. – 283
Müller, E. – 215
Muthsam, H. J. – 463

Palacios, A. – **175**
Piau, L. – **251**, **253**
Podsiadlowski, Ph. – 349
Poolamäe, R. – 41
Prialnik, D. – 245, 261
Pols, O. R. – **383**
Puls, j. – 283

Rauch, T. – **223**
Richard, O. – 289

Sądowski, A. – **399**
Sapar, A. – 41
Sapar, L. – 41
Shen, Z. Q. – 247
Siess, L. – **297**
Silver, I. M. – 157
Singh, H. P. – 43
Solheim – 157
Stein, R. F. – 253
Stępień, K. – **427**
Stökl, A. – **89**
Szczerba, R. – 263

Talon, S – 163
Tang, Y. K. – **115**
Turck-Chieze, S. – **257**

Vauclair, G. – **157**
Vauclair, S. – **97**
Verkoulen, M. – 365
Vick, M. – **289**
Vink, J. S. – **271**

Waldman, R. – **329**
Wang, H. – **263**
Wang, H. B. – 125
Wang, J. C. – **433**
Wang, Q. – **259**
Wang, W. – **333**
Wang, Y. P. – **345**
Webbink, R. – 419
Werner, K. – 223
Willson, L. A. – **189**, 259
Woitke, P. – **229**
Wood, M. A. – 157

Xiong, D. R. – **61**, 83

Yang, M. – **267**
Yang, W. M. – **259**
Yang, S. Z. – 437
Yang, Z. L. – **125**
Yaron, O. – 3–Z5, **261**
Yıldız, M. – **183**
Yoon, S,-Ch. – 365

Zahn, J.-P. – **47**, 245
Ziegler, M. – 223
Zhang, B. – 113, 119, 131, 339
Zhang, F. H. – 423, **435**
Zhang, H. W. – **127**
Zhang, J. – 339
Zhang, Q. S. – **133**
Zhang, S. N. – 343
Zhang, X. B. – **429**, 437
Zhang, Y. X. – **129**
Zhang, Z. B. – **431**
Zhao, G. – 127
Zhao, Y. H. – 129
Zhong, G. H. – 119, **131**
Ziółkowski, J. – 399
Žižňovský, J. – 347
Zvěřina, P. – 347
Zverko, J. – 347

Subject Index

Accretion: accretion disks 343
Atomic processes 1, 27
Atomic data 223
Astrometry 269
Astronomical data bases: miscellaneous 129, 223

black hole physics 343, 345

Cepheids 89
Convection 13, 43, 45, 61, 67, 75, 83, 89, 111, 133, 175, 183, 215, 253, 439, 463

Diffusion 123

Equation of state 1, 27

Gamma rays: bursts 431
Gamma rays: theory 431
galaxies: abundance 119
galaxies: clusters: general 435
galaxies: distances and redshifts 131
galaxies: elliptical and lenticular, cD 349
galaxies: evolution 119
galaxies: formation 359
galaxies: fundamental parameters 119, 131
galaxies: jets 345
galaxies: nuclei 343
galaxies: photometry 131
galaxies: spiral 119, 131
galaxies: starburst 119
galaxies: star clusters 121
galaxies: stellar content 131, 359
Galaxy: center 345
Galaxy: disk 265
Galaxy: evolution 265, 309, 359
Galaxy: globular clusters: general 269
Galaxy: kinematics and dynamics 265
Galaxy: open clusters and associations: individual M67 391
(Galaxy:) solar neighborhood 265

Hydrodynamics 1, 43, 45, 75, 89, 133, 163, 175, 215, 229, 245, 283, 341, 439

Instabilities 47, 163, 245
ISM: abundance 103, 333
ISM: HII regions 309

ISM: kinematics and hydrodynamics 309
Instrumentation: spectrographs 465

Line: formation 13, 35, 127
Line: identification 223
Line: profile 127
Local group 121

Masers 247
Methods: data analysis 129
Methods: numerical 43, 89, 463
Methods: statistical 129
MHD 255
Molecular processes 1

Neutrinos 157
Nucleosynthesis 113, 333, 339

Open clusters and associations: general 61

Planetary nebulae 197
Plasmas 27, 2-125

Radiative transfer 13, 41, 229

Stars: activity 423, 427
Stars: abundances 1, 61, 67, 97, 103, 163, 223, 245, 251, 289, 317, 365, 383, 425
Stars: AGB and post-AGE 103, 113, 197, 205, 223, 235, 259, 263, 297, 383
Stars: atmospheres 41, 75, 223, 347, 425
(Stars:) Binaries: close 349, 365, 379, 383, 399, 417, 419, 423, 427, 433, 437
(Stars:) Binaries: eclipsing 183, 429
(Stars:) Binaries: general 421, 433, 435, 437
(Stars:) Binaries: spectroscopic 425
Stars: blue stragglers 391, 417
Stars: chemically peculiar 347, 425
(stars:) circumstellar matter 247, 263
Stars: early-type 223, 283, 317, 341, 347
Stars: emission line, Be 317, 399
Stars: evolution 1,27, 47, 61, 75, 83, 97, 103, 117, 115, 123, 135, 163, 175, 183, 189, 197, 215, 223, 235, 243, 245, 249, 251, 261, 289, 297, 317,

329, 379, 383, 399, 405, 417, 419, 423, 425, 427, 433, 435, 439
Stars: fundamental parameters 149, 265
(Stars:) Hertzsprung-Russel diagram 249, 391
Stars: horizontal-branch 261
Stars: individual (PG 0122+200) 157
Stars: individual (α Cen, FL Lyr, V422 Cyg) 183, (AH Sco) 247, (Procyon, 61 Cyg) 405, (IU Per) 429
Stars: interiors 47, 117, 133, 183, 245, 255, 419, 439, 463
Stars: late-type 61, 183, 229, 427
Stars: magnetic fields 103, 205, 255, 347, 437
Stars: mass function 383
Stars: mass loss 189, 229, 259, 261, 263, 271, 283, 289, 317, 419
Stars: oscillations 75, 115, 123, 135, 149, 157, 267, 349, 429, 465
Stars: planetary systems 465
Stars: population II 253, 339
Stars: rotation 47, 103, 111, 163, 175, 243, 245, 271, 317, 365, 423
Stars: statistics 121
Stars: subdwarfs 349
Stars: supergiants 267, 309, 341

Stars: supernovae 271, 329
(Stars:) supernovae: general 329, 349, 379
Stars: variables: other 267, 347, 421
(Stars:) white dwarfs 157, 223, 379, 421
Stars: winds, outflow 229, 271, 437
Stars: Wolf-Rayet 365
Sun: abundances 13, 35, 61, 127, 135
Sun: activity 125
Sun: atmosphere 13
Sun: general 257
Sun: granulation 13
Sun: helioseismology 27, 135, 149
Sun: interior 45, 255
Sun: magnetic fields 125, 205, 255
Sun: photosphere 35
Sun: rotation 257

Techniques: high angular resolution 465
Techniques: interferometric 405
Techniques: miscellaneous 129
Turbulence 43, 67, 163, 289, 463

Waves 43, 163

x-rays: binaries 399